RELATIVITY IN FUNDAMENTAL ASTRONOMY:
DYNAMICS, REFERENCE FRAMES, AND DATA ANALYSIS

IAU SYMPOSIUM No. 261

COVER ILLUSTRATION: SERGEI A. KLIONER

This is an artist view of the research field of Applied Relativity. The picture shows the basic pillars of Applied Relativity, directly related to the main subjects of the Symposium: theoretical formulation of reference frames and theoretical (analytical and numerical) modelling of dynamics of celestial bodies and light rays, and the aspect of data processing.

INTERNATIONAL ASTRONOMICAL UNION

UNION ASTRONOMIQUE INTERNATIONALE

RELATIVITY IN FUNDAMENTAL ASTRONOMY

DYNAMICS, REFERENCE FRAMES, AND DATA ANALYSIS

PROCEEDINGS OF THE 261st SYMPOSIUM OF THE INTERNATIONAL ASTRONOMICAL UNION HELD IN VIRGINIA BEACH, VIRGINIA, USA APRIL 27 – MAY 1, 2009

Edited by

SERGEI A. KLIONER

Lohrmann-Observatorium, Technische Universität Dresden, Germany

P. KENNETH SEIDELMANN

University of Virginia, Charlottesville, VA, USA

and

MICHAEL H. SOFFEL

Lohrmann-Observatorium, Technische Universität Dresden, Germany

CAMBRIDGE UNIVERSITY PRESS
The Edinburgh Building, Cambridge CB2 8RU, United Kingdom
32 Avenue of the Americas, New York, NY 10013-2473, USA
477 Williamstown Road, Port Melbourne, VIC 3207, Australia
Ruiz de Alarcón 13, 28014 Madrid, Spain
Dock house, The Waterfront, Cape Town 8001, South Africa

First published 2010

Printed in the United Kingdom at the University Press, Cambridge

Typeset in System $\LaTeX\ 2_\varepsilon$

A catalogue record for this book is available from the British Library

Library of Congress Cataloguing in Publication data

This book has been printed on FSC-certified paper and cover board. FSC is an independent, non-governmental, not-for-profit organization established to promote the responsible management of the world's forests. Please see www.fsc.org for information.

ISBN 9780521764810 hardback
ISSN 1743–9213

Table of Contents

Section III. Time scales, clock and time transfer

Section IV. Equations of motion of astronomical bodies and light rays

Section V. Motion of astronomical bodies

Preface

The history of Einstein's theory of gravity (General Relativity Theory, GRT) can roughly be divided into three epochs. The first epoch started with Einstein's classical papers on the foundation of "General Relativity" (end of 1915), which soon opened up a vast opportunity for mathematicians and mathematical physicists. It was mainly a mathematically oriented discipline. The exceptions were related with the very first experimental tests of GRT. The central problem of celestial mechanics of the 19th century, namely Mercury's anomalous perihelion precession of $43''$ per century, could be explained.

The first measurements of the light deflection by the gravitational field of the Sun during the British expeditions to Sobral (Brazil) and Principe (Gulf of Guinea), taking photographic pictures of the solar vicinity during the solar eclipse on the 29th May, 1919, made Einstein famous.

The situation changed drastically in the second phase of testing, after about 1960. New technological innovations and techniques (atomic clocks, laser reflectors on dedicated satellites and on the lunar surface, radio interferometry, microwave techniques, etc.) not only allowed precise testing of the foundations of any physically reasonable theory of gravity (equivalence principles, gravitational redshift, etc.) and precise solar system tests of GRT, but also led to the rapid development of *relativistic astrophysics*, dealing with fantastic objects such as quasars, pulsars, black holes, gravitational lenses, and even the birth of our entire Universe some 14 billion years ago. The discovery of the binary pulsar PSR 1913+16 by Hulse and Taylor in 1974 revealed a new arena where theories of gravity can be tested. The existence of gravity waves, for the first time in history, was indirectly demonstrated. In parallel with that progress on the observational side, Kenneth Nordtvedt and Clifford Will came up with a parametrized post-Newtonian formalism that covers the post-Newtonian approximation of a great number of alternative theories of gravity. A certain set of PPN parameters, that can be determined experimentally together with realistic error bars, distinguishes these various limits. So far Einstein's theory of gravity has passed every experimental test with flying colors.

In the third present epoch, Einstein's theory has to be considered as an integral part of classical physics; nowadays it is employed to solve technologically oriented problems. Meanwhile, the stability of atomic clocks is of the order of a few times in 10^{-16}, with revolutionary consequences for the problem of navigation (GPS, GLONASS, etc.). VLBI measurements, as a basis for our present celestial reference system (ICRS) and the field of global geodynamics, presently aim at mm accuracies. Laser measurements to selected satellites (SLR) and retroreflectors on the lunar surface (LLR) have accuracies in the cm range. Consequently, solar system ephemerides, theories for time dissemination, clock synchronization, global geodynamics, light propagation, etc. have to be described in the framework of Einstein's theory of gravity. GRT has become the basis for astrometry, celestial mechanics, and metrology.

In the field of astrophysics, objects, that had been considered to be very exotic originally, such as gravitational lenses, pulsars, neutron stars, or black holes, have become quite common objects in the sky (though an ultimate proof of the existence of black holes is still overdue). Pulsars are being investigated as precise clocks that might provide an independent source of precise and stable time. Soon we might be in a position to test the theoretical 'no hair theorem' of black hole physics experimentally. After the first doubly imaged quasar was discovered in 1979, gravitational lensing became an observational science that has contributed significantly new results in areas as different as the cosmological distance scale, mass determination of galaxy clusters, physics of quasars, searches for dark matter, etc. Our present cosmological standard model based on GRT is now supported by numerous observations, especially those related with the anisotropies of the Cosmic Microwave Background Radiation. Detailed measurements of these anisotropies (e.g., by WMAP) lead to an observational determination of the basic cosmological parameters, including the age of the entire Universe.

Today, Applied General Relativity is a broad interdisciplinary field with various experts in different niches. Often they come from different branches of physics and astronomy and experience difficulties to communicate with each other because of their different languages. For that reason we felt the importance to bring such experts together, and to modify the various languages a bit in order to simplify communication with each other. We hope that our Symposium has contributed to this ambitious goal at least a little bit.

Sergei A. Klioner, P. Kenneth Seidelmann and Michael H. Soffel (Proceedings Editors)

THE ORGANIZING COMMITTEE

Scientific

Sergei A. Klioner (co-chair, Germany)
Nicole Capitaine (France)
Sylvio Ferraz Mello (Brasil)
Toshio Fukushima (Japan)
Michael Kramer (UK)
Andrea Milani (Italy)
Gérard Petit (France)
David Vokrouhlický (Czech Republic)

P. Kenneth Seidelmann (co-chair, USA)
Antonio Elipe (Spain)
William M. Folkner (USA)
Kenneth Johnston (USA)
François Mignard (France)
Wei-Tou Ni (China)
Michael Soffel (Germany)
Clifford Will (USA)

Local

M. Efroimsky (chair)
G. Kaplan
K. Marvel
A. Monet
W. Wooden

J. Bangert
B. Luzum
D. Matsakis
S. Urban
N. Zacharias

Acknowledgements

The symposium was sponsored and supported by the IAU Divisions I (Fundamental Astronomy) and X (Radio Astronomy); and by the IAU Commissions No. 4 (Ephemerides), No. 7 (Celestial Mechanics), No. 8 (Astrometry), No. 19 (Rotation of the Earth), No. 31 (Time), No. 33 (Structure and Dynamics of the Galactic Dynamics), and No. 52 (Relativity in Fundamental Astronomy); and by WG (Numerical Standards in Fundamental Astronomy), WG (Second Realization of International Celestial Reference Frame).

The Local Organizing Committee operated under the auspices of the
United States Naval Observatory.

Funding by the International Astronomical Union
and
National Science Foundation of USA
is gratefully acknowledged.

Administrative support was provided by
the American Astronomical Society and its staff.

Participants

John D. **Anderson**, Jet Propulsion Laboratory, Pasadena, CA, USA jdandy@earthlink.net
Guillem **Anglada-Escudé**, Carnegie Institution for Science, Washington D.C., USA guillem.anglada@gmail.com
E. Felicitas **Arias**, International Bureau of Weights and Measures, Sèvres, France farias@bipm.org
Neil **Ashby**, University of Colorado, Boulder, CO, USA ashby@boulder.nist.gov
Matthew **Bailes**, Swinburne University of Technology, Hawthorn, Australia matthew.bailes@gmail.com
Quentin G. **Bailey**, Embry Riddle Aeronautical University, Prescott, AZ, USA baileyq@erau.edu
Ronald **Beard**, U.S. Naval Research Laboratory, Washington, D.C., USA ronald.beard@nrl.navy.mil
Steven **Bell**, HM Nautical Almanac Office, Taunton, UK steve.bell@ukhc.gov.uk
Peter **Bender**, JILA, University of Colorado and NIST, Boulder, CO, USA pbender@jila.colorado.edu
Luc **Blanchet**, Institut d'Astrophysique de Paris, France blanchet@iap.fr
Stefanie **Bremer**, ZARM, University of Bremen, Germany bremer@zarm.uni-bremen.de
Nicoleta **Brinzei**, Transilvania University, Brasov, Romania nico.brinzei@rdslink.ro
Nicole **Capitaine**, SYRTE, Observatoire de Paris, France n.capitaine@obspm.fr
Ratana **Chhun**, ONERA, Chatillon, France ratana.chhun@onera.fr
Luis Filipe P. O. **Costa**, Centro de Física do Porto, Universidade do Porto, Portugal filipezola@hotmail.com
Mariateresa **Crosta**, Osservatorio Astronomico di Torino-INAF, Turin, Italy crosta@oato.inaf.it
Ignazio **Cuifolini**, University of Salento and INFN Sexione di Lecce, Lecce, Italy ignazio.ciufolini@unile.it
Jos **De Bruijne**, European Space Agency, Noordwijk, The Netherlands jdbruijn@rssd.esa.int
Michael **Efroimsky**, U.S. Naval Observatory, Washington D.C., USA me@usno.navy.mil
Frank **Eisenhauer**, Max-Planck-Institut für extraterrestrische Physik, Garching, Germany eisenhau@mpe.mpg.de
Thomas Marshall **Eubanks**, AMERICAFREE.TV, USA tme@multicasttech.com
Fernando **Fandino**, Universidad Nacional de Colombia, Bogotá, Colombia jffandilloc@unal.edu.co
Vincent **Fish**, MIT Haystack Observatory, Westford, MA, USA vfish@haystack.mit.edu
Willliam **Folkner**, Jet Propulsion Laboratory, Pasadena, CA, USA william.folkner@jpl.nasa.gov
Edward **Fomalont**, National Radio Astronomy Observatory, Charlottesville, VA, USA efomalon@nrao.edu
Toshio **Fukushima**, National Astronomical Observatory of Japan, Tokyo, Japan Toshio.Fukushima@nao.ac.jp
Bernard **Guinot**, SYRTE, Observatoire de Paris, France guinot.bernard@wanadoo.fr
Christine **Hackman**, U.S. Naval Observatory, Washington D.C., USA hackman.christine@usno.navy.mil
Aur/'elien **Hees**, Royal Observatory of Belgium, Bruxelles, Belgium aurelien.hees@oma.be
Robert **Heinkelmann**, Deutsches Geodaetisches Forschungsinstitut, Munich, Germany heinkelmann@dgfi.badw.de
Gregory **Hennessy**, U.S. Naval Observatory, Washington D.C., USA gsh@usno.navy.mil
Daniel **Hestroffer**, IMCCE, Observatoire de Paris, France hestro@imcce.fr
David **Hobbs**, Lund Observatory, Lund University, Lund, Sweden david@astro.lu.ed
George **Hobbs**, Australia Telescope National Facility, Epping, NSW, Australia george.hobbs@csiro.au
Catherine **Hohenkerk**, HM Nautical Almanac Office, Taunton, UK catherine.hohenkerk@ukho.gov.uk
Berry **Holl**, Lund Observatory, Lund University, Lund, Sweden berry@astro.lu.se
Luciano **Iess**, Universit'a La Sapienza, Rome, Italy luciano.iess@uniroma1.it
Robert **Jacobson**, Jet Propulsion Laboratory, Pasadena, CA, USA robert.a.jacobson@jpl.nasa.gov
George **Kaplan**, U.S. Naval Observatory (Ret.), Washington D.C., USA gk@gkaplan.us
Ramon **Khanna**, Springer, Germany Ramon.Khanna@springer.com
Sergei **Klioner**, Lohrmann-Observatorium, Dresden Technical University, Germany Sergei.Klioner@tu-dresden.de
Sergei **Kopeikin**, Univeristy of Missouri-Columbia, Columbia, MO, USA kopeikins@missouri.edu
Michael **Kramer**, Jodrell Bank Centre for Astophysics, Manchester, UK Michael.Kramer@manchester.ac.uk
Vladik **Kreinovich**, University of Texas at El Paso, El Paso, TX, USA vladik@utep.edu
Jacques **Laskar**, IMCCE, Observatoire de Paris, France laskar@imcce.fr
Mario **Lattanzi**, Osservatorio Astronomico di Torino-INAF, Turin, Italy lattanzi@oato.inaf.it
Christophe **Le Poncin-Lafitte**, SYRTE, Observatoire de Paris, France Christophe.Leponcin-Lafitte@obspm.fr
Lennart **Lindergren**, Lund Observatory, Lund University, Lund, Sweden lennart@astro.le.se
Meike **List**, ZARM, University of Bremen, Bremen, Germany list@zarm.uni-bremen.de
Jiacheng **Liu**, Nanjing University, P.R.China njuliujiacheng@gmail.com
Brian **Luzum**, U.S. Naval Observatory, Washington D.C., USA bjl@maia.usno.navy.mil
Valeri **Makarov**, NASA Exoplanet Science Institut, Caltech, Pasadena, CA, USA valeri.makarov@jpl.nasa.gov
Jean-Luc **Margot**, University of California, Los Angeles, CA, USA jlm@ess.ucla.edu
Dennis **McCarthy**, U.S. Naval Observatory, Washington D.C., USA dmc@maia.usno.navy.mil
Francois **Mignard**, Observatoire de la Côte d'Azur, Nice, France francois.mignard@obs-nice.fr
Andrea **Milani**, University of Pisa, Pisa, Italy milani@dm.unipi.it
Thomas **Murphy**, University of California, San Diego, CA, USA tmurphy@physics.ucsd.edu
Robert **Nelson**, Satellite Engineering Research Corporation, Bethesda, MD, USA robtnelson@aol.com
Anna **Nobili**, University of Pisa, Pisa, Italy nobili@dm.unipi.it
Xiaopei **Pan**, Jet Propulsion Laboratory, Pasadena, CA, USA xiaopei.pan@jpl.nasa.gov
Erricos **Pavlis**, JCET/UMBC - NASA Goddard Space Flight Center, Baltimore, MD, USA epavlis@umbc.edu
Gérard **Petit**, International Bureau of Weights and Measures, Sèvres, France gpetit@bipm.org
Sophie **Pireaux**, Royal Observatory of Belgium, Bruxelles, Belgium Sophie.Pireaux@oma.be
Elena **Pitjeva**, Institute of Applied Astronomy RAS, St.Petersburg, Russia evp@ipa.nw.ru
Dimitrios **Psaltis**, University of Arizona, Tucson, AZ, USA dpsaltis@physics.arizona.edu
Andreas **Quirrenbach**, ZAH, University of Heidelberg, Germany a.quirrenbach@lsw.uni-heidelberg.de
John **Ries**, University of Texas, Austin, TX, USA ries@csr.utexas.edu
Benny **Rievers**, ZARM, University of Bremen, Bremen, Germany Benny.Rievers@zarm.uni-bremen.de
Harald **Schuh**, Vienna University of Technology, Vienna, Austria harald.schuh@tuwien.ac.at
Bernard **Schutz**, Max Planck Institute for Gravitational Physics, Golm, Germany ute.schlichting@aei.mpg.de
P. Kenneth **Seidelmann**, University of Virginia, Charlottesville, VA, USA pks6n@virginia.edu
Kun **Shang**, Shanghai Astronomical Observatory, Shanghai, P.R.China shangkun@shao.ac.cn
Michael **Shao**, Jet Propulsion Laboratory, Pasadena, CA, USA mshao@huey.jpl.nasa.gov
Sergey **Siparov**, State University of Civil Aviation, St.Petersburg, Russia sergey@siparov.ru
Michael **Soffel**, Lohrmann-Observatorium, Dresden Technical University, Germany michael.soffel@tu-dresden.de
Ingrid **Stairs**, University of British Columbia, Vancouver, Canada stairs@astro.ubc.ca
E. Myles **Standish**, Jet Propulsion Laboratory, Pasadena, CA, USA ems@smyles.jpl.nasa.gov
Edilberto **Suarez**, Universidad Distrital, Bogotá, Colombia
Kai **Tang**, Shanghai Astronomical Observatory, Shanghai, P.R.China tangkai@shao.ac.cn
Pierre **Teyssandier**, SYRTE, Observatoire de Paris, France pierre.teyssandier@obspm.fr
Slava **Turyshev**, Jet Propulsion Laboratory, Pasadena, CA, USA turyshev@jpl.nasa.gov
Mauri **Valtonen**, Tuorla Observatory, University of Turku, Finland mvaltonen2001@yahoo.com

Alberto **Vecchiato**, Osservatorio Astronomico di Torino-INAF, Turin, Italy vecchiato@oato.inaf.it
Patrick **Wallace**, Rutherford Appleton Laboratory, Didcot, UK ptw@star.rl.ac.uk
Joachim **Wambsganss**, ZAH, University of Heidelberg, Germany jkw@ari.uni-heidelberg.de
Clifford **Will**, Washington University, St. Louis, MO, USA cmw@wuphys.wustl.edu
Carol **Williams**, University of South Florida (retired), Tampa, FL, USA cw@math.usf.edu
Peter **Wolf**, SYRTE, Observatoire de Paris, France peter.wolf@obspm.fr
Yi **Xie**, Univeristy of Missouri-Columbia, Columbia, MO, USA xiyi@missouri.edu
Chongming **Xu**, Shanghai Astronomical Observatory, Shanghai, P.R.China chongmingxu@hotmail.com
Vladimir **Zharov**, Sternberg State Astronomical Institute, Moscow, Russia zharov@sai.msu.ru
Shay **Zucker**, Tel Aviv University, Tel Aviv, Israel shayz@post.tau.ac.il

Relativity in Fundamental Astronomy
Proceedings IAU Symposium No. 261, 2009
S. A. Klioner, P. K. Seidelman & M. H. Soffel, eds.

doi:10.1017/S174392130999007X

Standard relativistic reference systems and the IAU framework

Michael Soffel
Lohrmann-observatory, TU Dresden
email: soffel@rcs.urz.tu-dresden.de

Abstract. The IAU framework for relativistic reference systems is based upon the work by Brumberg and Kopeikin and by Damour, Soffel and Xu (DSX). We begin with a brief introduction into the DSX-formalism. After that the various IAU Resolutions concerning relativistic astronomical reference systems are discussed. Finally, it is indicated how the expansion of the universe can be considered in the BCRS.

Keywords. Relativity, post-Newtonian approximation, astronomical reference systems.

1. Introduction

Soon after Einstein's seminal paper Einstein (1915) on General Relativity appeared it became obvious that solving Einstein's field equations for applications in the solar system becomes extremely complicated and one has to resort to approximation schemes. Consequently a slow-motion weak-field approximation called post-Newtonian approximation was worked out by Droste (1916), de Sitter (1916), Lorentz and Droste (1917) and then later by Fock (1959), Nordtvedt (1970) and Will (1993) who added a certain set of post-Newtonian parameters allowing for a violation of Einstein's theory of gravity in nature at the corresponding level of approximation. This classical post-Newtonian framework (e.g., Will 1993), however, has a certain number of drawbacks. First, one single coordinate system is used e.g., for the description of the gravitational N body problem. Second, matter is described as an ideal fluid (an exception can be found in Misner *et al.*, 1973). And third, after the paper by Blanchet and Damour (1989) it became obvious that the classical post-Newtonian center-of-mass and mass definitions should be improved. With respect to the coordinate system problem, it is obvious that for the description of local physics in the vicinity of a body A that is a member of some gravitational N-body problem, some local co-moving system, where the influence of external bodies is effaced and described in terms of tidal terms, should be employed.

2. The DSX-framework

In a series of papers Damour, Soffel and Xu (DSX) (Damour *et al.*, 1991–1994; see also Brumberg and Kopeikin 1988) laid the foundations of a new and improved relativistic celestial mechanics describing a system of N gravitationally interacting rotating bodies of arbitrary shape and composition at the first post-Newtonian approximation of Einstein's theory of gravity. No assumption is made on the internal composition of bodies. The basic matter variables of the DSX-formalism are the gravitational mass-energy

$$\sigma \equiv \frac{T^{00} + T^{ss}}{c^2}$$

where $T^{\mu\nu}$ are the components of the energy-momentum tensor and the gravitational mass-energy current density

$$\sigma^i \equiv \frac{T^{0i}}{c}.$$

In the gravitational N-body problem the DSX-framework employs a total of $N+1$ different coordinate systems: one global system with coordinates $(ct, \mathbf{x})$ that covers the entire model manifold and one local A-system with coordinates $(cT_A, \mathbf{X}_A)$ that is comoving with body A (Fig. 1). In each of the $N+1$ different coordinate systems the metric tensor is written in a special form. E.g., in the global system it is written as

$$\begin{aligned} g_{00} &= -1 + \frac{2w}{c^2} - \frac{2w^2}{c^4} + \ldots, \\ g_{0i} &= -\frac{4}{c^3} w^i + \ldots, \\ g_{ij} &= \delta_{ij}\left(1 + \frac{2w}{c^2}\right) + \ldots. \end{aligned} \tag{2.1}$$

Hence, the global metric tensor is completely determined by means of two potentials: a gravito-electric scalar potential w and a gravito-magnetic vector potential w^i. The gravito-electric potential merely generalizes the usual Newtonian potential U; the vector potential w^i describes the gravitational action of mass-energy currents (moving or rotating masses; Lense-Thirring effects). The corresponding Einstein field equations read:

$$\begin{aligned} \left(-\frac{1}{c^2}\frac{\partial^2}{\partial t^2} + \Delta\right) w &= -4\pi G\sigma + \mathcal{O}(c^{-4}), \\ \Delta w^i &= -4\pi G\sigma^i + \mathcal{O}(c^{-2}). \end{aligned} \tag{2.2}$$

In the local A-system the metric is written in the same form but with potentials W and W^a; there, the mass-energy density and corresponding current is written as Σ and Σ^a.

The formalism employs a spatial set of post-Newtonian mass- and spin-multipole moments (Blanchet-Damour moments) that is used to skeletonize the metric potentials in

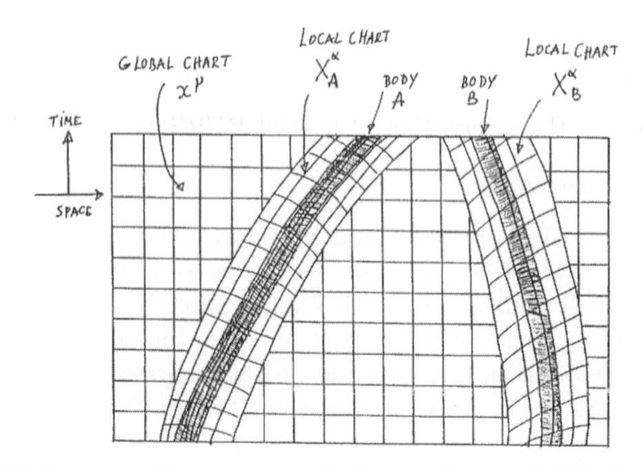

Figure 1. Various coordinate systems in the gravitational N-body system.

the local A-system, generated by body A itself outside of A:

$$\begin{aligned} M_L^A(T) &\equiv \int_A d^3X\hat{X}^L\Sigma + \frac{1}{2(2l+3)c^2}\frac{d^2}{dT^2}\int_A d^3X\hat{X}^L\mathbf{X}^2\Sigma \\ &\quad -\frac{4(2l+1)}{(l+1)(2l+3)c^2}\frac{d}{dT}\int_A d^3X\hat{X}^{aL}\Sigma^a \qquad (l \geqslant 0), \\ S_L^A(T) &\equiv \int_A d^3X\epsilon^{ab<c_l}\hat{X}^{L-1>a}\Sigma^b \qquad (l \geqslant 1). \end{aligned} \tag{2.3}$$

Here, the formalism of Cartesian STF- (symmetric and trace-free) tensors is used. The index L is a multi-index standing for l Cartesian indices: $L \equiv i_1 i_2 \dots i_l$ with each Cartesian index $i = 1, 2, 3 = x, y, z$. The hat as well as the sharp brackets indicate that the corresponding symmetric and trace-free part (see e.g., Damour *et al.*, 1991 for more details) has to be taken.

The heart of the DSX-formalism are the transformation rules for both the coordinates

$$(ct, \mathbf{x}) \longleftrightarrow (cT_A, \mathbf{X}_A)$$

and the metric potentials

$$(w, w^i) \longleftrightarrow (W, W^a)_A.$$

It is interesting to note that the Einstein-Infeld Hoffmann equations of motion that form the basis of any modern numerical ephemeris can simply be derived from

$$W_A = \frac{GM_A}{R_A}$$

and $W_A^a = 0$ by a transformation of the metric in the local A-systems into the global (BCRS) system as geodetic equations of the BCRS metric tensor (Damour *et al.*, 1991).

3. Standard astronomical reference systems and IAU resolutions

The IAU has recommended the use of two basic celestial reference systems: the Barycentric Celestial Reference System (BCRS) and the Geocentric Celestial Reference System (GCRS) (Soffel *et al.*, 2003). Both systems are needed to replace the Newtonian concept of a quasi-inertial space fixed system. The BCRS with coordinates $(ct, \mathbf{x})$ is defined by the form of the metric tensor as in (2.1) together with the condition for asymptotic flatness:

$$\lim_{r\to\infty} (w, w^i) = 0. \tag{3.1}$$

This definition leaves the orientation of spatial BCRS axes open. Later this was fixed by the ICRF. IAU 2006 Resolution B2 (Default orientation of the Barycentric Celestial Reference System (BCRS) and Geocentric Celestial Reference System (GCRS) reads:

.... Recommends

'that the BCRS definition is completed with the following: "For all practical applications, unless otherwise stated, the BCRS is assumed to be oriented according to the ICRS axes. The orientation of the GCRS is derived from the ICRS-oriented BCRS.'

Note, that the BCRS is used for solar system ephemerides (dynamical equations of motion for solar system bodies), for problems of interplanetary spacecraft navigation, high-precision astrometry, for a definition of proper motion or radial velocity, etc.

The GCRS was adopted by the IAU (2000) to model physical processes in the vicinity of the Earth. It is defined by the form of the metric tensor.

$$\begin{aligned} G_{00} &= -1 + \frac{2W}{c^2} - \frac{2W^2}{c^4} + \dots \\ G_{0a} &= -\frac{4}{c^3} W^a + \dots \\ G_{ab} &= \delta_{ab}\left(1 + \frac{2W}{c^2}\right) + \dots \end{aligned} \qquad (3.2)$$

In addition it is assumed that the spatial GCRS-coordinates are kinematically non-rotating with respect to the BCRS, i.e, they are locally non-inertial.

The quasi-linearity of the post-Newtonian field equations allows a unique split of the metric potentials W and W^a into three parts: i) internal, ii) inertial and iii) external parts. The internal parts result from the gravitational action of the Earth itself. In the absence of external bodies they present the post-Newtonian potentials of an isolated Earth, usually skeletonized by corresponding Blanchet-Damour moments. Outside the Earth these parts admit an expansion in terms of negative powers of R (the coordinate distance to the geocenter). For practical applications the internal parts of W, the gravitational potentials of the Earth itself, can be written as an expansion that is given by IAU2000 Resolution B1.4 (see below). Terms of first order in R are inertial terms. The corresponding term in W describes a deviation from free-fall due to the oblateness of the Earth, the one in W^a describes a Coriolis-force due to geodesic precession. A geocentric dynamically non-rotating system, where inertial forces are absent in dynamical equations of motion thus rotates with respect to the GCRS. The influence of external bodies is described as tidal terms, i.e., terms in the metric potentials that are at least quadratic in R. For more details the reader is referred to Soffel *et al.* (2003). The Geocentric Coordinate Time T is called TCG.

IAU 2000 Resolution B1.4, concerning the post-Newtonian expansion of the Earth's gravitational potentials reads:

The IAU .... Recommends

1. expansion of the post-Newtonian potential of the Earth in the Geocentric Celestial Reference System (GCRS) outside the Earth in the form

$$W_E = \frac{GM_E}{R}\left[1 + \sum_{l=2}^{\infty}\sum_{m=0}^{l}\left(\frac{R_E}{R}\right)^l P_{lm}(\cos\theta)(C_{lm}^E(T)\cos m\phi + S_{lm}^E(T)\sin m\phi\right],$$

where C_{lm}^E and S_{lm}^E are, to sufficient accuracy, equivalent to the post-Newtonian multipole moments introduced by Damour *et al.* (Damour *et al.*, Phys. Rev. D, **43**, 3273, 1991). θ and ϕ are the polar angles corresponding to the spatial coordinates X^a of the GCRS and $R = |\mathbf{X}|$, and

2. expression of the vector potential outside the Earth, leading to the well-known Lense-Thirring effect, in terms of the Earth's total angular momentum vector $\mathbf{S}_E$ in the form

$$W_E^a = -\frac{G}{2}\frac{(\mathbf{X}\times\mathbf{S}_E)^a}{R^3}.$$

Though the expansion for W_E looks completely Newtonian it is relativistic due to the use of BD-moments or their spherical equivalents (C_{lm}^E, S_{lm}^E). That implies that the DSX-formalism has been designed so that many relativistic terms are absorbed in the definitions of multipole moments. In practice they simply will be fit-parameters and a split into Newtonian- and post-Newtonian parts is superfluous, dangerous and confusing.

The GCRS is used to describe physical processes in the vicinity of the Earth, for the definitions of potentials coefficients (multipole moments), for satellite theory, for the description of dynamics of the Earth itself, for the definition of the CIP the CIO, etc, and for a direct relation to terrestrial systems and frames (ITRS and ITRF).

A set of auxiliary times scales has been linked to TCG and TCB. Originally Terrestrial Time was related with proper time of a clock located on the rotating geoid. However, IAU 2000 Resolution B1.9 (Re-definition of Terrestrial Time TT) recommends

that TT be a time scale differing from TCG by a constant rate: $\mathrm{dTT}/\mathrm{dTCG} = 1 - L_G$, where $L_G = 6.969290134 \times 10^{-10}$ is a defining constant. International Atomic Time TAI is related with TT by

$$\mathrm{TT} = \mathrm{TAI} + 32.184\,\mathrm{s}.$$

For barycentric ephemerides the times scale TDB was introduced. IAU 2006 Resolution B3 recommends:

that, in situations calling for the use of a coordinate time scale that is linearly related to Barycentric Coordinate Time (TCB) for an extended time span, TDB be defined as the following linear transformation of TCB:

$$\mathrm{TDB} = \mathrm{TCB} - L_B \times (\mathrm{JD}_{\mathrm{TCB}} - T_0) \times 86400 + \mathrm{TDB}_0,$$

where $T_0 = 2443144.5003725$, and $L_B = 1.550519768 \times 10^{-8}$ and $\mathrm{TDB}_0 = -6.55 \times 10^{-5}\,\mathrm{s}$ are defining constants. Note, that TDB was chosen to practically agree with the time argument of DE405.

4. The BCRS-metric and the expansion of the universe

So far the BCRS-metric was chosen to be asymptotically flat and all cosmological effects have been ignored, That is fine for the description of planetary motion and the propagation of light-rays in the solar system. However, at cosmological distances the expansion of the universe should be considered to deal with cosmological redshifts and the problem of various distance indicators (parallax-, luminosity-, angular diameter- and proper motion distance) that have to be distinguished for remote objects. Motivated also by the so-called Pioneer anomaly we included the expansion of the universe in the BCRS-metric and estimated orders of magnitude for corresponding effects in the solar system. Details can be found in Klioner and Soffel (2004). In the BCRS the expansion of the universe can be considered as a cosmic tidal force. The corresponding tidal acceleration grows with heliocentric distance; at Pluto's orbit it amounts to $2 \times 10^{-23}\ \mathrm{m/s^2}$ and points away from the Sun (Note, that the Pioneer anomaly amounts to $8.7 \times 10^{-10}\mathrm{m/s^2}$ and points towards the Sun). The perturbations of planetary orbits are completely negligible (the cosmic perihelion precession of Pluto's orbits is of order 10^{-5} μas/century).

References

Blanchet, J. & Damour, T. 1989, *Annales de l'institut Henri Poincare (A) Physique theorique*, 50, no. 4, 377

Brumberg, V. & Kopeikin, S. 1988, *Nuovo Cimento B*, 103, 63

Damour, T., Soffel, M., & Xu, C. 1991, *Phys. Rev. D*, 43, 3273 (DSX I); 1992, 45, 1017 (DSX II); 1993, 47, 3124 (DSX III); 1994, 49, 618 (DSX IV)

de Sitter, W. 1916, *Mon. Not. R. Astron. Soc.*, 76, 699 and 77, 155

Droste, J. 1916, *Versl. K. Akad. Wet. Amsterdam*, 19, 447

Einstein, A. 1915, *Preuss. Akad. Wiss. Berlin, Sitzber.*, 831

Fock, V. 1959 *The Theory of Space, Time and Gravitation*, Pergamon Press, Oxford

Klioner, S. & Soffel, M., 2004, *Refining the relativistic model for Gaia: cosmological effects in the BCRS*, Proc. of the GAIA meeting 4-7 October 2004, Paris (ESA SP-576), 305–308

Lorentz, H.A. & Droste, J., 1917 *Versl. K. Akad. Wet. Amsterdam*, 26, 392 (part I); 26, 649 (part II), English translation in H. A. Lorentz, Collected Papers, edited by P. Zeeman and A. D. Fokker (Nijhoff, The Hague, 1937), Vol. V, pp. 330–355

Misner, C., Thorne, K., & Wheeler, J.A., 1972, *Gravitation*, Freeman

Nordtvedt, K., 1970, *Astrophys. J.*, 161, 1059

Soffel, M., *et al.*, 2003, *Astron. J.*, 126, 2687

Will, C., 1993, *Theory and Experiment in gravitational physics*, Cambridge University Press, Cambridge

Relativity in Fundamental Astronomy
Proceedings IAU Symposium No. 261, 2009
S. A. Klioner, P. K. Seidelman & M. H. Soffel, eds.

doi:10.1017/S1743921309990081

Beyond the standard IAU framework

Sergei Kopeikin

Department of Physics & Astronomy, University of Missouri-Columbia,
Columbia, MO 65211, USA

Abstract. We discuss three conceivable scenarios of extension and/or modification of the IAU relativistic resolutions on time scales and spatial coordinates beyond the Standard IAU Framework. These scenarios include: (1) the formalism of the monopole and dipole moment transformations of the metric tensor replacing the scale transformations of time and space coordinates; (2) implementing the parameterized post-Newtonian formalism with two PPN parameters – β and γ; (3) embedding the post-Newtonian barycentric reference system to the Friedman-Robertson-Walker cosmological model.

Keywords. relativity, gravitation, standards, reference systems

---

1. Introduction

Fundamental scientific program of detection of gravitational waves by the space interferometric gravitational wave detectors like LISA is a driving motivation for further systematic developing of relativistic theory of reference frames in the solar system and beyond. LISA will detect gravitational wave sources from all directions in the sky. These sources will include thousands of compact binary systems containing neutron stars, black holes, and white dwarfs in our own Galaxy, and merging super-massive black holes in distant galaxies. However, the detection and proper interpretation of the gravitational waves can be achieved only under the condition that all coordinate-dependent effects are completely understood and subtracted from the signal. This is especially important for observation of gravitational waves from very distant sources located at cosmological distances because the Hubble expansion of our universe affects propagation of the waves.

New generation of microarcsecond astrometry satellites: SIM and a cornerstone mission of ESA – Gaia, requires a novel approach for an unambiguous interpretation of astrometric data obtained from the on-board optical instruments. SIM and Gaia complement one another. Both SIM and Gaia will approach the accuracy of 1 μas. Gaia will observe all stars ($\sim 10^9$) between magnitude 6 and 20. The accuracy of Gaia is about 5 μ as for the optimal stars (magnitude between 6 and 13). SIM is going to observe $\tilde{1}0000$ stars with magnitude up to 20. The accuracy of SIM is expected to be a few μas and can be reached for any object brighter than about 20 provided that sufficient observing time is allocated for that object. At this level the problem of propagation of light rays must be treated with taking into account relativistic effects generated by non-static part of the gravitational field of the solar system and binary stars (Kopeikin & Gwinn 2000). Astrometric resolution in 1 μas forces us to change the classic treatment of parallax, aberration, and proper motion of stars by switching to a more precise definition of reference frames on a curved space-time manifold (Kopeikin & Schäfer 1999, Kopeikin & Mashhoon 2000, Kopeikin *et al.* 2006). Advanced practical realization of an inertial reference frame is required for unambiguous physical interpretation of the gravitomagnetic precession of orbits of LAGEOS satellites (Ciufolini & Pavlis 2004) and GP-B gyroscope, which is measured relative to a binary radio star IM Pegasi that has large annual parallax and proper motion (Ransom *et al.* 2005).

Recent breakthroughs in technology of drag-free satellites, clocks, lasers, optical and radio interferometers and new demands of experimental gravitational physics (Lämmerzahl *et al.* 2001, Dittus *et al.* 2008) make it necessary to incorporate the parameterized post-Newtonian formalism (Will 1993) to the procedure of construction of relativistic local frames around Earth and other bodies of the solar system (Klioner & Soffel 2000, Kopeikin & Vlasov 2004). The domain of applicability of the IAU relativistic theory of reference frames (Soffel *et al.* 2003) should be also extended outside the boundaries of the solar system (Kopeikin & Gwinn 2000).

In what follows, Latin indices takes values 1,2,3, and the Greek ones run from 0 to 3. Repeated indices indicate the Einstein summation rule. The unit matrix is denoted $\delta_{ij} = \mathrm{diag}(1,1,1)$ and the fully anti-symmetric symbol ϵ_{ijk} is subject to $\epsilon_{123} = 1$. The Minkowski metric is $\eta_{\alpha\beta} = \mathrm{diag}(-1,1,1,1)$. Greek indices are raised and lowered with the Minkowski metric, Latin indices are raised and lowered with the unit matrix. Bold italic letters denote spatial vectors. Dot and cross between two spatial vectors denote the Euclidean scalar and vector products respectively. Partial derivative with respect to spatial coordinates x^i are denoted as $\partial/\partial x^i$ or $\vec{\nabla}$.

2. Standard IAU Framework

New relativistic resolutions on reference frames and time scales in the solar system were adopted by the 24-th General Assembly of the IAU in 2000 (Soffel *et al.* 2003, Soffel 2009). The resolutions are based on the first post-Newtonian approximation of general relativity. They abandoned the Newtonian paradigm of space and time and required corresponding change in the conceptual basis and terminology of the fundamental astronomy (Capitaine *et al.* 2006).

Barycentric Celestial Reference System (BCRS), $x^\alpha = (ct, \boldsymbol{x})$, is defined in terms of a metric tensor $g_{\alpha\beta}$ with components

$$g_{00} = -1 + \frac{2w}{c^2} - \frac{2w^2}{c^4} + O(c^{-5}), \tag{2.1}$$

$$g_{0i} = -\frac{4w^i}{c^3} + O(c^{-5}), \tag{2.2}$$

$$g_{ij} = \delta_{ij}\left(1 + \frac{2w}{c^2}\right) + O(c^{-4}). \tag{2.3}$$

Here, the post-Newtonian gravitational potential w generalizes the Newtonian potential, and w^i is a vector potential related to the gravitomagnetic effects (Ciufolini & Wheeler 1995). These potentials are defined by solving the field equations

$$\Box w = -4\pi G\sigma, \tag{2.4}$$

$$\Box w^i = -4\pi G\sigma^i, \tag{2.5}$$

where $\Box \equiv -c^{-2}\partial^2/\partial t^2 + \nabla^2$ is the wave operator, $\sigma = c^{-2}(T^{00} + T^{ss})$, $\sigma^i = c^{-1}T^{0i}$, and $T^{\mu\nu}$ are the components of the stress-energy tensor of the solar system bodies, $T^{ss} = T^{11} + T^{22} + T^{33}$.

Equations (2.4), (2.5) are solved by iterations

$$w(t,\boldsymbol{x}) = G\int \frac{\sigma(t,\boldsymbol{x}')d^3x'}{|\boldsymbol{x}-\boldsymbol{x}'|} + \frac{G}{2c^2}\frac{\partial^2}{\partial t^2}\int d^3x'\sigma(t,\boldsymbol{x}')|\boldsymbol{x}-\boldsymbol{x}'| + O(c^{-4}), \tag{2.6}$$

$$w^i(t,\boldsymbol{x}) = G\int \frac{\sigma^i(t,\boldsymbol{x}')d^3x'}{|\boldsymbol{x}-\boldsymbol{x}'|} + O(c^{-2}), \tag{2.7}$$

which are to be substituted to the metric tensor (2.1)–(2.3). Each of the potentials, w and w^i, can be linearly decomposed in two parts

$$w = w_E + \bar{w}, \tag{2.8}$$
$$w^i = w^i_E + \bar{w}^i, \tag{2.9}$$

where w_E and w^i_E are BCRS potentials depending on the distribution of mass and current only inside the Earth, and $\bar{w}_E$ and $\bar{w}^i_E$ are gravitational potentials of external bodies.

Geocentric Celestial Reference System (GCRS) is denoted $X^\alpha = (cT, \boldsymbol{X})$. GCRS is defined in terms of the metric tensor $G_{\alpha\beta}$ with components

$$G_{00} = -1 + \frac{2W}{c^2} - \frac{2W^2}{c^4} + O(c^{-5}), \tag{2.10}$$
$$G_{0i} = -\frac{4W^i}{c^3} + O(c^{-5}), \tag{2.11}$$
$$G_{ij} = \delta_{ij}\left(1 + \frac{2W}{c^2}\right) + O(c^{-4}). \tag{2.12}$$

Here $W = W(T, \boldsymbol{X})$ is the post-Newtonian gravitational potential and $W^i(T, \boldsymbol{X})$ is a vector-potential both expressed in the geocentric coordinates. They satisfy to the same type of the wave equations (2.4), (2.5).

The geocentric potentials are split into two parts: potentials W_E and W^i_E arising from the gravitational field of the Earth and external parts associated with tidal and inertial effects. IAU resolutions implied that the external parts must vanish at the geocenter and admit an expansion in powers of $\boldsymbol{X}$ (Soffel *et al.* 2003, Soffel 2009)

$$W(T, \boldsymbol{X}) = W_E(T, \boldsymbol{X}) + W_{\text{kin}}(T, \boldsymbol{X}) + W_{\text{dyn}}(T, \boldsymbol{X}), \tag{2.13}$$
$$W^i(T, \boldsymbol{X}) = W^i_E(T, \boldsymbol{X}) + W^i_{\text{kin}}(T, \boldsymbol{X}) + W^i_{\text{dyn}}(T, \boldsymbol{X}). \tag{2.14}$$

Geopotentials W_E and W^i_E are defined in the same way as w_E and w^i_E (see equations (2.6)–(2.7)) but with quantities σ and σ^i calculated in the GCRS. W_{kin} and W^i_{kin} are kinematic contributions that are linear in spatial coordinates $\boldsymbol{X}$

$$W_{\text{kin}} = Q_i X^i, \qquad W^i_{\text{kin}} = \frac{1}{4}\, c^2 \varepsilon_{ipq}(\Omega^p - \Omega^p_{\text{prec}})\, X^q, \tag{2.15}$$

where Q_i characterizes the minute deviation of the actual world line of the geocenter from that of a fiducial test particle being in a free fall in the external gravitational field of the solar system bodies (Kopeikin 1988)

$$Q_i = \partial_i \bar{w}(\boldsymbol{x}_E) - a^i_E + O(c^{-2}). \tag{2.16}$$

Here $a^i_E = dv^i_E/dt$ is the barycentric acceleration of the geocenter. Function Ω^a_{prec} describes the relativistic precession of dynamically non-rotating spatial axes of GCRS with respect to reference quasars

$$\Omega^i_{\text{prec}} = \frac{1}{c^2}\, \varepsilon_{ijk}\left(-\frac{3}{2}\, v^j_E\, \partial_k \bar{w}(\boldsymbol{x}_E) + 2\, \partial_k \bar{w}^j(\boldsymbol{x}_E) - \frac{1}{2}\, v^j_E\, Q^k\right). \tag{2.17}$$

The three terms on the right-hand side of this equation represent the geodetic, Lense-Thirring, and Thomas precessions, respectively (Kopeikin 1988, Soffel *et al.* 2003).

Potentials W_{dyn} and W^i_{dyn} are generalizations of the Newtonian tidal potential. For example,

$$W_{\text{dyn}}(T, \boldsymbol{X}) = \bar{w}(\boldsymbol{x}_E + \boldsymbol{X}) - \bar{w}(\boldsymbol{x}_E) - X^i \partial_i \bar{w}(\boldsymbol{x}_E) + O(c^{-2}). \tag{2.18}$$

It is easy to check out that a Taylor expansion of $\bar{w}(\boldsymbol{x}_E + \boldsymbol{X})$ around the point $\boldsymbol{x}_E$ yields

$W_{\rm dyn}(T, \boldsymbol{X})$ in the form of a polynomial starting from the quadratic with respect to $\boldsymbol{X}$ terms. We also note that the local gravitational potentials W_E and W_E^i of the Earth are related to the barycentric gravitational potentials w_E and w_E^i by the post-Newtonian transformations (Brumberg & Kopeikin 1989, Soffel *et al.* 2003).

3. IAU Scaling Rules and the Metric Tensor

The coordinate transformations between the BCRS and GCRS are found by matching the BCRS and GCRS metric tensors in the vicinity of the world line of the Earth by making use of their tensor properties. The transformations are written as (Kopeikin 1988, Soffel *et al.* 2003)

$$T = t - \frac{1}{c^2}\left[A + \boldsymbol{v}_E\cdot\boldsymbol{r}_E\right] + \frac{1}{c^4}\Big[B + B^i r_E^i + B^{ij} r_E^i r_E^j + C(t,\boldsymbol{x})\Big] + O(c^{-5}), \quad (3.1)$$

$$X^i = r_E^i + \frac{1}{c^2}\left[\frac{1}{2} v_E^i \boldsymbol{v}_E\cdot\boldsymbol{r}_E + \bar{w}(\boldsymbol{x}_E) r_E^i + r_E^i \boldsymbol{a}_E\cdot\boldsymbol{r}_E - \frac{1}{2} a_E^i r_E^2\right] + O(c^{-4}), \quad (3.2)$$

where $\boldsymbol{r}_E = \boldsymbol{x} - \boldsymbol{x}_E$, and functions $A, B, B^i, B^{ij}, C(t,\boldsymbol{x})$ are

$$\frac{dA}{dt} = \frac{1}{2} v_E^2 + \bar{w}(\boldsymbol{x}_E), \quad (3.3)$$

$$\frac{dB}{dt} = -\frac{1}{8} v_E^4 - \frac{3}{2} v_E^2\, \bar{w}(\boldsymbol{x}_E) + 4\, v_E^i\, \bar{w}^i + \frac{1}{2}\bar{w}^2(\boldsymbol{x}_E), \quad (3.4)$$

$$B^i = -\frac{1}{2} v_E^2\, v_E^i + 4\,\bar{w}^i(\boldsymbol{x}_E) - 3\, v_E^i\, \bar{w}(\boldsymbol{x}_E), \quad (3.5)$$

$$B^{ij} = -v_E^i Q_j + 2\partial_j \bar{w}^i(\boldsymbol{x}_E) - v_E^i \partial_j \bar{w}(\boldsymbol{x}_E) + \frac{1}{2}\delta^{ij}\dot{\bar{w}}(\boldsymbol{x}_E), \quad (3.6)$$

$$C(t,\boldsymbol{x}) = -\frac{1}{10} r_E^2\,(\dot{\boldsymbol{a}}_E\cdot\boldsymbol{r}_E). \quad (3.7)$$

Here again x_E^i, v_E^i, and a_E^i are the barycentric position, velocity and acceleration vectors of the Earth, the dot stands for the total derivative with respect to t. The harmonic gauge condition does not fix the function $C(t,\boldsymbol{x})$ uniquely. However, it is reasonable to fix it in the time transformation for practical reasons (Soffel *et al.* 2003).

Earth's orbit in BCRS is almost circular. This makes the right side of equation (3.3) almost constant with small periodic oscillations

$$\frac{1}{2} v_E^2 + \bar{w}(\boldsymbol{x}_E) = c^2 L_C + (\text{periodic terms}), \quad (3.8)$$

where the constant L_C and the periodic terms have been calculated with a great precision in (Fukushima 1995). For practical reason of calculation of ephemerides of the solar system bodies, the BCRS time coordinate was re-scaled to remove the constant L_C from the right side of equation (3.8). The new time scale was called TDB

$$t_{TDB} = t(1 - L_B), \quad (3.9)$$

where a new constant L_B is used for practical purposes instead of L_C in order to take into account the additional linear drift between the geocentric time T and the atomic time on geoid, as explained in (Brumberg & Kopeikin 1990, Irwin & Fukushima 1999, Klioner *et al.* 2009). Time re-scaling changes the Newtonian equations of motion. In order to keep the equations of motion invariant scientists doing the ephemerides also re-scaled spatial coordinates and masses of the solar system bodies. These scaling transformations are included to IAU2000 resolutions (Soffel *et al.* 2003) but they have never been explicitly associated with transformation of the metric tensor. It had led to a long-standing discussion about the units of measurement of time, space, and mass in astronomical measurements (Klioner *et al.* 2009).

The scaling of time and space coordinates is associated with a particular choice of the metric tensor corresponding to either TCB or TDB time. In order to see it, we notice that equation (2.15) is a solution of the Laplace equation, which is defined up to an arbitrary function of time Q. If one takes it into account, equations (2.15), (3.3) can be re-written as

$$W_{\rm kin} = Q + Q_i X^i, \tag{3.10}$$

$$\frac{dA(t)}{dt} = \frac{1}{2} v_E^2 + \bar{w}(\boldsymbol{x}_E) - Q, \tag{3.11}$$

and, if we chose $Q = c^2 L_C$, it eliminates the secular drift between times T and t without explicit re-scaling of time, which is always measured in SI units. It turns out that Blanchet-Damour (Blanchet & Damour 1989) relativistic definition of mass depends on function Q and is re-scaled in such a way that the Newtonian equations of motion remain invariant. Introduction of Q to function W brings about implicitly re-scales spatial coordinates as well. We conclude that introducing function $Q = c^2 L_C$ to the metric tensor without apparent re-scaling of coordinates and masses might be more preferable for IAU resolutions as it allows us to keep the SI system of units without changing coordinates and masses made ad hoc "by hands". Similar procedure can be developed for re-scaling the geocentric time T to take into account the linear drift existing between this time and the atomic clocks on geoid (Brumberg & Kopeikin 1990, Irwin & Fukushima 1999).

4. Parameterized Coordinate Transformations

This section discusses how to incorporate the parameterized post-Newtonian (PPN) formalism (Will 1993) to the IAU resolutions. This extends applicability of the resolutions to a more general class of gravity theories. Furthermore, it makes the IAU resolutions fully compatible with JPL equations of motion used for calculation of ephemerides of major planets, Sun and Moon.

These equations of motion depend on two PPN parameters, β and γ (Seidelmann 1992) and they are presently compatible with the IAU resolutions only in the case of $\beta = \gamma = 1$. Rapidly growing precision of optical and radio astronomical observations as well as calculation of relativistic equations of motion in gravitational wave astronomy urgently demands to work out a PPN theory of relativistic transformations between the local and global coordinate systems.

PPN parameters β and γ are characteristics of a scalar field which makes the metric tensor different from general relativity. In order to extend the IAU 2000 theory of reference frames to the PPN formalism we employ a general class of Brans-Dicke theories (Brans & Dicke 1961). This class is based on the metric tensor $g_{\alpha\beta}$ and a scalar field ϕ that couples with the metric tensor through function $\theta(\phi)$. We assume that ϕ and $\theta(\phi)$ are analytic functions which can be expanded in a Taylor series about their background values $\bar{\phi}$ and $\bar{\theta}$.

The parameterized theory of relativistic reference frames in the solar system is built in accordance to the same rules as used in the IAU resolutions. The entire procedure is described in our paper (Kopeikin & Vlasov 2004). The parameterized coordinate transformations between BCRS and GCRS are found by matching the BCRS and GCRS metric tensors and the scalar field in the vicinity of the world line of the Earth. The transformations have the following form (Kopeikin & Vlasov 2004)

$$T = t - \frac{1}{c^2}\left[A + \boldsymbol{v}_E \cdot \boldsymbol{r}_E\right] + \frac{1}{c^4}\left[B + B^i\, r_E^i + B^{ij}\, r_E^i\, r_E^j + C(t,\boldsymbol{x})\right] + O(c^{-5}), \quad (4.1)$$

$$X^i = r_E^i + \frac{1}{c^2}\left[\frac{1}{2} v_E^i v_E^j r_E^j + \gamma Q r_E^i + \gamma \bar{w}(\boldsymbol{x}_E) r_E^i + r_E^i a_E^j r_E^j - \frac{1}{2} a_E^i r_E^2\right] + O(c^{-4}) \quad (4.2)$$

where $\boldsymbol{r}_E = \boldsymbol{x} - \boldsymbol{x}_E$, and functions $A(t), B(t), B^i(t), B^{ij}(t), C(t,\boldsymbol{x})$ are

$$\frac{dA}{dt} = \frac{1}{2} v_E^2 + \bar{w} - Q(\boldsymbol{x}_E), \quad (4.3)$$

$$\frac{dB}{dt} = -\frac{1}{8} v_E^4 - \left(\gamma + \frac{1}{2}\right) v_E^2\, \bar{w}(\boldsymbol{x}_E) + 2(1+\gamma)\, v_E^i\, \bar{w}^i + \left(\beta - \frac{1}{2}\right) \bar{w}^2(\boldsymbol{x}_E), \quad (4.4)$$

$$B^i = -\frac{1}{2} v_E^2\, v_E^i + 2(1+\gamma)\, \bar{w}^i(\boldsymbol{x}_E) - (1+2\gamma)\, v_E^i\, \bar{w}(\boldsymbol{x}_E), \quad (4.5)$$

$$B^{ij} = -v_E^i Q_j + (1+\gamma)\partial_j \bar{w}^i(\boldsymbol{x}_E) - \gamma v_E^i \partial_j \bar{w}(\boldsymbol{x}_E) + \frac{1}{2}\delta^{ij}\, \dot{\bar{w}}(\boldsymbol{x}_E), \quad (4.6)$$

$$C(t,\boldsymbol{x}) = -\frac{1}{10} r_E^2\, (\dot{\boldsymbol{a}}_E \cdot \boldsymbol{r}_E). \quad (4.7)$$

These transformations depends explicitly on the PPN parameters β and γ and the scaling function Q, and should be compared with those (3.1)-(3.7) adopted in the IAU resolutions.

PPN parameters β and γ have a fundamental physical meaning in the scalar-tensor theory of gravity along with the universal gravitational constant G and the fundamental speed c. It means that if the parameterized transformations (4.1)-(4.7) are adopted by the IAU, the parameters β and γ must be included to the number of the astronomical constants which values must be determined experimentally. The program of the experimental determination of β and γ began long time ago and it makes use of various observational techniques. So far, the experimental values of β and γ are indistinguishable from their general-relativistic values $\beta = 1$, $\gamma = 1$.

5. Matching IAU Resolutions with Cosmology

BCRS assumes that the solar system is isolated and space-time is asymptotically flat. This idealization will not work at some level of accuracy of astronomical observations because the universe is expanding and its space-time is described by the Friedman-Robertson-Walker (FRW) metric tensor having non-zero Riemannian curvature (Misner *et al.* 1973). It may turn out that some, yet unexplained anomalies in the orbital motion of the solar system bodies (Anderson 2009, Krasinsky & Brumberg 2004, Lämmerzahl *et al.* 2006) are indeed associated with the cosmological expansion. Moreover, astronomical observations of cosmic microwave background radiation and other cosmological effects requires clear understanding of how the solar system is embedded to the cosmological model. Therefore, it is reasonable to incorporate the cosmological expansion of space-time to the IAU 2000 theory of reference frames in the solar system (Kopeikin *et al.* 2001, Ramirez *et al.* 2002).

Because the universe is not asymptotically-flat the gravitational field of the solar system can not vanish at infinity. Instead, it must match with the cosmological metric tensor. This imposes the cosmological boundary condition. The cosmological model is not unique has a number of free parameters depending on the amount of visible and dark matter, and on the presence of dark energy. We considered a FRW universe driven by a scalar field imitating the dark energy ϕ and having a spatial curvature equal to zero (Dolgov *et al.* 1990, Mukhanov 2005). The universe is perturbed by a localized distribution of matter of the solar system. In this model the perturbed metric tensor reads

$$g_{\alpha\beta} = a^2(\eta) f_{\alpha\beta}, \qquad f_{\alpha\beta} = \eta_{\alpha\beta} + h_{\alpha\beta}, \tag{5.1}$$

where perturbation $h_{\alpha\beta}$ of the background FRW metric tensor $\bar{g}_{\alpha\beta} = a^2\eta_{\alpha\beta}$ is caused by the presence of the solar system, $a(\eta)$ is a scale factor of the universe depending on the, so-called, conformal time η related to coordinate time t by simple differential equation $dt = a(\eta)d\eta$. In what follows, a linear combination of the metric perturbations

$$\gamma^{\alpha\beta} = h^{\alpha\beta} - \frac{1}{2}\eta^{\alpha\beta} h, \tag{5.2}$$

where $h = \eta^{\alpha\beta} h_{\alpha\beta}$, is more convenient for calculations.

We discovered a new cosmological gauge, which has a number of remarkable properties. In case of the background FRW universe with dust equation of state (that is, the background pressure of matter is zero (Dolgov *et al.* 1990, Mukhanov 2005) this gauge is given by (Kopeikin *et al.* 2001, Ramirez *et al.* 2002)

$$\gamma^{\alpha\beta}{}_{|\beta} = 2H\varphi\delta^{\alpha}_{0}, \tag{5.3}$$

where bar denotes a covariant derivative with respect to the background metric $\bar{g}_{\alpha\beta}$, $\varphi = \phi/a^2$, $H = \dot{a}/a$ is the Hubble parameter, and the overdot denotes a time derivative with respect to time η. The gauge (5.3) generalizes the harmonic gauge of asymptotically-flat space-time for the case of the expanding non-flat background universe.

The gauge (5.3) drastically simplifies the linearized Einstein equations. Introducing notations $\gamma_{00} \equiv 4w/c^2$, $\gamma_{0i} \equiv -4w^i/c^3$, and $\gamma_{ij} \equiv 4w^{ij}/c^4$, and splitting Einstein's equations in components, yield

$$\Box\chi - 2H\partial_\eta\chi + \frac{5}{2}H^2\chi = -4\pi G\sigma, \tag{5.4}$$

$$\Box w - 2H\partial_\eta w = -4\pi G\sigma - 4H^2\chi, \tag{5.5}$$

$$\Box w^i - 2H\partial_\eta w^i + H^2 w^i = -4\pi G\sigma^i, \tag{5.6}$$

$$\Box w^{ij} - 2H\partial_\eta w^{ij} = -4\pi G T^{ij}, \tag{5.7}$$

where $\partial_\eta \equiv \partial/\partial\eta$, $\Box \equiv -c^{-2}\partial^2_\eta + \nabla^2$, $\chi \equiv w - \varphi/2$, the Hubble parameter $H = \dot{a}/a = 2/\eta$, densities $\sigma = c^{-2}(T^{00} + T^{ss})$, $\sigma^i = c^{-1}T^{0i}$ with $T^{\alpha\beta}$ being the tensor of energy-momentum of matter of the solar system defined with respect to the metric $f_{\alpha\beta}$. These equations extend the domain of applicability of equations (2.4), (2.5) of the IAU standard framework (Soffel 2009) to the case of expanding universe.

First equation (5.4) describes evolution of the scalar field while the second equation (5.5) describes evolution of the scalar perturbation of the metric tensor. Equation (5.6) yields evolution of vector perturbations of the metric tensor, and equation (5.7) describes generation and propagation of gravitational waves by the isolated N-body system. Equations (5.4)–(5.7) contain all corrections depending on the Hubble parameter and can be solved analytically in terms of generalized retarded solution. Exact Green functions for these equations have been found in (Kopeikin *et al.* 2001, Ramirez *et al.* 2002, Haas & Poisson 2005). They revealed that the gravitational perturbations of the isolated system on expanding background depend not only on the value of the source taken on the past null cone but also on the value of the gravitational field inside the past null cone.

Existence of extra terms in the solutions of equations (5.4)–(5.7) depending on the Hubble parameter brings about cosmological corrections to the Newtonian law of gravity. For example, the post-Newtonian solutions of equations (5.5), (5.6) with a linear correction due to the Hubble expansion are

$$w(\eta,\boldsymbol{x})=G\int\frac{\sigma(\eta,\boldsymbol{x}')d^3x'}{|\boldsymbol{x}-\boldsymbol{x}'|}+\frac{G}{2c^2}\frac{\partial^2}{\partial\eta^2}\int d^3x'\sigma(\eta,\boldsymbol{x}')|\boldsymbol{x}-\boldsymbol{x}'| \\ -GH\int d^3x'\sigma(\eta,\boldsymbol{x}')+O\left(c^{-4}\right)+O\left(H^2\right), \tag{5.8}$$

$$w^i(\eta,\boldsymbol{x})=G\int\frac{\sigma^i(\eta,\boldsymbol{x}')d^3x'}{|\boldsymbol{x}-\boldsymbol{x}'|}+O\left(c^{-2}\right)+O\left(H^2\right). \tag{5.9}$$

Matching these solutions with those defined in the BCRS of the IAU 2000 framework in equations (2.6), (2.7) is achieved after expanding all quantities depending on the conformal time η in the neighborhood of the present epoch in powers of the Hubble parameter.

Current IAU 2000 paradigm assumes that the asymptotically-flat metric, $f_{\alpha\beta}$, is used for calculation of light propagation and ephemerides of the solar system bodies. It means that the conformal time η is implicitly interpreted as TCB in equations of motion of light and planets. However, the physical metric $g_{\alpha\beta}$ differs from $f_{\alpha\beta}$ by the scale factor $a^2(\eta)$, and the time η relates to TCB as a Taylor series that can be obtained after expanding $a(\eta)=a_0+\dot{a}\eta+\ldots$, in polynomial around the initial epoch η_0 and defining TCB at the epoch as TCB$=a_0\eta$. Integrating equation $dt=a(\eta)d\eta$ where t is the coordinate time, yields

$$t=\mathrm{TCB}+\frac{1}{2}\mathcal{H}\cdot\mathrm{TCB}^2+\ldots, \tag{5.10}$$

where $\mathcal{H}=H/a_0$ and ellipses denote terms of higher order in the Hubble constant $\mathcal{H}$. The coordinate time t relates to the atomic time TAI (the proper time of observer) by equation, which does not involve the scale factor $a(\eta)$ at the main approximation. It means that in order to incorporate the cosmological expansion to the equations of motion, one must replace TCB to the quadratic form

$$\mathrm{TCB}\longrightarrow\mathrm{TCB}+\frac{1}{2}\mathcal{H}\cdot\mathrm{TCB}^2. \tag{5.11}$$

Distances in the solar system are measured by radio ranging spacecrafts and planets. Equations of light propagation preserve their form if one keeps the speed of light constant and replace coordinates $(\eta,\boldsymbol{x})$ to (t,Ξ), where the spatial coordinates Ξ relate to coordinates $\boldsymbol{x}$ by equation $d\Xi=a(\eta)d\boldsymbol{x}$. Because one uses TAI for measuring time, the values of the spatial coordinates in the range measurements are given in terms of the capitalized coordinates Ξ. Therefore, the ranging measurements are not affected by the time transformation (5.10) in contrast to the measurement of the Doppler shift, which deals with time only. This interpretation may reconcile the "Quadratic Time Augmentation" model of the Pioneer anomaly discussed by Anderson *et al.* (2002) in equation (61) of their paper.

Acknowledgements

This work was supported by the Research Council Grant No. C1669103 and by 2009 Faculty Incentive Grant of the Alumni Organization of the University of Missouri-Columbia.

References

Anderson, J. D., Laing, P. A., Lau, E. L., Liu, A. S., Nieto, M. M., & Turyshev, S. G. 2002, *Phys. Rev.* D, **65**, 082004

Anderson, J. D. 2009, *Astrometric Solar-System Anomalies, this proceedings*, 189

Blanchet, L. & Damour, T. 1989, *Ann. Inst. H. Poincare, Phys. Theor.*, **50**, 377

Brans, C. H. & Dicke, R. H. 1961, *Phys. Rev.* D, **124**, 925

Brumberg, V. A. & Kopeikin, S. M. 1989, *Nuovo Cim. B*, **103**, 63

Brumberg, V. A. & Kopeikin, S. M. 1990, *Cel. Mech. Dyn. Astron.*, **48**, 23

Capitaine, N., Andrei, A. H., Calabretta, M., Dehant, V., Fukushima, T., Guinot, B., Hohenkerk, C., Kaplan, G., Klioner, S., Kovalevsky, J., Kumkova, I., Ma, C., McCarthy, D. D., Seidelmann, K., & Wallace, P. T. 2006, *Nomenclature, Precession and New Models in Fundamental Astronomy*, 26th IAU Gen. Assembbly, JD16, Prague, Czech Republic

Ciufolini, I. & Wheeler, J. A. 1995, *Gravitation and Inertia*, (Princeton Univ. Press: Princeton, NJ, 1995)

Ciufolini, I. & Pavlis, E. C. 2004, *Nature*, **431**, 958

Dittus, H., Lämmerzahl, C., Ni, W.-T., & Turyshev, S. (eds.) 2008, *Lasers, Clocks and Drag-Free: Technologies for Future Exploration in Space and Tests of Gravity* (Springer: Berlin, 2008)

Dolgov, A. D., Sazhin, M. V., & Zeldovich, Ya. B. 1990, *Basics of Modern Cosmology* (Editions Frontieres: Gif-sur-Yvette, France, 1990)

Fukushima, T. 1995, *Astron. Astrophys.*, **294**, 895

Haas, R. & Poisson, E. 2005, *Class. Quant. Grav.*, **22**, 739

Irwin, A. W. & Fukushima, T. 1999, *Astron. Astrophys.*, **348**, 642

Klioner, S. A. & Soffel, M. H. 2000, *Phys. Rev.* D, **62**, 024019

Klioner, S., Capitaine, N., Folkner, W., Guinot, B., Huang, T. Y., Kopeikin, S., Petit, G., Pitjeva, E., Seidelmann, P. K., & Soffel, M. 2009, *Units of Relativistic Time Scales and Associated Quantities, this proceedings*, 79

Kopeikin, S. M. 1988, *Cel. Mech.*, **44**, 87

Kopeikin, S. M. & Schäfer, G. 1999, *Phys. Rev.* D, **60**, 124002

Kopeikin, S. M. & Gwinn, C. R. 2000, In: *Towards Models and Constants for Sub-Microarcsecond Astrometry*, eds. K.J. Johnston, D.D. McCarthy, B.J. Luzum and G.H. Kaplan (USNO, Washington DC, 2000), pp. 303–307

Kopeikin, S. M., Ramirez, J., Mashhoon, B., & Sazhin, M. V. 2001, *Phys. Lett.* A, **292**, 173

Kopeikin, S. M. & Mashhoon, B. 2002, *Phys. Rev.* D, **65**, 064025

Kopeikin, S. & Vlasov, I. 2004, *Phys. Reports*, **400**, 209

Kopeikin, S., Korobkov, P., & Polnarev, A. 2006, *Class. Quant. Grav.*, **23**, 4299

Krasinsky, G. A. & Brumberg, V. A. 2004, *Cel. Mech. Dyn. Astron.*, **90**, 267

Lämmerzahl, C., Everitt, C. W. F., & Hehl, F. W. (eds.) 2001, *Gyros, Clocks, Interferometers: Testing Relativistic Gravity in Space*, Lecture Notes in Physics, Vol. 562 (Springer: Berlin, 2001)

Lämmerzahl, C., Preuss, O., & Dittus, H. 2006, In: *Lasers, Clocks, and Drag-Free: Technologies for Future Exploration in Space and Tests of Gravity* Proc. of the 359th WE-Heraeus Seminar, eds. H. Dittus, C. Lämmerzahl, W.-T. Ni, & S. Turyshev (Springer: Berlin, 2006)

Misner, C. W., Thorne, K. S., & Wheeler, J. A. 1973, *Gravitation* (Freeman: San Francisco, 1973)

Mukhanov, V. 2005, *Physical Foundations of Cosmology*, (Cambridge University Press: Cambridge, 2005)

Ramirez, J. & Kopeikin, S. 2002, *Phys. Lett.* B, **532**, 1

Ransom, R. R., Bartel, N., Bietenholz, M. F., Ratner, M. I., Lebach, D. I., Shapiro, I. I., & Lestrade, J.-F. 2005, In: *Future Directions in High Resolution Astronomy: The 10th Anniversary of the VLBA*, ASP Conf. Proc., **340**. Eds. J. Romney and M. Reid, pp. 506–510

Seidelmann, P. K. 1992, *Explanatory Supplement to the Astronomical Almanac* (University Science Books: Mill Valley, California, 1992) pp. 281–282

Soffel, M., Klioner, S. A., Petit, G., Wolf, P., Kopeikin, S. M., Bretagnon, P., Brumberg, V. A., Capitaine, N., Damour, T., Fukushima, T., Guinot, B., Huang, T.-Y., Lindegren, L., Ma, C., Nordtvedt, K., Ries, J. C., Seidelmann, P. K., Vokrouhlický, D., Will, C. M., & Xu, C. 2003, *Astron. J. (USA)*, **126**, 2687

Soffel, M., 2009, *Standard Relativistic Reference Systems and the IAU Framework, this proceedings*, 1

Will, C. M. 1993, *Theory and Experiment in Gravitational Physics*, (Cambridge University Press: Cambridge, 1993)

Relativity in Fundamental Astronomy
Proceedings IAU Symposium No. 261, 2009
S. A. Klioner, P. K. Seidelman & M. H. Soffel, eds.

doi:10.1017/S1743921309990093

Relativity in the IERS Conventions

Gérard Petit[1]

[1]Bureau International des Poids et Mesures
92312 Sèvres Cedex, France
email: gpetit@bipm.org

Abstract. In the last years, a fully general relativistic definition of reference systems and of their application to astronomy and geodesy has been passed into Resolutions of the scientific unions, following work of several working groups and of the community at large. In this community, the role of the International Earth Rotation and Reference systems Service (IERS) is to generate the terrestrial and celestial reference systems and the transformation between them, and the IERS Conventions provide the set of models and procedures used in the generation of IERS products. It is therefore essential that the IAU framework for relativity is introduced in the IERS Conventions, and that this is done consistently and completely throughout the document. The paper reviews relativistic aspects in the IERS Conventions and presents recent and on-going work aiming at providing a complete and consistent presentation for a new reference edition of the IERS Conventions, expected to appear in the next year.

Keywords. Reference systems, relativity, time

The International Earth Rotation and Reference systems Service (IERS) has been established in 1988 by the International Astronomical Union (IAU) and the International Union of Geodesy and Geophysics (IUGG). Its primary objectives are to serve the astronomical, geodetic and geophysical communities by providing realizations of the celestial (ICRF) and terrestrial (ITRF) reference systems and parameters allowing to transform between the two systems. In addition the IERS provides geophysical data to interpret time/space variations in the ICRF, ITRF or earth orientation parameters, and to model such variations. The IERS Conventions present the set of standards, constants, models and procedures to generate the above-mentioned products. They are formalized in successive reference versions, the IERS Standards (1992) (McCarthy, 1992), the IERS Conventions (1996) (McCarthy, 1996), and IERS Conventions (2003) (McCarthy & Petit, 2004).

We recall the relativistic framework built up by the set of IAU Resolutions in Section 1 and present in Section 2 the work in progress or already realized concerning all relativistic aspects in the IERS Conventions.

1. The relativistic framework

In order to describe observations in astronomy and geodesy, one has to choose the relativistic reference systems best suited to the problem at hand. A barycentric celestial reference system (BCRS) should be used for all experiments not confined to the vicinity of the Earth, while a geocentric celestial reference system (GCRS) is physically adequate to describe processes occurring in the vicinity of the Earth. These systems have been defined in a series of Resolutions passed by scientific Unions, mostly the IAU, in the past 20 years, see a more complete description of the work until year 2000 in (Soffel *et al.*, 2003).

1991 and the following years. The reference systems were first defined by the IAU Resolution A4 (1991) which contains nine recommendations, the first four of which are summarized below.

In the first recommendation, the metric tensor for space-time coordinate systems $(t, \mathbf{x})$ centered at the barycenter of an ensemble of masses is recommended in the form

$$\begin{aligned} g_{00} &= -1 + 2U(t, \mathbf{x})/c^2 + \mathcal{O}(c^{-4}), \\ g_{0i} &= \mathcal{O}(c^{-3}), \\ g_{ij} &= \delta_{ij}\left(1 + 2U(t, \mathbf{x})/c^2\right) + \mathcal{O}(c^{-4}), \end{aligned} \tag{1.1}$$

where c is the speed of light in vacuum ($c = 299792458$ m/s) and U is the Newtonian gravitational potential (here a sum of the gravitational potentials of the ensemble of masses, and of a external potential generated by bodies external to the ensemble, the latter potential vanishing at the origin). The recommended form of the metric tensor can be used, not only to describe the barycentric reference system of the whole solar system, but also to define the geocentric reference system centered in the center of mass of the Earth with U, now depending upon geocentric coordinates.

In the second recommendation, the origin and orientation of the spatial coordinate grids for the barycentric and geocentric reference systems are defined.

The third recommendation defines TCB (Barycentric Coordinate Time) and TCG (Geocentric Coordinate Time) as the time coordinates of the BCRS and GCRS, respectively, and, in the fourth recommendation, another time coordinate named TT (Terrestrial Time), is defined for the GCRS as

$$\mathrm{TT} = \mathrm{TCG} - \mathrm{L_G} \times (\mathrm{JD_{TCG}} - T_0) \times 86400, \tag{1.2}$$

where $\mathrm{JD_{TCG}}$ is the TCG Julian date, $T_0 = 2443144.5003725$ and where $L_G = U_G/c^2$ with U_G being the gravity potential on the geoid.

Note that the IUGG, in its Resolution 2 (1991), endorsed the IAU Recommendations and explicitly based its definition of Terrestrial Reference Systems on the IAU relativistic framework.

In 1997 the IAU has supplemented the framework by one more recommendation stating that no scaling of spatial axes should be applied in any reference system, even if scaled time coordinate like TT is used for convenience of an analysis (this is in relation with e.g. discussions on the VLBI model, see next Section).

2000 and the recent years. In the years following the adoption of the IAU'1991 Resolution, it became obvious that this set of recommendations was not sufficient, especially with respect to planned astrometric missions with μas-accuracies and with respect to the expected improvement of atomic clocks and the planned space missions involving such clocks and improved time transfer techniques. For that reason the IAU WG "Relativity for celestial mechanics and astronomy" together with the BIPM-IAU Joint Committee for relativity suggested an extended set of Resolutions that was finally adopted at the IAU General Assembly in Manchester in the year 2000 as Resolutions B1.3 to B1.5 and B1.9.

Resolution B1.3 concerns the definition of Barycentric Celestial Reference System (BCRS) and Geocentric Celestial Reference System (GCRS). The Resolution recommends to write the metric tensor of the BCRS in the form

$$\begin{aligned} g_{00} &= -1 + 2w/c^2 - 2w^2/c^4 + \mathcal{O}(c^{-5}), \\ g_{0i} &= -4/c^3 w^i + \mathcal{O}(c^{-5}), \\ g_{ij} &= \delta_{ij}\left(1 + 2w/c^2\right) + \mathcal{O}(c^{-4}). \end{aligned} \tag{1.3}$$

where w is a scalar potential and w^i a vector potential. This extends the form of the metric tensor given in (1.1), so that its accuracy is now sufficient for all applications foreseen in the next years, including those involving accurate space clocks. For the GCRS, Resolution B1.3 also adds that the spatial coordinates are kinematically non-rotating with respect to the barycentric ones.

Resolution B1.4 provides the form of the expansion of the post-Newtonian potential of the Earth to be used with the metric of Resolution B1.3.

Resolution B1.5 applies the formalism of Resolutions B1.3 and B1.4 to the problems of time transformations and realization of coordinate times in the solar system. Resolution B1.5 is based upon a mass monopole spin dipole model. It provides an uncertainty not larger than 5×10^{-18} in rate and, for quasi-periodic terms, not larger than 5×10^{-18} in rate amplitude and 0.2 ps in phase amplitude, for locations farther than a few solar radii from the Sun. The same uncertainty also applies to the transformation between TCB and TCG for locations within 50 000 km of the Earth.

Some shortcomings appeared in the definition of TT (1.2) when considering accuracy below 10^{-17}: the uncertainty in the determination of U_G is limited, the surface of the geoid is difficult to realize so that it is difficult to determine the potential difference between the geoid and the location of a clock, and in addition the geoid varies with time. Therefore it was decided to desociate the definition of TT from the geoid while maintaining continuity with the previous definition. The constant L_G was turned into a defining constant with its value fixed to $6.969290134 \times 10^{-10}$ in Resolution B1.9, which therefore removes the limitations mentioned above when realizing TT from clocks onboard terrestrial satellites.

Finally in 2006 it was decided to redefine the coordinate time TDB, which had been introduced by the IAU in 1976 as a dynamical time scale for barycentric ephemerides. As it had not been unambiguously defined, multiple realizations of TDB were possible. Because such realizations are still widely used for barycentric ephemerides, IAU Resolution B3 (2006) was passed to define TDB as the following linear transformation of TCB:

$$\mathrm{TDB} = \mathrm{TCB} - L_B \times (\mathrm{JD}_{\mathrm{TCB}} - T_0) \times 86400 + TDB_0, \tag{1.4}$$

where $\mathrm{JD}_{\mathrm{TCB}}$ is the TCB Julian date and where $L_B = 1.550519768 \times 10^{-8}$ and $TDB_0 = -6.55 \times 10^{-5} s$ are defining constants. Figure 1 shows graphically the relationships between the time scales following the IAU Resolutions of 1991, 2000 and 2006.

2. Relativistic aspects in the IERS Conventions

In the work to update the IERS Conventions (2003), relativistic aspects cover three topics. The first one is to review the nomenclature used throughout the document in order to ensure its consistency, both internally and with the Unions' recommendations. The second one concerns chapter 10 (models for space-time coordinates and equations of motion) where, in a recent update (see http://tai.bipm.org/iers/convupdt/convupdt.html) the transformation from proper time to coordinate time in the vicinity of the Earth is treated and numerical examples are provided for the different terms in the relativistic expression for the acceleration of an Earth satellite. The third topic concerns chapter 11 (models for signal propagation) and covers models for VLBI and (radio and laser) ranging techniques.

Nomenclature. Nomenclature issues in the IERS Conventions can be loosely classified in two categories, although several issues are interconnected. The first type, and the most important for what concerns relativistic aspects, is about the designation of coordinates and coordinate quantities; the second type relates to the definition and realization of reference systems and to the transformation between celestial and terrestrial reference systems.

In the first category, the wording used to designate coordinates and coordinate quantities (e.g. space coordinates, gravitational constants GM, etc...) has to be reviewed. The IAU Commission "Relativity in fundamental astronomy" (RIFA) has proposed a conventional wording (Klioner, 2008) that can be summarized by two main rules:

• All quantities intended for use with time scale XX (e.g. TDB) should be called "XX-compatible quantities" and the corresponding values "XX-compatible values". In the case of constants having the same value in BCRS and GCRS (e.g. mass parameters $\mu = GM$ of celestial bodies) the value can be called "unscaled".

• Avoid attaching any adjectives to the names of the units second and meter for numerical values of these quantities. For example, the expression "The interval is xx seconds of TDB" should be written as "The TDB interval is xx seconds".

In the Conventions (2003) a variety of wordings are used, which can be classified into three types:

• One makes use of the word "unit", like in "[so-called] TDB unit" (for cases when a scaled coordinate time is used, here TDB), or in "TCB (SI) units" TCB (SI) units (for cases when an unscaled coordinate time is used).

• One uses the word "scale", like in "ITRF ... uses the TT scale".

• One uses a full set of words in a sentence, like in " ... coordinates consistent with TDB".

The first two types may be corrected following the above-mentioned rules, while expressions of the third type are acceptable in the proposed nomenclature.

The second category of nomenclature issues mostly concerns Chapter 5 (transformation between celestial and terrestrial systems) and Chapter 4 (Terrestrial Reference System). Both chapters have been revised (see http://tai.bipm.org/iers/convupdt/convupdt.html) following the work of the IAU Division I Working Group "Nomenclature for Fundamental Astronomy" (NFA), see http://syrte.obspm.fr/iauWGnfa/NFA_Glossary.html and (Capitaine, 2008).

Chapter 10: Models for space-time coordinates and the equations of motion. This chapter has been updated in 2008 and the presentation of coordinate time scales now accounts for all IAU Resolutions (see section 1). The relationship between all time scales used in this context is shown on Figure 1, taken from this chapter in the IERS Conventions.

In addition, a new section covers the transformation between proper time and coordinate time in the vicinity of the Earth (typically up to geosynchronous orbit or slightly above). Evaluating the contributions of the higher order terms in the metric (1.3) applied to the geocentric reference system GCRS, it is found that the IAU'1991 metric (1.1) is sufficient for time and frequency applications in the GCRS in the light of present clock accuracies.

When considering TT as coordinate time, the proper time of a clock A located at the GCRS coordinate position $\mathbf{x}_A(t)$, and moving with the coordinate velocity $\mathbf{v}_A$, is

$$\frac{d\tau_A}{d\mathrm{TT}} = 1 + L_G - 1/c^2 \left[\mathbf{v}_A^2/2 + U_\mathrm{E}(\mathbf{x}_A) + U_\mathrm{ext}(X_\mathrm{A}) - U_\mathrm{ext}(X_\mathrm{E}) - x_A^i \partial_i U_\mathrm{ext}(X_\mathrm{E})\right] \quad (2.1)$$

Here, U_E denotes the Newtonian potential of the Earth at the position $\mathbf{x}_A$ of the clock in the geocentric frame, and U_ext is the sum of the Newtonian potentials of the other bodies (mainly the Sun and the Moon) computed at a location X in barycentric coordinates, either at the position X_E of the Earth center of mass, or at the clock location X_A. The

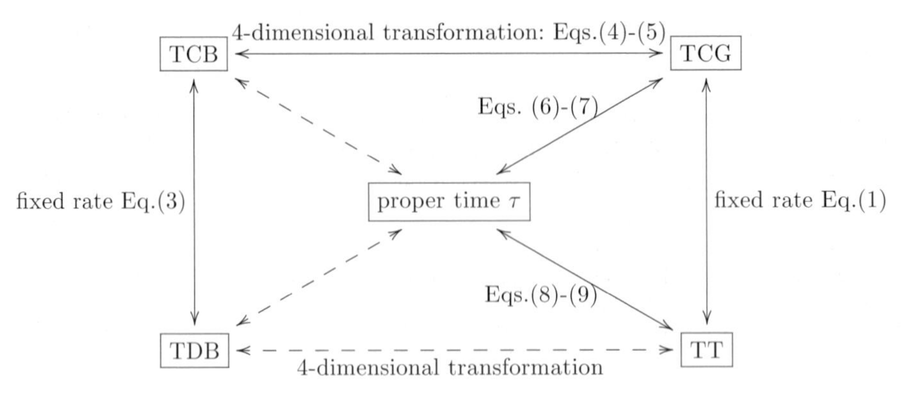

Figure 1. Various relativistic time scales and their relations. Each of the coordinate time scales TCB, TCG, TT and TDB can be related to the proper time τ of an observer, provided that the trajectory of the observer in the BCRS and/or GCRS is known. See chapter 10 in the IERS Conventions web site at http://tai.bipm.org/iers/convupdt/convupdt.html for reference to the transformations listed in the Figure.

last three terms are tidal terms and their contribution will be limited to below 1×10^{-15} in frequency and a few ps in time amplitude up the GPS orbit, so they may be skipped depending on the uncertainty required. Nevertheless, some care needs to be taken when evaluating the Earth's potential $U_{\rm E}$ at the location of the clock as the uncertainty in $U_{\rm E}$ should be consistent with the uncertainty expected on (2.1). Analytical formulas may be specified e.g. for GPS (Ashby, 2003; Kouba, 2004), however a numerical integration of equation (2.1) using the proper development for the potential is always worth using. This is specially the case for low Earth orbit satellites (see *e.g.* Larson *et al.*, 2007), where analytical expressions may be significantly in error or even completely misleading.

Chapter 11: Models for signal propagation. In the Conventions (2003), chapter 11 describes the relativistic model for VLBI time delay and for Laser ranging.

In the short term, no change is expected in the VLBI model which provides an accuracy below 1 ps, but it should be reviewed in the next years in view of the planned improvement in VLBI observations (Behrend *et al.*, 2008). However some cosmetic changes may be introduced, linked to the nomenclature. In order to describe the possible effect of an incorrect use of coordinates and coordinate quantities, it is worth reminding the past history of this model since its introduction in 1990. In 1990, the so-called "Consensus model" was adopted at a USNO workshop. This was in an era before the adoption of the IAU'1991 Resolution so that, although relativity was carefully accounted for, the currently agreed notations did not exist at that time. The model appeared first in the IERS Standards (1992), but was modified in the IERS Conventions (1996), erroneously intending to comply with the IAU/IUGG'1991 Resolutions, on the basis that "as the time argument is now based on TAI ..., distance estimates from these conventions will now be consistent 'in principle' with physical distances". However, the change so introduced in the Conventions (1996) would have produced space coordinates which would have differed from the usual TT-compatible space coordinates by a scale change of 1.4×10^{-9}. Furthermore it must be recognized that the goal stated in the Conventions (1996) is not achievable as no coordinate quantity can be consistent with a physical (proper) quantity over the extension of the Earth. The change was however never implemented by analysis centers and the model was eventually restored in its original form in the IERS Conventions (2003) with additional explanations. Indeed, the consensus model can provide either TT-compatible space coordinates when used with raw VLBI

(TT-compatible) delays (as is the usual case in VLBI analysis), or it could provide TCG-compatible space coordinates if used with delays transformed to be TCG-compatible. Such issues, which may result in scale differences in case of misinterpretation, are re-examined whenever the scale of the terrestrial reference frame is discussed, however it should be stressed that no ambiguity exists in the present model of the Conventions.

Finally, the section on laser ranging is to be re-examined in order to cover all ranging techniques by electromagnetic signals in the vicinity of the Earth (up to the Moon). As it has been shown (Klioner, 2007) that post-post Newtonian terms are not required in view of the present uncertainty, no significant model change is expected.

3. Conclusions

The relativistic framework specified by IAU Resolutions in 1991 and 2000, and supplemented by additional Recommendations, is now complete and adapted to the current and planned applications in astrometry and space geodesy. Work remains to be done to apply it in all fields and, in some cases, a conventionally adopted nomenclature is still not widely used. This work is under way in IAU working groups (NSFA) and commissions (RIFA) and in the IERS Conventions center. It should be put into application in the next reference edition of the IERS Conventions expected in the near future.

References

Ashby, N. 2003, *Living Rev. Relativity*, 6, 1, http://www.livingreviews.org/lrr-2003-1

Behrend D. *et al.* 2008, *IAG Symposia series (Springer Verlag)*, 133, Part 5, 833

Capitaine, N. 2008, *Proc. Journées SRST*, Lohrmann-Observatorium and Observatoire de Paris, 46

Klioner, S. A. 2007, *IERS Workshop on Conventions*, http://www.bipm.org/en/events/iers

Klioner, S. A. 2008, *A&A*, 478-3, 951

Kouba, J. 2004, *GPS Solutions*, 8,3, 170

Larson, K. M., Ahsby, N., Hackman, C., & Bertiger, W. 2007, *Metrologia*, 44, 484

McCarthy, D. D. (ed.) 1992, *IERS TN13*, Observatoire de Paris, 150 p.

McCarthy, D. D. (ed.) 1996, *IERS TN21*, Observatoire de Paris, 95 p.

McCarthy, D. D. & Petit, G. (eds.) 2004, *IERS TN32*, Verlag des BKG, 127 p.

Soffel, M., Klioner, S., Petit, G., Wolf, P., *et al.* 2003, *AJ*, 126(6), 2687

Relativity in Fundamental Astronomy
Proceedings IAU Symposium No. 261, 2009
S. A. Klioner, P. K. Seidelman & M. H. Soffel, eds.

doi:10.1017/S174392130999010X

The global positioning system, relativity, and extraterrestrial navigation

Neil Ashby[1] **and Robert A. Nelson**[2]

[1]Dept. of Physics, University of Colorado
Boulder, CO 80309-0390 USA
National Institute of Standards & Technology Affiliate
email: ashby@boulder.nist.gov

[2]Satellite Engineering Research Corporation
7710 Woodmont Ave., Suite 1109, Bethesda, MD 20814 USA
email: robtnelson@aol.com

Abstract. Relativistic effects play an important role in the performance of the Global Positioning System (GPS) and in world-wide time comparisons. The GPS has provided a model for algorithms that take relativistic effects into account. In the future exploration of space, analogous considerations will be necessary for the dissemination of time and for navigation. We discuss relativistic effects that are important for a navigation system such as at Mars. We describe relativistic principles and effects that are essential for navigation systems, and apply them to navigation satellites carrying atomic clocks in orbit about Mars, and time transfer between Mars and Earth. It is shown that, as in the GPS, relativistic effects are not negligible.

Keywords. relativity, navigation, reference systems, time, GPS

---

1. Introduction: relativity principles in the GPS

The earth and its satellites are in free fall. The principle of equivalence implies that over some region near earth's center of mass, the gravitational field strength due to external bodies is cancelled by an apparent equal but opposite field arising from acceleration. Studies of this cancellation have shown that, when transforming from barycentric coordinates to local inertial coordinates, the source of the cancellation is a time derivative of the synchronization term in the time transformation (Ashby, N. & Bertotti, B. (1986), Nelson, R. A. (1987)). In the local freely falling frame external bodies can be ignored approximately. A single coordinate time variable occurs in the scalar invariant ds^2. Coordinate time has the property that an event occurs at a unique coordinate time for all observers.

The basis for computation of relativistic effects in the GPS is the following expression for ds^2:

$$ds^2 = g_{\mu\nu} dx^\mu dx^\nu = -\left(1 + \frac{2V}{c^2}\right)(cdt)^2 + \left(1 - \frac{2V}{c^2}\right)(dx^2 + dy^2 + dz^2). \tag{1.1}$$

Here there are no PPN parameters; the spatial coordinates are isotropic; only leading terms of order c^{-2} are kept; the effects of external bodies are neglected, and the potential of the earth V is modelled keeping only monopole and quadrupole terms. Even with these simplifications, numerous relativistic concepts must be employed in understanding the GPS. These include time dilation, gravitational frequency shifts, the Sagnac effect, constancy of the speed of light, the principle of equivalence, relativity of simultaneity, coordinate speed of light, and local inertial frames.

The proper time elapsed on an atomic clock depends on the clock's history. The increment of proper time may be computed from the scalar ds;

$$d(c\tau) = d(ct)\sqrt{1 + \frac{2V}{c^2} - \left(1 - \frac{2V}{c^2}\right)\frac{dx^2 + dy^2 + dz^2}{c^2 dt^2}} \approx \left(1 + \frac{V}{c^2} - \frac{v^2}{2c^2}\right)d(ct). \quad (1.2)$$

In the GPS this result is exploited to synchronize atomic clocks to coordinate time by tracking the clock's coordinate time in terms of the proper time:

$$\Delta t = \int_{\text{path}} d\tau \left(1 - \frac{V}{c^2} + \frac{v^2}{2c^2}\right). \quad (1.3)$$

When the potential is due only to the earth, the coordinate time is very close to Geocentric Coordinate Time (TCG). For a clock at rest on earth's surface, the gravitational potential is approximately:

$$V = -\frac{GM}{r}\left(1 - \frac{J_2 a_1^2}{2r^2}(3\cos^2\theta - 1)\right), \quad (1.4)$$

and the coordinate time increment will be

$$dt = d\tau\left(1 + \frac{GM}{c^2 r}\left(1 - \frac{J_2 a_1^2}{2r^2}(3\cos^2\theta - 1)\right) + \frac{\omega^2 r^2 \sin^2\theta}{2c^2}\right), \quad (1.5)$$

where $GM = 3.986004418 \times 10^{14}$ m^3/s^2; $a_1 = 6.378137 \times 10^6$ m is earth's equatorial radius, $J_2 = 1.08268 \times 10^{-3}$ is the quadrupole moment coefficient, $\omega = 7.292115 \times 10^{-5}$ s^{-1} and θ is the the geocentric colatitude, measured down from the north pole. Here we quote values of the constants that define the WGS-84 system, the basis for navigation in the GPS.

The coefficient of $d\tau$ in Eq. (1.5) has very nearly a constant value on earth's geoid, a surface of constant effective potential in the rotating frame in which the last term in Eq. (1.5) contributes a centripetal potential. The constant can be evaluated on the equator, and the result is

$$\begin{aligned}&\frac{GM}{c^2 a_1} + \frac{GMJ_2}{2c^2 a_1} + \frac{\omega^2 a_1^2}{2c^2} \\ &\quad = (6.95349 + .00376 + .01203) \times 10^{-10} = 6.96928 \times 10^{-10}. \quad (1.6)\end{aligned}$$

2. GPS time

The constant estimated in Eq. (1.6) is, to within the limitations of the model potential, the same as the defined constant $L_G = W_0/c^2 = 6.969290134 \times 10^{-10}$ that gives the rate change between TCG and TAI:

$$d(t_{\text{TAI}}) = d(t_{\text{TCG}})(1 - L_G). \quad (2.1)$$

Noise on individual clocks can be mitigated by averaging over many clocks; Eq. (2.1) can be interpreted as a definition of the TCG time scale in terms of TAI, which is an averaged time scale maintained by the BIPM that incorporates hundreds of atomic clocks distributed around the world. The USNO maintains a time scale, UTC(USNO), based on its own ensemble of atomic clocks, and is a major contributor to TAI (or terrestrial time TT). To change the coordinate time t of Eq. (1.2) so that it represents TT, let $t \to t_{\text{TT}}/(1 - L_G)$. Solving to leading order in c^{-2} for the elapsed coordinate time with the new scale,

$$\Delta t_{\rm TT} = \int_{\rm path} d\tau \left(1 - \frac{V}{c^2} - \frac{W_0}{c^2} + \frac{v^2}{2c^2}\right). \tag{2.2}$$

Clocks in GPS satellites are offset so that they beat at the same rate as the references on earth's geoid; also the altitude is so great that the orbits are nearly Keplerian, earth's quadrupole potential being very small. For such clocks, conservation of energy gives

$$\frac{1}{2}v^2 - \frac{GM}{r} = -\frac{GM}{2a}, \tag{2.3}$$

where a is the semi-major axis. Combining Eqs. (2.2-2.3) and rearranging,

$$\begin{aligned}\Delta t &= \int_{\rm path} d\tau \left(1 + \frac{3GM}{2ac^2} - \frac{W_0}{c^2} + \frac{2GM}{c^2}\left(\frac{1}{a} - \frac{1}{r}\right)\right) \\ &= \Delta\tau(1 - 4.4647 \times 10^{-10}) + \frac{2\sqrt{GMa}}{c^2} e \sin E + \text{const},\end{aligned} \tag{2.4}$$

where E is the eccentric anomaly. The constant rate offset, -4.4647×10^{-10} contains a handful of relativistic corrections:

$$\frac{3GM}{2ac^2} - \frac{W_0}{c^2} = -4.4647 \times 10^{-10}. \tag{2.5}$$

GPS satellite clocks are given this offset before launch so that in orbit, they will beat at the correct coordinate rate. GPS clocks are steered (without leap seconds) to the UTC(USNO) time scale so they provide a realization of the coordinate time scale TT, except that the last term in Eq. (2.4) must be implemented in all GPS receivers.

A mysterious "break" in satellite clock frequencies occurred when satellite orbits were adjusted, moving them up or down between their assigned slots and parking orbits. This was explained in about 2000 in terms of changes in the satellite semimajor axis:

$$\delta\left(\frac{3GM}{2c^2 a}\right) \approx -\frac{3GM\delta a}{2c^2 a^2}. \tag{2.6}$$

This term is a combination of time dilation and gravitational frequency shifts. A change of 20 km in the semi-major axis results in a frequency change of a few parts in 10^{13}. This correction is currently implemented by hand in the GPS.

3. Earth rotation and the Sagnac effect

To be useful for navigation, GPS clocks are synchronized in a freely falling, locally inertial frame (the ECI frame) whose origin is at earth's center of mass. If earth rotation were ignored when synchronizing clocks on earth's surface, the results would be inconsistent because light does not travel in a straight line with uniform speed c in an earth-fixed, rotating reference frame. If we ignore the gravitational potential, the metric of special relativity in cylindrical coordinates is

$$ds^2 = -(cdt)^2 + dr^2 + r^2 d\phi^2 + dz^2. \tag{3.1}$$

If we transform to an earth-centered, earth-fixed (ECEF) reference frame by making the replacement

$$\phi \rightarrow \phi + \omega t, \tag{3.2}$$

where ω is earth's angular velocity of rotation, then the metric becomes

$$ds^2 = -\left(1 - \frac{\omega^2 r^2}{c^2}\right)(c^2 dt^2) + dr^2 + 2\omega r^2 d\phi d(ct) + dz^2. \tag{3.3}$$

The coordinate time is still t (or effectively TT). Solving for dt to leading order in ω for a slowly moving clock,

$$\Delta t = \int_{\text{path}} d\tau + \frac{2\omega}{c^2}\int_{\text{path}} dA_z, \tag{3.4}$$

where A_z is the area swept out by a vector from the rotation axis to the clock, projected onto a plane parallel to earth's equator. This correction can be several hundred nanoseconds and is an important correction that must be accounted for in long-distance clock comparisons such as when using GPS in common view from laboratories in Europe and the U.S., or in TWSTFT (two-way satellite time and frequency transfer). The same effect occurs if electromagnetic signals are used to synchronize clocks.

4. Navigation with the GPS

We henceforth assume that clocks in GPS satellites are synchronized (or equivalently, that their biases are known). Suppose that at coordinate time t a receiver at position $\mathbf{r}$ detects time ticks originating at times t_j from satellites at positions $\mathbf{r}_j$. The navigation equations are then

$$c(t - t_j) = |\mathbf{r} - \mathbf{r}_j|, \quad j = 1, 2, 3, 4... \tag{4.1}$$

Given four such equations, in principle these non-linear equations can be solved for the unknown time t and position $\mathbf{r}$ of the receiver's detection event. However, the GPS is designed so that the satellite positions are transmitted in the ECEF! This is a source of much confusion. The satellite positions must be transformed to some common ECI frame before solving, then the solution transformed to the ECEF frame that exists at time t.

The basic measurement of the detector is accomplished by alignment of a code unique to each transmitter, that is impressed on the signal by phase reversals, with a replica of the code generated within the receiver. The receiver carries a relatively inexpensive (and noisy) oscillator for code generation and timing. Suppose the clock in the receiver has an error Δt_r. The time difference between transmitted time t_j and time $t + \Delta t_r$ on the receiver oscillator is called the "pseudorange." Then the pseudorange PR_j is

$$PR_j = c(t+\Delta t_r - t_j) = \sqrt{(x - x_j)^2 + (y - y_j)^2 + (z - z_j)^2} + c(\Delta t_r - \Delta t_t + \Delta t_{\text{rel}} + \Delta t_{\text{atm}}), \tag{4.2}$$

where Δt_{rel} is the eccentricity correction, Δt_t is the transmitter error, and Δt_{atm} are atmospheric delay corrections. This is illustrated in Figure 1.

The time bias on the oscillator is continually being recalibrated to within a few nanoseconds by internal solutions of the navigation equations, but inexpensive receivers do not make such precision available to the user.

Signals from the GPS satellites are right circularly polarized, in which the electric and magnetic fields oscillate in phase. At places in the wave train where a phase reversal is imposed, all electromagnetic fields will pass through zero. At a physical point where all fields are zero, the fields are zero in every reference system; these invariant zeros sweep through free space with speed c.

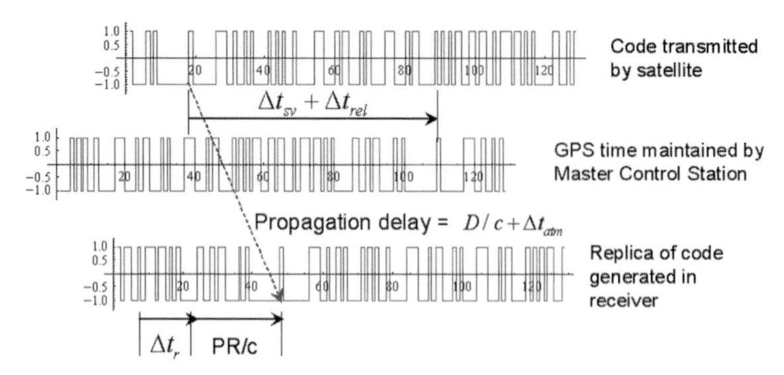

Figure 1. Pseudorange is the true range D with corrections due to relativity, clock biases, and atmospheric delays. The carrier is not shown; the plotted code corresponds to phase reversals impressed on the carrier.

5. Other satellite navigation systems

Several other navigation systems have been partially deployed or are in various stages of planning. The Russion GLONASS system is very similar to GPS. Satellites orbit 17 times while GPS orbits 16 times. A clock frequency offset, which is not quite as large as that for GPS, is applied to the satellite clocks. A full constellation of 24 satellites was planned but was never completely implemented; currently the constellation is being replenished and a full constellation is in the works.

GALILEO is a European satellite navigation system which at present has two satellites in orbit. Specifications for this system call for all relativistic effects to be the responsibility of the receiver. No frequency offsets will be applied in hardware to the orbiting clocks, which will necessarily run faster than TT by a few parts in 10^{10}. It will be up to software in the receiver to correct the transmitted time ticks for this large time drift.

The Chinese Republic has plans for a global satellite navigation system involving over 30 satellites; some of these will be geosynchronous and more than one such satellite has already been launched.

Several augmentation systems such as WAAS, EGNOS, and QZSS monitor GPS signals and upload data to geostationary satellites for retransmission to users. This mitigates problems with GPS due to outages, multipath, ionospheric delays, and other problems that make GPS somewhat unreliable.

6. Effects not currently accounted for

Solar and lunar tidal potentials. At the origin of a locally inertial, freely falling reference frame such as the ECI frame, the principle of equivalence implies that the gravitational field strength due to external bodies is cancelled by the induced field strength due to acceleration. Such a cancellation has been overlooked time and again by no small number of individuals claiming that relativity has not been accounted for properly in the GPS. The cancellation is a consequence of the relativity of simultaneity. To summarize the calculation briefly, consider a time transformation from barycentric or heliocentric coordinates to a locally inertial, freely falling frame, that has a resynchronization term of the form

$$\frac{\mathbf{v} \cdot \mathbf{r}}{c^2} \tag{6.1}$$

where $\mathbf{v}$ is the velocity of the local frame's origin and $\mathbf{r}$ is the displacement of the observation point from the origin. The tensor transformation of the 00-component of the metric tensor will then produce contributions including terms

$$-\left(1+\frac{\partial \mathbf{v}\cdot\mathbf{r}/c^2}{\partial t}\right)^2 = -\frac{2\mathbf{a}\cdot\mathbf{r}}{c^2}+... \tag{6.2}$$

where $\mathbf{a}$ is the acceleration of the freely falling frame. Let the contributions to g_{00} of external bodies be expanded in a Taylor series about the origin of the local frame. The first term will be the same for all clocks in the neighborhood of the local frame origin and can be removed by absorbing the term into the local time scale. The second term will be $-2\nabla V\cdot\mathbf{r}/c^2$ and these two terms together will be

$$-\frac{2}{c^2}\left(\mathbf{a}+\nabla V\right)\cdot\mathbf{r}=0, \tag{6.3}$$

because in free fall, $\mathbf{a}=-\nabla V$. The cancellation does not arise from second-order Doppler effects as some have claimed. The first terms that contribute are tidal terms, second derivatives of the external potential evaluated at the origin of the local frame. The net fractional frequency shift of orbiting GPS clocks due to the moon has an amplitude of only 7×10^{-16} and the sun contributes about half that amount.

Earth's quadrupole potential. Earth's quadrupole moment produces a small, predictable periodic effect on orbiting clocks, that is not currently accounted for. The time correction that should be applied to the orbiting clock is (Ashby (2003))

$$\Delta t_{\text{oblateness}} = \frac{GMJ_2a_1^2}{2c^2a^3}\left(1-\frac{3}{2}\sin^2 i\right)\Delta\tau+\sqrt{\frac{GM}{a^3}}\frac{J_2a_1^2\sin^2 i}{2c^2}\sin(2\omega+2f). \tag{6.4}$$

where i is the orbital inclination, ω is the altitude of perigee, f is the satellite's true anomaly, and a is the mean semi-major axis. It is coincidental that the chosen inclination $i=55^\circ$ so that the secular term is nearly zero. The remaining term has a period of nearly 6 hours and an amplitude of 24 ps, contributing almost a centimeter to navigation error.

Coordinate speed of light. For signals traveling from earth's surface at radius R to or from a GPS satellite, at an elevation angle E, solving the equation of the null geodesic $ds^2=0$ for the propagation time gives

$$c\Delta t=(1-L_G)\left(\sqrt{r_{SV}^2-R^2\cos^2 E}-R\sin E\right)+$$
$$\frac{2GM}{c^2}\log\frac{r_{SV}+R+\sqrt{r_{SV}^2-R^2\cos^2 E}-R}{r_{SV}+R-\sqrt{r_{SV}^2-R^2\cos^2 E}+R}, \tag{6.5}$$

where r_{SV} is the satellite's radial coordinate. The correction factor L_G multiplying the geometric range very nearly cancels the usual logarithmic Shapiro delay term (Petit & Wolf (1994)). The isotropic coordinate gauge has been selected and the product GM for earth is determined as part of the definition of the WGS-84 reference frame used for navigation in the GPS. Also, c is defined so GM/c^2 has fixed units of length. Further scaling of the length coordinate would be inconsistent since the first-order potential term $GM/(c^2r)$ is unitless and at a given physical point must have the same numerical value in all coordinate systems. Figure 1 plots both corrections to the delay.

7. Extension to extraterrestrial navigation

Consider extending the analysis of relativistic effects to the vicinity of another celestial body such as Mars (Ares) (Nelson (2007)). An equipotential surface exists in the Mars-centered, Mars-fixed rotating frame (ACAF frame) in which atomic clocks at rest beat

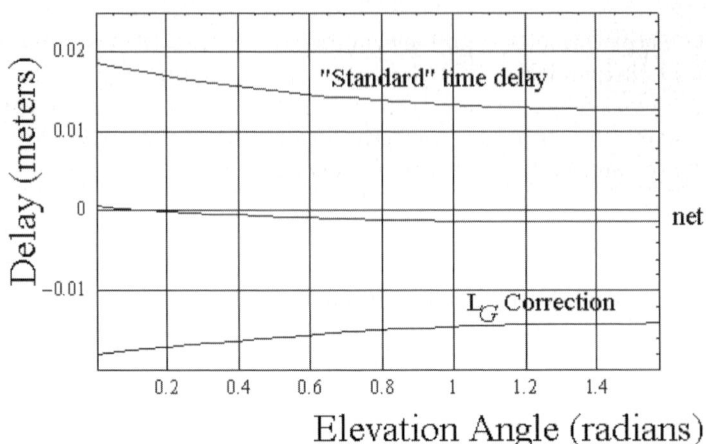

Figure 2. Plot of relativistic corrections to coordinate time delay. The log term is nearly cancelled by the L_G correction to the geometric distance.

at the same rate. Ignoring all masses except Mars, the rate of Aretian coordinate time relative to clocks at "infinity" may be expressed in terms of a constant like L_G.

$$\begin{aligned} L_A &= \frac{GM_A}{c^2 a_{1A}} + \frac{GM_A J_{2A}}{2c^2 a_{1A}} + \frac{\omega_A^2 a_{1A}^2}{2c^2} \\ &\approx (1.40362 + .00128 + .00322) \times 10^{-10} \\ &= 1.40812 \times 10^{-10} \qquad (21\% \text{ of } L_G). \end{aligned} \tag{7.1}$$

Then the frequency offset of an atomic clock in orbit around Ares can be calculated when the semimajor axis is known. For a circular orbit,

$$\frac{\Delta f}{f} = \frac{3GM_A}{2a_A c^2} - L_A. \tag{7.2}$$

For an areosynchronous orbit, $a_A = 2.04277 \times 10^7$ km, and

$$\frac{\Delta f}{f} = -1.05738 \times 10^{-10}. \tag{7.3}$$

Figure 3 shows an areosynchronous navigation system poised over a possible base near the Arean equator. The satellites are given slightly inclined orbits; five satellites are needed in order to provide continuous navigation coverage. The reason for this can be visualized as follows. Think of the navigation error associated with signals from one satellite in terms of a narrow space between two concentric shells of nearly equal radii. Two such shells that intersect in general give a thin torus of position, but if the shells happen to be tangent such a torus spreads into a disc and the position is not well determined. This can happen if the centers of the shells are in line with the point of tangency, and this is the reason that poor navigation precision could result if all the satellites were in the equatorial plane. Imagine any two of the five satellites, not in equatorial orbits. The plane generated by vectors from the center of Ares to the satellites will be wobbling in space and will eventually intersect one of the remaining satellites. When that happens, with

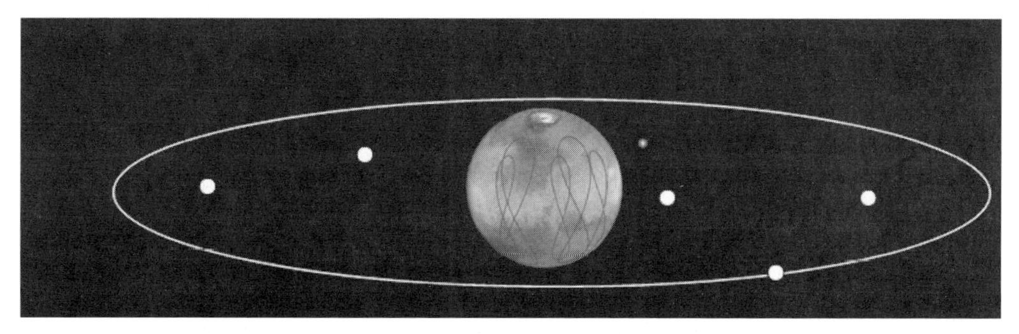

Figure 3. Notional areosynchronous navigation system with five satellites in slightly inclined orbits

three satellites in line, the geometrical situation is poor for navigation; one of the three satellites is relatively useless; in general one needs five satellites.

For continuous communication coverage between the Arean surface and the satellites, twelve satellites distributed in three orbital planes will suffice; however such a configuration is not sufficient for navigation, where continuous view of at least four satellites from points on the surface are needed. A navigational system with 24 satellites similar to the GPS configuration would be visible most of the time at a radius of about 4.2 planetary radii: if $a_A = 14,259$ km, the orbital period would be 51,695 s, and the fractional frequency offset relative to clocks on the surface would be

$$\frac{3GM_A}{2c^2 a_A} - L_A = 1.90941 \times 10^{-10}. \tag{7.4}$$

Barycentric Coordinate Time (TCB). Time transfer between an extraterrestrial time system and a terrestrial one calls for a common time system which overlaps both systems. We consider here TCB according to the IAU definitions (IAU/IUGG 1991). The elapsed Barycentric Coordinate Time on a moving clock anywhere within the solar system is

$$\Delta t_{\mathrm{TCB}} = \int_{\mathrm{path}} d\tau \left(1 - \frac{V(\mathbf{r})}{c^2} + \frac{1}{2}\frac{v^2}{c^2}\right), \tag{7.5}$$

where $\mathbf{r}$ is the position of the clock and v its velocity. All solar system bodies are included in the potential $V(\mathbf{r})$. To obtain terrestrial time in terms of TCB, The vector $\mathbf{r}$ and the corresponding velocity are split up into into vectors from the barycenter to earth, and relative to earth:

$$\mathbf{r} = \mathbf{r}_{\mathrm{earth}} + \mathbf{R}, \qquad \mathbf{v} = \mathbf{v}_{\mathrm{earth}} + \dot{\mathbf{R}}. \tag{7.6}$$

where $\mathbf{R}$ is the displacement of the clock relative to earth's center of mass. The potential is split into a contribution from the earth and an external potential that is expanded in a Taylor series for small displacements:

$$V = V_{\mathrm{earth}} + V_{\mathrm{ext}} \approx V_{\mathrm{earth}}(\mathbf{R}) + V_{\mathrm{ext}}(\mathbf{r}_{\mathrm{earth}}) + \nabla V_{\mathrm{ext}} \cdot \mathbf{R} + \ldots \tag{7.7}$$

Then after substitution and rearrangement, and replacing the integration variable τ by t in first-order correction terms,

$$\Delta t_{\mathrm{TCB}} = \Delta\tau + \int_{t_0}^{t_1} dt \left(-\frac{V_{\mathrm{ext}}(\mathbf{r}_{\mathrm{earth}})}{c^2} + \frac{1}{2}\frac{v_{\mathrm{earth}}^2}{c^2}\right)$$
$$+ \int_{t_0}^{t_1} dt \left(-\frac{V_{\mathrm{earth}}(\mathbf{R})}{c^2} + \frac{1}{2}\frac{\dot{\mathbf{R}}^2}{c^2}\right) + \left.\frac{\mathbf{R}\cdot\mathbf{v}_{\mathrm{earth}}}{c^2}\right|_{t_0}^{t_1}. \tag{7.8}$$

The last term in Eq. (7.8) represents the relativity of simultaneity of earth's ECI frame moving relative to the barycenter. Only tidal terms from the external potential survive, because of the principle of equivalence. The first integral in Eq. (7.8) gives hundreds of correction terms; the principal ones are similar to the corrections discussed in Eq. (2.4) for a GPS clock:

$$\int_{t_0}^{t_1} dt\left(-\frac{V_{\rm ext}(\mathbf{r}_{\rm earth})}{c^2}+\frac{1}{2}\frac{v_{\rm earth}^2}{c^2}\right) \approx \frac{3GM_\odot}{2c^2 a_{\rm earth}}+\frac{2\sqrt{GM_\odot a_{\rm earth}}}{c^2}e_{\rm earth}\sin E_{\rm earth}+\ldots \quad (7.9)$$

The second integral in Eq. (7.8) depends on the position and velocity of the clock relative to the earth, and is just the TCG of the event:

$$\int_{t_0}^{t_1} dt\left(-\frac{V_{\rm earth}(\mathbf{R})}{c^2}+\frac{1}{2}\frac{\dot{\mathbf{R}}^2}{c^2}\right) = t_{\rm TCG} = (1+L_G)t_{\rm TT}. \quad (7.10)$$

The relation between Arean time and TCB is analogous. The time of an event near Ares can then be transformed to TCB and thence to TT.

8. Conclusions

Numerous relativistic effects must be accounted for in global navigation systems. A distributed network of clocks, synchronized in the ECI frame, provides a realization of coordinate time used for many purposes other than simple navigation. Trends toward improving the precision of navigation algorithms will entail the incorporation of additional relativistic effects.

Clocks on GPS satellites would run fast by 38 microseconds per day relative to terrestrial clocks if they were not offset prior to launch. Also, the navigation solution in the GPS could be in error if the residual effects of time dilation and gravitational redshift due to orbit eccentricity were not corrected in the receiver (14 meters error if $e = 0.02$).

Similarly, the times registered by clocks on Mars will require relativistic corrections. If signals from a clock on the planetary surface were used for navigation by an approaching spacecraft, then serious errors would result if the physics of relativity were not considered. Also, any precisely timed astronomical event measured from the vicinity of the planet must be calibrated correctly in terms of the times on Earth, and the gravitational delay in the propagation of an electromagnetic signal must be considered.

The fundamental notion is that time is the reading of a clock. All time comparisons are made between clocks, whose readings represent "proper time." In any coordinate system, the relation between coordinate time and proper time is given by the invariant spacetime interval. The choice of coordinate time is arbitrary and is based on convenience.

The paradigm for relativistic time transfer has been successfully applied in the GPS. It is important that analogous relativistic effects be recognized in the synchronization of clocks for future applications in the exploration of the solar system. They are not merely of theoretical scientific interest. Relativity has entered into the realm of engineering practice.

References

Ashby, N. & Bertotti, B. 1986 *Phys. Rev.*, D34, 2246-2258

Nelson, R. A. 1987 *J. Math. Phys.*, 28, 2379-2383; *ibid.* 35, 6224-6225

Ashby, N. 2003 *http://relativity.livingreviews.org/Articles/lrr-2003-1*

Ashby, N. 2002 *Phys. Today*, 55(5) 41-47

Petit, G. & Wolf, P. 1994 *Astron. Astrophys.* 286, 971-977

Nelson, R. A. 2007 "Relativistic Time Transfer in the Solar System", Proc. EFTF/IEEE-FCS, Geneva, 1278-1283

Relativity in Fundamental Astronomy
Proceedings IAU Symposium No. 261, 2009
S. A. Klioner, P. K. Seidelman & M. H. Soffel, eds.

doi:10.1017/S1743921309990111

Reference frames and the physical gravito-electromagnetic analogy

L. Filipe O. Costa[1] **and Carlos A. R. Herdeiro**[2]

[1,2]Centro de Física do Porto e Departamento de Física da Universidade do Porto
Rua do Campo Alegre 687, 4169-007 Porto, Portugal
[1]email: filipezola@fc.up.pt, [2]email: crherdei@fc.up.pt
Illustrations by Rui Quaresma (quaresma.rui@gmail.com)

Abstract. The similarities between linearized gravity and electromagnetism are known since the early days of General Relativity. Using an exact approach based on tidal tensors, we show that such analogy holds only on very special conditions and depends crucially on the reference frame. This places restrictions on the validity of the "gravito-electromagnetic" equations commonly found in literature.

Keywords. Gravitomagnetism, Frame Dragging, Papapetrou equation

1. Gravito-electromagnetic analogy based on tidal tensors

The topic of the gravito-electromagnetic analogies has a long story, with different analogies being unveiled throughout the years. Some are purely formal analogies, like the splitting of the Weyl tensor in electric and magnetic parts, e.g. Maartens-Basset 1998; but others (e.g Damour *et al.* 1991, Costa-Herdeiro 2008, Jantzen *et al.* 1992, Natário 2007, Ruggiero-Tartaglia 2002) stem from certain physical similarities between the gravitational and electromagnetic interactions. The linearized Einstein equations (see e.g. Damour *et al.* 1991, Ruggiero-Tartaglia 2002, Ciufolini-Wheeler 1995), in the harmonic gauge $\bar{h}_{\alpha\beta}{}^{;\beta} = 0$, take the form $\Box\bar{h}^{\alpha\beta} = -16\pi T^{\alpha\beta}$, similar to Maxwell equations in the Lorentz gauge: $\Box A^{\beta} = -4\pi j^{\beta}$. That suggests an analogy between the trace reversed time components of the metric tensor $\bar{h}_{0\alpha}$ and the electromagnetic 4-potential A_{α}. Defining the 3-vectors usually dubbed gravito-electromagnetic fields, the time components of these equations may be cast in a Maxwell-like form, e.g. eqs (16)–(22) of Ruggiero-Tartaglia 2002. Furthermore (on certain special conditions, see section 2) geodesics, precession and forces on gyroscopes are described in terms of these fields in a form similar to their electromagnetic counterparts, e.g. Ruggiero-Tartaglia 2002, Ciufolini-Wheeler 1995. Such analogy may actually be cast in an exact form using the 3+1 splitting of spacetime (see Jantzen *et al.* 1992, Natário 2007).

These are analogies comparing physical quantities (electromagnetic forces) from one theory with inertial gravitational forces (i.e. fictitious forces, that can be gauged away by moving to a freely falling frame, due to the equivalence principle); it is clear that (non-spinning) test particles in a gravitational field move with zero acceleration $DU^{\alpha}/d\tau = 0$; and that the spin 4-vector of a gyroscope undergoes Fermi-Walker transport $DS^{\alpha}/d\tau = S_{\sigma}U^{\alpha}DU^{\sigma}/d\tau$, with no real torques applied on it. In this sense the gravito-electromagnetic fields are pure coordinate artifacts, attached to the observer's frame.

However, these approaches describe also (not through the "gravito-electromagnetic" fields themselves, but through their derivatives; and, again, under very special conditions)

tidal effects, like the force applied on a gyroscope. And these are covariant effects, implying physical gravitational forces.

Herein we will discuss under which precise conditions a similarity between gravity and electromagnetism occurs (that is, under which conditions the physical analogy $\bar{h}_{0\mu} \leftrightarrow A_\mu$ holds, and Eqs. like (16)–(22) of Ruggiero-Tartaglia 2002 have a physical content). For that we will make use of the tidal tensor formalism introduced in Costa-Herdeiro 2008. The advantage of this formalism is that, by contrast with the approaches mentioned above, it is based on quantities which can be covariantly defined in both theories — tidal forces (the only physical forces present in gravity) — which allows for a more transparent comparison between the electromagnetic (EM) and gravitational (GR) interactions.

Table 1. The gravito-electromagnetic analogy based on tidal tensors.

Electromagnetism		Gravity	
Worldline deviation:		Geodesic deviation:	
$\frac{D^2\delta x^\alpha}{d\tau^2} = \frac{q}{m}E^\alpha{}_\beta \delta x^\beta, \quad E^\alpha{}_\beta \equiv F^\alpha{}_{\mu;\beta}U^\mu$	(1a)	$\frac{D^2\delta x^\alpha}{d\tau^2} = -\mathbb{E}^\alpha{}_\beta \delta x^\beta, \quad \mathbb{E}^\alpha{}_\beta \equiv R^\alpha{}_{\mu\beta\nu}U^\mu U^\nu$	(1b)
Force on magnetic dipole:		Force on gyroscope:	
$F^\beta_{EM} = \frac{q}{2m}B_\alpha{}^\beta S^\alpha, \quad B^\alpha{}_\beta \equiv \star F^\alpha{}_{\mu;\beta}U^\mu$	(2a)	$F^\beta_G = -\mathbb{H}_\alpha{}^\beta S^\alpha, \quad \mathbb{H}^\alpha{}_\beta \equiv \star R^\alpha{}_{\mu\beta\nu}U^\mu U^\nu$	(2b)
Maxwell Equations:		Eqs. Grav. Tidal Tensors:	
$E^\alpha{}_\alpha = 4\pi\rho_c$	(3a)	$\mathbb{E}^\alpha{}_\alpha = 4\pi\left(2\rho_m + T^\alpha{}_\alpha\right)$	(3b)
$E_{[\alpha\beta]} = \frac{1}{2}F_{\alpha\beta;\gamma}U^\gamma$	(4a)	$\mathbb{E}_{[\alpha\beta]} = 0$	(4b)
$B^\alpha{}_\alpha = 0$	(5a)	$\mathbb{H}^\alpha{}_\alpha = 0$	(5b)
$B_{[\alpha\beta]} = \frac{1}{2}\star F_{\alpha\beta;\gamma}U^\gamma - 2\pi\epsilon_{\alpha\beta\sigma\gamma}j^\sigma U^\gamma$	(6a)	$\mathbb{H}_{[\alpha\beta]} = -4\pi\epsilon_{\alpha\beta\sigma\gamma}J^\sigma U^\gamma$	(6b)

$\rho_c = -j^\alpha U_\alpha$ and j^α are, respectively, the charge density and current 4-vector; $\rho_m = T_{\alpha\beta}U^\alpha U^\beta$ and $J^\alpha = -T^\alpha_\beta U^\beta$ are the mass/energy density and current (quantities measured by the observer of 4-velocity U^α); $T_{\alpha\beta} \equiv$ energy-momentum tensor; $S^\alpha \equiv$ spin 4-vector; $\star \equiv$ Hodge dual. We use $\tilde{\epsilon}_{0123} = -1$.

The tidal tensor formalism unveils a new gravito-electromagnetic analogy, summarized in Table 1, based on exact and covariant equations. These equations make clear key differences, and under which conditions a similarity between the two interactions may occur.

Eqs. (1) are the worldline deviation equations yielding the relative acceleration of two neighboring particles (connected by the infinitesimal vector δx^α) with the *same* 4-velocity U^α (and the same q/m ratio, in the electromagnetic case). These equations manifest the physical analogy between electric tidal tensors: $\mathbb{E}_{\alpha\beta} \leftrightarrow E_{\alpha\beta}$.

Eq. (2a) yields the electromagnetic force exerted on a magnetic dipole moving with 4-velocity U^α, and is the covariant generalization of the usual 3-D expression $\mathbf{F_{EM}} = \nabla(\mathbf{S}.\mathbf{B})q/2m$ (valid only in the dipole's proper frame); Eq. (2b) is exactly the Papapetrou-Pirani equation for the gravitational force exerted on a spinning test particle. In both (2a) and (2b), Pirani's supplementary condition $S_{\mu\nu}U^\nu = 0$ is assumed (c.f. Costa-Herdeiro 2009). These equations manifest the physical analogy between magnetic tidal tensors: $B_{\alpha\beta} \leftrightarrow \mathbb{H}_{\alpha\beta}$.

Taking the traces and antisymmetric parts of the EM tidal tensors, one obtains Eqs. (3a)-(6a), which are explicitly covariant forms for each of Maxwell equations. Eqs. (3a) and (6a) are, respectively, the time and space projections of Maxwell equations $F^{\alpha\beta}{}_{;\beta} = 4\pi j^{\alpha}$; i.e., they are, respectively, covariant forms of $\nabla\cdot\mathbf{E} = 4\pi\rho_c$ and $\nabla\times\mathbf{B} = \partial\mathbf{E}/\partial t + 4\pi\mathbf{j}$; Eqs. (4a) and (5a) are the space and time projections of the electromagnetic Bianchi identity $\star F^{\alpha\beta}{}_{;\beta} = 0$; i.e., they are covariant forms for $\nabla\times\mathbf{E} = -\partial\mathbf{B}/\partial t$ and $\nabla\cdot\mathbf{B} = 0$. These equations involve only tidal tensors and sources, which can be seen substituting the following decomposition (or its Hodge dual) in (4a) and (6a):

$$F_{\alpha\beta;\gamma} = 2U_{[\alpha}E_{\beta]\gamma} + \epsilon_{\alpha\beta\mu\sigma}B^{\mu}{}_{\gamma}U^{\sigma}. \tag{1.1}$$

It is then straightforward to obtain the *physical* gravitational analogues of Maxwell equations: one just has to apply the same procedure to the gravitational tidal tensors, i.e., write the equations for their traces and antisymmetric parts (that is more easily done decomposing the Riemann tensor in terms of the Weyl tensor and source terms, see Costa-Herdeiro 2007 sec. 2), which leads to Eqs. (3b) - (6b). Underlining the analogy with the situation in electromagnetism, Eqs. (3b) and (6b) turn out to be the time-time and and time-space projections of Einstein equations $R_{\mu\nu} = 8\pi(T_{\mu\nu} - \frac{1}{2}g_{\mu\nu}T^{\alpha}{}_{\alpha})$, and Eqs. (4b) and (5b) the time-space and time-time projections of the algebraic Bianchi identities $\star R^{\gamma\alpha}{}_{\gamma\beta} = 0$.

1.1. *Gravity vs Electromagnetism*

Charges — the gravitational analogue of ρ_c is $2\rho_m + T^{\alpha}_{\ \alpha}$ ($\rho_m + 3p$ for a perfect fluid) $\Rightarrow$ in gravity, pressure and all material stresses contribute as sources.

Ampere law — in stationary (in the observer's rest frame) setups, $\star F_{\alpha\beta;\gamma}U^{\gamma}$ vanishes and equations (6a) and (6b) match up to a factor of 2 $\Rightarrow$ currents of mass/energy source gravitomagnetism like currents of charge source magnetism.

Symmetries of Tidal Tensors — The GR and EM tidal tensors do not generically exhibit the same symmetries, signaling fundamental differences between the two interactions. In the general case of fields that are time dependent in the observer's rest frame (that is the case of an intrinsically non-stationary field, or an observer moving in a stationary field), the electric tidal tensor $E_{\alpha\beta}$ possesses an antisymmetric part, which is the covariant derivative of the Maxwell tensor along the observer's worldline; there is also an antisymmetric contribution $\star F_{\alpha\beta;\gamma}U^{\gamma}$ to $B_{\alpha\beta}$. These terms consist of time projections of EM tidal tensors (cf. decomposition 1.1), and contain the laws of electromagnetic induction. The gravitational tidal tensors, by contrast, are symmetric (in vacuum, in the magnetic case) and spatial, manifesting the absence of analogous effects in gravity.

Gyroscope vs. magnetic dipole — According to Eqs. (2), both in the case of the magnetic dipole and in the case of the gyroscope, it is the magnetic tidal tensor, *as seen by the test particle* (U^{α} in Eqs. (2) is the gyroscope/dipole 4-velocity), that determines the force exerted upon it. Hence, from Eqs. (6), we see that the forces can be similar only if the fields are stationary (besides weak) in the gyroscope/dipole frame, i.e., when it is at "rest" in a stationary field. Eqs. (2) also tell us that in gravity the angular momentum S plays the role of the magnetic moment $\mu = S(q/2m)$; the relative minus sign manifests that masses/charges of the same sign attract/repel one another in gravity/electromagnetism, as do charge/mass currents with parallel velocity.

2. Linearized Gravity

If the fields are stationary in the observer's rest frame, the GR and EM tidal tensors have the same symmetries, which by itself does not mean a close similarity between the two interactions (note that despite the analogy in Table 1, EM tidal tensors are linear, whereas the GR ones are not). But in two special cases a matching between tidal tensors occurs: ultrastationary spacetimes (where the gravito-magnetic tidal tensor is linear, see Costa-Herdeiro 2008 Sec. IV) and linearized gravitational perturbations, which is the case of interest for astronomical applications.

Consider an arbitrary electromagnetic field $A^\alpha = (\phi, \mathbf{A})$ and arbitrary perturbations around Minkowski spacetime in the form†

$$ds^2 = -c^2\left(1 - 2\frac{\Phi}{c^2}\right)dt^2 - \frac{4}{c}\mathcal{A}_j dt dx^j + \left[\delta_{ij} + 2\frac{\Theta_{ij}}{c^2}\right]dx^i dx^j . \qquad (2.1)$$

Tidal effects. — The GR and EM tidal tensors from these setups will be in general very different, as is clear from equations (3-6), and as one may check from the explicit expressions in Costa-Herdeiro 2008.

But if one considers time independent fields, and a static observer of 4-velocity $U^\mu = c\delta^\mu_0$, then the *linearized* gravitational tidal tensors match their electromagnetic counterparts identifying $(\phi, A^i) \leftrightarrow (\Phi, \mathcal{A}^i)$ (in expressions below colon represents partial derivatives; $\epsilon_{ijk} \equiv$ Levi Civita symbol):

$$\mathbb{E}_{ij} \simeq -\Phi_{,ij} \overset{\Phi\leftrightarrow\phi}{=} E_{ij}, \quad \mathbb{H}_{ij} \simeq \epsilon_i^{\ lk}\mathcal{A}_{k,lj} \overset{\mathcal{A}\leftrightarrow A}{=} B_{ij}. \qquad (2.2)$$

This suggests the physical analogy $(\phi, A^i) \leftrightarrow (\Phi, \mathcal{A}^i)$, and defining the "gravito-electromagnetic fields" $\mathbf{E_G} = -\nabla\Phi$ and $\mathbf{B_G} = \nabla\times\boldsymbol{\mathcal{A}}$, in analogy with the electromagnetic fields $\mathbf{E} = -\nabla\phi$, $\mathbf{B} = \nabla\times\mathbf{A}$. In terms of these fields we have $\mathbb{E}_{ij} \simeq (E_G)_{i,j}$ and $\mathbb{H}_{ij} \simeq (B_G)_{i,j}$, in analogy with the electromagnetic tidal tensors $E_{ij} = E_{i,j}$ and $B_{ij} = B_{i,j}$.

The matching (2.2) means that a gyroscope at rest (relative to the static observer) will feel a force F^α_G similar to the electromagnetic force F^α_{EM} on a magnetic dipole, which in this case take the very simple forms (time components are zero):

$$\mathbf{F_{EM}} = \frac{q}{2mc}\nabla(\mathbf{B.S}); \qquad F^j_G = -\frac{1}{c}\mathbb{H}^{ij}S_i \approx -\frac{1}{c}(B_G)^{i,j}S_i \;\Leftrightarrow\; \mathbf{F_G} = -\frac{1}{c}\nabla(\mathbf{B_G.S}). \quad (2.3)$$

Had we considered gyroscopes/dipoles with different 4-velocities, not only the expressions for the forces would be more complicated, but also the gravitational force would significantly differ from the electromagnetic one, as one may check comparing Eqs. (12) with (17)-(20) of Costa-Herdeiro 2008. This will be exemplified in section 2.1.

The matching (2.2) also means, by similar arguments, that the relative acceleration between two neighboring masses $D^2\delta x^i/d\tau^2 = -\mathbb{E}^{ij}\delta x_j$ is similar to the relative acceleration between two charges (with the same q/m): $D^2\delta x^i/d\tau^2 = E^{ij}\delta x_j(q/m)$, *at the instant* when the test particles have 4-velocity $U^\alpha = c\delta^\alpha_0$ (i.e., are *at rest* relative to the static observer $\mathcal{O}$).

Gyroscope precession. — The evolution of the spin vector of the gyroscope is given by the Fermi-Walker transport law, which, for a gyroscope at rest reads $DS^i/d\tau = 0$; hence, we have, in the coordinate basis, Eq. (2.4a). The last term of Eq. (2.4a) vanishes if we express $\mathbf{S}$ in the local orthonormal tetrad $e^{\hat{\alpha}}$: $S^i = S^{\hat{i}}e^i_{\hat{i}}$, where to linear order $e^i_{\hat{i}} = \delta^i_{\ \hat{i}} - \Theta^i_{\ \hat{i}}/c^2$; in this fashion we obtain Eq. (2.4b), which is similar to the precession of a magnetic dipole in a magnetic field $d\mathbf{S}/dt = q\mathbf{S}\times\mathbf{B}/2mc$:

$$\frac{dS^i}{dt} = -c\Gamma^i_{0j}S^j = -\frac{1}{c}\left[(\mathbf{S}\times\mathbf{B_G})^i + \frac{1}{c}\frac{\partial\Theta^{ij}}{\partial t}S_j\right] \; (a); \qquad \frac{dS^{\hat{i}}}{dt} = -\frac{1}{c}(\mathbf{S}\times\mathbf{B_G})^{\hat{i}} \; (b). \quad (2.4)$$

† In the previous sections we were putting $c = 1$. In this section we re-introduce the speed of light in order to facilitate comparison with relevant literature.

Thus, in the special case of gyroscope precession, the linear gravito-electromagnetic analogy holds even if the fields vary with time.

Geodesics. — The space part of the equation of geodesics $U^{\alpha}{}_{,\beta}U^{\beta} = -\Gamma^{\alpha}_{\beta\gamma}U^{\beta}U^{\gamma}$ is given, to first order in the perturbations and in test particle's velocity by ($a^i \equiv d^2x^i/dt^2$):

$$\mathbf{a} = \nabla\Phi + \frac{2}{c}\frac{\partial \boldsymbol{\mathcal{A}}}{\partial t} - \frac{2}{c}\mathbf{v} \times (\nabla \times \boldsymbol{\mathcal{A}}) - \frac{1}{c^2}\left[\frac{\partial \Phi}{\partial t}\mathbf{v} + 2\frac{\partial \Theta^i{}_j}{\partial t}v^j \mathbf{e_i}\right]. \tag{2.5}$$

Comparing with the electromagnetic Lorentz force:

$$\mathbf{a} = \frac{q}{m}\left[-\nabla\phi - \frac{1}{c}\frac{\partial \mathbf{A}}{\partial t} + \frac{\mathbf{v}}{c} \times (\nabla \times \mathbf{A})\right] = \frac{q}{m}\left[\mathbf{E} + \frac{\mathbf{v}}{c} \times \mathbf{B}\right], \tag{2.6}$$

these equations do not manifest, in general, a close analogy. Note that the last term of (2.5), which has no electromagnetic analogue, is, for the problem at hand (see next section), of the same order of magnitude as the second and third terms. But when one considers stationary fields, then (2.5) takes the form $\mathbf{a} = -\mathbf{E_G} - 2\mathbf{v} \times \mathbf{B_G}/c$ analogous to (2.6).

Note the difference between this analogy and the one from the tidal effects considered above: in the case of the latter, the similarity occurs only when the *test particle* sees time independent *fields* (fields $\equiv$ derivatives of potentials/of metric perturbations); for geodesics, it is when *the observer* (not the test particle!) sees a time independent *potential* (ϕ)*/metric perturbations*(Φ, Θ_{ij}).

2.1. *Translational vs. Rotational Mass Currents*

The existence of a similarity between gravity and electromagnetism thus relies on the time dependence of the mass currents: if the currents are (nearly) stationary, for example from a rotating celestial body, the gravitational field generated is analogous to a magnetic field; such is the field detected on LAGEOS data by Ciufolini *et al.*. But when the currents seen by the observer vary with time — e.g. the ones resulting from translation of the celestial body, considered in Soffel *et al.* 2008 — then the dynamics differ significantly.

Rotational Currents. — We will start by the well known analogy between the electromagnetic field of a spinning charge (charge Q, magnetic moment μ) and the gravitational field (in the far region $r \to \infty$) of a rotating celestial body (mass m, angular momentum J), see Fig. 1.

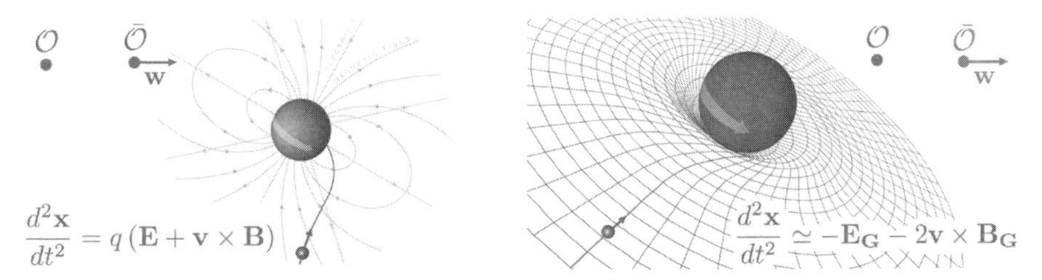

Figure 1. Spinning charge vs. spinning mass

The electromagnetic field of the spinning charge is described by the 4-potential $A^{\alpha} = (\phi, \mathbf{A})$, given by (2.7a). The spacetime around the spinning mass is asymptotically described by the linearized Kerr solution, obtained by putting in (2.1) the perturbations (2.7a):

$$\phi = \frac{Q}{r}, \quad \mathbf{A} = \frac{1}{c}\frac{\boldsymbol{\mu} \times \mathbf{r}}{r^3} \quad (a); \qquad \Phi = \frac{M}{r}, \quad \boldsymbol{\mathcal{A}} = \frac{1}{c}\frac{\mathbf{J} \times \mathbf{r}}{r^3}, \quad \Theta_{ij} = \Phi\delta_{ij} \quad (b). \tag{2.7}$$

For the observer at rest $\mathcal{O}$ the gravitational tidal tensors asymptotically match the electromagnetic ones, identifying the appropriate parameters:

$$\mathbb{E}_{ij} \simeq \frac{M}{r^3}\delta_{ij} - \frac{3Mr_ir_j}{r^5} \overset{M \leftrightarrow Q}{=} E_{ij}; \quad \mathbb{H}_{ij} \simeq \frac{3}{c}\left[\frac{(\mathbf{r}.\mathbf{J})}{r^5}\delta_{ij} + 2\frac{r_{(i}J_{j)}}{r^5} - 5\frac{(\mathbf{r}.\mathbf{J})r_ir_j}{r^7}\right] \overset{J \leftrightarrow \mu}{=} B_{ij}$$

(all the time components are zero for this observer). This means that $\mathcal{O}$ will find a similarity between *physical* (i.e., tidal) gravitational forces and their electromagnetic counterparts: the gravitational force $F^i_G = -\mathbb{H}^{ji}S_j/c$ exerted on a gyroscope carried by $\mathcal{O}$ is similar to the force $F^i_{EM} = qB^{ji}S_j/2mc$ on a magnetic dipole; and the worldline deviation $D^2\delta x^i/d\tau^2 = -\mathbb{E}^{ij}\delta x_i$ of two masses dropped from rest is similar to the deviation between two charged particles with the same q/m.

Moreover, observer $\mathcal{O}$ will see test particles moving on geodesics described by equations analogous to the electromagnetic Lorentz force (see Fig. 1).

Translational Currents. — For the observer $\bar{\mathcal{O}}$ moving with velocity $\mathbf{w}$ relative to the mass/charge of Fig. 1, however, the electromagnetic and gravitational interactions will look significantly different. For simplicity we will specialize here to the case where $\mathbf{J} = \boldsymbol{\mu} = 0$, so that the mass/charge currents seen by $\bar{\mathcal{O}}$ arise solely from translation. To obtain the electromagnetic 4-potential $A^{\bar{\alpha}}$ in the frame $\bar{\mathcal{O}}$, we apply the boost $A^{\bar{\alpha}} = \Lambda^{\bar{\alpha}}{}_{\alpha}A^{\alpha} = (\bar{\phi}, \bar{\mathbf{A}})$, where $\Lambda^{\bar{\alpha}}{}_{\alpha} \equiv \partial\bar{x}^{\bar{\alpha}}/\partial x^{\alpha}$, using the expansion of Lorentz transformation (as done in e.g. Will-Nordtvedt 1972):

$$t = \bar{t}\left(1 + \frac{w^2}{2c^2} + \frac{3w^4}{8c^4}\right) + \left(1 + \frac{w^2}{2c^2}\right)\frac{\bar{\mathbf{x}}.\mathbf{w}}{c^2}; \quad \mathbf{x} = \bar{\mathbf{x}} + \frac{1}{2c^2}(\bar{\mathbf{x}}.\mathbf{w})\mathbf{w} + \left(1 + \frac{w^2}{2c^2}\right)\mathbf{w}\bar{t}, \quad (2.8)$$

yielding, to order c^{-2}, $A^{\bar{\alpha}} = (\bar{\phi}, \bar{\mathbf{A}})$, with $\bar{\phi} = Q(1 + w^2/2c^2)/r$ and $\bar{\mathbf{A}} = -Q\mathbf{w}/rc$. To obtain $A^{\bar{\alpha}}$ in the coordinates $(\bar{x}^i, \bar{t})$ of $\bar{\mathcal{O}}$, we must also express r (which denotes the distance between the source and the point of observation, in the frame $\mathcal{O}$) in terms of $R \equiv |\bar{\mathbf{r}} + \mathbf{w}\bar{t}|$, i.e., the distance between the source and the point of observation in the frame $\bar{\mathcal{O}}$. Using transformation (2.8), we obtain: $r^{-1} = R^{-1}[1 - (\mathbf{w}.\mathbf{R})^2/(2R^2c^2)]$, and finally the electromagnetic potentials seen by $\bar{\mathcal{O}}$:

$$\bar{\phi} = \frac{Q}{R}\left(1 + \frac{w^2}{2c^2} - \frac{(\mathbf{w}.\mathbf{R})^2}{4R^2c^2}\right); \quad \bar{\mathbf{A}} = -\frac{1}{c}\frac{Q}{R}\mathbf{w}. \quad (2.9)$$

The metric of the spacetime around a point mass, in the coordinates of $\bar{\mathcal{O}}$, is also obtained using transformation (2.8), which is accurate to Post Newtonian order, by an analogous procedure. First we apply the Lorentz boost $g_{\bar{\alpha}\bar{\beta}} = \Lambda^{\alpha}{}_{\bar{\alpha}}\Lambda^{\beta}{}_{\bar{\beta}}g_{\alpha\beta}$ to the metric (2.7) (with $\mathcal{A} = 0$); then, expressing r in terms of R, we finally obtain (note that, although we are not putting the bars therein, indices $\alpha = 0, i$ in the following expressions refer to the coordinates of $\bar{\mathcal{O}}$):

$$g_{00} = -1 + 2\frac{M}{Rc^2} + \frac{4Mw^2}{Rc^4} - \frac{M(\mathbf{w}.\mathbf{R})^2}{c^4R^3} \equiv -1 + \frac{2\Phi}{c^2};$$
$$g_{0i} = \frac{4Mw_i}{Rc^3} \equiv -\frac{2\bar{\mathcal{A}}_i}{c^2}; \quad g_{ij} = \left[1 + 2\frac{M}{Rc^2}\right]\delta_{ij} \equiv \left[1 + 2\frac{\bar{\Theta}}{c^2}\right]\delta_{ij}, \quad (2.10)$$

where we retained terms up to c^{-4} in g_{00}, up to c^{-3} in g_{i0} and c^{-2} in g_{ij}, as usual in Post-Newtonian approximation. This matches, to linear order in M, Eqs. (5) of Soffel *et al.* 2008 for the case of one single source; or e.g. Eqs. (11) of Nordtvedt 1988 (in the case of the latter, an additional gauge choice, Eq. (19) of Will-Nordtvedt 1972, was made). The metric (2.10), like the electromagnetic potential (2.9), is now time dependent, since $\mathbf{R}(\bar{t}) = \bar{\mathbf{r}} + \mathbf{w}\bar{t}$.

The gravitational tidal tensors seen by $\bar{\mathcal{O}}$ are ($\mathbb{E}_{\alpha 0} = \mathbb{E}_{0\alpha} = \mathbb{H}_{\alpha 0} = \mathbb{H}_{0\alpha} = 0$):

$$\begin{aligned}
\mathbb{E}_{ij} =& -\bar{\Phi}_{,ij} - \frac{2}{c}\frac{\partial}{\partial \bar{t}}\bar{\mathcal{A}}_{(i,j)} - \frac{1}{c^2}\frac{\partial^2}{\partial \bar{t}^2}\bar{\Theta}\delta_{ij} \\
=& \frac{M\delta_{ij}}{R^3}\left[1 + \frac{3w^2}{c^2} - \frac{9}{2}\frac{(\mathbf{R}.\mathbf{w})^2}{c^2 R^2}\right] - \frac{3MR_iR_j}{R^5}\left[1 + \frac{2w^2}{c^2} - \frac{5(\mathbf{R}.\mathbf{w})^2}{2c^2R^2}\right] \\
& - \frac{3Mw_iw_j}{c^2R^3} + \frac{6Mw_{(i}R_{j)}(\mathbf{R}.\mathbf{w})}{c^2R^5}; \qquad (2.11)
\end{aligned}$$

$$\mathbb{H}_{ij} = \epsilon_i{}^{lk}\bar{\mathcal{A}}_{k,lj} - \frac{1}{c}\epsilon_{ij}{}^{l}\frac{\partial \bar{\Theta}_{,l}}{\partial \bar{t}} = \frac{M}{cR^3}\left[3\epsilon_{ij}{}^{k}w_k - \frac{3}{R^2}(\mathbf{R}.\mathbf{w})\epsilon_{ij}{}^{k}R_k - \frac{6}{R^2}(\mathbf{R}\times\mathbf{w})_iR_j\right], \qquad (2.12)$$

which significantly differ from the electromagnetic ones ($E_{0\alpha} = B_{0\alpha} = 0$):

$$\begin{aligned}
E_{ij} =& -\bar{\phi}_{,ij} - \frac{1}{c}\frac{\partial}{\partial \bar{t}}\bar{A}_{i;j} = E_{i,j} \\
=& \frac{Q\delta_{ij}}{R^3}\left[1 + \frac{w^2}{2c^2} - \frac{3}{4}\frac{(\mathbf{R}.\mathbf{w})^2}{c^2R^2}\right] - \frac{3QR_iR_j}{R^5}\left[1 + \frac{w^2}{2c^2} - \frac{5(\mathbf{R}.\mathbf{w})^2}{4c^2R^2}\right] \\
& - \frac{Qw_iw_j}{2c^2R^3} + \frac{3Qw_{[i}R_{j]}(\mathbf{R}.\mathbf{w})}{c^2R^5}; \qquad (2.13)
\end{aligned}$$

$$E_{i0} = -\frac{1}{c}\frac{\partial}{\partial \bar{t}}\bar{\phi}_{;i} - \frac{1}{c^2}\frac{\partial^2 \bar{A}_i}{\partial \bar{t}^2} \equiv \frac{1}{c}\frac{\partial E_i}{\partial \bar{t}} = \frac{Q}{cR^3}\left[w_i - \frac{3(\mathbf{R}.\mathbf{w})R_i}{R^2}\right]; \qquad (2.14)$$

$$B_{ij} = \epsilon_i{}^{lm}\bar{A}_{m;lj} \equiv B_{i,j} = \frac{Q}{cR^3}\left[\epsilon_{ij}{}^{k}w_k - \frac{3}{R^2}(\mathbf{R}\times\mathbf{w})_iR_j\right]; \qquad (2.15)$$

$$B_{i0} = \frac{1}{c}\frac{\partial B_i}{\partial \bar{t}} = -\frac{3Q}{c^2R^5}(\mathbf{R}.\mathbf{w})(\mathbf{R}\times\mathbf{w})_i. \qquad (2.16)$$

Note in particular that, unlike their gravitational counterparts, $E_{\alpha\beta}$ and $B_{\alpha\beta}$ are not symmetric, and have non-zero time components. The antisymmetric parts $E_{[ij]} = E_{[i,j]}$ and $B_{[ij]} = B_{[i,j]}$ above are (vacuum) Maxwell equations $\nabla\times\mathbf{E} = -(1/c)\partial\mathbf{B}/\partial t$ and $\nabla\times\mathbf{B} = (1/c)\partial\mathbf{E}/\partial t$, implying that a time varying electric/magnetic field endows the magnetic/electric tidal tensor with an antisymmetric part. For instance, a time varying electric field will always induce a force on a magnetic dipole. The fact that $\mathbb{E}_{\alpha\beta}$ and $\mathbb{H}_{\alpha\beta}$ are symmetric reflects the absence of analogous gravitational effects. The time component B_{i0} means that the force on a magnetic dipole (magnetic moment $\mu = q/2m$) will have a time component $(F_{EM})_0 = (1/c)\boldsymbol{\mu}.\partial\mathbf{B}/\partial t$, which (see Costa-Herdeiro 2009 sec. 1.2) is minus the power transferred to the dipole by Faraday's law of induction (and is reflected in the variation of the dipole's proper mass $m = -P^\alpha U_\alpha/c^2$). Again, this is an effect which has no gravitational counterpart: $\mathbb{H}_{\alpha 0} = \mathbb{H}_{0\alpha} = 0$, thus $(F_G)_0 = 0$, and the proper mass of the gyroscope is a constant of the motion.

The space part of the geodesic equation for a test particle of velocity $\mathbf{v}$ is:

$$\begin{aligned}
\mathbf{a} =& \nabla\bar{\Phi} + \frac{2}{c}\frac{\partial\bar{\boldsymbol{\mathcal{A}}}}{\partial \bar{t}} - 2\mathbf{v}\times(\nabla\times\bar{\boldsymbol{\mathcal{A}}}) - \frac{3}{c^2}\frac{\partial}{\partial \bar{t}}\left(\frac{M}{R}\right)\mathbf{v} \qquad (2.17) \\
=& -\frac{M}{R^3}\left[1 + \frac{2w^2}{c^2} - \frac{3(\mathbf{R}.\mathbf{w})^2}{2c^2R^2}\right]\mathbf{R} + \frac{3M(\mathbf{R}.\mathbf{w})}{c^2R^3}\mathbf{w} - \frac{4M}{c^2R^3}\mathbf{v}\times(\mathbf{R}\times\mathbf{w}) + \frac{3}{c^2}\frac{M}{R^3}(\mathbf{R}.\mathbf{w})\mathbf{v},
\end{aligned}$$

which matches equation (10) of Soffel *et al.* 2008, or (7) of Nordtvedt 1973, again, in the special case of only one source, and keeping therein only linear terms in the perturbations and test particle's velocity $\mathbf{v}$.

Comparing with its electromagnetic counterpart

$$\left(\frac{m}{q}\right)\mathbf{a} = \mathbf{E} + \frac{\mathbf{v}}{c}\times\mathbf{B} = \frac{Q}{R^3}\left[1 + \frac{w^2}{2c^2} - \frac{3(\mathbf{R}.\mathbf{w})^2}{4c^2R^2}\right]\mathbf{R} - \frac{1}{2}\frac{Q(\mathbf{R}.\mathbf{w})}{c^2R^3}\mathbf{w} + \frac{Q}{c^2R^3}\mathbf{v}\times(\mathbf{R}\times\mathbf{w})$$

we find them similar to a certain degree (up to some factors), except for the last term of (2.17). That term signals a difference between the two interactions, because it means that there is a velocity dependent acceleration which is parallel to the velocity; that is in contrast with the situation in electromagnetism, where the velocity dependent accelerations arise from magnetic forces, and are thus always perpendicular to $\mathbf{v}$.

As expected from Eqs. (2.4) (and by contrast with the other effects), the precession of a gyroscope carried by $\bar{\mathcal{O}}$, Eq. (2.18b) takes a form analogous to the precession of a magnetic dipole, Eq. (2.18a), if we express $\mathbf{S}$ in the local orthonormal tetrad $e^{\hat{i}}$, non rotating relative to the inertial observer at infinity, such that $S^i = (1 - M/R)S^{\hat{i}}$:

$$\frac{d\mathbf{S}}{d\bar{t}} = \frac{q}{2m}\frac{Q}{c^2 R^3}\left[\mathbf{S} \times (\mathbf{R} \times \mathbf{w})\right] \quad (a); \qquad \frac{dS^{\hat{i}}}{d\bar{t}} = \frac{2M}{c^2 R^3}\left[(\mathbf{R} \times \mathbf{w}) \times \mathbf{S}\right]^{\hat{i}} \quad (b). \tag{2.18}$$

If instead of the gyroscope comoving with observer $\bar{\mathcal{O}}$ (with constant velocity $\mathbf{w}$), we had considered a gyroscope moving in a circular orbit, then an additional term would arise in analogy with Thomas precession for the magnetic dipole; for a circular geodesic that term amounts to $-1/4$ of expression (2.18b), and we would obtain the well known equation for geodetic precession (e.g. Ciufolini-Wheeler 1995).

3. Conclusion

We conclude our paper by discussing some of the implications of our conclusions in the approaches usually found in literature. In the framework of linearized theory, e.g. Ruggiero-Tartaglia 2002, Ciufolini-Wheeler 1995, Einstein equations are often written in a Maxwell-like form; likewise, geodesics, precession and gravitational force on a spinning test particle are cast (in terms of 3-vectors defined in analogy with the electromagnetic fields $\mathbf{E}$ and $\mathbf{B}$) in a form similar to, respectively, the Lorentz force on a charged particle, the precession and the force on a magnetic dipole.

We have concluded that the actual physical similarities between gravity and electromagnetism (on which the physical content of such approaches relies) occur only on very special conditions. For tidal effects, like the forces on a gyroscopes/dipoles, the analogy manifest in Eqs. (2.3) holds only when the *test particle* sees time independent *fields*. In the example of analogous systems considered in section 2.1, this means that the center of mass of the gyroscope/dipole must not move relative to the central body. In the case of the analogy between the equation of geodesics and the Lorentz force law (see Fig. 1), as manifest in equation (2.5), it is in the *potentials/metric perturbations*, as seen by *the observer* (not the test particle!), that the time independence is required. The latter condition is not as restrictive as the one of the tidal effects: consider for instance observers moving in circular orbits around a static mass/charge; such observers see an unchanging spacetime, and unchanging electromagnetic potentials, so, for them, the equation of geodesics and Lorentz force take similar forms (such analogy may actually be cast in an exact form, see Natário 2007, Jantzen *et al.* 1992). However, those observers see a time-varying electric field $\mathbf{E}$ (constant in magnitude, but varying in direction), which, by means of equations (4) and (6), implies that the tidal tensors are not similar to the gravitational ones†.

† The electromagnetic field $F^{\alpha\beta}$ is not constant along the worldline of an observer moving in a circular orbit (radius R, angular velocity $\mathbf{\Omega}$, velocity $\mathbf{w} = \mathbf{\Omega} \times \mathbf{R}$) around a point charge. Its variation endows the magnetic tidal tensor with an antisymmetric part, and the electric tidal tensor with a time component: $dF^{0i}/d\tau = Qw^i/cR^3 = -2E^{[i0]} = -\epsilon^{ijk}B_{[jk]}$. This means that they significantly differ from the GR tidal tensors seen by an observer in circular motion around a point mass. Note that both the GR and the EM tidal tensors for these analogous problems can be obtained from, respectively, Eqs. (2.11)-(2.12) and (2.13)-(2.16), making therein $\mathbf{R}.\mathbf{w} = 0$ (corresponding to circular motion), despite the fact that these expressions were originally derived for an observer with constant velocity. This is because, as can be seen from their definitions in Table 1, it is the 4-velocity U^α (regardless of the way it varies), at the given point, that determines the tidal tensors.

Finally, as a consequence of this analysis, a distinction, from the point of view of the analogy with electrodynamics, between effects related to (stationary) rotational mass currents, and those arising from translational mass currents, becomes clear: albeit in the literature both are dubbed "gravitomagnetism", one must note that, while the former are clearly analogous to magnetism, in the case of the latter the analogy is not so close.

References

Maartens, R. & Basset, B. 1998, *Class. Q. Grav.*, 15, 705

Damour, T., Soffel, M., & Xu, C. 1991, *Phys. Rev. D.*, 43, 3273

Ruggiero, M. & Tartaglia, A. 2002, *Nuovo Cim.*, 117B 743 [arXiv:gr-qc/0207065].

Costa, L. Filipe, O., & Herdeiro, Carlos, A. R. 2008, *Phys. Rev. D*, 78, 024021

Costa, L. Filipe, & Herdeiro, Carlos, A. R. 2007 [arXiv:gr-qc/0612140]

Costa, L. Filipe, O., & Herdeiro, Carlos, A. R. 2009, *Int. J. Mod. Phys. A*, 24, 1695

Ciufolini, I. & Pavlis, E. 2004, *Nature* 431, 958; Ciufolini I., Pavlis E., Peron R., *New Astron.* 11, 527-550 (2006).

Ciufolini, I. & Wheeler, J. 1995, *Gravitation and Inertia* (Princeton Univ. Press)

Soffel, M., Klioner, S., Muller, J., & Biskupek, L. 2008, *Phys. Rev. D*, 78, 024033

Jantzen, R., Carini, P., & Bini, D. 1992, *Ann. Phys.*, 215, 1 [arXiv:gr-qc/0106043].

Natário, J. 2007, *Gen. Rel. Grav.*, 39, 1477 [arXiv:gr-qc/0701067].

Nordtvedt, K. 1973, *Phys. Rev. D*, **7**, 2347

Will, C. & Nordtvedt, K. 1972, *APJ*, 177, 757

Nordtvedt, K. 1988, *Int. J. Theoretical Physics*, 27, 2347

Relativity in Fundamental Astronomy
Proceedings IAU Symposium No. 261, 2009
S. A. Klioner, P. K. Seidelman & M. H. Soffel, eds.

doi:10.1017/S1743921309990123

Reference frames, gauge transformations and gravitomagnetism in the post-Newtonian theory of the lunar motion

Yi Xie[1,2] and Sergei Kopeikin[2]†

[1]Astronomy Department, Nanjing University,
Nanjing, Jiangsu 210093, China

[2]Department of Physics & Astronomy, University of Missouri-Columbia,
Columbia, MO 65211, USA

Abstract. We construct a set of reference frames for description of the orbital and rotational motion of the Moon. We use a scalar-tensor theory of gravity depending on two parameters of the parametrized post-Newtonian (PPN) formalism and utilize the concepts of the relativistic resolutions on reference frames adopted by the International Astronomical Union in 2000. We assume that the solar system is isolated and space-time is asymptotically flat. The primary reference frame has the origin at the solar-system barycenter (SSB) and spatial axes are going to infinity. The SSB frame is not rotating with respect to distant quasars. The secondary reference frame has the origin at the Earth-Moon barycenter (EMB). The EMB frame is local with its spatial axes spreading out to the orbits of Venus and Mars and not rotating dynamically in the sense that both the Coriolis and centripetal forces acting on a free-falling test particle, moving with respect to the EMB frame, are excluded. Two other local frames, the geocentric (GRF) and the selenocentric (SRF) frames, have the origin at the center of mass of the Earth and Moon respectively. They are both introduced in order to connect the coordinate description of the lunar motion, observer on the Earth, and a retro-reflector on the Moon to the observable quantities which are the proper time and the laser-ranging distance. We solve the gravity field equations and find the metric tensor and the scalar field in all frames. We also derive the post-Newtonian coordinate transformations between the frames and analyze the residual gauge freedom of the solutions of the field equations. We discuss the gravitomagnetic effects in the barycentric equations of the motion of the Moon and argue that they are beyond the current accuracy of lunar laser ranging (LLR) observations.

Keywords. gravitation, relativity, astrometry, reference systems

---

The tremendous progress in technology, which we have witnessed during the last 30 years, has led to enormous improvement of precision in measuring time and distances within the boundaries of the solar system. Observational techniques like lunar and satellite laser ranging, radar and Doppler ranging, very long baseline interferometry, high-precision atomic clocks, gyroscopes, etc. have made it possible to start probing the kinematic and dynamic effects in motion of celestial bodies to unprecedented level of fundamental interest. Current accuracy requirements make it inevitable to formulate the most critical astronomical data-processing procedures in the framework of Einstein's general theory of relativity. This is because major relativistic effects are several orders of magnitude larger than the technical threshold of practical observations and in order to interpret the results of such observations, one has to build physically-adequate relativistic models. The future projects will require introduction of higher-order relativistic models

† email: kopeikins@missouri.edu

supplemented with the corresponding parametrization of the relativistic effects, which will affect the observations.

The dynamical modeling for the solar system (major and minor planets), for deep space navigation, and for the dynamics of Earth's satellites and the Moon must be consistent with general relativity. LLR measurements are particularly crucial for testing general relativistic predictions and advanced exploration of other laws of fundamental gravitational physics. Current LLR technologies allow us to arrange the measurement of the distance from a laser on the Earth to a corner-cube reflector (CCR) on the Moon with a precision approaching 1 millimeter (Battat *et al.* 2007 and Murphy *et al.* 2008).

At this precision, the LLR model must take into account all the classical and relativistic effects in the orbital and rotational motion of the Moon and Earth. Although a lot of effort has been made in constructing this model, there are still many controversial issues, which obscure the progress in better understanding of the fundamental principles of the relativistic model of the Earth-Moon system.

The theoretical approach used for construction of the JPL ephemeris accepts that the post-Newtonian description of the planetary motions can be achieved with the Einstein-Infeld-Hoffmann (EIH) equations of motion of point-like masses (Einstein *et al.* 1938), which have been independently derived by Petrova (1949) and Fock (1959) for massive fluid balls as well as by Lorentz & Droste (1917) under assumptions that the bodies are spherical, homogeneous and consist of incompressible fluid. These relativistic equations are valid in the barycentric frame of the solar system with time coordinate t and spatial coordinates $x^i \equiv \mathbf{x}$.

However, due to the covariant nature of the general theory of relativity the barycentric coordinates are not unique and are defined up to the space-time transformation (Brumberg 1972, Brumberg 1991, and Soffel 1989)

$$t \quad \mapsto \quad t - \frac{1}{c^2}\sum_B \nu_B \frac{GM_B}{R_B}(\mathbf{R}_B \cdot \mathbf{v}_B), \tag{1}$$

$$\mathbf{x} \quad \mapsto \quad \mathbf{x} - \frac{1}{c^2}\sum_B \lambda_B \frac{GM_B}{R_B}\mathbf{R}_B, \tag{2}$$

where summation goes over all the massive bodies of the solar system ($B = 1, 2, ..., N$); G is the universal gravitational constant; c is the fundamental speed in the Minkowskian space-time; a dot between any spatial vectors, $\mathbf{a} \cdot \mathbf{b}$ denotes an Euclidean dot product of two vectors $\mathbf{a}$ and $\mathbf{b}$; M_B is mass of body B; $\mathbf{x}_B = \mathbf{x}_B(t)$ and $\mathbf{v}_B = \mathbf{v}_B(t)$ are coordinates and velocity of the center of mass of the body B; $\mathbf{R}_B = \mathbf{x} - \mathbf{x}_B$; ν_B and λ_B are constant, but otherwise free parameters being responsible for a particular choice of the barycentric coordinates. These parameters can be chosen arbitrary for each body B of the solar system. Standard textbooks (Brumberg 1972, Brumberg 1991, Soffel 1989, and section 4.2 in Will 1993) assume that the coordinate parameters are equal for all bodies. This simplifies the choice of coordinates and their transformations, and allows one to identify the coordinates used by different authors. For instance, $\nu = \lambda = 0$ corresponds to harmonic or isotropic coordinates (Fock 1959), $\lambda = 0$ and $\nu = 1/2$ realizes the standard coordinates used in the book of Landau & Lifshitz (1975) and in PPN formalism (Will 1993). The case of $\nu = 0, \lambda = 2$ corresponds to the Gullstrand-Painlevé coordinates (Painlevé 1921, Gullstrand 1922), but they have not been used so far in relativistic celestial mechanics of the solar system. We prefer to have more freedom in transforming EIH equations of motion and do not equate the coordinate parameters ν_B, λ_B for different massive bodies.

If the bodies in the N-body problem are numbered by indices B, C, D, etc., and the coordinate freedom is described by equations (1), (2), EIH equations have the following form (equation 88 in Brumberg 1972)

$$a^i_B = F^i_N + \frac{1}{c^2} F^i_{EIH}, \tag{3}$$

where the Newtonian force

$$F^i_N = -\sum_{C\neq B} \frac{GM_C R^i_{BC}}{R^3_{BC}}, \tag{4}$$

the post-Newtonian perturbation

$$\begin{aligned}
F^i_{_{EIH}} = &-\sum_{C\neq B} \frac{GM_C R^i_{BC}}{R^3_{BC}} \Bigg\{ (1+\lambda_C) v_B^2 - (4+2\lambda_C)(\mathbf{v}_B \cdot \mathbf{v}_C) + (2+\lambda_C) v_C^2 \\
&-\frac{3}{2}\left(\frac{\mathbf{R}_{BC}\cdot\mathbf{v}_C}{R_{BC}}\right)^2 - 3\lambda_C \left[\frac{\mathbf{R}_{BC}\cdot\mathbf{v}_{BC}}{R_{BC}}\right]^2 - (5-2\lambda_B)\frac{GM_B}{R_{BC}} - (4-2\lambda_C)\frac{GM_C}{R_{BC}} \\
&-\sum_{D\neq B,C} GM_D \Bigg[\frac{1}{R_{CD}} + \frac{4-2\lambda_D}{R_{BD}} - \left(\frac{1+2\lambda_C}{2R^3_{CD}} - \frac{\lambda_C}{R^3_{BD}} + \frac{3\lambda_D}{R_{BD}R^2_{BC}} - \frac{3\lambda_D}{R_{CD}R^2_{BC}}\right) \\
&\times(\mathbf{R}_{BC}\cdot\mathbf{R}_{CD})\Bigg]\Bigg\} - \sum_{C\neq B}\Bigg\{\frac{GM_C v^i_{CB}}{R^3_{BC}}\Big[(4-2\lambda_C)(\mathbf{v}_B\cdot\mathbf{R}_{BC}) - (3-2\lambda_C) \\
&\times(\mathbf{v}_C\cdot\mathbf{R}_{BC})\Big] + \frac{GM_C}{R_{BC}} \sum_{D\neq B,C} GM_D R^i_{CD}\left(\frac{7-2\lambda_C}{2R^3_{CD}} + \frac{\lambda_C}{R^3_{BD}} + \frac{\lambda_D}{R_{CD}R^2_{BC}}\right. \\
&\left.-\frac{\lambda_D}{R_{BD}R^2_{BC}}\right)\Bigg\},
\end{aligned} \tag{5}$$

and $\mathbf{v}_B = \mathbf{v}_B(t)$ is velocity of the body B, $\mathbf{a}_B = \dot{\mathbf{v}}_B(t)$ is its acceleration, $\mathbf{R}_{BC} = \mathbf{x}_B - \mathbf{x}_C$, $\mathbf{R}_{CD} = \mathbf{x}_C - \mathbf{x}_D$ are relative distances between the bodies, and $\mathbf{v}_{CB} = \mathbf{v}_C - \mathbf{v}_B$ is a relative velocity.

Barycentric coordinates $\mathbf{x}_B$ and velocities $\mathbf{v}_B$ of the center of mass of body B are adequate theoretical quantities for description of the world-line of the body with respect to the center of mass of the solar system. However, the barycentric coordinates are global coordinates covering the entire solar system. Therefore, they have little help for efficient physical decoupling of the post-Newtonian effects existing in the description of the local dynamics of the orbital motion of the Moon around Earth (Brumberg & Kopeikin 1989). The problem originates from the covariant nature of EIH equations and the gauge freedom of the general relativity theory. Its resolution requires a novel approach based on introduction of a set of local coordinates associated with the barycenter of the Earth-Moon system, the Earth and the Moon (Kopeikin & Xie 2009).

The gauge freedom is already seen in the post-Newtonian EIH force (5) as it explicitly depends on the choice of spatial coordinates through the gauge-fixing parameters λ_C, λ_D. Each term, depending explicitly on λ_C and λ_D in equation (5), has no direct physical meaning because it can be eliminated after making a specific choice of these parameters. In many works on experimental gravity and applied astronomy (including JPL ephemerides) researches fix parameters $\lambda_C = \lambda_D = 0$, which corresponds to working in harmonic coordinates. Harmonic coordinates simplify EIH equations to large extent but one has to keep in mind that they have no physical privilege anyway, and that a separate term or a limited number of terms from EIH equations of motion can not be measured if they are gauge-dependent (Brumberg 1991).

This opinion was recently confronted in publications by Murphy *et al.* (2007a,b), Soffel *et al.* (2008), Williams *et al.* (2004), who followed Nordtvedt (1988). They separated EIH equations (3)-(5) to the form being similar to the Lorentz force in electrodynamics

$$a_B^i = \sum_{C \neq B} \left[E_{BC}^i + \frac{4 - 2\lambda_C}{c} (\mathbf{v}_B \times \mathbf{H}_{BC})^i - \frac{3 - 2\lambda_C}{c} (\mathbf{v}_C \times \mathbf{H}_{BC})^i \right] \tag{6}$$

where E_{BC}^i is called the "gravitoelectric" force, and the terms associated with the cross products $(\mathbf{v}_B \times \mathbf{H}_{BC})^i$ and $(\mathbf{v}_C \times \mathbf{H}_{BC})^i$ are referred to as the "gravitomagnetic" force (Nordtvedt 1988). The "gravitomagnetic" field is given by equation

$$H_{BC}^i = -\frac{1}{c} (\mathbf{v}_{BC} \times \mathbf{E}_{BC})^i = \frac{GM_C}{c} \frac{(\mathbf{v}_{BC} \times \mathbf{R}_{BC})^i}{R_{BC}^3}, \tag{7}$$

and is proportional to the Newtonian force multiplied by the factor of v_{BC}/c, where v_{BC} is the relative velocity between two gravitating bodies.

The gravitomagnetic field is of paramount importance for theoretical foundation of general relativity (Ciufolini & Wheeler 1995). Therefore, it is not surprising that the acute discussion has started about whether LLR can really measure the "gravitomagnetic" field H_{BC}^i (Murphy *et al.* 2007a, Kopeikin 2007, Murphy *et al.* 2007b, Ciufolini 2007, Soffel *et al.* 2008). It is evident that equation (6) demonstrates a strong dependence of the "gravitomagnetic" force of each body on the choice of the barycentric coordinates. For this reason, by changing the coordinate parameter λ_C one can eliminate either the term $(\mathbf{v}_B \times \mathbf{H}_{BC})^i$ or $(\mathbf{v}_C \times \mathbf{H}_{BC})^i$ from EIH equations of motion (6). In particular, the term $(\mathbf{v}_B \times \mathbf{H}_{BC})^i$ vanishes in the Painlevé coordinates, making the statement of Murphy *et al.* (2007a,b) about its "measurement" unsupported, because the strength of the factual "gravitomagnetic" force is coordinate-dependent. Notice that the barycentric (SSB) frame remains the same. We eliminate the "gravitomagnetic" force by changing the spatial coordinate only. In particular, the Lorentz transformation does not play any role. Hence a great care should be taken in order to properly interpret the LLR "measurement" of such gravitomagnetic terms in consistency with the covariant nature of the general theory of relativity and the theory of astronomical measurements in curved space-time. We keep up the point that the "gravitomagnetic" field (7) is unmeasurable with LLR due to its gauge-dependence that is not associated with the transformation from one frame to another but with the coordinate transformation (2).

Nevertheless, the observable LLR time delay is gauge invariant. This is because the gauge transformation changes not only the gravitational force but the solution of the equation describing the light ray propagation. For this reason, the gauge parameter λ_C appears in the time delay *explicitly*

$$\begin{aligned} t_2 - t_1 \quad &= \quad \frac{R_{12}}{c} + 2 \sum_C \frac{GM_C}{c^3} \ln \left[\frac{R_{1C} + R_{2C} + R_{12}}{R_{1C} + R_{2C} - R_{12}} \right] \\ &\quad + \sum_C \lambda_C \frac{GM_C}{c^3} \frac{(R_{1C} - R_{2C})^2 - R_{12}^2}{2R_{1C} R_{2C} R_{12}} (R_{1C} + R_{2C}). \end{aligned} \tag{8}$$

At the same time the "Newtonian" distance R_{12} depends on the parameter λ_C *implicitly* through the solution of EIH equations (3)-(5). This implicit dependence of the right side of (8) is exactly compensated by the explicit dependence of (8) on λ_C, making the time delay gauge-invariant.

Papers (Murphy *et al.* 2007a,b, Williams *et al.* 2004, Soffel *et al.* 2008) do not take into account the explicit gauge-dependence of the light time delay on λ_C. If the last term in (8) is omitted but EIH force is taken in form (6), the equations (6) and (8) become theoretically incompatible. In this setting LLR "measures" only the consistency of the EIH equations with the expression for time delay of the laser pulse. However, this is not a test of gravitomagnetism, which actual detection requires more precise measurement of the gauge-invariant components of the Riemann tensor associated directly either with the spin multipoles of the gravitational field of the Earth (Ciufolini 2008, Ciufolini & Pavlis 2004) or with the current-type multipoles of the tidal gravitational field of external bodies (Kopeikin 2008).

In order to disentangle physical effects from numerous gauge dependent terms in the equations of motion of the Moon we need a precise analytic theory of reference frames in the lunar motion that includes several reference frames: SSB, GRF, SRF and EMB. This gauge-invariant approach to the lunar motion has been initiated in our paper (Kopeikin & Xie 2009) to which we refer the reader for further particular details.

Acknowledgements

Y. Xie is thankful to the Department of Physics & Astronomy of the University of Missouri-Columbia for hospitality and accommodation. The work of Y. Xie was supported by the China Scholarship Council Grant No. 2008102243 and IAU travel grant. The work of S. Kopeikin was supported by the Research Council Grant No. C1669103 of the University of Missouri-Columbia and by 2009 faculty incentive grant of the Arts and Science Alumni Organization of the University of Missouri-Columbia.

References

Battat, J., Murphy, T. W., Adelberger, E., *et al.* 2007, *APS Meeting Abstracts*, 12.003
Brumberg, V. A. 1972, *Relativistic Celestial Mechanics*, Moscow: Nauka (in Russian)
Brumberg, V. A. 1991, *Essential Relativistic Celestial Mechanics*, New York: Adam Hilger
Brumebrg, V. A. & Kopeikin, S. M. 1989, *Nuovo Cimento B*, 103, 63
Ciufolini, I. & Wheeler, J. A. 1995, *Gravitation and Inertia*, NJ: Princeton University Press
Ciufolini, I. 2007, arXiv:0704.3338v2
Ciufolini, I. 2008, arXiv:0809.3219v1
Ciufolini, I. & Pavlis, E. C. 2004, *Nature*, 431, 958
Einstein, A., Infeld, L., & Hoffmann, B. 1938, *The Annals of Mathematics*, 39, 65
Fock, V. A. 1959, *The Theory of Space, Time and Gravitation*, New York: Pergamon Press
Gullstrand, A. 1922, *Arkiv. Mat. Astron. Fys.*, 16, 1
Kopeikin, S. M. 2007, *Phys. Rev. Lett.*, 98, 229001
Kopeikin, S. M. 2008, arXiv:0809.3392v1
Kopeikin, S. M. & Xie, Y. 2009, arXiv:0902.2416v2
Landau, L. D. & Lifshitz, E. M. 1975, *The classical theory of fields*, Oxford: Pergamon Press
Lorentz, H. A. & Droste, J. 1937, *Versl. K. Akad. Wet. Amsterdam*, 26, 392 (part I) and 649 (part II)
Murphy, Jr., T. W., Nordtvedt, K., & Turyshev, S. G. 2007a, *Phys. Rev. Lett.*, 98, 071102
Murphy, Jr., T. W., Nordtvedt, K., & Turyshev, S. G. 2007b, *Phys. Rev. Lett.*, 98, 229002
Murphy, T. W., Adelberger, E. G., Battat, J. B. R., *et al.* 2008, *Publ. Astron. Soc. Pacific*, 120, 20
Nordtvedt, K. 1988, *Int. J. Theor. Phys.*, 27, 1395
Painlevé, P. 1921, *C. R. Acad. Sci. (Paris)*, 173, 677
Petrova, N. M. 1949, *Zh. Exp. Theor. Phys.*, 19, 989
Soffel, M. H. 1989, *Relativity in Astrometry, Celestial Mechanics and Geodesy*, Berlin: Springer
Soffel, M., Klioner, S., Müller, J., & Biskupek, L. 2008, *Phy. Rev. D*, 78, 024033
Will, C. M. 1993, *Theory and Experiment in Gravitational Physics*, Cambridge University Press
Williams, J. G., Turyshev, S. G., & Murphy, T. W. 2004, *IJMPD*, 13, 567

Relativity in Fundamental Astronomy
Proceedings IAU Symposium No. 261, 2009
S. A. Klioner, P. K. Seidelman & M. H. Soffel, eds.

doi:10.1017/S1743921309990135

Relativistic description of astronomical objects in multiple reference systems

Chongming Xu[1] & Zhenghong Tang[2]

Shanghai Astronomical Observatory, Chinese Academy of Sciences, P.R.China;
email: [1]cmxu1938@gmail.com, [2]zhtang@shao.ac.cn

Abstract. Many astronomical systems require for their description in the frame of Einstein's theory of gravity not just one but several reference systems. In the first post-Newtonian approximation the Damour-Soffel-Xu (DSX) formalism presents a new and improved treatment of celestial mechanics and astronomical reference systems for the gravitational N-body problem. In the DSX-formalism the astronomical bodies are characterized by their Blanchet-Damour (BD) mass- and spin-multipole moments. However, the time dependence of these moments requires additional dynamical equations, usually local flow equations describing the internal motions inside the bodies or additional assumptions about them. In this article the internal motion of astronomical bodies will be adressed within the 1st post-Newtonian approximation to Einstein's theory of gravity. A concept of quasi-rigid bodies will be introduced; after that, astronomical fluid and elastic bodies will be discussed.

Keywords. GRT, relativistic description of astronomical objects, post-Newtonian approximation

1. Introduction

We are at the point where astronomical measurements will soon reach incredible accuracies; e.g., the astrometric satellite Gaia will measure angular distances with a precision of about $10\,\mu$as. At this level of accuracy not only the various astronomical reference systems, but also the internal motion of astronomical bodies have to be described in the framework of relativity, at least at the first post-Newtonian approximation. The problem of relativistic astronomical reference systems and the applications of the standard IAU framework have been discussed by Soffel (2009) in detail. In this paper we discuss the problem of internal motion of astronomical bodies. The discussion will be such that several reference systems should be used to describe the motion of the whole system relativistically assuming e.g., that the astronomical body is member of an N-body system. A (DSX) framework dealing with the celestial mechanics of N rotating bodies of arbitrary shape and composition at the first post-Newtonian approximation to Einstein's theory of gravity has recently been introduced by Damour, Soffel & Xu (1991, 1992, 1993, 1994). This framework employs a total of $N+1$ different coordinate (reference) systems: one global one that covers the entire model manifold to describe the motion of the whole N-body system and one local system co-moving with each of the N bodies to describe the local physics, especially the internal motion of the body under consideration. In the DSX-framework the gravitational potentials (one scalar and one vector potential) of some body in the outside region is described by a set of mass- and a set of spin-multipole moments (Blanchet-Damour moments) in the corresponding local comoving system. If $(cT, \boldsymbol{X})$ denote the coordinates of some local A-system these BD-moments are defined by

$$M_L^A(T) \equiv \int_A d^3X\hat{X}^L\Sigma + \frac{1}{2(2l+3)c^2}\frac{d^2}{dT^2}\left[\int_A d^3XX^LX^2\Sigma\right]$$
$$-\frac{4(2l+1)}{(i+1)(2l+3)c^2}\frac{d}{dT}\left[\int_A d^3XX^{aL}\Sigma^a\right] + O(4) \ \ (l \geqslant 0),$$
$$S_L^A(T) \equiv \int_A d^3X\epsilon^{ab<c_l}\hat{X}^{L-1>a}\Sigma^b + O(2) \ \ (l \geqslant 1),$$

where L is a multi-index of l Cartesian indices: $L = i_1i_2\cdots i_l$; both, the mass moments M_L and the spin moments S_L are assumed to be symmetric and trace-free (STF); Σ and Σ^a are the active gravitational mass and active mass current density which are defined as $\Sigma = c^{-2}(T^{00} + T^{bb})$ and $\Sigma^a = c^{-1}T^{0a}$ in the local coordinate system. $T^{\alpha\beta}$ are the contravariant components of the stress-energy tensor. The order symbol, $O(n)$, in the equation indicates that terms of order c^{-n} have been neglected.

The two gravitational potentials, W_A and W_A^a, appearing in the canonical form of the local metric tensor (e.g., Damour, Soffel & Xu 1991) outside of body A can easily be expressed in terms of the BD moments of A. For astronomical bodies of almost spherical shape this expansion will converge rapidly.

In the DSX-framework the time dependence of the BD-moments has not been treated in detail. It depends upon the internal dynamics of the astronomical body, which is the main topic of this article.

For many astronomical bodies (e.g., stars) it is sufficient to assume the internal material to be fluid. Corresponding Newtonian hydrodynamical and thermodynamical equations can be found in many textbooks. Relativistic hydrodynamical and thermodynamical equations have been discussed, e.g., by Chandrasekhar (1971), who, however, employs a single coordinate system. Wu and Xu (2001) have discussed such equations in the problem of multiple reference systems which might be relevant for close binary or multiple stellar systems.

For other astronomical bodies, such as the Earth or neutron stars, the elastic components cannot be described by a fluid picture. Here, the formalism of relativistic elastomechanics can be employed. Within the DSX-framework such a formalism has been worked out by Xu, Wu & Soffel (2001) and Xu *et al.*, (2003, 2005). This formalism is based on a displacement-field as is the ell known Newtonian formalism of elastomechanics. For applications in the field of geodynamics post-Newtonian equations for the displacement field have been derived that generalize the well known Jeffreys-Vicente (Jeffreys & Vicence 1957) equations from Newtonian physics.

Both, the fluid and elastic material of astronomical bodies will be treated below. In Newtonian celestial mechanics, however, dynamical equations are often drastically simplified by assuming the astronomical bodies to be rigid (only three internal degrees of freedom). For the construction of a solar system ephemeris such an assumption is a good starting point. E.g., for the problem of Earth rotation in the Newtonian framework the well known SMART solution (Bretagnon *et al.*, 1997), which might serve as excellent basis for treating the real problem of geodynamics, is based on 'rigid bodies'. It is well known that rigid bodies with incompressible material elements do not exist in relativity since in reality the sound speed always has to be finite. Nevertheless, quasi-rigid bodies may be introduced by auxiliary conditions that simplify the post-Newtonian formalism drastically. Such quasi-rigid relativistic bodies have been treated in Xu & Tao (2004) and Tao & Xu (2003).

2. Quasi-rigid bodies

For our model of a quasi-rigid astronomical body a certain relation for the internal motion is formulated in the local system that is co-moving with the body under consideration. We can define the quantities $\bar{\Sigma}$ and $\bar{\Sigma^a}$ by

$$\begin{aligned}
\overline{\Sigma} &\equiv \Sigma + \frac{\Sigma_{\rm PN}}{c^2}, \quad \text{where} \quad \Sigma_{\rm PN} \equiv \frac{11}{42} X^2 \ddot{\Sigma}, \\
\overline{\Sigma}^a &\equiv \Sigma^a + \frac{\Sigma^a_{\rm self}}{c^2} + \frac{\Sigma^c_{\rm ext}}{c^2}, \\
\Sigma^a_{\rm self} &\equiv \Sigma \left(\frac{7}{2} \epsilon_{ade} \Omega^d \partial_e Z^+ + \frac{1}{2} \epsilon_{edf} \Omega^d X^f \partial_{ae} Z^+ \right), \\
Z^+ &\equiv G \int_A d^3 X' \Sigma(T_A, \mathbf{X}') |\mathbf{X} - \mathbf{X}'|, \\
\Sigma^a_{\rm ext} &\equiv \sum_{l \geqslant 0} \frac{\Sigma}{l!} \Bigg[4 \epsilon_{ade} \Omega^d X^e X^{<L>} G_L(T) + \frac{1}{l+2} \epsilon_{aed} X^{<dL>} H_{eL} \\
&\quad - \frac{l+10}{2(l+2)(2l+5)} \hat{X}^L X^2 \dot{G}_{cL} + \frac{l+10}{2(l+2)(2l+5)} \partial_T (\ln \Sigma) \hat{X}^L X^2 G_{cL} \Bigg],
\end{aligned}$$

where G_L and H_L are gravito-electric and gravito-magnetic tidal moments (e.g., Damour, Soffel & Xu 1991). Now, a model of a 1PN quasi-rigid body can be constructed by means of the following constraint:

$$\overline{\Sigma}^a + \frac{1}{2c^2} X^a \left[\partial_T T^{bb} - \Sigma \partial_T W + \partial_T \Sigma_{\rm PN} + \partial_a (\Sigma^a_{\rm self} + \Sigma^a_{\rm ext}) \right] = \epsilon_{abc} \Omega^b X^c \overline{\Sigma} + O(4).$$

In this equation Ω^a might be considered as a formal parameter of our quasi-rigid body, whereas in the Newtonian approximation Ω^a is the angular velocity. Our constraint equation has the following consequences: The 1PN spin vector (defined in Damour, Soffel & Xu 1993) is given by

$$S_a^{\rm PN} = I_{ab} \Omega^b + O(4),$$

where the 1PN moment of inertia tensor reads

$$I_{ab} = I_{ba} = \int_A d^3 X (\delta_{ab} \mathbf{X}^2 - X^a X^b) \overline{\Sigma} + O(4).$$

The post-Newtonian MacCullagh relations then simply read

$$M_{ab} = -I_{ab} + \frac{1}{3} \delta_{ab} I_{cc} + O(4).$$

As is well known such MacCullagh relations between components of the inertia tensor and potential coefficients are very useful in geodynamics. Actually, with our constraint equation all relations for 'Newtonian rigid bodies' are valid even at the first post-Newtonian level except for one:

$$\dot{M}_L = \epsilon_{pq<a_l} M_{L-1>q} \Omega^p, \quad l > 2.$$

This difference, however, might not be relevant in most situations since the relativistic part of M_L with $L > 2$ usually can be neglected.

3. Elastic and fluid bodies

Our formalism that describes the dynamics of elastic material based upon a displacement field presents an application of the Carter and Quintana (Carter & Quintana 1972) formalism. Let s^a denote the components of the displacement field that describes deviations from an equilibrium configuration (the elastomechanical ground state). The post-Newtonian Jeffreys-Vicente (J-V) equation for an almost spherical body then reads (Xu, Wu & Soffel 2001):

$$\rho^* \frac{D^2 s_a}{D\tau^2} = \overline{W}_{G,a}\delta\rho^* + \rho^*(\delta\overline{W})_{,a} - \frac{1}{2}\rho^*(\overline{V}^2)_{,ab}\dot{s}^b - (\delta p)_{,a} + (2\mu s^b{}_a)_{;b}$$
$$+\frac{1}{c^2}\left\{\rho^*\left[-\frac{1}{2}(\overline{V}^2)_{,ab}s^b\overline{V}^2 - \overline{V}^a(\overline{V}^b\ddot{s}^b) + 4W_{G,[a}\overline{V}^{b]}\dot{s}^b - (\delta\overline{W})_{,\overline{T}}\overline{V}^a\right.\right.$$
$$\left.\left.+4(\delta\overline{W}_a)_{,\overline{T}} + \overline{W}_{,a}\overline{W}_{,b}s^b - 8\overline{V}^b_{,(c}\overline{W}_{[a),b]}s^c\right] - \delta\dot{p}\overline{V}^a\right\} + O(4),$$

where the variables with a bar are in rotating coordinates (angular velocity: Ω^a), ρ is the mass-energy density ($\rho^* = \rho + pc^{-2}$), μ denotes the shear modulus, $\overline{W}_G$ is the PN geopotential $\overline{W}_G = \overline{W} + \frac{1}{2}\overline{V}^2$, $\overline{V}^a$ the rotation velocity, $\overline{W}$ and $\overline{W}^a$ the scalar potential and vector potential of the metric tensor; δp, $\delta\rho$, δW are the Eulerian variation of p, ρ and W, respectively. For $c \to \infty$, the above equation reduces to the Newtonan J-V equation. Boundary and junction conditions of this relativistic J-V equation have been discussed in Xu *et al.* (2003). The PN J-V equation and related equations have been expanded in terms of generalized spherical harmonics so that the partial differential equations become a set of ordinary differential equations (Xu *et al.*, 2005) to simplify further calculations.

A fluid body might be viewed as a special case of an elastic body. Corresponding hydrodynamical equations have been discussed by Xu and Wu (2001) within the DSX-formalism for applications to multiple stellar systems. In the DSX-framework, the hydrodynamic equations (energy equation and Euler equation) take the form

$$\frac{\partial}{\partial T}\Sigma + \frac{\partial}{\partial X^a}\Sigma^a = \frac{1}{c^2}\frac{\partial}{\partial T}T^{bb} - \frac{1}{c^2}\Sigma\frac{\partial}{\partial T}W + O(4),$$
$$\frac{\partial}{\partial T}\left[\left(1+\frac{4W}{c^2}\right)\Sigma^a\right] + \frac{\partial}{\partial X^b}\left[\left(1+\frac{4W}{c^2}\right)T^{ab}\right] = F^a(T, X^a) + O(4),$$

where

$$F^a = \Sigma E_a + \frac{1}{c^2}B_{ab}\Sigma^b,\ E_a = \partial_a W + \frac{4}{c^2}\partial_T W_a,\ B_{ab} = -4(\partial_a W_b - \partial_b W_a).$$

In the case of a non-perfect fluid the stress-energy tensor in local coordinates reads

$$T^{\alpha\beta} = \epsilon U^\alpha U^\beta + ph^{\alpha\beta} - \frac{1}{3}\beta\theta h^{\alpha\beta} - \lambda\sigma^{\alpha\beta} + \frac{2}{c^2}Q^{(\mu}U^{\nu)},$$

where ϵ is the density of total mass-energy, p the isotropic pressure, λ and β are the coefficients of shear and bulk viscosity, U^α is the 4-velocity, $h^{\alpha\beta}$ the usual projection operator into the fluid's rest space, Q^α the heat flux vector, $\sigma^{\alpha\beta}$ the shear tensor and $\theta = U^\mu_{;\mu}$ the expansion scalar. When $\lambda = \beta = \kappa = 0$, $T^{\alpha\beta}$ describes a perfect fluid. ϵ, θ, U^α, $h^{\alpha\beta}$ and $\sigma^{\alpha\beta}$ can then be explicitly expressed in terms of Σ, Σ^a, W and W^a. Q^α also depends upon temperature T. If the heat flux is taken into account then $T^{\alpha\beta}$ depends upon Σ, Σ^a, W, W^a, T, λ, β and κ. For reasons of brevity, here we only show the hydrodynamic equations for a perfect fluid

$$\dot{\Sigma} + \Sigma^a_{,a} = \frac{1}{c^2}\left(\frac{2\dot{\Sigma}^d\Sigma^d}{\Sigma} - \frac{\dot{\Sigma}\Sigma^d\Sigma^d}{\Sigma^2} + 3\dot{p} - \Sigma\dot{W}\right) + O(4),$$

$$\frac{\partial}{\partial T}\left[\left(1+\frac{4W}{c^2}\right)\Sigma^a\right]+\frac{\partial}{\partial X^b}\left[\left(1+\frac{4W}{c^2}\right)\left(\frac{\Sigma^a\Sigma^b}{\Sigma}\left(1+\frac{\Sigma^c\Sigma^c}{\Sigma^2c^2}+\frac{2p}{\Sigma c^2}\right)\right.\right.$$
$$\left.\left.+p\delta^{ab}\left(1-\frac{2W}{c^2}\right)\right)\right]=\Sigma\partial_a W+\frac{4}{c^2}\left[\Sigma\partial_T W_a-\Sigma^b\left(\partial_a W_b-\partial_b W_a\right)\right]+O(4).$$

The post-Newtonian expansion scalar reads

$$\theta=U^\mu_{;\mu}=\frac{\partial V^a}{\partial X^a}+\frac{1}{c^2}\left[\frac{\partial V^a}{\partial X^a}\left(W+\frac{V^2}{2}\right)+3\frac{dW}{dT}+\frac{1}{2}\frac{d}{dT}V^2\right]+O(4).$$

Note that $\theta=0$ does not imply that $\partial V^a/\partial X^a=0$ (incompressible fluids only exists in Newtonian hydrodynamics, not in the general relativistic hydrodynamics). For more details the reader is referred to Xu and Wu (2001).

4. Discussion

Three different models for the dynamics of an astronomical body for an extension of the DSX-framework have been introduced: elastic bodies, fluid bodies and quasi-rigid bodies. Such models have been introduced in the local coordinate system that is co-moving with the body under consideration and can be used in situations where multiple reference systems should be introduced. Several problems still have to be treated, e.g., for the problem of geodynamics an expansion of relevant functions in terms of generalized spherical harmonics has not yet been done for an oblate ground state, which is important for a relativistic theory of precession-nutation.

Acknowledgements

The authors thank the editor and the referee for improving the text and the English. This work was supported by the Natural Science Foundation of China under the grants 10673026 and 10873014.

References

Bretagnon, P., Rocher, P., & Simon, J. L. 1997, *A&A*, 319, 305
Carter, B. & Quintana, H. 1972, *Proc. Roy. Soc. London*, A331, 57
Chandrasekhar, S. 1971, *ApJ*, 164, 569
Damour, T., Soffel, M., & Xu, C. 1991, *Phys. Rev. D* 43, 3273
Damour, T., Soffel, M., & Xu, C. 1992, *Phys. Rev. D* 45, 1017
Damour, T., Soffel, M., & Xu, C. 1993, *Phys. Rev. D* 47, 3124
Damour, T., Soffel, M., & Xu, C. 1994, *Phys. Rev. D* 49, 618
Jeffereys, H. & Vicence, R. O. 1957, *MNRAS*, 117, 142
Soffel, M. 2009, *this proceedings*, 1
Tao, J. & Xu, C. 2003, *Int. J. Mod. Phys. D* 12, 811
Wahr, J. M. 1982, *Geo-dynamics*, No. 41, 327
Xu, C. & Wu, X. 2001, *Phys. Rev. D* 63, 064001
Xu, C., Wu, X., & Soffel, M. 2001, *Phys. Rev. D* 63, 043002
Xu, C., Wu, X., Soffel, M., & Klioner, S. 2003, *Phys. Rev. D* 68, 064009
Xu, C. & Tao, J. 2004, *Phys. Rev. D* 69, 024003
Xu, C., Wu, X., & Soffel, M. 2005, *Phys. Rev. D* 71, 024030

Relativity in Fundamental Astronomy
Proceedings IAU Symposium No. 261, 2009
S. A. Klioner, P. K. Seidelman & M. H. Soffel, eds.

doi:10.1017/S1743921309990147

The celestial reference frame stability and apparent motions of the radio sources

V. E. Zharov[1], M. V. Sazhin[2], V. N. Sementsov[2], K. V. Kuimov[2], O. S. Sazhina[2] and N. T. Ashimbaeva[2]

[1]Physics Departement of Moscow State University,
Universitetsky pr. 13, Moscow, Russia
email: VladZh2007@yandex.ru

[2]Sternberg State Astronomical Institute of Moscow State University,
Universitetsky pr. 13, Moscow, Russia
email: valera@sai.msu.ru

Abstract. Time series of the coordinates of the ICRF radio sources were analyzed. It was shown that part of radio sources, including even the so-called "defining" sources, show a the significant apparent motion. Corrections for their a priori coordinates are time functions. The celestial reference frame stability is provided by the no-net-rotation condition applied to the selected subset of sources, which leads in our case to a rotation of the frame axes with time. Parameters of this rotation were calculated for different subsets of sources.

To improve stability of the celestial reference frame new methods of selection of the extragalactic radio sources were suggested. The first one was called "cosmological" and the second one "kinematical". It was shown that a selected subset of the ICRF sources, according to cosmological criteria, determines the most stable coordinate system during next decade.

1. Introduction

The International Celestial Reference System (ICRS) is based on the positions of 608 selected compact extragalactic radio sources (quasars, active galactic nuclei (AGN), and blazars) (Ma *et al.*, 1998). Stability of the system axes is guaranteed by positions of the "defining" radio sources. One assumes that their coordinates are known as precisely as possible. These sources are unresolved with VLBI baselines comparable to the Earth diameter, and it was assumed that variations of their coordinates are negligible.

To prepare a new catalogue of radio sources (new realization of ICRF-2) one has first to analyze the observation data during 1980-2008 years. The first aim of this analysis was comparison of the time series of the sources' coordinates calculated by different groups which use different software for reduction and analysis of VLBI data. The second aim was to introduce a selection of the sources into "stable" and "unstable", instead of "defining", "candidates", and "others". Time series used for analysis are listed in table 1.

We used a method of approximation of time series of coordinates by a polynomial model. A linear model with respect to β_i (i = 0, 1, 2, 3) regression polynomial coefficients is

$$y(t) = \beta_0 + \beta_1 t + \beta_2 t^2 + \beta_3 t^3 + \varepsilon(t), \tag{1.1}$$

where t is time, $y(t)$ are corrections ($\Delta\alpha\cos\delta$, $\Delta\delta$) to the ICRF coordinates (right ascension or declination) of a source, and ε is a stochastic value residual.

Table 1. The list of the time series of the sources' coordinates and codes

Name of series	Code	Number of session	Observation interval
aus002a	OCCAM 6.2	3554	1979-04.2007
bkg00g	CALC/SOLVE 10.0	3466	1984-07.2007
dgf000g	OCCAM 6.1	2981	1984-08.2007
gsf001a	CALC/SOLVE 10.0	4389	1979-11.2007
iaa000b	QUASAR	4202	1979-05.2007
mao00a	Steelbreeze	3548	1980-05.2007
opa002a	CALC/SOLVE 10.0	3750	1984-12.2007
sai000b	ARIADNA	3209	1984-12.2007
usn000g	CALC/SOLVE 10.0	4170	1979-05.2007

The coefficients of the polynomials were found out by regression analysis. The power of the polynomial was determined by a R^2 statistic, where

$$R^2 = \frac{\sum(\hat{y}_j - \bar{y})^2}{\sum(y_j - \bar{y})^2} = 1 - \frac{\sum(y_j - \hat{y}_j)^2}{\sum(y_j - \bar{y})^2}. \tag{1.2}$$

Here y_j is the correction of right ascension or declination at the moment $t = t_j, j = 1, 2, \ldots, N$, and $\hat{y}_j$ is the estimation of polynomial function at t_j, and $\bar{y}$ is the average value of the series over whole span interval. The value R depends on the correlation between y and $\hat{y}$ (Draper & Smith, 1998). Obviously, if the polynomial model is correct, that is values of $\hat{y}_j$ are equal to y_j, the coefficient $R = 1$. Actually, $\hat{y}_j \neq y_j$ and $R < 1$, but the maximal value of R corresponds to the best fitting model.

As an example, one can see from Fig.1 and Fig.2 that the polynomial model of coordinate variations of the ICRF source $1404 + 286$ has power larger then 3. It is well known that when one increases the number of parameters in the fitting function the residual decreases, but the predicted confidence intervals increase enormously. The choice of the model with a small residual and large prediction error, or with a precise prediction and a large residual is determined by the general problem under consideration. At the same time one has to estimate the significance level of regression, i.e. is an increasing R significant or not.

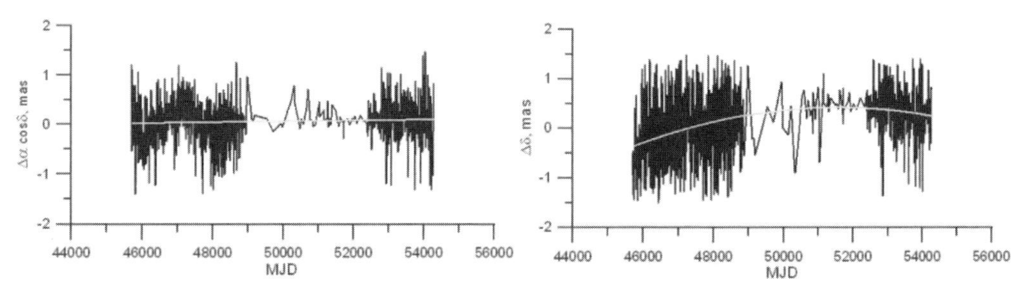

Figure 1. Right ascension variation (left curve) and declination (right curve) of the ICRF source 1404+286 as function of time. The best fit of α is linear model ($\beta_1 = 3 \pm 1$ μas/year), while the best fit of δ is quadratic polynomial ($\beta_2 = (-3 \pm 2) \times 10^{-4} \mu$as/year^2).

We have to estimate the stability of the radio source positions. The stability of the celestial reference frame is determined by sources with negligible apparent velocities. It means that the coefficients β_i, $i = 1, 2, 3$ in (1.1) have to be close to zero. But analysis shows that many of the ICRF sources have apparent motion, i.e. coefficients β_i are significantly larger than the errors. Therefore, this fact requests physical explanation.

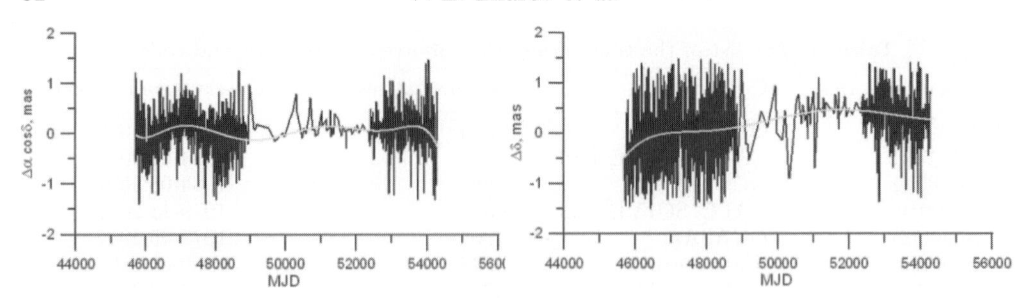

Figure 2. Right ascension variation (left curve) and declination (right curve) of the ICRF source 1404+286 as function of time. Functions α and δ were fitted by polynomials of 7-th power.

To choose the stable ICRF sources we propose the following method of source selection. First of all we consider kinematical characteristics of sources. We can predict the value of y_{N+1} and its confidence intervals outside the observing data span when we have a well-fitted model of $y(t)$.

We call stable sources that show small apparent motion, i.e. the confidence intervals of the predicted corrections to right ascension or declination include a zero value. Otherwise, if the model shows a significant difference of correction to α or δ from zero, we can call this source unstable (at a corresponding confidence level).

The analysis of data (table 1) shows that many of the ICRF sources reveal significant linear motion, their confidence intervals increase rapidly, and do not include zero. Therefore, we must consider them as unstable in according to the above-mentioned criteria. Actually, one can subtract a well predicted linear trend, and then the confidence intervals include zero and this source can be considered as "stable".

2. Blandford–Rees model of extragalactic radio sources

Let us consider a physical mechanism which can explain this apparent motion (Zharov *et al.*, 2009), (Sazhin *et al.*, 2009).

To explain this phenomena we choose the Blandford–Rees (BR) model (Blandford & Königl, 1979), (Begelman, Blandford & Rees, 1984). The main idea of this hypothesis is that the quasars and AGN-objects (most of the ICRF sources are quasar or AGN's) represent the system of a massive black hole and jets (see Fig. 3). The optical radiation is formed in black hole's accretion disk while the radio emission is formed into the jet, at some distance from the optical source. Below we will call the radio source a "jet-core" instead of the optical core which coincides with position of massive black hole.

In papers of Zacharias *et al.* (1999), Assafin *et al.* (2003) the catalogue of optical positions of 172 ICRF sources was composed. In the first paper a significant spacial difference between optical and radio quasar components was already pointed out. From our point of view this difference indicates that the BR model is appropriate for our purposes. The physical distance between optical and radio components can be estimated for sources with known redshifts and measured angular distances. The mean distance is 300-500 pc (Fig. 4). The uncertainty of distance is connected with uncertainties of the optical positions and the unknown value of the angle between the precession axis and the line of sight. It means that the radio component moves relative to the optical one with the angular velocity $1 - 10\,\mu$as/year.

We assume that the linear apparent motion can be explained by the precession of the jet while the quadratic apparent motion can be explained by the stochastic process

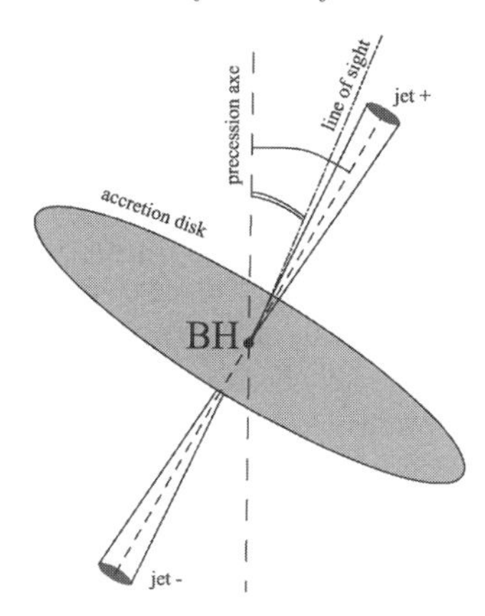

Figure 3. This figure represents the Blandford–Rees model. Central black hole (BH) is surrounded by accretion disk, and two jets from polar regions. The "jet+" is directed to observer. The small black ellipse ending the "+" cone represents a "jet core".

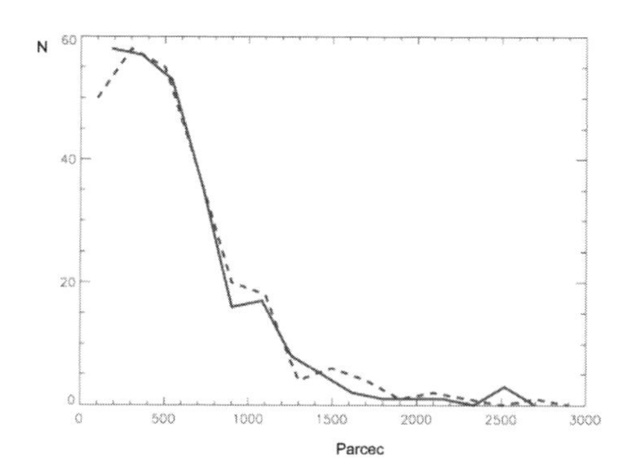

Figure 4. Observed distribution (dashed line) of distances between optical and radio component in quasars according the BR model. The solid line is simulated distribution calculated by the multiplication of two random functions: gamma and sine of a uniformly distributed angle between the precession axis and the line of sight.

of interaction of jet particles with interstellar clouds. The period of jet precession is expected to be $10^3 \div 10^6$ years (Zharov *et al.*, 2009). As long as the precession periods of the jets are significantly larger than the time of observations, the source motions can be treated as linearly stable and predictable with a high accuracy for the time interval of VLBI observations $\sim$ 30 years, while the quadratic motion is stochastic and unpredictable.

As a result we restrict our consideration by two models

$$y(t) = \beta_0 + \beta_1 t + \varepsilon(t), \quad y(t) = \beta_0 + \beta_2 t^2 + \varepsilon(t),$$

and the decision of which model is valid is taken with the following criteria. We calculated the function R_1^2 (1.2) for linear model and R_2^2 (1.2) for a quadratic model. If and only if $R_2^2/R_1^2 > 5$, we accept the quadratic model of apparent motion. Approximately two-thirds of the sources show linear and one-third show quadratic motion.

As long as linear motion is predictable and stationary for a long time interval, one can subtract the linear trend of the data and work with these "residual" data.

3. Cosmological criterion for ICRF sources choice.

All other motions inside the radio source represent a noise component of astrometric observation. These motions occur inside some linear scale. The shorter scale the smaller angular displacement as seen by an observer.

Thereby to decrease astrometric noise and to improve the coordinate system stability we have to choose the most remote sources. It is correct in the Euclidean space: the more remote a source the less the angular scale of its apparent motion. In the Friedman model of expanding Universe it is not correct. Extragalactic objects have to be considered in expanding space-time, and in framework of the Standard Cosmological Model.

According to this model the apparent angular size of the source has a minimum for an redshift $z = 1.63$. An object located at this distance with a physical size of about 1 pc has an angular size equals to $\theta = 116$ μas. This is a minimal angular size of an object, and it will increase for $z < 1.63$ and for $z > 1.63$. It was shown that the interval of redshift $0.8 \leqslant z \leqslant 3.0$ is the most favorable in terms that the physical shift inside such sources corresponds to minimal apparent angular shift of a "jet core". Details of this calculation can be found in (Sazhin *et al.*, 2009).

After the selection of sources as "unstable" and "stable" according to the "kinematical" and "cosmological" criteria we obtained the final list of 137 sources (see Table 1 in (Sazhin *et al.*, 2009)).

4. The ICRF system instability.

As was pointed out that the main purpose of the selection of "stable" sources is the stability of the celestial reference frame which is connected with the predictability of the source motion.

The variation of the ICRF source coordinates leads to a small rotation of reference system. To estimate the stability of the system three small angles θ_1, θ_2, θ_3, which describe a small rotation were calculated:

$$\mathbf{s}(t) = \begin{pmatrix} 1 & -\theta_3 & \theta_2 \\ \theta_3 & 1 & -\theta_1 \\ -\theta_2 & \theta_1 & 1 \end{pmatrix} \mathbf{s}(t_0)$$

where $\mathbf{s}(t)$, $\mathbf{s}(t_0)$ are unity vectors of a source at moments t and $t_0 = J2000.0$.

In the Fig. 5 the angles θ_1, θ_2, θ_3, calculated for different subsets of the ICRF sources ("defining", "stable" and for "cosmological") are shown.

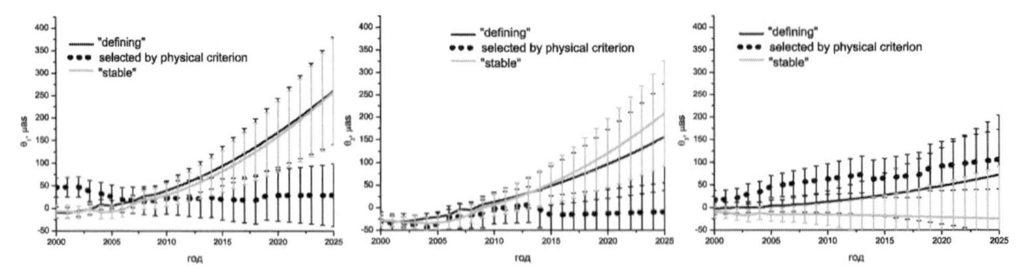

Figure 5. The rotation angles θ_1(a), θ_2(b), θ_3(c) calculated for different subsets of the ICRF sources: defining (black), the Feissel's stable (gray) and our list based on physical criteria (dotted).

Obviously the subset formed on bases of the cosmological criteria makes the stability of the ICRF better.

5. Conclusions

The physical basis of "cosmological" and "kinematical" criteria is based on the assumption that apparent motion of quasars is connected with real motion inside quasars. Therefore apparent angular motion corresponds to a real physical shift of a "jet core" inside a radio source. The interval of redshift $0.8 \leqslant z \leqslant 3.0$ is the most favorable in terms that physical shift inside such sources corresponds to a minimal apparent angular shift of a "jet core". The method of "cosmological" selection improves stability of the ICRF over next decade.

Acknowledgements

Authors thank Dr. D. Gabuzda, Prof. K. Postnov, Prof. B. Somov, and Dr. O. Titov for fruitful discussion. This work has been supported by Russian Foundation for Basic Research grants 07-02-01034 and 08-02-00971, grant of the President of RF MK-2503.2008.2 (O.S.).

References

Assafin, M., Zacharias, M., Rafferty, T. J. *et al.*, 2003, *AJ*, **125**, 2728

Begelman, M. C., Blandford, R. D., & Rees, M. J., 1984, *Rev. Mod. Phys*, **56**, 255

Blandford, R. D. & Königl, A., 1979, *AJ*, **232**, 34

Draper, N. R. & Smith, H. Applied Regression Analysis. 3-rd edition. John Wiley & Sons, Inc. 1998.

Ma, C., Arias, E. F., Eubanks, T. M., *et al.*, *AJ*, **116**, 516, 1998.

Sazhin, M. V., Sementsov, V. N., Zharov, V. E., Kuimov, K. V., Ashimbaeva, N. T., & Sazhina, O. S., 2009, arXiv:0904.2146v1

Zacharias, N., Zacharias, M. I., Hall, D. M. *et al.*, 1999, *AJ*, **118**, 2511

Zharov, V. E., Sazhin, M. V., Sementsov, V. N., Kuimov, K. V., & Sazhina, O. S., *AZh*, **86**, 527, 2009 (in russian); *Astron. Rep.*, **53**, 579, 2009.

Relativity in Fundamental Astronomy
Proceedings IAU Symposium No. 261, 2009
S. A. Klioner, P. K. Seidelman & M. H. Soffel, eds.

doi:10.1017/S1743921309990159

Astronomical tests of relativity: beyond parameterized post-Newtonian formalism (PPN), to testing fundamental principles

Vladik Kreinovich

University of Texas at El Paso, El Paso, TX 79968, USA
email: vladik@utep.edu

Abstract. By the early 1970s, the improved accuracy of astrometric and time measurements enabled researchers not only to experimentally compare relativistic gravity with the Newtonian predictions, but also to compare different relativistic gravitational theories (e.g., the Brans-Dicke Scalar-Tensor Theory of Gravitation). For this comparison, Kip Thorne and others developed the Parameterized Post-Newtonian Formalism (PPN), and derived the dependence of different astronomically observable effects on the values of the corresponding parameters.

Since then, all the observations have confirmed General Relativity. In other words, the question of which relativistic gravitation theory is in the best accordance with the experiments has been largely settled. This does not mean that General Relativity is the final theory of gravitation: it needs to be reconciled with quantum physics (into quantum gravity), it may also need to be reconciled with numerous surprising cosmological observations, etc. It is, therefore, reasonable to prepare an extended version of the PPN formalism, that will enable us to test possible quantum-related modifications of General Relativity.

In particular, we need to include the possibility of violating fundamental principles that underlie the PPN formalism but that may be violated in quantum physics, such as scale-invariance, T-invariance, P-invariance, energy conservation, spatial isotropy violations, etc. In this paper, we present the first attempt to design the corresponding extended PPN formalism, with the (partial) analysis of the relation between the corresponding fundamental physical principles.

Keywords. Gravitation, relativity, celestial mechanics, etc.

1. Introduction

One of the main motivations for the development of General Relativity was the discrepancy between the astronomical observations and the predictions of Newton's theory: namely, the 43 sec/100 years difference in the perihelion of Mercury. The first confirmation of General Relativity also came from astronomy, as the 1919 eclipse observations of near-solar objects that confirmed the gravity-based bending the light paths. Until the early 1970s, astronomical and time measurements have been used to compare different predictions of General Relativity with the Newtonian ones – and in all the case, General relativity was confirmed.

By the early 1970s, the improved accuracy of astrometric and time measurements enabled the researchers not only to experimentally compare relativistic gravity with the Newtonian predictions, but also to compare different relativistic gravitational theories (e.g., the Brans-Dicke Scalar-Tensor Theory of Gravitation). For this comparison, Kip Thorne (Thorne & Will, 1971) and others developed the Parameterized Post-Newtonian Formalism (PPN), and derived the dependence of different astronomically observable effects on the values of the corresponding parameters; see, e.g., Brumberg (1991), Will (1993).

Since then, all the observations have confirmed General Relativity. In other words, the question of which relativistic gravitation theory is in the best accordance with the experiments has been largely settled. This does not mean that General Relativity is the final theory of gravitation: it needs to be reconciled with quantum physics (into quantum gravity), it may also need to be reconciled with numerous surprising cosmological observations, etc. It is, therefore, reasonable to prepare an extended version of the PPN formalism, that will enable us to test possible quantum-related modifications of General Relativity.

In particular, we need to include the possibility of violating fundamental principles that underlie the PPN formalism but that may be violated in quantum physics, such as scale-invariance, T-invariance, P-invariance, energy conservation, spatial isotropy violations, etc. We present the first attempt to design the corresponding extended PPN formalism, with the (partial) analysis of the relation between the corresponding fundamental physical principles.

2. Possible Violations of T-Invariance

Derivation of possible terms. One of the assumptions behind most terms of the PPN formalism is T-invariance, i.e., invariance with respect to changing time direction $t \to -t$. The largest non-T-invariant terms usually considered in celestial mechanics are radiation c^{-5} terms in binary systems (such as pulsars) Will (2001).

However, it is well known that quantum physics is not T-invariant: interaction experiments have shown that weak interactions are not T-invariant. It is, therefore, reasonable to consider possible effects of T-non-invariance on c^{-3} and c^{-4} terms in celestial mechanics. As usual, the c^{-4} terms in $ds^2 = g_{\alpha\beta} dx^\alpha dx^\beta$ mean c^{-4} terms in g_{00}, c^{-3} terms in g_{0i} $(1 \leqslant i \leqslant 3)$, and c^{-2} terms in g_{ij}.

Following the general ideas of PPN (see, e.g., Will (2001)), we assume that the terms g_{ij} analytically depend on the masses m_a of the celestial bodies (of order c^{-2}), on their velocities $\mathbf{v}_a$ (of order c^{-1}), on the inverse distances r_a^{-1} between the current point and the a-th body, r_{ab}^{-1} between the bodies, on the corresponding unit vectors $\mathbf{e}_a$ and $\mathbf{e}_{ab}$ (and on other terms like pressure p). The terms should be dimensionless and rotation-invariant.

The only T-non-invariant quantity is $\mathbf{v}_a$ of order c^{-1}. Every term must contain masses (since it must tend to 0 when $m_a \to 0$), so with m_a of order c^{-2} and $\mathbf{v}_a$, we have c^{-3}. In g_{ij}, we look for c^{-2} terms, so there are no T-non-invariant terms there.

Similarly, the values g_{0i} can contain $\mathbf{v}_a$ at most linearly – else they are $\sim c^{-4}$. Terms containing $\mathbf{v}_a$ linearly are T-invariant, so the new terms must contain no velocities at all. These terms should contain m_a and no other relativistic terms – else they would be c^{-4}. We need to add r_a^{-1} to make these terms dimensionless and $\mathbf{e}_a$ to make it a vector, so we get

$$\delta g_{0i} = \delta_2 \cdot \sum \frac{m_a \cdot e_{a,i}}{r_a}$$

for some parameter δ_2.

The values g_{00} can contain $\mathbf{v}_a$ at most quadratically. Quadratic terms are T-invariant, so new terms must be linear in $\mathbf{v}_a$. Adding r_a^{-1} to make it dimensionless and multiplying by $\mathbf{e}_a$ to make it a scalar, we get

$$\delta g_{00} = \delta_1 \cdot \sum \frac{m_a \cdot (\mathbf{e}_a \cdot \mathbf{v}_a)}{r_a}$$

for some parameter δ_1.

Effect on light. Light is determined by c^{-2} terms in $g_{\alpha\beta}$, so the new terms have no effect on light.

Additional coordinate transformations. Are the new terms coordinate-invariant? To find out, we need to add the possibility of T-non-invariant coordinate transformation $x^{\alpha'} = x^\alpha + \xi^\alpha(x^\beta)$ to the usual PPN transformations Will (2001). These transformations lead to $\delta g_{\alpha\beta} = -\xi_{\alpha,\beta} - \xi_{\beta,\alpha}$. To maintain PPN approximation, we must consider terms up to c^{-2} in x^i and up to c^{-3} in x^0. The term ξ^i contains m_a of order c^{-2}, so it cannot contain any non-T-invariant terms $\mathbf{v}_a$ (which would add the order c^{-1}); thus, the ξ^i terms are T-invariant. The ξ^0 terms must contain $\mathbf{v}_a$ at most linearly; linear terms are T-invariant, so the only new terms do not contain $\mathbf{v}_a$ at all. To make the resulting terms in $g_{\alpha\beta}$ dimensionless, we must multiply m_a by $\log(r_a)$ (to get r_a^{-1} in the derivative). Thus, we get the additional coordinate transformation $x_0' = x_0 + \xi_0$, with

$$\xi_0 = \alpha \cdot \sum_a m_a \cdot \ln(r_a),$$

for some parameter α.

This transformation leads to terms

$$\delta g_{00} = -2\xi_{0,0} = -2\alpha \cdot \sum \frac{m_a \cdot (\mathbf{e}_a \cdot \mathbf{v}_a)}{r_a}$$

and

$$\delta g_{0i} = -\xi_{0,i} = -\alpha \cdot \sum \frac{m_a \cdot e_{a,i}}{r_a},$$

i.e., to $\delta_1' = \delta_1 - 2\alpha$ and $\delta_2' = \delta_2 - \alpha$. One can easily conclude that the necessary and sufficient condition for a metric to be T-invariant in some coordinate system (i.e., to have α for which $\delta_1' = \delta_2' = 0$) is $\delta_1 = 2\delta_2$.

The coordinate-invariant combination of new parameters is $\delta_1' \stackrel{\text{def}}{=} \delta_1 - 2\delta_2$.

The existence of a Lagrange function. When can these new terms come from a Lagrange function L? The new terms must contain m_a (else they do not tend to 0 as $m_a \to 0$), they must contain at least one other m_b – else there are no distances to make them dimensionless, and they must be of order $\geqslant c^{-6}$. Thus, they can contain $\mathbf{v}_a$ at most quadratically. Quadratic terms are T-invariant, so the only possible non-T-invariant terms contain $\mathbf{v}_a$ linearly. Using dimensionless-ness and rotation-invariance, we conclude that the only possible term is

$$\sum_{a \neq b} \frac{m_a \cdot m_b \cdot (\mathbf{v}_a \cdot \mathbf{e}_{ab})}{r_{ab}},$$

but this term is a full time derivative of the expression

$$\sum_{a \neq b} m_a \cdot m_b \cdot \ln(r_{ab})$$

and therefore, does not affect the Lagrange equations of motion.

Thus, the Lagrange function exists if and only if the metric is T-invariant.

Lorentz-invariance. By applying Lorentz transformation, we can see that the new terms are Lorentz-invariant if and only if $\delta_1 = 2\delta_2$, i.e., if the metric is T-invariant.

Thus, T-non-invariant effects are ether-dependent, i.e., depend on the velocity $\mathbf{w}$ of the system's center of mass with respect to the stationary system.

Effects on the restricted 2-body problem. Following Brumberg (1991), we find the additional term

$$\delta L = -\frac{1}{2} \cdot \delta_1' \cdot \frac{m \cdot (\mathbf{r} \cdot \mathbf{w})}{r^2}$$

in the Lagrange function of the restricted 2-body problem, then compute the average $[\delta L]$ over fast changing angular variables:

$$[\delta L] = \frac{1}{2} \cdot \delta_1' \cdot \frac{m}{a} \cdot \frac{1 - \sqrt{1 - e^2}}{e} \cdot E,$$

where

$$E = w_x[\cos\Omega \sin\omega - \sin\Omega \cos i \sin\omega] + w_y[\sin\Omega \cos\omega - \cos\Omega \cos i \sin\omega] + w_z \sin i \sin\omega$$

Thus, we get the following formulas for the osculating elements:

$$\frac{da}{dt} = 0; \quad \frac{de}{dt} = \delta_1' \cdot \frac{\sqrt{1 - e^2} \cdot (1 - \sqrt{1 - e^2}) \cdot m}{2 \cdot n \cdot a^3 \cdot e^2} \cdot E_e,$$

where

$$E_e = w_x[\cos\Omega \sin\omega + \sin\Omega \cos i \cos\omega] - w_y[\sin\Omega \sin\omega + \cos\Omega \cos i \cos\omega] - w_z \sin i \cos\omega$$

$$\frac{di}{dt} = -\delta_1' \cdot \frac{\cot i \cdot \left(1 - \sqrt{1 - e^2}\right) \cdot m}{2 \cdot n \cdot a^3 \cdot \sqrt{1 - e^2} \cdot e} E_e - \delta_1' \cdot \frac{\left(1 - \sqrt{1 - e^2}\right) \cdot m}{2 \cdot n \cdot a^3 \cdot \sqrt{1 - e^2} \cdot e \cdot \sin i} \cdot E_i,$$

where

$$E_i = w_x[\sin\Omega \sin\omega + \cos\Omega \cos i \cos\omega] - w_y[\sin\Omega \sin\omega + \cos\Omega \cos i \cos\omega] - w_z \sin i \cos\omega$$

etc.

In particular, for the perihelion shift, we get

$$\frac{d\pi}{dt} = k \cdot \delta_1' \cdot w \cdot \frac{m}{n \cdot a^3},$$

with $k \approx 1$, so the shift per orbit cycle $\Delta\pi \sim k \cdot \delta_1' \cdot w$ does not depend on the distance a to the Sun. For the accuracy of $0.01''$ per 100 years, and with $w \approx 700$ km/s, we get $|\delta_1'| \leqslant 3 \cdot 10^{-5}$.

Comment. In the Lunar motion, the effect of new terms is also negligible.

T-non-invariance without scale-invariance. If we do not assume that the metric is dimensionless (scale-invariant), then for the gravitational acceleration $\mathbf{a}$ of a body we get a general formula $\mathbf{a} = \mathbf{f}(m_a, \mathbf{r}, \mathbf{r}_a, \mathbf{v}, \mathbf{v}_a)$. The only requirements are that the formula is rotation-invariant and that $\mathbf{f} = 0$ when all $m_a = 0$.

From the physical viewpoint, it is natural to add a requirement of energy conservation, i.e., that it is impossible to have a closed cycle and gain some work while returning all the bodies to their original locations with original velocities.

Under this additional assumption, our first conclusion is that the radial motion in a central field is T-invariant. Indeed, if it was not T-invariant, then we could reverse $\mathbf{v}$ and get a different acceleration. Then, by letting the body go closer to the center and back (or vice versa), we would be able to gain energy.

Our second conclusion is that under P-invariance (under $\mathbf{x} \to -\mathbf{x}$), circular motion in a central field is T-invariant. Indeed, if it was not, then we could reverse the velocities and get a different acceleration. Then, by letting a body go first in one circular direction and then back, we would gain energy.

Since for planets, orbits are almost circular, we can thus conclude that under energy conversation, P-invariance is equivalent to T-invariance (modulo eccentricity e). If we additionally assume that $\mathbf{f}$ is analytical with respect to m_a, $\mathbf{v}$, and $\mathbf{v}_a$, and Lorentz-covariant, then we see that the smallest non-T-invariant terms are of order c^{-5} – as in radiation effect.

3. Possible Violations of P-Invariance

Similarly to the case of T-non-invariance, one can show that in the PPN-order approximation, the only P-non-invariant term is

$$\delta g_{0i} = \varepsilon \cdot \sum \frac{m_a}{r_a^2} \cdot (\mathbf{v}_a \times \mathbf{r}_a)_i.$$

This expression has been described before: it is the Chern-Simons term in Yunes & Pretorius (2009) coming from supersymmetry; we show that this is the only possible P-non-invariant terms of PPN order.

The above formula is in agreement with the above conclusion that all P-asymmetric terms are T-invariant, and that, therefore, PT-invariance implies P- and T-invariance.

Here, no new coordinate transformations are possible. The Lagrange function for an N-body problem exists if and only if the metric is P-invariant, and the motion is Lorentz-invariant if and only if it is P-invariant.

The secular effects in the 2-body problem are (here, $\mathbf{w}$ is the same as above):

$$\frac{da}{dt} = \frac{de}{dt} = \frac{d\mathcal{M}}{dt} = 0; \quad \frac{di}{dt} = \varepsilon \cdot \frac{m}{a^2\sqrt{1-e^2}} \cdot (w_x \cdot \cos(\Omega) + w_y \cdot \sin\Omega);$$

$$\frac{d\Omega}{dt} = -\varepsilon \cdot \frac{m}{a^2\sqrt{1-e^2}} \cdot (\cot(i)(w_x \cdot \sin(\Omega) - w_y \cdot \cos(\Omega)) - w_z);$$

$$\frac{d\omega}{dt} = \varepsilon \cdot \frac{m}{a^2\sqrt{1-e^2}} \cdot (\cot(i) \cdot \cos(i) \cdot (w_x \sin\Omega - w_y \cos\Omega) - w_z \cdot \cos(i)).$$

The effects are of the usual form m/a^2. Thus, $\varepsilon \leqslant$ accuracy of measuring perihelion shift, i.e., $|\varepsilon| \leqslant 0.01$.

4. Possible Violations of Equivalence Principle and Their Relation to Non-Conservation of Energy

In general, we can distinguish between the inertial mass m^I, and active m^A and passive m^P gravitational masses. Under this distinction, the force $\mathbf{F}_1$ with which the 2nd body attracts the 1st one is equal to

$$\mathbf{F}_1 = m_1^I \cdot \mathbf{a}_1 = -G \cdot \frac{m_1^P \cdot m_2^A}{r_{12}^3} \cdot \mathbf{r}_{12}.$$

What if we assume that energy is preserved? First, we can connect two bodies with an elastic rod. In general, the resulting 2-body system moves with the force

$$\mathbf{F} = \mathbf{F}_1 + \mathbf{F}_2 \sim (m_1^P \cdot m_2^A - m_2^P \cdot m_1^A).$$

If $\mathbf{F} \neq 0$, we can get the immobile combination moving and thus, get energy out of nothing. Thus, if energy is preserved, we have $\mathbf{F} = 0$ and hence, $m^A/m^P = \text{const}$

(i.e., $m_a \propto m_P$), and

$$\mathbf{a}_1 = -G \cdot \frac{m_1^A}{m_1^I} \cdot \frac{m_2^A}{r_{12}^3} \cdot \mathbf{r}_{12}.$$

We know that every particle a annihilates with its antiparticle $\widetilde{a}$ into a pair of photons: $a + \widetilde{a} \leftrightarrow 2\gamma$. It is reasonable to assume that gravity is C-invariant, i.e., that $m_a = m_{\widetilde{a}}$. Thus, if $m^I \not\propto m^A$, we can make the following experiment with an originally immobile combination of a and $\widetilde{a}$:

- first, let this combination move towards the gravity source;
- after a while, annihilate a and $\widetilde{a}$, turn the photons into $b + \widetilde{b}$, and move the new combination $b + \widetilde{b}$ back to the original location;
- once in the original location, annihilate b and $\widetilde{b}$, and turn the photons into $a + \widetilde{a}$.

If $m_a^A / m_a^I \neq m_b^A / m_b^I$, the accelerations are different, so the system gains velocity (hence energy). In other words, if $m^I \not\propto m^A$, energy is not preserved.

Thus, in the presence of C-invariance, energy conservation implies the equivalence principle.

5. Other Possible Effects

Other possible effects include cosmology, Finsler (non-Riemannian) space-time, etc.; e.g., for torsion $S^\alpha_{\beta\gamma}$, instead of the formula $T^{\alpha\beta}_{;\beta} = 0$ (with which the derivation of relativistic celestial mechanic effects starts (Brumberg, 1991)), we have $T^{\alpha\beta}_{;\beta} + S_\beta T^{\alpha\beta} = 0$, where $S_\beta \stackrel{\text{def}}{=} S^\alpha_{\alpha\beta}$. The general PPN-type dependence is

$$S_0 = \beta_T \cdot \sum \frac{m_a \cdot (\mathbf{e}_a \cdot \mathbf{v}_a)}{r_a^2} \text{ and } S_i = \beta_T \cdot \sum \frac{m_a \cdot e_{ai}}{r_a^2};$$

additional T- and P-non-invariant terms are also possible. Interestingly, we now have a class of theories including Newton's gravity and intermediate theories. The fact that one of the terms is Newtonian simplifies the computations of the celestial mechanical effects of torsion.

Acknowledgements

This work was supported in part by National Science Foundation grant HRD-0734825. The author is greatly thankful to all the participants of IAU Symposium 261 for valuable suggestions, and to the anonymous referees for their help.

References

Brumberg, V. A. 1991, *Essential Relativistic Celestial Mechanics* (New York: Taylor and Francis).

Thorne, K. S. & Will, C. M. 1971, Thoretical frameworks for testing relativistic gravity. I. Foundations. *Astroph. J.*, 1971, 163, 595–610.

Will, C. M. 1993, *Theory and Experiment in Gravitational Physics* (London: Cambridge University Press).

Will, C. M. 2001, The confrontation between General Relativity and experiment, *Living Reviews in Relativity.*

Yunes, P. & Pretorius, F. 2009, Dynamical Chern-Simons modified gravity: spinning black holes in the slow-rotation approximation, *Phys. Rev. D*, 79(8), 084043.

Relativity in Fundamental Astronomy
Proceedings IAU Symposium No. 261, 2009
S. A. Klioner, P. K. Seidelman & M. H. Soffel, eds.

doi:10.1017/S1743921309990160

Units of measurement in relativistic context

Bernard Guinot

Observatoire de Paris, 61 avenue de l'Observatoire, F-75014 Paris, France

Abstract. In the Newtonian approximation of General Relativity, employed for the dynamical modelling in the solar system, the coordinates have the dimension of time and length. As these coordinates are close to their Newtonian counterpart, the adherence to the rules of the Quantity Calculus does not raise practical difficulties: the second and the metre should be used as their units, in an abstract conception of these units. However, the scaling of coordinate times, applied for practical reasons, generates controversies, because there is a lack of information about the metrics to which they pertain. Nevertheless, it is not satisfactory to introduce specific units for these scaled coordinate times.

---

1. Introduction

We consider in this paper the units to be used in the post-Newtonian approximations in the dynamical modeling of the solar system: the coordinates and other quantities appearing in the components of the metric.

In textbooks on general relativity, one can find a diversity of statements on the dimensions and units of coordinates, which can perplex the reader. However, their goal is often didactic: it is to explain that coordinates are not concretely measurable with clocks and rods as the Newtonian coordinates are supposed to be.

Misner, *et al.* (1973) wrote that relativistic coordinates are pure numbers (dimensionless), the "telephone numbers of events". In applications, this point of view leads to the introduction of dimensionless graduation units which bring unnecessary complications, as noted by Klioner (2008). It is much simpler to adhere to the rules of Quantity Calculus and to apply these rules in a symbolic manner, as recommended in discussions between the author and eminent metrologists (J. De Boer, T.J. Quinn).

The Quantity Calculus, which has its roots in the work of Maxwell, is not familiar outside the field of metrology. A magisterial paper 'On the History of Quantity Calculus and the International System' has been written by J. de Boer (1994/1995), designated by (JdB) in the following. Of particular interest for us, are also two recent papers by W. H. Emerson (2005) and (2008), although they refer to classical physics. As the subtleties of these papers make their reading rather difficult, I tried to retain what is essential for our purpose: the unit of coordinates and of quantities that appear in the Newtonian approximations of general relativity used in dynamical astronomy. We deal with the mechanical units only, the second, the metre and the kilogram, when we use the International System of Units (SI). Moreover, the mass Mdoes not appear explicitly, but only through its product by the gravitational constant G, GM, which has the dimension $(\text{length})^3/(\text{time})^2$. Thus, we avoid the additional difficulties of electrostatic and electromagnetic quantities and of thermodynamics.

2. Proper time and coordinates

In General Relativity, proper time is supposed to be directly measurable. We postulate that the count of periods of an atomic transition of an unperturbed, freely falling atom provides a good measure of the theoretical proper time along its world line. In an atomic clock, the corrections (relativistic or not) are applied so that a conventional number of periods provides the unit of proper time at a specified connector of the clock. This is the basis of the so-called "definition of the second" of the SI adopted in 1967: "The second is the duration of 9 192 631 770 periods of the radiation corresponding to the transition between the two hyperfine levels of the caesium 133 atom". This definition is rather a recipe to realize a second of proper time. Thus, locally, proper time is a measurable quantity, by comparison with a standard, in a concrete manner. Note that the definition (recipe) could change within a decade in order to reach a better accuracy (presently the inaccuracy is about 10^{-15} in relative value).

The metrological problems arise in relativity with the coordinates, which are implicitly determined by the adopted form of the components of the metric, after fixing the unit of proper time and their origins (conventional for coordinate time). There is theoretically a quasi-total freedom to select the most convenient coordinates for the problem at hand and their metrological dimension is not imposed. However, in the post-Newtonian approximation, the signature of the metric tensor provides a clear distinction between time and space. This is made explicit by the coordinates which are linked to proper time by factors which are close to unity and to $1/c$ (c, velocity of light). These coordinates are measurable quantities, which, according to the *International Vocabulary of basic and general terms of metrology (VIM),* are defined in the following terms: (*measurable) quantity, attribute of a phenomenon, body or substance that may be distinguished qualitatively and determined quantitatively.* However, they cannot be measured in the concrete conception of measurement, by direct use of a physical standard (Brumberg, 1991). Following JdB, we call them "abstract quantities". The problem we have to solve is then the designation of their units.

3. Quantity Calculus and units

Klioner (2008) has already exposed the essential features of the "Quantity Calculus" for our purpose. We recall them briefly.

A quantity Q is conceptual and is independent of any system of units. It may be concrete, as the length of an object, or abstract, as a relativistic space coordinate. *Using physical quantities gives a representation* [of the laws of physics], which *is invariant with respect of the choice of units*(JdB, p. 406). For application of this representation, Q is expressed, according to ISO notations, as the product

$$Q = \{Q\} \cdot [Q], \tag{3.1}$$

where $\{Q\}$ is the numerical value of the quantity and $[Q]$ its unit in some stated system of units. Here we consider the International System SI.

If, for example, we define quantity D by a product of powers of quantities A, B, C,

$$D = A^a\, B^b\, C^c, \tag{3.2}$$

it is evident that the use of a coherent derived unit $[A]^a\,[B]^b[C]^c$ ensures that the same relation exists between quantities and between their numerical values.

This is also true in the case of sums and differences of quantities under the essential conditions that these quantities are of the same kind and that they are expressed with the same unit.

These rules are straightforward in the case of concrete measurements. For abstract quantities, their application requires an abstract (or symbolic) conception of units. This will be first illustrated by an example.

4. Concrete and abstract points of view on units

Consider two tables of lengths L_1 and L_2 measured in meters. Their difference of lengths $L_2 - L_1$ is measured in meters, concretely, after juxtaposition of the extremities of the tables.

Now consider two time scales (relativistic or not) T_1 and T_2. In conformity with a well-established usage, the same abbreviation designates a time scale and its reading, but the reading is a quantity written in Italics, while the designation is in Roman. The numerical values of their readings at some event E are $\{T_1\}$ and $\{T_2\}$, and the difference of these values is

$$T_2 - T_1 = (\{T_2\} - \{T_1\}) \text{ seconds at event E,} \qquad (4.1)$$

where $(\{T_2\} - \{T_1\})$ is a pure number.

This notation was popularized in the 1950's and was called *algebraic notation of time differences* (advocated, in particular, by W. Markowitz). It was adopted by the IAU in 1967 (Commission 31, Resolution 2) and by the International Consultative Committee of the Radiocommunications in 1970 (Recommendation 459). It is now employed let us say 'instinctively' without any difficulties although its interpretation is puzzling. If we refer to the definition of the second, the second is a duration. However, there is no duration involved in eq. (4.1); it expresses an instantaneous measurement. The reason of this paradox is that on the left of (4.1) we use "second" as an abstract unit, in a *symbolic* conception of the unit of time, both for T_1 and T_2, while on the right we refer to the concrete definition of the second of some particular quantity (supposed to be an interval of proper time). The interpretation of eq. (4.1) requires that "second" be seen as symbolic (the symbol "s") in both sides of the equal sign.

This symbolic use of SI units may be difficult to accept. It is, nevertheless, unconsciously applied, as shown in the example of data on UT1 provided by the International Earth Rotation and References System Service.

Historically, the SI was born from the need of unification of units for trade, engineers, and laboratory scientists, all of them needing concrete units that can be represented by standards (étalons). In other terms, the units were seen as quantities of the same kind as the measurands. However, it appears that quantities of a different kind may have the same expression for their units when applying the algebra as in (3.2) to the base units. For example, the moment of a force and the energy are both expressed in $\mathrm{m}^2\ \mathrm{kg}\ \mathrm{s}^{-2}$. Even in that case of concrete quantities the unit expressed by $\mathrm{m}^2\ \mathrm{kg}\ \mathrm{s}^{-2}$should be considered as symbolic and designates two different units for two different quantities. The definition of a quantity imposes the expression of its unit. However, the inverse is not true, the unit does not bring unambiguous information on the quantity that it measures. Quantities must always be defined.

Suppose that one makes a measurement of the moment of a force. He uses a dynamometer and a rule graduated in meters. These instruments bring the uncertainties of their calibration. The result of the measurement is not expressed in an ideal meter, kilogram and second that conform to their definition. Nevertheless, it is expressed in units of m^2 $\mathrm{kg}\ \mathrm{s}^{-2}$. It is, of course, possible to cover this inconsistency by the statement of an uncertainty of the numerical value. However, the uncertainty is not always mentioned and, even in the case of a concrete measurement, the expression of the unit is also symbolic.

The name "second" and the letter "s" represent symbolically (or conventionally, if one prefers) the unit for all quantities having the dimension of time, quantities which have to be precisely defined and to be referred to by their name. These quantities can be, for example (a) a duration of proper time, (b) a reading of a given time scale, (c) a difference of readings of such a time scale, (d) the difference of readings of two time scales at some event. In these four examples, we have quantities that are not of the same *nature:* the reading of a time scale has no extension in time, while duration has. The distinction is made in the usual language: Oxford dictionary says *Six o'clock is a point of time; six hours is a period of time.* However, in science, we use the same name of unit "second" and the same symbol "s" for these quantities. Emerson (2008) gives another interesting example, with thermodynamic temperature expressed in kelvin, K. Here the definition of thermodynamic temperature fixes exactly the zero of the scale and this is implied by the expression of a temperature in kelvin. However, in a difference of temperatures, the zero disappears. This difference has not the nature of a thermodynamic temperature. Nevertheless, the symbol K is used. This symbol is also used for the International Temperature Scale of 1990 (ITS-90), which is a different quantity.

The fact that units may be used symbolically is underlined by their writing in Roman characters, not in Italics as quantities. More convincing is to observe, as did Emerson, that they cannot always represent a concrete quantity. For example, we can conceive that m^3 represents the volume equal to that of a cube of side one meter. However, who can conceive a s^2?

5. Barycentric Celestial Reference System (BCRS)

The metric recommended for the BCRS by IAU Resolution B1.3 (2000) has the form

$$d\tau^2 = \left[-g_{00}(t, \mathbf{x}) - \frac{2}{c}\, g_{0i}(t, \mathbf{x})\, \dot{x}^i - \frac{1}{c^2}\, g_{ij}(t, \mathbf{x})\, \dot{x}^i\, \dot{x}^j \right] dt^2 \tag{5.1}$$

with the usual conventions for indices and summation. The components $g_{\mu\nu}$ of the metric are dimensionless quantities. Proper time τ is concretely measured in seconds. The velocity of light, a physical constant, is expressed in units of proper time and length, m/s. The homogeneity of the units of (5.1) requires that t (designating here Barycentric Coordinate Time, TCB) be reckoned in seconds and x^i in meters in a symbolic use of these units.

Suppose that we introduce a "graduation unit of TCB" and "a graduation unit of space coordinates". Then to keep the homogeneity of units with proper time τ, we would have to introduce factors such as second/(graduation unit of TCB). This complication serves no useful purpose. It is much easier to consider that the second and the meter, seen as symbolic, are the graduation units (Klioner, 2008).

6. Geocentric Celestial Reference System (GCRS)

Similarly, in the GCRS, the Geocentric Coordinate Time TCG and geocentric space coordinates are expressed in seconds and meters.

Expression of the difference TCB – TCG, as given by IAU Resolution B1.5(2000) would have no meaning and could not be applied to numerical values, if different units were employed for TCB and TCG.

The symbolic use of the SI units in these cases seems to be well accepted. However, the adjective "symbolic" (or equivalently "abstract") may be controversial. One must stress that they do not qualify new units, but indicate the way we use only SI units. For brevity sake we are tempted to speak of "symbolic seconds", for example, but it is more correct to say "symbolic use of the second".

7. Units of and Terrestrial Time TT and Barycentric Dynamical Time TDB

Difficulties of terminology appeared with the scaling of the coordinate times leading to the definition of TT and TDB, as linear functions of TCG and TCB. The expressions TT units and TDB units are sometimes employed.

7.1. *Terrestrial Time and International Atomic Time (TAI)*

Let us recall that IAU 2000 B1.9 recommends "that TT be a time scale differing from TCG by a constant rate: $dTT/dTCG = 1 - L_G$ where $L_G = 6.969290134 \times 10^{-10}$ is a defining constant."

The recommendation does not define explicitly TT as a new time scale, as in the case of TDB (see 7.2). However the words "differing from TCG", and the analogy with TDB, implies it.

The value of L_G was chosen so that the rate of TT is very close to the rate of proper time of a clock at any fixed point on the rotating geoid, operating in conformity with the definition of the second. The International Atomic Time TAI is a realization of TT + 32.184 s (with a constant time offset for historical reasons). TAI has, therefore, the nature of a coordinate time. However, its dissemination provides directly the unit of proper time, the second, at any fixed point on the ground (until the top of the Everest!) at better than 1×10^{-12} in relative value. Maybe this close agreement with proper time is one reason why the use of "second" as a unit of TT and TAI raised no objections. However, in orbit modeling it leads to the introduction of new quantities differing from those associated with TCG and sometimes expressed in so-called TT-units, similarly as in the case of TDB discussed below.

7.2. *Barycentric Dynamical Time*

The IAU 2006 Resolution B3 recommends

"That in situations calling for the use of a coordinate time scale that is linearly related to Barycentric Coordinate Time (TCB) and, at the geocenter, remains close to Terrestrial Time (TT) for an extended time span, TDB be defined as the following transformation of TCB

$$TDB = TCB - L_B \times (JD_{TCB} - T_0) \times 86400 + TDB_0 \tag{7.1}$$

where $T_0 = 2443144.5003725$, and $L_B = 1.550519768 \times 10^{-8}$ and $TDB_0 = -6.55 \times 10^{-5}$ s are defining constants."

Note 1 states: "JD_{TCB} is the TCB Julian date. Its value is $T_0 = 2443144.5003725$ for the event 1977 January 1 00h 00m 00s TAI at the geocenter, and it increases by one for each 86400 s of TCB".

Is equation (7.1) an equation between quantities or between numerical values?

(a) The wording "...TDB be defined..." as well as the indication "are defining constants" which follows the numerical values of these constants, indicates that TDB is to be considered as a quantity different from TCB.

(b) If nevertheless one considers that (7.1) is a relation between numerical values, that leaves the possibility to consider that it expresses a change of unit.

Let us consider the consequences of these options.

In the case (a) a new coordinate time is defined. Equation (7.1) shows that the unit of TDB is the second, seen as symbolic, as the unit of TCB. The use of TDB should imply the definition of new space coordinates expressed in metres. In order to accomplish this change of coordinates correctly, a new form of the components of the metric should be provided, introducing new quantities, to be also expressed in SI units. A difficulty (and may be some misunderstandings) arises from the fact that this metric has not been officially stated, although "natural" definitions are mentioned in (Klioner, 2008).

In the case (b), whichever be the adopted metric, according to the rules of the quantity calculus, the use of TDB units would require that the proper time be expressed in proper TDB seconds. In particular, if the same form of the metric as for TCB is retained (which is the main advantage of this option), the TDB second for proper time would be longer than the SI second by $1.55\ldots \times 10^{-8}$. There is a possibility to obviate this unacceptable change of unit for proper time by introduction of factors of the form (SI Unit/TDB unit) in the metric. This is like the introduction of graduation units mentioned in 1. It would not be a satisfactory solution.

8. Related problems

8.1. *Scale units*

The contact between the symbolic and the concrete point of view on units cannot be avoided. For example, it is often necessary to consider in a concrete manner the interval between two consecutive second markers of a time scale. This occurs in relativity when one needs to evaluate the rate in proper time, at some stated event, of a coordinate time. A non-relativistic example is the evaluation of the so-called "length of the day of UT1 (LOD)" in "TAI second" (note the impropriety of terms). In these cases, the astronomers often use expressions such as "the second of UT1, of TAI, of sidereal time (when considered as a time scale), the ephemeris second, etc." which appear to be different seconds. In the symbolic point of view, there is only the "second" for different quantities

Discussions on this problem took place at the 1980 meeting of the Comité Consultatif pour la Définition de la Seconde (CCDS). In the "Déclaration" of 1980 on the definition of TAI, appeared the expression "durée de l'intervalle unitaire de TAI" ("duration of the unitary interval of TAI"). However, later intervalle unitaire and unitary interval were replaced by "unité d'échelle" and "scale unit". IAU Resolution A4, 1991 uses also this terminology. As "scale unit" is widely used and well understood, I believe that we could retain its use and extend it to space coordinates.

8.2. *Some practical limitations to the symbolic point of view on units*

The symbolic use of the SI unit for quantities having the same dimension, but different definitions, is in practice limited to quantities which are nearly equal, or, in the case of scales, which have scale units nearly equal. In particular, the time scales are defined so that their rates differ by a small amount. Even in the case of sidereal time, when considered as a time scale (not an angle), the unit second has always been used without objection, although its scale unit duration differs by about 1/365 in relative value from that of UT1, TAI, etc.

8.3. *Notation of duration and dates*

In many publications, appears a typographical distinction between the notation of the unit for the date and the unit for duration. For example, the Astronomer's Handbook (1966) says "Une heure doit être notée : $4^{\text{h}}\ 39^{\text{s}}$ de préférence à 4 h 39 min". According to ISO, this distinction should not be made, and the symbol of the units should not be written in superscript. The symbol of the unit must appear after the full numerical value, including its decimal part if any.

9. Conclusion

In numerical applications of theoretical modeling, the requirement that the same algebraic relations exist between quantities and their numerical values imposes that there is only one unit per dimension: this is the basis of the Quantity Calculus. It is essential to note that the information on a quantity is contained it its definition, not in its unit.

In the case of quantities that cannot be measured concretely by comparison with a standard, this leads to the concept of abstract quantities and units whose logical interpretation may appear difficult. It is then possible to consider that it is a mere convention, which is simple and efficient. In particular, there is no reason to make exceptions for coordinate times and space coordinates. Their units must be the second and the meter, without adjective or qualifiers, even in the case where scaling factors are applied for convenience.

References

Astronomer's Handbook, Transactions of the IAU Vol. XIIC (1966), Academic Press.

Brumberg, V. A., 1991, Essential relativistic celestial mechanics, Adam Hilger, Bristol, Philadelphia and New York, 263 p.

De Boer, J. 1994/95, On the history of quantity calculus and International System, Metrologia, 32, 405-429.

Emerson, W. H. 2005, On the concept of dimension, Metrologia, 42, L21-L22.

Emerson, W. H. 2008, On quantity calculus and units of measurement, Metrologia, 45, 134-138.

Klioner, S. A. 2008, Relativistic scaling of astronomical quantities and the system of astronomical units, A&A, 478, 951-958.

Misner, C. W., Thorne K. S., & Wheeler J. A., 1973, Gravitation, Freeman, New York, 1280 p.

Relativity in Fundamental Astronomy
Proceedings IAU Symposium No. 261, 2009
S. A. Klioner, P. K. Seidelman & M. H. Soffel, eds.

doi:10.1017/S1743921309990172

Models and nomenclature in Earth rotation

Nicole Capitaine

SYRTE, Observatoire de Paris, CNRS, UPMC, 61 Av. de l'Observatoire, 75014, Paris, France
email: n.capitaine@obspm.fr

Abstract. The celestial Earth's orientation is required for many applications in fundamental astronomy and geodesy; it is currently determined with sub-milliarcsecond accuracy by astro-geodetic observations. Models for that orientation rely on solutions for the rotation of a rigid Earth model and on the geophysical representation of non-rigid Earth effects. Important IAU 2000/2006 resolutions on reference systems have been passed (and endorsed by the IUGG) that recommend a new paradigm and high accuracy models to be used in the transformation from terrestrial to celestial systems. This paper reviews the consequences of these resolutions on the adopted Earth orientation parameters, IAU precession-nutation models and associated nomenclature. It summarizes the fundamental aspects of the current IAU precession-nutation models and reports on the consideration of General Relativity (GR) in the solutions. This shows that the current definitions and nomenclature for Earth's rotation are compliant with GR and that the IAU precession-nutation is compliant with the IAU 2000 definition of the geocentric celestial reference system in the GR framework; however, the underlying Earth's rotation models basically are Newtonian.

Keywords. standards; astrometry; ephemerides; reference systems; time; Earth

1. Introduction

Earth's rotation is a diurnal rotation that exhibits fluctuations with time, about an axis that is moving both in space due to precession-nutation and within the Earth due to polar motion. The knowledge of that motion is essential for representing the coordinate transformation between the celestial and terrestrial systems that is required for many applications in fundamental astronomy and geodesy. Models for representing this motion are based on IAU and IUGG standards, on IAU models for precession-nutation and geophysical models for the part of polar motion and fluctuations in speed that are predictable. Earth Orientation parameters (EOP) can be estimated from observations by Very Long Baseline Interferometry (VLBI) of extragalactic radio sources, laser ranging of artificial satellites (SLR) and the Moon (LLR), observations of the GNSS systems, and observations with the DORIS system; each of these techniques has a specific potential for Earth orientation and reference systems determination.

Observing and modeling Earth rotation with a submilliarcsecond accuracy rely on a huge international effort. Observations and their analyses are coordinated at an international level by the International Earth Rotation and Reference systems Service (IERS). The IERS products, i.e. the realizations of the International Terrestrial and Celestial Reference Systems (ITRS and ICRS) and the EOP, are based on data provided by the international services (IVS for VLBI observations, ILRS for SLR and LLR observations, IGS for GNSS observations and IDS for DORIS observations). There also has been a continuing scientific effort for improving the models and the geophysical interpretation.

Several IAU resolutions on reference systems have been passed in 2000 and 2006, and endorsed by the IUGG in 2003 and 2007, respectively. This paper reviews the current status of the parameters for Earth's rotation, the IAU/IUGG adopted models and the associated nomenclature. It also summarizes the fundamental aspects of the IAU precession-nutation models and reports on the consideration of General Relativity in the current solutions.

2. The observed Earth orientation parameters

Three angles would be sufficient to specify the Earth's orientation in the celestial reference system. However, for practical reasons, five parameters are traditionally used in order to describe and observe the various fluctuations in Earth's rotation. The observed EOP consist of polar motion, celestial pole offsets and Universal Time, UT1. Polar motion represents variations in the terrestrial direction of the pole (see Fig. 1); it is quasi-periodic and essentially unpredictable. It includes a free, nearly circular motion of period of about 435 d (Chandler term) with a variable amplitude, an annual elliptical motion forced by the seasonal displacement of air and water masses, a small drift, and diurnal and semi-diurnal variations with amplitudes of a fraction of milliarcsecond (mas) that are due to the oceanic tides. Celestial pole offsets are corrections to the predicted celestial direction of the pole (see Fig. 3 and Sect. 4.2); they include corrections to the IAU precession-nutation (see Sect. 5) and the motion in space with a period of about 430 d (see Fig. 1) corresponding to the free retrograde diurnal motion of the Earths' axis with respect to the Earth (i.e. free core nutation, or FCN).

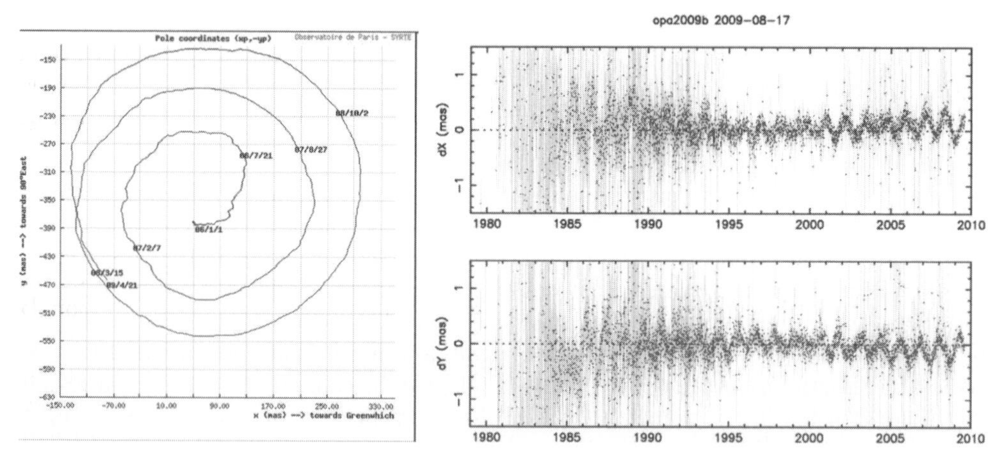

Figure 1. Spiral motion of the pole as seen from the Earth (left graph) and VLBI estimated corrections to the IAU 2006/2000 precession-nutation of the pole (right graph); unit: milliarcsecond; credit: IERS EOP Product Center and IVS OPA Analysis Center, Observatoire de Paris.

Variations in UT1, which reflects the fluctuations in the Earth's speed, are generally represented by variations in the length of day (LOD). They amount to several parts in 10^{-8}, including a secular variation, decadal variations, as well as tidal and seasonal variations (see Fig. 2).

IERS determination of polar motion mainly results from GPS observations that rely on a very dense network on the Earth, while celestial pole offsets and UT1 result mainly from VLBI that provides an accurate realization of the ICRS. The current uncertainty in the IERS determination of the EOP is a few tens of microarcseconds (μas) in the terrestrial and celestial pole directions, and a few μs for UT1.

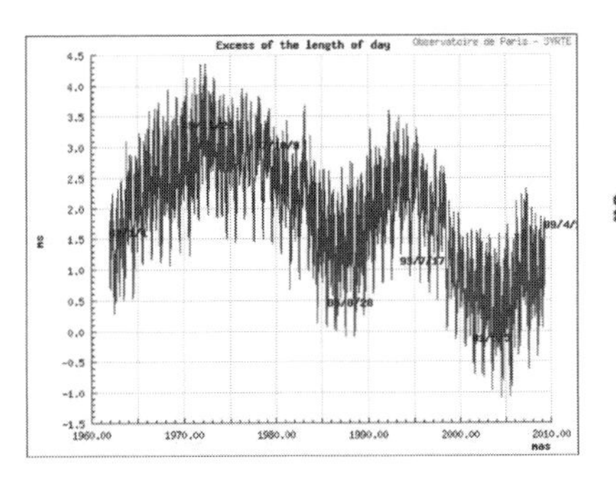

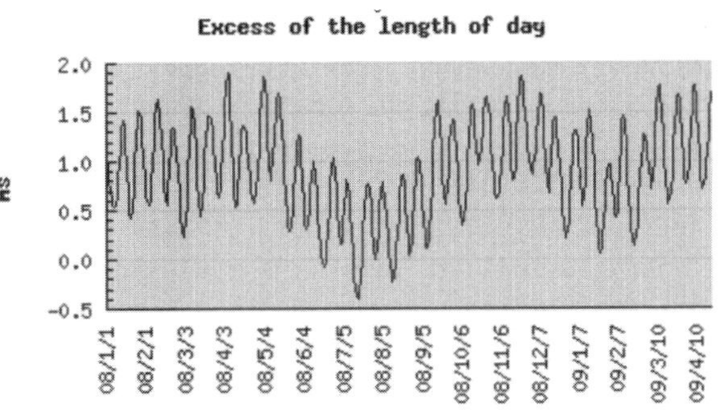

Figure 2. The variation in the length of day (unit: ms): variations over five years (left graph) and seasonal variations (right graph); credit: IERS EOP Product Center, Observatoire de Paris.

3. The IAU 2000/2006 Resolutions on reference systems

The IAU 2000 resolutions adopted by the XXIVth IAU General Assembly (August 2000) and endorsed by the XXIIIrd IUGG General Assembly (July 2003), have important consequences on the reference systems, the concepts, the parameters, and the models for Earth's rotation. IAU 2000 Resolution B1.3 specifies the systems of space-time coordinates for the solar system and the Earth within the framework of General Relativity and provides clear procedures for theoretical and computational developments of those space-time coordinates, and especially the transformation between the barycentric and geocentric coordinates (see Soffel *et al.* 2003). IAU 2000 Resolution B1.6 recommends the adoption of the IAU 2000 precession-nutation. IAU 2000 Resolution B1.7 defines the pole of the nominal rotation axis, while IAU 2000 Resolution B1.8 defines new origins on the equator, the Earth Rotation Angle (ERA) and UT1. The latter resolution also recommends a new paradigm for the terrestrial-to-celestial coordinate transformation. IAU 2000 Resolution B1.9 provides a re-definition of Terrestrial time (TT).

The IAU 2006 resolutions, adopted by the XXVIth IAU General Assembly (August 2006) and endorsed by the XXIVth IUGG General Assembly (July 2007), supplement the IAU 2000 resolutions on reference systems. IAU 2006 Resolution B1 recommends a new precession model as a replacement to the IAU 2000 precession in order to be consistent with both dynamical theory and the IAU 2000A nutation. IAU 2006 Resolution B2 addresses definition, terminology or orientation issues relative to reference systems that needed to be specified after the adoption of the IAU 2000 resolutions (e.g. that for all practical applications, unless otherwise stated, the BCRS (and hence GCRS) is assumed to be oriented according to the ICRS axes). IAU 2006 Resolution B3 provides a re-definition of Barycentric Dynamical Time (TDB).

The new terminology associated with the IAU 2000/2006 resolutions, along with some additional definitions related to them, were recommended by the 2003–2006 IAU Working Group on "Nomenclature for Fundamental Astronomy" (IAU NFA WG) (Capitaine *et al.* 2007) and endorsed by IAU 2006 Resolutions B2 and B3.

4. The parameters and nomenclature for Earth's rotation

4.1. *The IAU 2000/2006 space-time coordinates for Earth's rotation*

As specified by IAU 2000 Resolution B1.3, the Barycentric Celestial Reference System (BCRS), as a global coordinate system for the solar system, should be used with Barycentric Coordinate Time (TCB) for planetary ephemerides. In contrast, the Geocentric Celestial Reference System (GCRS), as a local coordinate system for the Earth, should be used with Geocentric Coordinate Time (TCG) for the Earth's rotation and precession-nutation of the equator. The spatial orientation of the GCRS is derived from that of the BCRS. Consequently, the GCRS is "kinematically non-rotating" so that Coriolis terms (that come mainly from geodesic precession) have to be considered when dealing with equations of motion in that system.

The IUGG 2003/2007 resolutions have endorsed the IAU 2000/2006 resolutions on reference systems and have additionally defined (IUGG 2007 Resolution 2) a Geocentric Terrestrial Reference System (GTRS) in agreement with IAU 2000 Resolution B1.3, and the International Terrestrial Reference System (ITRS) as the specific GTRS for which the orientation is operationally maintained in continuity with past international agreements.

The IAU 2000/2006 resolutions have clarified the definitions of both the Terrestrial Time (TT) and Barycentric Dynamical Time (TDB). The new definitions are such that TT is a time scale differing from TCG by a constant rate, which is a defining constant. In a very similar way, the new TDB is a linear transformation of TCB, the coefficients of which are defining constants. The consequence is that TT (or TDB), which may be for some practical applications of more convenient use than TCG (or TCB), can be used with the same rigorous approach. This applies in particular to the solutions of the Earth's rotational equations expressed in TT and the solar system ephemerides (necessary for computing the luni-solar and planetary torque acting on Earth's rotation) expressed in TDB.

4.2. *The IAU 2000/2006 definition and use of the Earth orientation parameters*

IAU 2000 Resolution B1.7, specifies that the pole of the nominal rotation axis is the *Celestial Intermediate Pole* (CIP), which is defined as being the intermediate pole, in the ITRS to GCRS transformation, separating nutation from polar motion by a specific convention in the frequency domain. The convention defining the CIP is such that (i) the GCRS CIP motion includes all the terms with periods greater than 2 days in the GCRS (i.e. frequencies between −0.5 cycles per sidereal day (cpsd) and +0.5 cpsd); (ii) the ITRS CIP motion, includes all the terms outside the retrograde diurnal band in the ITRS (i.e. frequencies less than −1.5 cpsd or greater than −0.5 cpsd).

IAU 2000 Resolution B1.8 recommends using the "non-rotating origins" (Guinot 1979) as origins on the CIP equator in the GCRS and ITRS; they were re-named *Celestial and Terrestrial Intermediate Origins* (CIO and TIO), respectively by IAU 2006 Resolution B2. Their kinematical property provides a very straightforward definition of the Earth's diurnal rotation based on the *Earth Rotation Angle* (ERA) between those two origins. The definition of UT1 has been refined as being linearly proportional to the ERA through the following conventional transformation (Capitaine *et al.* 2000):

$$\begin{aligned}\mathrm{ERA(UT1)} = 2\pi[&0.7790572732640 \\ &+ 1.00273781191135448\ (\text{Julian UT1 date} - 2451545.0)].\end{aligned} \tag{4.1}$$

According to IAU 2000 Resolution B1.8, the ITRS to GCRS transformation should be specified by the position of the CIP in the GCRS, the position of the CIP in the ITRS, and the ERA. The GCRS direction of the CIP unit vector (which includes precession, nutation and the frame bias) thus replaces the classical precession and nutation quantities (see Fig. 3). The CIO (σ) is at present very close to the GCRS x-origin, Σ_0, and almost stationary in longitude, while the equinox (γ) to which Greenwich sidereal time, GST, refers is moving at about $50''$/year in longitude. The CIO based procedure allows a clear separation between precession-nutation and the ERA, which is not model-dependent. In contrast, precession and nutation are mixed up with Earth's rotation into the expression

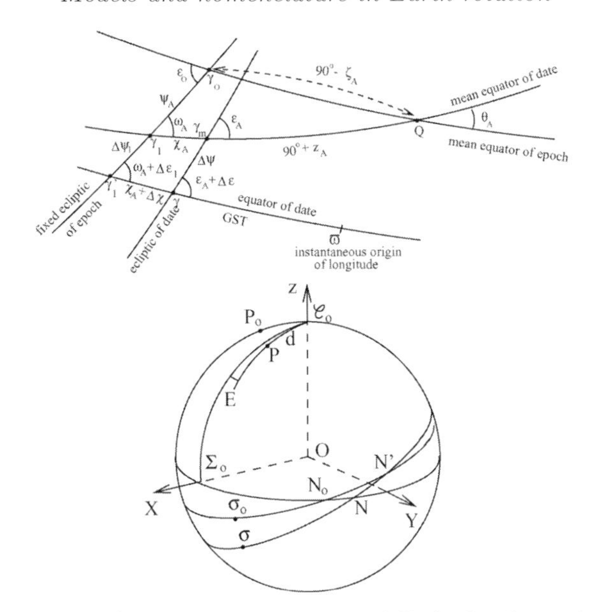

Figure 3. Precession-nutation of the equator in the GCRS: the classical precession-nutation angles (upper part) versus the CIP coordinates, d and E, and the CIO, σ (lower part).

for GST = ERA(UT1) − EO(TT), where EO, represents the accumulated precession and nutation in right ascension.

4.3. *The IAU 2000/2006 Nomenclature for Earth's rotation*

The IAU NFA WG made a number of recommendations on terminology (see Capitaine *et al.* 2007). It also produced the "IAU 2006 Glossary" including a set of detailed definitions (compliant with GR) that best explain all the terms required for implementing the IAU 2000 resolutions, as well as new definitions proposed by the WG, including those formally endorsed by the IAU in 2006 and the IUGG in 2007. The IAU 2000/2006 resolutions have provided the appropriate terminology for the pole, the Earth's angle of rotation, the longitude origins and the related reference systems. The IAU 2006 NFA Glossary (2006) includes in particular definitions for the celestial and terrestrial reference systems ICRS, BCRS, GCRS, ITRS and the *Celestial and Terrestrial Intermediate Reference Systems*, the *intermediate equator*, as the equator of the CIP, the origins CIO and TIO, the *CIO and TIO locators*, s and s', for positioning those origins in the GCRS, the *equation of the origins* (EO), as the distance between the CIO and the equinox along the intermediate equator, and the time scales TCB, TDB, TCG and TT.

5. The IAU 2006/2000 precession-nutation

5.1. *The adoption of the current IAU model*

IAU 2000 Resolution B1.6 recommends the adoption of the new precession-nutation model that is designated IAU 2000 (version A corresponding to the model of Mathews *et al.* (2002), denoted MHB2000, of 0.2 mas accuracy, and version B corresponding to its shorter version (McCarthy & Luzum 2002) with an accuracy of 1 mas). The precession part of the IAU 2000A model consists only in corrections $\delta\psi_A = -0''.29965$/century and $\delta\omega_A = -0''.02524$/century to the precession rates (in longitude and obliquity referred to the J2000.0 ecliptic), of the IAU 1976 precession and hence does not correspond to a dynamical theory. The second step in improving the IAU precession model was the endorsement by IAU 2006 Resolution B1 of the recommendation of the 2003–2006 IAU Working Group on "Precession and the Ecliptic" (Hilton *et al.* 2006) to adopt the P03 Precession (Capitaine *et al.* 2003) as a replacement for the precession part of the IAU 2000A precession-nutation, beginning on 1 January 2009.

The procedures, data and software for implementing the IAU 2000/2006 space-time coordinates, parameters and paradigm, nomenclature and models for Earth's rotation have been made available by Chapter 5 of the IERS Conventions 2003 that was updated in 2009 (see at: `http://tai.bipm.org/iers/convupdt/convupdt.html`) and the Standards Of Fundamental Astronomy (SOFA) (Wallace 1998).

5.2. *Main features of the IAU 2000A Nutation*

The IAU 2000A nutation is based on the REN2000 rigid Earth nutation of Souchay *et al.* (1999) for the axis of figure. The latter is expressed as a series of luni-solar and planetary nutations in longitude $\Delta\psi$ and obliquity $\Delta\epsilon$ referred to the ecliptic of date, composed of "in-phase" and "out-of-phase" components with their time variations, as follows:

$$\begin{aligned}\Delta\psi &= \textstyle\sum_{i=1}^{N}(A_i + A_i't)\sin(\text{ARGUMENT}) + (A_i'' + A_i'''t)\cos(\text{ARGUMENT}),\\ \Delta\epsilon &= \textstyle\sum_{i=1}^{N}(B_i + B_i't)\cos(\text{ARGUMENT}) + (B_i'' + B_i'''t)\sin(\text{ARGUMENT}),\end{aligned} \tag{5.1}$$

where t is measured in Julian centuries of TT from epoch J2000.0 and ARGUMENT is a function of the fundamental arguments of the nutation theory.

The rigid Earth nutation was transformed to the non-rigid Earth nutation by applying the MHB2000 "transfer function" to the REN2000 series of the corresponding prograde and retrograde nutations. The sub-diurnal terms due to the imperfect axial symmetry of the Earth are not part of the solution, so that the axis of reference of the nutation model is compliant with the definition of the CIP. The MHB "transfer function" is based on the solution of the linearized dynamical equation of the wobble-nutation problem. Seven "Basic Earth Parameters" (BEP) were treated as adjustable for fitting the theoretical outputs to the VLBI. This improves the IAU 1980 theory of nutation by taking into account the effect of mantle anelasticity, ocean tides, electromagnetic couplings produced between the fluid outer core and the mantle as well as between the solid inner core and fluid outer core, and the consideration of nonlinear terms. The axis of reference is the axis of maximum moment of inertia of the Earth ignoring time-dependent deformations.

The IAU 2000A nutation series that is expressed in the form of Eq (5.1), includes, as the REN2000 series, 678 lunisolar terms and 687 planetary terms. The resulting nutation is expected to have an accuracy of about 10 μas for most of its terms. On the other hand, the FCN (see Sect. 2), being a free motion which cannot be predicted rigorously, is not considered a part of the IAU 2000A model, which limits the accuracy in the computed direction of the celestial pole in the GCRS to about 0.3 mas.

The IAU 2000A nutation includes the geodesic nutation contributions to the annual, semiannual and 18.6-year terms from Fukushima (1991); these contributions to the nutations in longitude and obliquity are in μas:

$$\begin{aligned}\Delta\psi_g &= -153\ \sin l' - 2\ \sin 2l' + 3\ \sin\Omega,\\ \Delta\epsilon_g &= 1\ \cos\Omega,\end{aligned} \tag{5.2}$$

where l' is the Sun's mean anomaly and Ω the Moon's longitude of the ascending node.

5.3. *Main features of the IAU 2006 precession*

The IAU 2006 precession (Capitaine *et al.* 2003) provides improved polynomial expressions up to the 5th degree in time t, both for the precession of the ecliptic and the precession of the equator.

The precession of the equator was derived from the dynamical equations expressing the motion of the mean pole about the ecliptic pole. Consequently, the IAU 2006 precession is consistent with a dynamical theory. The convention for separating precession from nutation, as well as the integration constants used in solving the equations, has been chosen in order to be consistent with the IAU 2000A nutation. This includes corrections for the perturbing effects in the observed quantities.

In particular, the IAU 2006 value for the precession rate in longitude is such that the the corresponding Earth's dynamical flattening is consistent with the MHB value for that parameter. This required applying a multiplying factor to the IAU 2000 precession rate of $\sin\epsilon_{\mathrm{IAU2000}}/\sin\epsilon_{\mathrm{IAU2006}} = 1.000000470$ in order to compensate for the change (by 42 mas) of the J2000 mean obliquity of the IAU 2006 model with respect to the IAU 2000 value (i.e. the IAU 1976 value). Moreover, the IAU 2006 precession includes the Earth's J_2 rate effect (*i.e.* $\dot{J}_2 = -3 \times 10^{-9}$/century), mostly due to the post-glacial rebound, which was not taken into account in the IAU precession models previously.

The contributions to the IAU 2006 precession rates for the 2nd order effects, the J_3 and J_4 effects of the luni-solar torque, the J_2 and planetary tilt effects, as well as the tidal effects are from Williams (1994), and the non-linear terms are from MHB2000.

The geodesic precession is from Brumberg *et al.* (1992), i.e. $p_g = 1''\!.919883$/century. It is important to note that including the geodesic precession and geodesic nutation in the precession-nutation model ensure that the GCRS is without any time-dependent rotation with respect to the BCRS.

5.4. *IAU 2006 adjustments to the IAU 2000A nutation*

The difference between IAU 2006 and IAU 2000 lies essentially in the precession part, though very small changes are needed in a few of the IAU 2000A nutation amplitudes in order to ensure compatibility with the IAU 2006 values for ϵ_0 and the J_2 rate:

- the amplitudes of the nutation in longitude have to be adjusted in order to compensate for the change from the IAU 2000 to the IAU 2006 value for ϵ_0, the largest term being of the order of 10 μas for the 18.6-yr nutation (note that no such adjustment is needed in the case of the coordinate $X \approx \psi \sin\epsilon_0$);

- introducing the IAU 2006 J_2 rate value gives rise to additional Poisson terms in nutation, the coefficients of which are proportional to $\dot{J}_2/J_2$ (*i.e.* -2.7774×10^{-6}/century); the largest effects for the corresponding changes in the X, Y series are of the order of a few tens of μas after a century (Capitaine & Wallace 2006).

Whenever these small adjustments are included in the periodic terms, the notation "IAU $2000\mathrm{A}_{\mathrm{R06}}$" can be used to indicate that the nutation has been revised for use with the IAU 2006 precession. These adjustments are taken into account in the SOFA implementation of the IAU 2006/2000A precession-nutation.

5.5. *The IAU 2000/2006 expressions for the GCRS coordinates of the CIP*

Expressions for the coordinates X and Y of the CIP in the GCRS have been derived from the IAU $2006/2000\mathrm{A}_{\mathrm{R06}}$ expressions for the precession and nutation quantities referred to the J2000 ecliptic and the relationships between the X and Y coordinates and those quantities. The developments for X and Y include polynomial expressions up to the 5th degree in time t that are mainly due to precession plus the frame biases, and a periodic part, with a form similar to Eq. (5.1), but with Poisson terms up to the 4th degree.

Those expressions for X, Y, as well as the procedures for implementing the IAU 2006/2000 precession-nutation, have been provided by Capitaine & Wallace (2006) and Wallace & Capitaine (2006); they have been implemented in SOFA.

5.6. *Comparisons between models and observations*

Comparisons of the IAU 2006/2000 precession-nutation model with VLBI observations, once corrected for an empirical model for the FCN (see Fig. 1 and Sect. 5.2), show residuals with a w.r.m.s of about 130 μas. Those residuals can be empirically modeled in a variety of ways (see Capitaine *et al.* (2009) for more details). The fit of a combination of linear and 18.6-yr terms to the residuals show that the residuals would be compatible with corrections of a few tens of μas to the 18.6-yr nutation. Note that this would correspond to small corrections to the estimates for a couple of the BEP of the MHB model. This result is consistent with independent fits to the LLR celestial pole offsets with respect to the same IAU precession-nutation model (Zerhouni *et al.* 2008).

A comparison over 400 years has also been made between the INPOP06 (Fienga *et al.* 2008) numerical integration of the GCRS motion of the axis of angular momentum and the IAU 2006 precession plus the IAU 2000A nutation for the axis of angular momentum (i.e. the REN 2000 solution for that axis). The INPOP06 solution corresponds to an external torque modified by the non-rigid Earth; it uses the J_2 rate value of IAU 2006 and has been fitted to the IAU 2006 linear term in longitude. The only differences appearing in the comparison are one Fourier and Poisson term at the 18.6-yr period in X with a coefficient of about 50 μas/century and a linear term in Y of about 200 μas/century, which is less than the expected accuracy in that term. This provides another external check of the precision of the IAU 2006/2000A precession-nutation.

6. Theoretical basis for the models

The basic dynamical equation for Earth's rotation is the equation of angular momentum balance with the luni-solar and planetary torques acting on the oblate Earth. This can be developed in various forms. The classical Euler equations in the terrestrial system have an appropriate form for best considering the non-rigid Earth effects. Such equations have been developed by Sasao *et al.* (1980) for a non-rigid Earth with fluid core (designated SOS equations). The MHB transfer function on which the IAU 2000A nutation is based results from a generalization of the SOS equations to an Earth model including an inner core, with dissipative phenomena and BEP parameters fitted to VLBI (see Sect. 5.2). That transfer function should be applied to a rigid Earth solution.

The rotational equations for a rigid Earth can be written in the celestial reference system using various formalisms. The REN2000 analytical solution is based on an Hamiltonian formalism; it is the sum of solutions corresponding to each part of the second member for successive orders of approximation and different contributions (luni-solar effects, planetary effects, etc.). Other forms of equations and resolutions have been used.

The equations as functions of the two first Euler angles, ψ and ω, (e.g. Woolard 1953, Bretagnon *et al.* 1997) can be expressed as:

$$\begin{cases} -\ddot{\omega} + \sigma\,\dot{\psi}\sin\omega + \dot{\psi}^2\sin\omega\cos\omega = L_1/A \\ \sin\omega\,\ddot{\psi} + \sigma\,\dot{\omega} + 2\,\dot{\psi}\,\dot{\omega}\cos\omega = M_1/A, \end{cases} \tag{6.1}$$

where A and C are the Earth's principal moments of inertia, $\sigma = (C/A)\,\Omega$, is the frequency of the Euler free motion in the GCRS, Ω being the mean angular velocity of the Earth. L_1 and M_1 are the components of the external torque in the equatorial reference system defined by the CIP and the intersection of the CIP equator with the J2000 ecliptic.

The equations as functions of the GCRS CIP coordinates, X, Y can be written with similar notations as (Capitaine *et al.* 2005):

$$\begin{cases} -\ddot{Y} + \sigma\dot{X} = L/A + F'' \\ \ddot{X} + \sigma\dot{Y} = M/A + G'', \end{cases} \tag{6.2}$$

where L and M are the components of the torque in the equatorial reference system defined by the CIP and the point of the CIP equator that is distant from the CIO by the quantity s (see Sect. 4.3). F'' and G'' are functions, of the second order, of the X and Y variables and their first and second time derivatives. This form is best appropriate for expressing all the quantities, including the nutation arguments, in the GCRS.

Eq. (6.1) and (6.2), applied to a semi-analytical expression for the external torque, provide a semi-analytical solution for the parameters (i.e. Euler angles, or X, Y) using the method of variations of parameters (see Woolard 1953) and successive iterations. Non-rigid effects expressed in space can also be introduced in the second member.

7. Summary

The consequences of IAU 2000/2006 resolutions for Earth's rotation are the following:

– the definition of the celestial and terrestrial reference systems that are essential for Earth rotation theory and observations are compliant with General Relativity (GR),

– the definition of the Earth orientation parameters (EOP) have been clarified thanks to the use of the Celestial intermediate pole and origin that is compliant with GR,

– the nomenclature associated with the new concepts and quantities has been specified,

– the IAU precession-nutation model, including the geodesic precession-nutation, is compliant with the "kinematically non-rotating" definition of the GCRS.

The current definitions and nomenclature for Earth's rotation are thus compliant with General Relativity and the current IAU precession-nutation is compliant with the GCRS definition in the GR framework. However, it should be noted that the IAU precession-nutation does not result from a rigorous GR treatment. Firstly, both the development of the equations and the transformation between BCRS and GCRS coordinates of the Moon, Sun and planets used for computing the torque, were considered in a Newtonian framework. Secondly, the relativistic rotation of the dynamical geocentric celestial reference system with respect to the GCRS, was taken into account by adding the geodesic precession-nutation to the dynamical solution, while the rigorous way would be to consider an additional torque in the second member of the angular momentum equation. Thirdly, TT has been used instead of TDB in the semi-analytical expressions of the solutions. The latter effect can be shown to be less than $10^{-8}{}''$ in the CIP location. In contrast, according to Brumberg & Simon (2004), the effect of the relativistic part of the BCRS-to-GCRS transformation can reach 150 μas on the precession-nutation solution after one century. A complete GR treatment is, therefore, required in order that the precession-nutation models can achieve a microarcsecond accuracy (see Klioner *et al.* 2009).

References

Bretagnon, P., Rocher, P., & Simon, J.-L., 1997, *A&A* 319, pp. 305

Brumberg, V. A., Bretagnon, P., & Francou, G., 1992, Proceedings of the Journées 1991 "Systèmes de référence spatio-temporels", N. Capitaine (ed), Observatoire de Paris, pp. 141–148

Brumberg, V. A. & Simon, J.-L., 2004, Proceedings of the Journées 2003 "Systèmes de référence spatio-temporels", A. Finkelstein & N. Capitaine (eds), pp. 302–313

Capitaine, N., Guinot, B., & McCarthy, D. D., 2000, *A&A* 355, 398

Capitaine, N., Wallace, P. T., & Chapront, J., 2003, *A&A* 412, 567

Capitaine, N., Folgueira, M., & Souchay, J., 2005, *A&A* 445, 347

Capitaine, N. & Wallace, P. T., 2006, *A&A* 450, 855

Capitaine & IAU NFA WG, 2007, in *Transactions of the IAU* XXVIB, van derHucht, K. A. (ed), 14, pp. 474–475

Capitaine, N., Mathews, P. M., Dehant, V., Wallace, P. T., & Lambert, S. B., 2009, *Celest. Mech. Dyn. Astr.* 103, 179

Fienga, A., Manche, H., Laskar, J., & Gastineau, M., 2008, *A&A* 477, 315

Guinot, B., 1979, in *Time and the Earth's Rotation*, D. D. McCarthy and J. D. Pilkington (eds), D. Reidel Publishing Company, 7

Hilton, J., Capitaine, N., Chapront, J., *et al.*, 2006, *Celest. Mech. Dyn. Astr.* 94, 3, 351

IAU 2000, *Transactions of the IAU* XXIVB; Manchester, Rickman. H. (ed), Astronomical Society of the Pacific, Provo, USA, 2001, pp. 34–58

IAU 2006, *Transactions of the IAU* XXVIB; van der Hucht, K. A. (ed)

IAU 2006 NFA Glossary of the IAU Working Group on "Nomenclature for Fundamental Astronomy", `http://syrte.obspm.fr/iauWGnfa/NFA Glossary.html`

IUGG 2007, IUGG Resolutions, `http://www.iugg.org/resolutions/perugia07.pdf`

IERS Conventions (2003), *IERS Technical Note 32*, D. D. McCarthy and G. Petit (eds), Frankfurt am Main: Verlag des desamts für Kartographie und Geodäsie, 2004

Klioner, S., Gerlach, E., Soffel, M., 2009, *this proceedings*, 112

Mathews, P. M., Herring, T. A., & Buffett B. A., 2002, *J. Geophys. Res.* 107, B4, 10.1029/2001JB000390

McCarthy, D. D. & Luzum, B. J., 2003, *Celest. Mech. Dyn. Astr.* 85, 37

Sasao, T., Okubo, S., & Saito, M., 1980, Proceedings of the IAU Symposium 78, E. P. Federov, M. L. Smith, P. L. Bender (eds), Dordrecht, D. Reidel Publishing Co., 1980, p. 165–183

Soffel, M., Klioner, S. A., Petit *et al.*, 2003, *AJ* 126, 6, 2687

Souchay, J., Loysel, B., Kinoshita, H., & Folgueira, M., 1999, A&AS 135, 111

Wallace, P. T., 1998, in *Highlights of Astronomy* Vol. 11A, J. Andersen (ed), Kluwer Academic Publishers, 11, 191

Wallace, P. T. & Capitaine, N., 2006, *A&A* 459, 3, 981

Williams, J. G., 1994, *AJ* 108 (2), 711

Woolard, E. W., 1953, *Astr. Pap. Amer. Ephem. Naut. Almanach* XV, I, 1–165

Zerhouni, W., Capitaine, N., & Francou, G., 2009, Proceedings of the Journées 2008 "Systèmes de référence spatio-temporels", M. Soffel & N. Capitaine (eds), pp. 186–189

Relativity in Fundamental Astronomy
Proceedings IAU Symposium No. 261, 2009
S. A. Klioner, P. K. Seidelman & M. H. Soffel, eds.

doi:10.1017/S1743921309990184

Units of relativistic time scales and associated quantities

S. A. Klioner[1], N. Capitaine[2], W. M. Folkner[3], B. Guinot[2], T.-Y. Huang[4], S. M. Kopeikin[5], E. V. Pitjeva[6], P. K. Seidelmann[7] and M. H. Soffel[1]

[1]Lohrmann Observatory, Dresden Technical University, 01062 Dresden, Germany
[2]SYRTE, Observatoire de Paris, CNRS, UPMC, 61 Av. de l'Observatoire, 75014 Paris, France
[3]JPL m/s 301-150, 4800 Oak Grove Drive, Pasadena CA 91109, USA
[4]Nanjing University, Astronomy Department, 22 Hankou Road, 210093 Nanjing, China PR
[5]Department of Physics & Astronomy, University of Missouri, Columbia, Missouri 65211, USA
[6]Institute of Applied Astronomy RAS, Nab Kutuzova 10, 191187 St Petersburg, Russia
[7]University of Virginia, 129 Fontana Ct, Charlottesville VA 22911-3531, USA

Abstract. This note suggests nomenclature for dealing with the units of various astronomical quantities that are used with the relativistic time scales TT, TDB, TCB and TCG. It is suggested to avoid wordings like "TDB units" and "TT units" and avoid contrasting them to "SI units". The quantities intended for use with TCG, TCB, TT or TDB should be called "TCG-compatible", "TCB-compatible", "TT-compatible" or "TDB-compatible", respectively. The names of the units second and meter for numerical values of all these quantities should be used without any adjectives. This suggestion comes from a special discussion forum created within IAU Commission 52 "Relativity in Fundamental Astronomy".

1. Introduction

In the current literature one can read different, sometimes contradictory and illogical statements about the units associated with the values of various astronomical parameters. One sees wording like "TDB units" and "TT units" that is often contrasted to "SI units". Such terminology is often a source of confusion: no serious discussion of how those "TDB units" and "TT units" are defined can be found in the literature. The present note puts forward the case for using clear and consistent wording concerning the units and values of various astronomical quantities to be used with all standard astronomical reference systems (BCRS and GCRS) and time scales (TT, TDB, TCB and TCG). It is the result of a special discussion forum created within IAU Commission 52 "Relativity in Fundamental Astronomy".

2. Quantities, values and units

For the purposes of the present note, it is important to distinguish clearly between quantities and their numerical values. According to ISO (1993, definition 1.1), *quantity is an attribute of a phenomenon, body or substance that may be distinguished qualitatively and determined quantitatively. A value (of a quantity)* is defined as *the magnitude of a particular quantity generally expressed as a unit of measurement multiplied by a number* (ISO 1993, definition 1.18). The numerical values of quantities are pure numbers that appear when quantities are expressed using some units. For any quantity A one has

$$A = \{A\}\,[A], \tag{2.1}$$

where $\{A\}$ is the numerical value (a pure number) of quantity A and $[A]$ is the corresponding unit. Notations $\{A\}$ and $[A]$ for the numerical value and unit of a quantity A, respectively, are recommended in ISO 31-0 (ISO 1992).

The official definition of the concept of "unit" is given by ISO (1993, definition 1.7): *a unit (of measurement) is a particular quantity, defined and adopted by convention, with which other quantities of the same kind are compared in order to express their magnitudes relative to that quantity.* Therefore, a unit is a sort of recipe of how an observer can realize a specific physical quantity. The observer can then express numerically all other quantities which have the same physical dimensionality by comparing them with that specific quantity called "unit".

3. SI second as the unit of proper time

The official definition of the SI second can be found in (BIPM 2006):

The second is the duration of 9 192 631 770 periods of the radiation corresponding to the transition between the two hyperfine levels of the ground state of the caesium 133 atom.

In the relativistic framework it is very important to realize that this definition does not contain any hints on how the observer realizing the second should move, or where (in what gravitational field) that observer should be situated. From the relativistic point of view this is the only correct approach to define a physical unit of time. A physical unit of time can be realized only by physical observations. One of the fundamental principles of General Relativity, the Einstein Equivalence Principle, in combination with the so-called locality hypothesis claims, in particular, that an observer using only its proper time (the reading of an ideal clock moving together with him) cannot judge how he is moving and how strong the gravitation field along his trajectory is. Therefore, the concrete second, called in the following the "SI second", can be realized by any observer: a clock on the surface of the Earth, or on Mars or on board a space vehicle far away from any planet. In this sense, the SI second is the same for any observer and represents a recipe (as for any unit of measurement: see Section 2 above) to be executed in order to realize unit time intervals locally. Thus, the SI second is the unit of proper time and, as for proper time itself, it can and should be realized only locally (but by any observer at an arbitrary location and in an arbitrary gravitational field). Hence, it is clear that wording "SI seconds on the geoid" used in the original definition of TAI does not mean that SI seconds are defined on the geoid or can be realized only on the geoid. Such a wording actually refers to the proper time of an observer on the geoid expressed in SI seconds.

The official definition of the SI meter reads BIPM (2006):

The metre is the length of the path travelled by light in vacuum during a time interval of 1/299 792 458 of a second.

The SI meter is, therefore, based on the SI second and on the specific defining value of the speed of light $c = 299\,792\,458$ m/s (assumed here to be constant according to Special Relativity).

In the framework of General Relativity, one should distinguish between observable (or measurable) and coordinate quantities. A measurable quantity has dimension, a unit, and gets a numerical value after comparison with its unit. Its value is independent of the choice of theory and reference systems.

A coordinate quantity has dimension, cannot be measured directly but can get a numerical value after computation from observables with proper theoretical (relativistic) relations. Its numerical value is usually followed by "second", "meter" or some combination according to its dimension and the system of units used for the observables. Its value depends on the choice of theory (General Relativity in present IAU Resolutions) and reference systems.

In practice, all quantities not resulting directly from measurements are coordinate quantities, such as time and space coordinates, orbital elements, distances between remote points, and so on.

4. Unit time intervals of different observers

It is a common mistake to believe that intervals of proper time $\Delta\tau_1$ and $\Delta\tau_2$ measured by different observers can be "uniquely" and "naturally" compared to each other. The only way to do so in General Relativity is to define a 4-dimensional relativistic reference system having coordinate time t, establish a relativistic procedure of coordinate synchronization of clocks with respect to t, and convert the intervals of proper time $\Delta\tau_1$ and $\Delta\tau_2$ of each observer into corresponding intervals of coordinate time Δt_1 and Δt_2. These two intervals of coordinate time can indeed be compared directly. If both observers use SI seconds to measure their proper times, and $\{\Delta\tau_1\} = 1$ and $\{\Delta\tau_2\} = 1$ (i.e. both proper time intervals have length of 1 SI second as realized by the corresponding observer), the values of coordinate time intervals $\{\Delta t_1\}$ and $\{\Delta t_2\}$ are in general different. It does not mean, however, that the observers use different units of time. Only if the same units are used for $\Delta\tau$ and Δt, numerical values $\{\Delta t_1\}$ and $\{\Delta t_2\}$ are related to each other according to the standard formulas of special- and general-relativistic time dilations.

5. Proper time, and coordinate times TCB and TCG

Along with the proper times of individual observers, coordinate times are indispensable for relativistic modelling of physical processes. Coordinate times together with 3 spatial coordinates constitute relativistic 4-dimensional reference systems. Full definition of a reference system can be achieved only by fixing its metric tensor, as is done by the IAU (IAU 2001; Rickman 2001) for the standard reference systems BCRS and GCRS. Physical and mathematical details of this definition can be found in Soffel *et al.* (2003).

A relativistic reference system can be associated with a set of rules allowing one to label any phenomena or physical events with 4 real numbers. One of these numbers is called coordinate time and the other three are called spatial coordinates. Any coordinate time is a coordinate and, therefore, cannot be measured directly. They can only be *computed,* from the readings of real clocks together with additional parameters and information. For this computation one should use the theoretical relation between the proper time of an observer and coordinate time in General Relativity:

$$\frac{d\tau}{dt} = \left(-g_{00}(t,\boldsymbol{x}_{\mathrm{obs}}(t)) - \frac{2}{c}\,g_{0i}(t,\boldsymbol{x}_{\mathrm{obs}}(t))\,\dot{x}^i_{\mathrm{obs}}(t) - \frac{1}{c^2}\,g_{ij}(t,\boldsymbol{x}_{\mathrm{obs}}(t))\,\dot{x}^i_{\mathrm{obs}}(t)\,\dot{x}^j_{\mathrm{obs}}(t)\right)^{1/2}, \tag{5.1}$$

where t is the coordinate time of a reference system having a metric tensor with components g_{00}, g_{0i} and g_{ij} (i and j running from 1 to 3, and each component of the metric tensor being a function of coordinate time t and spatial position $\boldsymbol{x}$), and τ is the proper time of an observer having position $\boldsymbol{x}_{\mathrm{obs}}(t)$ and velocity $\dot{\boldsymbol{x}}_{\mathrm{obs}}(t)$ with respect to this reference system. Einstein's implicit summation is used in the above formula. In order to be useful this formula needs an initial condition of the form

$$\tau(t_0) = \tau_0, \tag{5.2}$$

where t_0 and τ_0 are constants to be determined from the procedure of clock synchronization (if there is only one observer these two constants can be taken to be zero). In the form written above, Eq. (5.1) allows one to compute τ if t, $\boldsymbol{x}_{\rm obs}(t)$, $\dot{\boldsymbol{x}}_{\rm obs}(t)$ and the components of the metric tensor $g_{\alpha\beta}$ are given. This formula can be inverted (e.g. numerically) in order to compute t for a given τ.

We should note that Eq. (5.1) is a relation between quantities τ and t. By analogy with the rules of quantity calculus, the values of τ and t can be also related by this formula if and only if the same units are used for both τ and t. This implies that if proper time τ is expressed in SI seconds, then values of t computed from numerical inversion of Eq. (5.1) should be also expressed in SI seconds. A semantic difficulty here is that SI seconds can only be realized for physically measurable proper time, whereas t is a non-measurable coordinate quantity related to the measurable τ by Eqs. (5.1)–(5.2). To accommodate this objection, one can agree to call the unit of time t "SI-induced second". These "SI-induced seconds" can be realized only through the proper time of an observer. In the following we will call both "SI seconds" (used for proper times) and "SI-induced seconds" (used for coordinate times) simply "seconds".

All the comments and arguments given above are equally correct for both TCG and TCB.

Spatial coordinates $\boldsymbol{x}$ and $\boldsymbol{X}$ of BCRS and GCRS, respectively, are also defined by the metric tensors of these reference systems. The standard formulas of Special and General Relativity assume that the locally measured light speed in vacuum is equal to a constant quantity c that enters equations in many ways. In practical calculations the specific value of c from the definition of the SI meter is always used. Therefore, if t is expressed in seconds, x is expressed in meters.

Applying the same arguments of inheritance of the units to the equations of motion for celestial bodies (Newtonian equations of motion or post-Newtonian EIH equations) we conclude that mass parameters $\mu = GM$ of celestial bodies should also be expressed in the units of the SI.

We note that the views expressed above are closely related to the idea of symbolic or abstract quantities and units developed in metrology (see, de Boer (1994) and Emerson (2008)). In particular the "SI-induced second" discussed above appears as a symbolic second (see e.g., de Boer 1994). A detailed discussion of these concepts in the relativistic framework can be found in Guinot (1997).

6. Scaled time scales TDB and TT

The reasons for and the mathematical details of the relativistic scaling of BCRS and GCRS, and in particular their coordinate times, are summarized by Klioner (2008). TT and TDB are conventional linear functions of TCG and TCB, respectively. The definition of TT (given by IAU 2000 Resolution B1.9 and IAU 1991 Resolution A4) can be written as

$$TT = F_G\, TCG \tag{6.1}$$

where $F_G = 1 - L_G$, and $L_G = 6.969290134 \times 10^{-10}$ is a defining constant. TDB defined in IAU 2006 Resolution 3 is related to TCB as

$$TDB = F_B\, TCB + TDB_0 \tag{6.2}$$

where $F_B = 1 - L_B$, and $L_B = 1.550519768 \times 10^{-8}$ and TDB_0 are defining constants. Let us note here that TAI is a physical realization of TT with a shift of -32.184 s for historical reasons. Therefore, TAI is a realization of coordinate time "TT−32.184 s". The difference between TAI and an ideal realization of "TT−32.184 s" is only due to imperfections of the participating clocks and the clock synchronization procedures.

The slopes F_G and F_B of both linear functions have the same purpose: both TT and TDB should show no linear drift with respect to proper times of observers situated on the rotating geoid, i.e. close to the surface of the Earth. Since this property depends on our model of the solar system (i.e. on a planetary ephemeris and on a number of astronomical and geodetic constants), the latter requirement cannot be satisfied exactly. Therefore, some conventional constants have been chosen in the definitions of TT and TDB so that the requirement is satisfied approximately, but with an accuracy totally sufficient for practical purposes. Similarly, the constant TDB_0 was chosen merely to keep $TDB - TT$ approximately centered on zero.

Eqs. (6.1) and (6.2) define two new quantities: coordinate time scales TT and TDB. As discussed in Klioner (2008), for practical reasons (keeping equations of motion of celestial bodies and photons invariant) these scalings of coordinate time are accompanied by the corresponding scalings of spatial coordinates $\boldsymbol{x}_{TDB} = F_B\,\boldsymbol{x}$ and $\boldsymbol{X}_{TT} = F_G\,\boldsymbol{X}$ and mass parameters μ of celestial bodies $\mu_{TDB} = F_B\,\mu$ and $\mu_{TT} = F_G\,\mu$.

The scaled coordinate times and spatial coordinates can be thought of as defining two new reference systems: those with coordinates $(TT, \boldsymbol{x}_{TT})$ and $(TDB, \boldsymbol{x}_{TDB})$. These new reference systems can be characterized by their own metric tensors, different from those of the BCRS and GCRS. Formulas $\mu_{TDB} = F_B\,\mu$ and $\mu_{TT} = F_G\,\mu$ for the mass parameters follow from the requirement to have the same form of the equations of motion with both unscaled and scaled coordinates. This is an additional requirement that does not immediately follow from the scaling of time and spatial coordinates.

In combination with (5.1), Eqs. (6.1) and (6.2) define how TT and TDB are related to the proper time of any observer. The proper time can be considered a function of TT and TDB in a similar way to when we considered it as functions of TCG and TCB above. Therefore, the same arguments as in Section 5 can be used to demonstrate that if proper times are expressed in SI seconds, both TT and TDB are by inheritance expressed in SI-induced seconds or simply in seconds. Here "by inheritance" simply means that the formulas linking TT and TDB to the other timescales, and ultimately to proper times, provide a formal connection back to SI seconds.

Similarly, scaled spatial coordinates are expressed in SI-induced meters or simply in meters. The same arguments allow us to conclude that the scaled mass parameters are also expressed in the units of the SI.

7. Suggested terminology

All these arguments allow us to suggest the following nomenclature:

– Avoid using the wording "TDB units" ("TDB seconds/meters"), "TT units" ("TT seconds/meters") and avoid contrasting these terms with "SI units" ("SI seconds/meters").

– All quantities intended for use with TDB should be called "TDB-compatible quantities" and corresponding values "TDB-compatible values".

– All quantities intended for use with TT should be called "TT-compatible quantities" and corresponding values "TT-compatible values".

– All quantities intended for use with TCB or TCG should be called "TCB-compatible quantities" or "TCG-compatible quantities" and the corresponding values "TCB-compatible values" or "TCG-compatible values", respectively. In the case of constants having the same value in BCRS and GCRS (e.g. mass parameters $\mu = GM$ of celestial bodies) the value can be called "unscaled". Note that it is misleading to describe these values as "SI-compatible" or "in SI units" since this does not distinguish unscaled values from TT- and TDB-compatible values. Such wording should be avoided.

– Consider that the numerical values of all above-mentioned quantities (TT-compatible, TCG-compatible, TDB-compatible and TCB-compatible) are expressed in the usual units of the SI. Avoid attaching any adjectives to the names of the units second and meter for numerical values of these quantities † ‡ .

Acknowledgement

The authors gratefully acknowledge discussions with George Kaplan, Gérard Petit, and Patrick Wallace who also contributed to the text of this paper and to the proposed terminology.

References

Emerson, W. H. 2008, Metrologia, 45, 134

de Boer, J. 1994/1995, Metrologia, 32, 405

Guinot, B. 1997, Metrologia, 34, 261

IAU, Information Bulletin, 88 (2001) (errata in IAU Information Bulletin, 89)

IAU Resolutions adopted at the XXVIth General Assembly of the IAU (2006), Resolution 3 "Redefinition of Barycentric Dynamical Time, TDB", `http://www.iau.org/Resolutions_at_GA-XXVI.340.0.html`

IERS Conventions 2003, ed. D. D. McCarthy & G. Petit, IERS Technical Note No. 32 (Frankfurt am Main: Verlag des Bundesamtes für Kartographie und Geodäsie)

International standard ISO 31-0: Quantities and units - Part 0: General principles. International Organization for Standardization, Geneva, 1992

The International System of Units (SI), 8th edition. Bureau International des Poids et Mesures, 2006

International Vocabulary of Basic and General Terms in Metrology. International Organization for Standardization, Geneva, 1993

Klioner, S. A. 2008, *A*stron.Astrophys., **478**, 951

Rickman, H. 2001, Reports on Astronomy, Trans. IAU, XXIV B

Soffel, M., Klioner, S. A., Petit, G. *et al.* , 2003, *A*stron. J. **126**, 2687

† So, for example, it would be improper to say "The interval is xx seconds of TDB"; the correct wording is "The TDB interval is xx seconds".

‡ An example of numerical values for a quantity is as follows: (i) the TCB/TCG-compatible value for GM_E (the mass parameter of the Earth) is $3.986004418 \times 10^{14}$ $\mathrm{m^3 s^{-2}}$; (ii) the TT-compatible value for GM_E is $3.986004415 \times 10^{14}$ $\mathrm{m^3 s^{-2}}$; (iii) the TDB-compatible value for GM_E is $3.986004356 \times 10^{14}$ $\mathrm{m^3 s^{-2}}$.

Relativity in Fundamental Astronomy
Proceedings IAU Symposium No. 261, 2009
S. A. Klioner, P. K. Seidelman & M. H. Soffel, eds.

doi:10.1017/S1743921309990196

Overview of current precision clocks and future prospects

R. L. Beard

U. S. Naval Research Laboratory, Washington, D. C., U.S.A.

Abstract. Today's time and frequency standards range from the most sophisticated reference standards to the smallest oscillator for handheld radios. The technical requirements and technologies needed are different for the various applications but they derive from similar physical concepts. These different technologies can be categorized into four major areas, reference standards, mobile systems, man-portable (handheld) and space systems. These areas are the core areas of time and frequency standard applications and different areas of technology are needed to address them. This presentation discusses the time and frequency standards used or available for these areas.

1. Reference Standards

Clocks and oscillators within this area are of the type needed by reference timescale centers such as the U.S. Naval Observatory(USNO) Master Clock. This specialized area requires the most highly stable and accurate time standards that are maintained under controlled conditions. Their outputs are processed with special ensembling algorithms designed to produce an absolute reference for all systems. For example, the current suite of clocks used at USNO consists of commercial cesium beam frequency standards and hydrogen masers. These clocks are physically separated and operated in tightly controlled environment chambers. Size, weight and power are not issues pertinent for these clocks, primary emphasis is on performance, mostly in the long term.

2. Mobile Systems

Clocks in this area are typically crystal oscillator based devices and small atomic clock/oscillator, used for positioning, communications or internal subsystems. Although small, the requirements for mobile devices are typically not demanding or rigorous. Clock technology to produce smaller, lower power devices is slowly developing primarily through government support.

3. Man-Portable

Devices in this area are the most demanding in terms of size, weight and power. The most commonly used are quartz crystal oscillator devices. However, in recent years there have been several government sponsored efforts to develop extremely small atomic standards. These devices offer better accuracy and stability than crystal oscillators in an extremely small package. Although their performance exceeded that of crystal oscillator based devices, they have yet to perform as well as their larger mobile or timing center devices.

4. Space Qualified Atomic Standards

Space qualified atomic clocks have been essential for the development and deployment of the Global Positioning System (GPS). GPS is the dominant user of high precision and stable space qualified atomic clocks, the only other user, are the few MILSTAR satellites containing lesser performing Rubidium standards. These clocks provide high stability for navigation performance and a large part of the development of these devices for space was to provide high stability reliably. GPS user equipment, and the timing capability resulting from the atomic clocks in the GPS system, are producing an inexpensive alternative to high precision atomic clocks for many systems. By displacing higher cost, higher performing atomic clocks, GPS User Equipment receivers or timing receivers with low quality clocks are being deployed in a wide variety of systems. Naval tactical and strategic systems have utilized hundreds of these units. Larger ships may have multiple cesium standards on board. Secondary standards such as rubidium vapor cells and crystal oscillators are being used extensively in aircraft, shipboard and man portable applications, since virtually every system has a clock or oscillator of some quality contained in it.

5. Current Standards

The primary frequency standard for the U.S. is the laser cooled cesium fountain at the National Institute for Standards and Technology. This type of standard is the metrological reference standard in frequency and a number of similar units are in use through the world. There are some 21 centers using cesium fountain clocks, although they are not commercially available. Each center has virtually built their our fountain clock. The performance is determined by comparison and coordination with the Bureau de Poids et Mesure. For other than timing centers, such as telecommunications centers, the most prevalent standard is the commercial cesium standard. Second is the active hydrogen maser that is in limited commercial availability. These devices are expensive with the hydrogen maser being about an order of magnitude more expensive.

The space qualified atomic clocks in the GPS operational satellites have stability requirements ranging from 2×10^{-13}/day for cesium and 1×10^{-14}/day for the rubidium on the later satellites. On-orbit performance has provided better than expected stabilities. The use of hydrogen masers for the ground stations and eventually in spacecraft was considered even before the beginning of the GPS Program. Efforts at that time were based on adapting the active hydrogen maser design initially developed by the Smithsonian Astrophysical Observatory. Their Gravity Probe One unit, built for a NASA sponsored relativity experiment, was launched in the mid-1970's and demonstrated potential for operation in orbiting spacecraft. To reduce the size of the active hydrogen maser a compact passive physics unit design was developed by U.S. Naval Research Laboratory for GPS. The final selected approach was the Hughes Q-enhanced design with a small magnetron cavity. This maser design approach reduced the overall unit size roughly to that of a GPS space qualified cesium clock. The European development of a navigation satellite system, to be known as GALILEO, has also developed rubidium and hydrogen maser clocks for spacecraft. The latest technology satellite, known a Giove B, has a space qualified hydrogen maser on-board that has thus far shown very good results.

6. Potential Future Clocks

Microwave standards are a mature technology and have good potential for further significant improvements. For instance, a "juggling" rubidium fountain clock that launches multiple "balls" of atoms in rapid succession could greatly improve the Signal to Noise Ratio (SNR) and result in short term fractional frequency stability in the high $10^{-15's}$ at one second while still maintaining excellent long term systematics well below 10^{-16}. This stability requires an local oscillator (LO) with better performance than an Ovenized Crystal Oscillator (OCXO). The Time and Frequency Group at the Jet Propulsion Lab (JPL) in Pasadena has built a cryo-cooled sapphire-loaded ruby oscillator that achieves 3×10^{-15} performance from 1 to 1000 seconds, thus meeting the local oscillator requirements for an advanced fountain. It is possible that further refinements to the fountain concept could bring that device into the low $10^{-15's}$ at a second. Laser-cooled neutral atom microwave standards based on rubidium have been under development by USNO, and they are in the process of incorporating them into their operations.

Laser-cooled microwave ion standards are expected to have an exceptional long term systematic noise floor. It is likely that the main limitation will be magnetic field sensitivity, which is largely an engineering problem of providing good shielding while still maintaining good optical access. However, while the systematic floor is likely to be in the low $10^{-17's}$, the short term stability is probably limited to the low $10^{-13's}$ due to the low SNR inherent in a device with only a few ions. As a result, a microwave laser-cooled ion trap device is unlikely to meet the stated goals.

Buffer-gas-cooled ion standards have already demonstrated a stability of 3×10^{-14} at 1 second. These devices have large signals (many ions), but only a moderate SNR due to large background signals. A factor of 3–10 improvement in SNR could be achieved with better detection schemes to reduce background. This almost certainly means using lasers instead of lamps, as is the current practice. One of these ion standards coupled with an advanced LO (such as the cryo-cooled LO already discussed) could get close to the short term stability goal, but the systematic floor is unlikely to be below 10^{-16} (larger numbers of ions at higher temperatures means both exposure to higher rf fields and larger Doppler shifts). Nevertheless, this type of approach should not be dismissed too quickly, since this frequency stability still allows several ps timing stability at one day. The buffer-gas-cooled ion standard with laser interrogation would require the fewest technological advances and would be the simplest to implement.

The JPL Time and Frequency Group has developed a new technology standard known as the Linear Ion Trap Standard. Operational versions of these units are being deployed in the NASA Deep Space Network as replacements for the large active hydrogen masers currently in use. A spacecraft version of these units is being investigated and offers the potential of very small size and power for potentially high stability. The physics package has been shown to be capable of small design, but since it is a passive device, a high quality local oscillator is needed to gain the full potential of these devices. The potential performance gain using a modest performance local oscillator and the adaptability to digital implementation of the electronics could be a major step in space qualified atomic clock technology.

The next step in atomic clock evolution is to move from microwave "clock" frequencies to optical frequencies. With frequencies measured in the 10^{15} Hz range instead of 10^{10} Hz, optical clocks have a potentially huge gain in Q (the ratio of the oscillator frequency to the uncertainty in that frequency). Since short-term stability is inversely proportional to Q, it too improves. An ion trap clock based on an optical transition then combines very good short-term stability due to the high Q of the optical transition with an exceptionally low systematic noise floor.

There are two technologies that are critical to optical clock progress. The first is Octave-wide Optical Comb Generation (OOCG). This is the phase-coherent spanning of a factor of two in optical frequencies by a frequency "comb" with radio frequency (rf) (100 to

1000 MHz) spacing. The OOCG makes it possible to link optical frequencies coherently down to rf frequencies, where timing information is usually generated, transferred and analyzed. The first successful OOCGs have been demonstrated, but significant research needs to be performed to map out their characteristics and capabilities. This is a huge step for optical clocks, since previous chains linking optical to rf frequencies required man-years of highly skilled work to build and maintain. With advent of OOCGs the amount of work has been reduced by several orders of magnitude. The second critical technology is laser frequency stabilization. To take full advantage of the optical line Q, the "clock" laser, which is now the LO, must have a frequency uncertainty on the order of 1 Hz or less. This is difficult to achieve but offers great potential for future development of clock technology.

Relativity in Fundamental Astronomy
Proceedings IAU Symposium No. 261, 2009
S. A. Klioner, P. K. Seidelman & M. H. Soffel, eds.

doi:10.1017/S1743921309990202

Time ephemeris and general relativistic scale factor

Toshio Fukushima

National Astronomical Observatory of Japan,
181-8588, Mitaka, Tokyo, Japan
email: Toshio.Fukushima@nao.ac.jp

Abstract. Time ephemeris is the location-independent part of the transformation formula relating two time coordinates such as TCB and TCG (Fukushima 1995). It is computed from the corresponding (space) ephemerides providing the relative motion of two spatial coordinate origins such as the motion of geocenter relative to the solar system barycenter. The time ephemerides are inevitably needed in conducting precise four dimensional coordinate transformations among various spacetime coordinate systems such as the GCRS and BCRS (Soffel *et al.* 2003). Also, by means of the time average operation, they are used in determining the information on scale conversion between the pair of coordinate systems, especially the difference of the general relativistic scale factor from unity such as L_C. In 1995, we presented the first numerically-integrated time ephemeris, TE245, from JPL's planetary ephemeris DE245 (Fukushima 1995). It gave an estimate of L_C as $1.4808268457(10) \times 10^{-8}$, which was incorrect by around 2×10^{-16}. This was caused by taking the wrong sign of the post-Newtonian contribution in the final summation. Four years later, we updated TE245 to TE405 associated with DE405 (Irwin and Fukushima 1999). This time the renewed vale of L_C is $1.48082686741(200) \times 10^{-8}$ Another four years later, by using a precise technique of time average, we improved the estimate of Newtonian part of L_C for TE405 as $1.4808268559(6) \times 10^{-8}$ (Harada and Fukushima 2003). This leads to the value of L_C as $L_C = 1.48082686732(110) \times 10^{-8}$. If we combine this with the constant defining the mean rate of TCG-TT, $L_G = 6.969290134 \times 10^{-10}$ (IAU 2001), we estimate the numerical value of another general relativistic scale factor $L_B = 1.55051976763(110) \times 10^{-8}$, which has the meaning of the mean rate of TCB-TT. The main reasons of the uncertainties are the truncation effect in time average and the uncertainty of asteroids' perturbation. The former is a natural limitation caused by the finite length of numerical planetary ephemerides and the latter is due to the uncertainty of masses of some heavy asteroids. As a compact realization of the time ephemeris, we prepared HF2002, a Fortran routine to compute approximate harmonic series of TE405 with the RMS error of 0.446 ns for the period 1600 to 2200 (Harada and Fukushima 2003). It is included in the IERS Convention 2003 (McCarthy and Petit 2003) and available from the IERS web site; http://tai.bipm.org/iers/conv2003/conv2003_c10.html.

Keywords. general relativity, ephemerides, reference systems, time

---

1. Concept of Time Ephemeris

Consider a four dimensional spacetime coordinate transformation, $x^\mu = f^\mu(X^\alpha)$, where x^μ is the four dimensional coordinates of an event in a certain four dimensional coordinate system, which we call the background coordinate system, while X^α is the four dimensional coordinates of the same event in another four dimensional coordinate system, which we call the target coordinate system. If we expand the above coordinate transformation around the space coordinate origin of the target coordinate system, its time part is written as

$$t = f(T) + f_k(T)X^k + \cdots \tag{1.1}$$

where t and T are the time coordinates of the background and target coordinate systems, respectively. Then, we define the time ephemeris as a function expressed (Fukushima 1995) as

$$\Delta\tau(t) \equiv t - f^{-1}(t). \tag{1.2}$$

In case of the Earth, the BCRS is the background coordinate system, the GCRS is the target coordinate system, TCB is t, and TCG is T (IAU 1992, Seidelmann and Fukushima 1992, IAU 2001). Namely the above time-time relation is rewritten (Fukushima 1995) as

$$\mathrm{TCB} - \mathrm{TCG} = \Delta\tau_E(\mathrm{TCB}) + \frac{\mathbf{v}_E \cdot \mathbf{X}}{c} + \cdots \tag{1.3}$$

where c is the speed of light in vacuum.

2. General Relativistic Scale Factor

The time average of the time ephemerides has an important meaning in the unit conversion between different coordinate systems (Fukushima *et al.* 1986). To make the description more understandable, hereafter, we deal with the case of the Earth only.

Assume that the background and target coordinate systems use different unit systems in length and in time as $[\mathrm{m}_B, \mathrm{s}_B]$ for the BCRS and $[\mathrm{m}_G, \mathrm{s}_G]$ for the GCRS. Adopt a convention such that the numerical values of c are the same in both coordinate systems as

$$c = 299792458\ \mathrm{m}_B/\mathrm{s}_B = 299792458\ \mathrm{m}_G/\mathrm{s}_G. \tag{2.1}$$

then the ratios of the length units and time units must be the same as

$$\frac{\mathrm{m}_G}{\mathrm{m}_B} = \frac{\mathrm{s}_G}{\mathrm{s}_B}. \tag{2.2}$$

The latter quantity is nothing but the time average of the differential ratio of the time scales, which is rewritten (Fukushima *et al.* 1986) as

$$\frac{\mathrm{s}_G}{\mathrm{s}_B} = \frac{< d\mathrm{TCG} >}{< d\mathrm{TCB} >} = 1 - L_C, \tag{2.3}$$

where

$$L_C = \left\langle \frac{d\Delta\tau_E}{dt} \right\rangle, \tag{2.4}$$

is the general relativistic scale factor of the Earth. Once this factor is obtained, the numerical values of all the physical quantities measured in each coordinate system have the proportional relations as

$$\frac{M_G}{M_B} = \frac{R_G}{R_B} = \frac{P_G}{P_B} = 1 - L_C, \tag{2.5}$$

where M, R, and P denote the mass, the radius, and the period of any kind.

3. Computation of Time Ephemeris

The relation between the two time scales are computed by assuming that the space coordinate origin of the target coordinate system, the geocenter in the case of the Earth, follows a geodesic in the background coordinate system, the solar system barycentric coordinate system in the case of the Earth. Adopt Einstein's general theory of relativity as the general relativistic theory to be based. Then, the time development equation of $\Delta\tau$ is obtained from the equation of geodesic, or that of proper time more specifically, as

$$\frac{d\Delta\tau_E}{dt} = \frac{\mathbf{v}_E^2 + U_E}{2c^2} + \frac{\mathbf{v}_E^4 + 12\mathbf{v}_E^2 U_E - 4U_E^2 - 32\mathbf{v}_E \cdot \mathbf{w}_E - 4W_E}{8c^4} + \cdots \tag{3.1}$$

where $\vec{v}_E$ is the velocity of the geocenter, U_E is the Newtonian gravitational potential acting on the Earth (and excluding the self-gravitational potential of the Earth itself), and $\vec{w}_E$ and W_E represent the post-Newtonian contributions in general (Soffel *et al.* 2003). In the EIH metric, they are expressed (Fukushima 1995) as

$$U_E \equiv \sum_{J \neq E} U_{EJ}, \quad \mathbf{w}_E \equiv \sum_{J \neq E} U_{EJ}\mathbf{v}_J,$$

$$W_E \equiv \sum_{J \neq E} U_{EJ} \left[4\mathbf{v}_J^2 - \left(\frac{\mathbf{r}_{EJ} \cdot \mathbf{v}_J}{r_{EJ}} \right)^2 + \sum_{K \neq J} U_{JK} \left(2 + \frac{\mathbf{r}_{EJ} \cdot \mathbf{r}_{JK}}{r_{JK}^2} \right) \right], \tag{3.2}$$

where

$$U_{JK} \equiv \frac{GM_K}{r_{JK}}, \quad \mathbf{r}_{JK} \equiv \mathbf{x}_J - \mathbf{x}_K, \quad r_{JK} \equiv |\mathbf{r}_{JK}|. \tag{3.3}$$

The right hand side of Equation (3.1) is independent on $\Delta\tau_E$ itself. Thus, it is a pure function of t if the motion of major celestial bodies in the solar system is known, i.e. if the planetary/lunar ephemerides is provided. In this sense, we may obtain the time ephemeris simply by the quadrature of the right hand side of the above equation.

4. Realizations of Time Ephemeris

There have been several analytical time ephemerides. Moyer's pioneer works (Moyer 1981a, Moyer 1981b) are based on the Keplerian approximation of planetary/lunar orbits. All of the later computations (Hirayama *et al.* 1987, Fairhead *et al.* 1988, Fairhead and Bretagnon 1990) are based on the analytical planetary ephemeris, VSOP82 (Bretagnon 1982) and the analytical lunar ephemeris, ELP2000 (Chapront-Touze and Chapront 1982). Since the ephemerides are expressed as Fourier series, it is easily to conduct the quadrature.

On the other hand, numerical time ephemerides are obtained twice (Fukushima 1995, Irwin and Fukushima 1999), all of which are based on the JPL numerical ephemerides, DE102, DE200, DE245, and DE405 (Newhall 1989, Standish *et al.* 1992, Standish 1998a, Standish 1998b). In this case, the quadrature was executed by the Romberg method (Press *et al.* 2007). Taking the same number of DE ephemerides used, we named the time ephemerides as TE102, TE200, TE245, and TE405, respectively.

The numerical representation of time ephemerides are, as the same as in case of numerical lunar/planetary ephemerides, usually the resulting numerical tables themselves or their Chebyshev polynomial representation. These are appropriate for fast computation. However, the periodic features are difficult to find out. Also, the full implementation requires an expertise on its installation and some disk storages.

In order to complement these weak points, we presented the harmonic decomposition of TE405, the latest time ephemeris of the Earth (Harada and Fukushima 2003). The used approach is a nonlinear method of harmonic analysis (Harada 2003), an excerpt of which is reported in Appendix B of our analysis of the planetary precession derived from

Table 1. Main Terms of Fourier Series Expression of TE405

S (ns)	C (ns)	Period (days)
+505079.2018	−1551857.1407	365.2652622182
+21856.7326	−23134.7679	365.22102337
+20733.1083	−8526.5271	398.88401884
−11108.6620	−8369.7220	182.62982594
−3405.2830	−3354.5797	4333.21415

Note. Listed are the largest five Fourier terms of the harmonic decomposition of TE405 (Harada and Fukushima 2003).

DE405 (Harada and Fukushima 2004). The expression is in the form

$$\Delta\tau_E \approx \sum_{j=0}^{2} P_j \xi^j + \sum_{j=1}^{J} \left[S_j \sin(\omega_j \xi_j) + C_j \cos(\omega_j \xi_j) \right] + \sum_{j=1}^{K} \xi_j \left[S'_j \sin(\omega_j \xi_j) + C'_j \cos(\omega_j \xi_j) \right], \tag{4.1}$$

where

$$\xi \equiv \frac{\mathrm{JD} - 2414949.0}{54749.25}, \tag{4.2}$$

and P_j, S_j, C_j, ω_j, S'_j, and C'_j are certain constants. For the full period of TE405, i.e. from 1600 until 2200, the RMS of the residual of the approximation is 0.446 ns and the absolute maximum difference is 2.95 ns. For the shorter period from 1960 to 2020, the maximum difference reduces to 1.58 ns and most of the differences are less than 1 ns.

Table 1 shows the main Fourier terms. The full result contains a quadratic polynomial, 463 Fourier terms, and 36 mixed secular terms. Namely $J = 473$ and $K = 36$ in the above expression. The published article (Harada and Fukushima 2003) contains only the full coefficients of S'_j and C'_j, some of P_j, and the first five coefficients of Fourier terms.

For the full expression, refer the Fortran routine HF2002 and its parameter file included in the IERS Convention 2003 (McCarthy and Petit 2003). They are available from the IERS web site;

`http://tai.bipm.org/iers/conv2003/conv2003_c10.html`

5. Determination of General Relativistic Scale Factor

Let us return to the issue of scale factor. The factor L_C is split into the sum of three parts;

$$L_C = L_C^N + L_C^{PN} + L_C^A, \tag{5.1}$$

where the superscripts N and PN denote the Newtonian and the post-Newtonian contribution by the Earth's velocity in the BCRS and by the Newtonian gravitational potential of the Sun and major planets, while the superscript A does the Newtonian effect by asteroids (Fukushima 1995). Table 2 shows the dilated contribution of the first two parts for the case of TE245. Note that the asteroid part is too small to be listed.

The numerical value of L_C^N significantly differ ephemeris by ephemeris. See Table 3. The uncertainties shown here are basically caused by the finiteness of the effective period of the lunar/planetary ephemeris used. In fact, all the practical ephemerides whether being numerical or analytical are limited. Let us explain this situation more plainly.

Table 2. Contribution to General Relativistic Scale Factor, L_C

Source	Contribution (10^{-17})
Sun	987062583
velocity	493530342
Jupiter	182856
Saturn	29647
Moon	14191
Venus	2877
Uranus	2250
Neptune	1741
Mars	240
Mercury	171
post-Newtonian	11

Note. Listed are the contribution of each source to the value of L_C for the case of TE245 (Fukushima 1995). Note that all the contributions including the post-Newtonian one are positive. The asteroids' contribution, which is dropped from the list, is 0.45 in the unit of table.

Table 3. Estimated Values of Main Newtonian Parts of General Relativistic Scale Factor, L_C^N

L_C^N (10^{-17})	Time Ephemeris	Reference
1480826869.80±0.5	TE102	Fukushima 1995
57.13±0.5	TE200	Fukushima 1995
56.21±0.5	TE245	Fukushima 1995
55.94±1.0	TE405	Irwin and Fukushima 1999
55.90±0.6	TE405	Harada and Fukushima 2003

Assume that the ephemeris contain a very long period term of the frequency Ω. The associated Fourier terms are expanded as

$$\cos \Omega t \approx 1 - \frac{(\Omega t)^2}{2} + \cdots, \quad \sin \Omega t \approx \Omega t - \frac{(\Omega t)^3}{6} + \cdots, \tag{5.2}$$

Therefore, for the finite time period such as $|t| < T$, we cannot discriminate the cosine and sine terms of the frequency Ω with a constant offset, 1, and a linear trend, Ωt, if the leading residual terms, $(\Omega T)^2/2$ or $(\Omega T)^3/6$, are sufficiently small. See the detailed discussion in our reports (Fukushima 1995, Irwin and Fukushima 1999, Harada and Fukushima 2003).

On the other hand, the values of the last two parts do not differ significantly so that we may fix them by the value of TE245 (Fukushima 1995) as

$$L_C^{PN} = (10.97 \pm 0.01) \times 10^{-17}, \quad L_C^A = (0.45 \pm 0.50) \times 10^{-17}. \tag{5.3}$$

The uncertainty of the post-Newtonian term comes from that of the PPN parameters, β and γ. Meanwhile, the large uncertainty of the asteroid effect is due to their mass uncertainty.

At any rate, let us calculate the final value. Table 4 shows the summed value of L_C for TE405 based on the latest determination of L_C^N (Harada and Fukushima 2003). Using this, we calculate another scale factor, L_B, as

$$L_B \equiv L_C + L_G - L_C L_G = (1550519767.63 \pm 1.1) \times 10^{-17}, \tag{5.4}$$

which determines the mean rate of TCB-TT and becomes a key factor to convert the numerical values of physical quantities obtained from the astronomical observations in the

Table 4. General Relativistic Scale Factors

Constant	Meaning	Value (10^{-17})	σ (10^{-17})
L_C^N	Main Newtonian Part of L_C	1480826855.90	0.60
L_C^{PN}	post-Newtonian Part of L_C	10.97	0.01
L_C^A	Asteroid Part of L_C	0.45	0.50
L_C	Mean Rate of TCB-TCG	1480826867.32	1.1
L_G	Mean Rate of TCG-TT	69692901.34	
L_B	Mean Rate of TCB-TT	1550519767.63	1.1

solar system and those determined from the experimental measurements at laboratories on the Earth. Here

$$L_G \equiv 69692901.34 \times 10^{-17}, \tag{5.5}$$

is a defining constant to specify the mean rate of TCG-TT (McCarthy and Petit 2003).

References

Bretagnon, P., 1982, Astron. Astrophys., 114, 278

Chapront-Touzé, M., Chapront, J., 1983, Astron. Astrophys., 124, 50

Fairhead, L., Bretagnon, P., Lestrade, J. F., 1988, Proc. IAU Symp, 128, 419

Fairhead, L. & Bretagnon, P., 1990, Astron. Astrophys., 229, 240

Fukushima, T., 1995, Astron. Astrophys., 294, 895

Fukushima, T., Fujimoto, M.-K., Aoki, Sh., & Kinoshita, H., 1986, Celestial Mechanics, 36, 215

Harada, W., 2003, M. Sc. Thesis, Univ. Tokyo

Harada, W. & Fukushima, T., 2003, Astron. J., 126, 2557

Harada, W. & Fukushima, T., 2004, Astron. J., 127, 531

Hirayama, Th., Fujimoto, M.-K., Kinoshita, H., & Fukushima, T., 1987, Proc. IAG Symposia at IUGG XIX General Assembly, Tome I, 91

International Astronomical Union, 1992, in Proc. 21st General Assembly Buenos Aires 1991, Trans. of IAU XXIB, IAU, Paris

International Astronomical Union, 2001, in Proc. 24th General Assembly Manchester 2000, Trans. of IAU XXIVB, IAU, Paris

Irwin, A. W. & Fukushima, T., 1999, Astron. Astrophys., 348, 642

McCarthy, D. D. & Petit G., 2003, IERS Convention (2003), IERS Tech. Note 32, Obs. Paris, Paris

Moyer, T. D., 1981a, Celest. Mech. 23, 33

Moyer, T. D., 1981b, Celest. Mech. 23, 57

Newhall, X. X, 1989, Celest. Mech. 45, 305

Press, W. H., Teukolsky, S. A., Vetterling, W. T., & Flannery, B. P., 2007, Numerical Recipes: the Art of Scientific Computing, 3rd ed., Cambridge Univ. Press, Cambridge

Seidelmann, P. K. & Fukushima, T., 1992, Astron. Astrophys., 265, 833

Soffel, M., Klioner, S. A., Petit, G., Wolf, P., Kopeikin, S. M., Bretagnon, P., Brumberg, V. A., Capitaine, N., Damour, T., Fukushima, T., Guinot, B., Huang, T.-Y., Lindegren, L., Ma, C., Nordtvedt, K., Ries, J. C., Seidelmann, P. K., Vokrouhlicky, D., Will, C. M., & Xu, C., 2003, Astron. J., 126, 2687

Standish, E. M., 1998a, JPL Planetary and Lunar Ephemerides, DE405/LE405, JPL interoffice memorandum 312.F-98-048

Standish, E. M., 1998b, Astron. Astrophys., 336, 381

Standish, E. M., Newhall, X. X., Williams, J. G., & Yeomans D. K., 1992, Orbital Ephemerides of the Sun, Moon, and Planets. In: Seidelmann, P. K. (ed.) Explanatory Supplement to the Astronomical Almanac, University Science Books, Mill Valley, CA

Relativity in Fundamental Astronomy
Proceedings IAU Symposium No. 261, 2009
S. A. Klioner, P. K. Seidelman & M. H. Soffel, eds.

doi:10.1017/S1743921309990214

Current and future realizations of coordinate time scales

E. Felicitas Arias[1,2]

[1]International Bureau of Weights and Measures,
Sèvres, France
email: farias@bipm.org

[2]Associated astronomer at the Paris Observatory,
Paris, France

Abstract. Two atomic time scales maintained at the International Bureau of Weights and Measures (BIPM) are realizations of terrestrial time: International Atomic Time (TAI) and TT(BIPM). They are calculated from atomic clocks realizing proper time in national laboratories. The algorithm for the calculation of TAI has been designed to optimize the frequency stability and accuracy of the time scale. Plans for the future improvement of the reference time scales are presented.

Keywords. coordinate time scales, atomic time scales

1. Introduction

Time and frequency metrology provides time references for experiments confined to a small environment a well as applications in astronomy and Earth sciences in an extended environment, where time constitutes a coordinate in an arbitrarily chosen space time coordinate system.

While a clock realizes proper time and provides a reference for observations in its vicinity, only a coordinate time scale can form the basis of a world-wide time reference. We adopt a rotating geocentric reference system where the time-coordinate is the geocentric coordinate time. Such a time scale is to be constructed from an ensemble of clocks, each realizing a proper time. An algorithm has to be designed to produce the world time reference from these individual clock proper times. The choice of the algorithm will depend on the interval over which the frequency stability is to be assured and a compromise between frequency stability and accuracy, and should be adapted to the characteristics of the standards involved and the techniques used to compare them. This algorithm should produce a time scale that is more stable and accurate than any of the individual participating clocks. To make the best use of the time and frequency standards operating in national laboratories today, time scale algorithms should apply in the framework of general relativity.

International Atomic Time (TAI) is a realization of terrestrial time maintained since 1988 at the International Bureau of Weights and Measures (BIPM). It uses data from about 350 atomic clocks and a dozen primary frequency standards operating in national metrology laboratories and scientific institutes world-wide. TAI is calculated from thirty-day data batches and is published with a latency of about 15 days after the last date of data. The frequency stability of TAI, over 30-40 days is better than 4 parts in 10^{16}, and its frequency accuracy is a few parts in 10^{16}. However, TAI has some long-term instabilities that make it unsuitable for applications such as pulsar timing. For this purpose, another atomic time scale is calculated at the BIPM under the acronym TT(BIPMYY) using an algorithm that eliminates the instabilities present in TAI.

TAI forms the basis for realizing a number of time scales used in dynamics, for modelling the motions of artificial and natural celestial bodies, and in the exploration of the solar system, tests of theories, geodesy, geophysics, studies of the environment. In all these applications, relativistic effects are important.

2. The SI unit of time

The atomic second is the SI unit of time, defined by the 13th General Conference on Weights and Measures (Terrien 1968) based on the hyperfine transition of caesium 133. The definition of the second should be understood as the definition of the unit of proper time; it applies in a small spatial domain which shares the motion of the caesium atom used to realize the definition. In a laboratory sufficiently small to allow the effects of the non-uniformity of the gravitational field to be neglected when compared to the uncertainties of the realization of the second (a few parts in 10^{16} for caesium fountains), the proper second is obtained after application of the special relativistic correction for the velocity of the atom in the laboratory.

3. General relativity and atomic time; definition of TAI

In 1980, the Consultative Committee for the Definition of the Second (CCDS, now Consultative Committee for Time and Frequency, CCTF) declared that TAI is a coordinate time defined in a geocentric reference frame, and that its scale unit is the SI second as realized on the rotating geoid. The IAU endorsed this definition only when a global treatment of space-time reference systems was recommended in the framework of general relativity. After much controversy this finally happened in 1991 with the adoption of a specified metric (IAU 1991). In 2000 the IAU adopted a metric extended to higher order terms (IAU 2000). In these developments, several theoretical coordinate times were defined, for use in the vicinity of the Earth and for the dynamics of the solar system. All these coordinate times are realized on the basis of TAI after relativistic transformations. TAI itself appears as a realization of an ideal Terrestrial Time TT. Terrestrial Time is obtained from a Geocentric Coordinate Time TCG by a linear transformation chosen so that the mean rate of TT is close to the mean rate of the proper time of an observer located on the rotating geoid.

As seen above, TAI is the reference time scale, defined in the context of general relativity. Since 1988 it has been calculated at the BIPM as the result of international cooperation. An algorithm developed at the Bureau International de l'Heure (BIH) in the 1970s, denoted ALGOS (BIH 1974), (Guinot & Thomas 1988), (Audoin & Guinot 2001) has fixed the principles of the construction of TAI. After numerous tests and various improvements, it remains the basis of the present calculation at the BIPM.

Coordinated Universal Time (UTC) is derived from TAI by the application of leap seconds. The dates of the insertion of leap seconds in UTC are decided and announced by the International Earth Rotation and Reference Systems Service (IERS). At then time of writing and at least until 31 December 2009, the difference between TAI and UTC amounts to 34 s.

Whereas TAI is the uniform time scale that provides a precise reference for scientific applications, UTC is the time scale of practical use that serves for international coordination in time keeping and is the basis of many legal national times.

4. Metrological quality of a reference time scale: the case of TAI

A time scale to be used as a reference should be continuous. It is characterized by its reliability, frequency stability and accuracy, and accessibility. The algorithm ALGOS produces a continuous non-stepped time scale. TAI and UTC have the same metrological quality, with the exception that UTC is intentionally stepped by the application of leap seconds to compensate for the irregular rate of rotation of the Earth.

The *reliability* of a time scale is closely linked to the reliability of the clocks involved in its construction. Reliability is also associated with redundancy; in the case of TAI, a large number of clocks are required; this number is today about 350, most of them high-performance caesium atomic standards and active auto-tuned hydrogen masers.

The *frequency stability* of a time scale is its capacity to maintain a fixed ratio between its unitary scale interval and its theoretical counterpart. One measure of the frequency stability of a time scale is its Allan variance (Allan *et al.* 1988), which is the two-sample variance designed for the statistical analysis of time series, and depends on the sampling interval.

The *frequency accuracy* of a time scale is the aptitude of its unitary scale interval to reproduce its theoretical counterpart. After the calculation of a time scale on the basis of an algorithm conferring the requested frequency stability, frequency accuracy can be improved by comparing the frequency of the time scale with that of primary frequency standards, and by applying, if necessary, frequency corrections.

The *accessibility* of a world-wide time scale is its aptitude to provide a means of dating events for everyone. This depends on the precision which is required. In the case of TAI the ultimate precision requires a delay of a few tens of days in order to reach the long-term frequency stability required for a reference time scale. In addition, the process needs to be designed in such a way that the measurement noise is eliminated or at least minimized, which requires a minimum number of data sampling intervals.

The frequency instability of TAI, estimated today as 4 parts in 10^{16} for averaging times of 30 to 40 days (Petit 2008), is obtained by processing clock and clock comparison data at 5-day intervals over a monthly analysis, with a delay to publication of about 10 days after the last date of reported data. In the very long term, over a decade, the stability is maintained by primary frequency standards and is limited by the accuracy at the level of parts in 10^{15} assuming that the present performances are constant.

5. The algorithm for TAI and UTC

The time laboratories realize a stable local time scale using individual atomic clocks or a clock ensemble. Clock readings are then combined at the BIPM through an algorithm designed to optimize the frequency stability and accuracy, and increase the reliability of the time scale above the level of performance that can be realized by any individual clock in the ensemble. The calculation of UTC is carried out in three successive steps:

a) The free atomic time scale EAL (Echelle Atomique Libre) is computed as a weighted average of free-running atomic clocks spread world-wide. A clock weighting procedure has been designed to optimize the long-term frequency stability of the scale. No constraint is imposed to match the interval unit of EAL to the SI second.

b) The frequency of EAL is steered to maintain agreement with the definition of the SI second, and the resulting time scale is International Atomic Time TAI. The steering correction is determined by comparing the EAL frequency with that of primary frequency standards.

c) Leap seconds are inserted to maintain agreement with the time derived from the rotation of the Earth. The resulting time scale is UTC.

Different algorithms can be considered depending on the requirements of the scale; for an international reference such as UTC, the requirement is extreme reliability and long-term frequency stability. UTC therefore relies on the largest possible number of atomic clocks of different types, located in different parts of the world and connected in a network that allows precise time comparisons between remote sites.

The algorithm ALGOS for defining EAL is structured in three parts, each one associated with an algorithm: the *weighting algorithm* optimized to guarantee the long-term stability of the time scale (Azoubib 2001); the *prediction algorithm* used to avoid time and frequency jumps due to different clock ensembles being used in consecutive calculation periods (Thomas *et al.* 1994), and the *steering algorithm* used to improve the time scale accuracy (Arias & Petit 2005).

In time-scale algorithms, clock weights are generally chosen as the reciprocals of a statistical quantity which characterizes their frequency stability, such as a frequency variance (classical variance, Allan variance, etc.). If strictly applied, this gives a time scale which is more stable than any contributing element. In the EAL computation, the weight attributed to each clock is the reciprocal of the individual classical variance computed from the frequencies of the clock, relative to EAL, estimated over the current 30-day interval and over the past eleven consecutive 30-day intervals. The weight determination thus uses clock measurements covering a full year. This reduces the weight of both clocks that are highly sensitive to seasonal changes and hydrogen masers that show a large frequency drift. It thus helps to improve the long-term stability of EAL.

The weight of a clock is considered as constant during the 30-day period of computation and continuity with the previous period is assured by clock frequency prediction, a procedure that renders the scale insensitive to changes in the set of participating clocks. The algorithm is able to detect abnormal behaviour of clocks, and to disregard them if necessary; this is done in an iterative process that starts by the weights obtained in the previous month, and serves as an indicator of the behaviour of the clock in the month of computation.

To limit individual clock contributions, preventing domination of the scale by a small number of very stable clocks, a maximum value for the weight is chosen for each month of calculation, expressed as a fraction between 0 and 1. In ALGOS the maximum weight is chosen as a function of the total number of clocks in a period of calculation.

In the generation of a time scale, the prediction of the atomic clock behavior plays an important role; in fact the prediction is useful to avoid or minimize frequency jumps of the time scale when a clock is added or removed from the ensemble or when its weight changes. Considering two successive one-month intervals of TAI calculation we impose several constraints on the prediction term at the boundary date to avoid or minimize time and frequency jumps in the resulting time scale. In the case of commercial caesium clocks, for averaging times around 30 days the predominant noise is random walk frequency modulation. All clocks in TAI are treated with this same linear frequency prediction model, but a revision appears to be necessary to take into account the increasing number of participating hydrogen masers, for which the predominant frequency noise is a linear drift. Hydrogen masers represent today about 12% of the total weight of clocks in TAI.

TAI accuracy is assured from measurements of a small number of primary frequency standards (PFS) developed by a few metrology laboratories. The frequency of EAL is compared with that of the primary frequency standards using all available data over a one-year interval, and a frequency shift (frequency steering correction) is applied to EAL to ensure that the frequency of TAI conforms to its definition. Changes to the steering correction are expected to ensure accuracy without degrading the long-term (several months) stability of TAI. The value of this correction varies, with a maximum fixed at 6×10^{-16} in a month of calculation. The accuracy of TAI therefore depends on PFS measurements, which are reported more or less regularly to the BIPM. Data from several PFS are combined to estimate the duration of the scale unit of TAI (Azoubib *et al.* 1977), (Arias & Petit 2005). As at June 2009, twelve primary frequency standards, including nine caesim fountains, provide the best representation of the SI second with uncertainties of a few parts in 10^{16}, and contribute to improving the accuracy of TAI.

6. TT(BIPM)

An algorithm similar to that used to evaluate the frequency of EAL is used in post-processing to calculate another time scale strictly identified by the acronym TT(BIPMYY), where yy indicates the last two digits of the year of computation. TT(BIPM) is also a realization of the ideal Terrestrial Time TT (Petit 2003), (Petit 2008). TT(BIPM) is calculated every year and its frequency is steered using all available measurements of primary frequency standards reported to the BIPM by national laboratories. TT(BIPM) is a time scale optimized for frequency accuracy. The accuracy of TT(BIPM) for 2008 is estimated to be 5×10^{-16}.

TT(BIPM) provides a stable and accurate reference for characterizing the performance of the frequency of EAL and the frequency drift of the H-masers and caesium clocks. The frequency of EAL, when compared to that of TT(BIPM), presents a drift of 4×10^{-16} whose origin is under study.

7. Clocks in TAI

As at June 2009, 68 time laboratories from about 50 countries participate in the calculation of TAI at the BIPM. They contribute data each month from about 350 clocks. About 83% of clocks are either commercial caesium clocks of the Symmetricom/HP/Agilent 5071A-type or active, auto-tuned hydrogen masers. Commercial caesium clocks with high-performance tubes realize the atomic second with a relative accuracy in frequency of 1×10^{-13}, and they have an excellent long-term frequency stability. Active hydrogen masers also benefit from high-frequency stabilities of the order of 10^{-15} over 1 day.

8. Further improvement and perspectives

The present frequency stability of TAI is estimated to be 0.4×10^{-15} over one month. This is obtained through the procedures for clock weighting and frequency prediction described in the previous sections. An improvement in the stability would be possible if an increased number of more stable clocks participated in the formation of TAI. Such progress do not rely on the BIPM. However, improving the algorithm is the responsibility of the BIPM.

The effect of the linear prediction algorithm of the clock frequency has been studied at the BIPM for different types of clocks in TAI (Panfilo & Arias 2009). ALGOS predicts the clock frequency with a linear model that is well adapted to the caesium clocks, but not to the hydrogen maser clocks which represent 12% of the total weight in EAL. A test version of EAL without hydrogen masers has been calculated to evaluate the effects of the equal modelling of the clock frequencies. A new mathematical expression for the prediction of the hydrogen maser frequency is proposed taking into account the drift. Tests over a 3-year period have been performed applying the linear prediction to the caesium clocks and a quadratic prediction to the hydrogen masers. A version of EAL on the basis of the proposed frequency prediction for hydrogen masers, but with the classical clock weighting, has been evaluated. When all clocks are predicted with the linear model, a drift of about 4 parts in 10^{16} is observed between EAL and TT(BIPM). The results seem to indicate that inappropriate modelling of the frequency drift of hydrogen masers could be responsible for 20% of this drift. In this test one month of past data has been used to evaluate frequency drift between EAL and TT(BIPM); a longer period will also be tested. EAL still shows a significant drift after having introduced a quadratic model for the hydrogen maser clocks; further work needs to be done, and in particular a revision of the clock weighting algorithm.

Since 1971 the definition of the second has been based on the transition between two hyperfine levels of the ground state of the 133-caesium atom. Primary frequency standards (in particular caesium fountains) realize the definition of the second with an uncertainty of one part in 10^{15} (BIPM 2009) . In some laboratories a new generation of fountains is under study with the goal of reducing the uncertainty to parts in 10^{16}.

New devices make use of the properties of other radiations. In the microwave region, rubidium has been used for the construction of a double Cs-Rb fountain (Guéna *et al.* 2008). Frequency standards based on optical radiations (ytterbium, mercury, strontium) have been constructed in several metrology institutes (Dubé *et al.* 2006), (Fouche *et al.* 2007), (Tamm *et al.* 2007) (the list of references is not exhaustive). The accuracy of these realizations is expected to exceed that of caesium fountains, and some have already been recommended by the International Committee for Weights and Measures (CIPM) for use as secondary representations of the second (Gill & Riehle 2006).

The current techniques and methods of time transfer are not yet able to perform frequency transfer at the level of the optical standards accuracy, limiting the possibility of their remote comparison. Work is under way within the time metrology community aimed at overcoming this difficulty. The CCTF has established in 2006 a working group for coordinating activities on highly accurate time and frequency transfer.

9. Conclusion

Two realizations of Terrestrial Time TT are maintained at the BIPM; TAI is the continuous, atomic time scale which serves as the basis for constructing UTC. The instability of TAI over 30 to 40 day intervals is of 4 parts in 10^{16}; however, long-term instabilities of 1 to 2 parts in 10^{15} mean that TAI cannot be used as a reference for applications requiring long-term stability. TT(BIPM) is computed yearly based on all available measurements of primary frequency standards; its long-term instability is better that that of TAI in a factor at between two and three (Petit 2003), and its accuracy is of 5 parts in 10^{16}.

The algorithm used for the clock frequency prediction is under revision. The frequency of caesium clocks is well predicted by a linear model, but this is not the case for hydrogen masers where a quadratic model seems to better represent their drift. Preliminary studies indicate that the drift observed in the last years between EAL and TT(BIPM) could be partially explained by the effects of the inappropriate modelling of the hydrogen masers.

The 133-caesium atom provides the definition of the SI second as well as its unique realization. Progress in fundamental physics applied to the development of new standards for metrology has led the CIPM to recommend a list of radiations for providing secondary representations of the second. However, studies on highly accurate frequency transfer are still necessary to allow the comparison of these standards at the best of their performances. In view of this evolution, the metrological community has started discussions on a possible new definition and realization of the second which could arrive in the next decade.

The SI unit of length, the meter, is realized through a list of radiations approved for practical realization of the meter. With the establishment of a list of appropriate radiations for realizing the second, we could imagine that access to the SI second could be achieved in the future similar to the meter, with different levels of uncertainty, through the various entries in this list.

References

Allan D. W., Hellwig, H., Kartaschoff, P., Vanier, J., Vig, J., Winkler, G. M. R., & Yannoni, N. F. 1988, *Proc. of the* 42nd *Annual IFCS*, 419

Arias E. F. & Petit G. 2005, *Proc. Joint IEEE FCS and PTTI*, 244

Audoin, C & Guinot, B. 2001, *The measurement of time*, Cambridge University Press

Azoubib, J., Graveaud, M., & Guinot, B. 1977, *Metrologia*, 13, 87

Azoubib, J. 2001, *Proc.* 32nd *PTTI*, 195

Bureau International de l'Heure 1974, *BIH Annual Report for 1973*, Observatoire de Paris

Bureau International des Poids et Mesures 2009, *BIPM Annual Report on Time Activities for 2008*, 26

Dubé, P., Madej, A. A., Bernard, J. E., & Shiner, A. D. 2006, *Proc. of the 2006 IEEE IFCS*, 409

Fouche M., Le Targat, R., Baillard, X., Brusch, A., Tcherbakoff, F., Rovera, G. D., & Lemonde, P. 2007, *IEEE Trans. Instr. Meas*, 56, 236

Gill, P. & Rihele, F. 2006, *Proc.* 20th *EFTF*, 282

Guna, J., Chapelet, F., Rosenbusch, P., Laurent, P., Abgrall, M., Rovera, G. D., Santarelli, G., Tobar, M., E., Bize, S., & Clairon, A. 2008, *Proc. of the* 22nd *EFTF*, CD-ROM

Guinot, B. & Thomas, C. 1988, *Annual Report of the BIPM Time Section*, 1, D1-D22

International Astronomical Union 1991, *Proceedings of the* 21st *General Assembly*, Resolution A4, Kluwer, Dordrecht, The Nederlands

International Astronomical Union 2000, *Proceedings of the* 24th *General Assembly*, Resolutions B1.3 and B1.9, Astronomical Society of the Pacific, Provo, USA.

Panfilo, G. & Arias, E. F. 2009, *Special issue of UFFC*, submitted

Petit, G. 2003, *Proc.* 35th *PTTI*, 307

Petit, G. 2008, *Proc.* 7th *Symposium on Freqency standards and metrology*, in press

Tamm, Ch., Lipphardt, B., Schnatz, H., Wynands, R., Weyers, S., Schneider, T., & Peik, E. 2007, *IEEE Trans. Instrum. Meas.*, 56, 601

Terrien, J. 1968, *Metrologia*, 4, 43

Thomas, C, Wolf, P., & Tavella, P. 1994, *IEEE Trans. Instrum. Meas.*, 56, 601

Thomas, C. & Azoubib, J. 1996, *Metrologia*, 33, 227

Relativity in Fundamental Astronomy
Proceedings IAU Symposium No. 261, 2009
S. A. Klioner, P. K. Seidelman & M. H. Soffel, eds.

doi:10.1017/S1743921309990226

Relativistic equations of motion of massive bodies

Luc Blanchet
Institut D'Astrophysique de Paris
email: blanchet@iap.fr

Abstract. Highly relativistic equations of motions will play a crucial role for the detection and analysis of gravitational waves emitted by inspiralling compact binaries in detectors LIGO/ VIRGO on ground and LISA in space. Indeed these very relativistic systems (with orbital velocities of the order of half the speed of light in the last orbital rotations) require the application of a high-order post-Newtonian formalism in general relativity for accurate description of their motion and gravitational radiation [1]. In this contribution the current state of the art which has reached the third post-Newtonian approximation for the equations of motion [2–6] and gravitational waveform [7–9] has been described (see [10] for an exhaustive review). We have also emphasized the successful matching of the post-Newtonian templates to numerically generated predictions for the merger and ring-down in the case of black-hole binaries [11].

References

[1] Cutler, C., Apostolatos, T., Bildsten, L. Finn, L., Flanagan, E., Kennefick, D. Markovic, D. Ori, A. Poisson, E. Sussman, G., & Thorne, K. 1993, The last three minutes: Issues in gravitational-wave measurements of coalescing compact binaries, *Phys.Rev.Lett.*, 70, 2984
[2] Jaranowski, P. & Schaefer, G. 1998, Third post-Newtonian higher order ADM Hamilton dynamics for two-body point-mass systems, *Phys.Rev.D*, 57, 7274
[3] Damour, T., Jaranowski, P., & Schaefer, G. 2001, Dimensional regularization of the gravitational interaction of point masses, *Phys.Lett.B*, 513, 147
[4] Blanchet, L. & Faye, G. 2001, General relativistic dynamics of compact binaries at the third post-Newtonian order, *Phys.Rev.D*, 63, 062005
[5] Andrade, V., de. Blanchet, L., & Faye, G. 2001, Third post-Newtonian dynamics of compact binaries: Noetherian conserved quantities and equivalence between the harmonic-coordinate and ADM-Hamiltonian formalisms, *Class.Quant.Grav.*, 18, 753
[6] Itoh, Y. & Futamase, T. 2003, New derivation of a third post-Newtonian equation of motion for relativistic compact binaries without ambiguity, *Phys.Rev.D*, 68, 121501R
[7] Blanchet, L., Faye, G., Iyer, B., & Joguet, B. 2002, Gravitational-wave inspiral of compact binary systems to 7/2 post-Newtonian order, *Phys.Rev.D*, 65, 061501R
[8] Blanchet, L., Damour, T., Esposito-Farese, G., & Iyer, B. 2004, Gravitational radiation from inspiralling compact binaries completed at the third post-Newtonian order, *Phys.Rev.Lett.*, 93, 091101
[9] Blanchet, L., Faye, G., Iyer, B., & Sinha, S. 2008, The third post-Newtonian gravitational wave polarisations and associated spherical harmonic modes for inspiralling compact binaries in quasi-circular orbits, *Class.Quant.Grav.*, 25, 165003
[10] Blanchet, L. 2006, Gravitational radiation from post-Newtonian sources and inspiralling compact binaries, *Liv.Rev. in Relat.*, 9, 4
[11] Buonanno, A., Cook, G., & Pretorius, F. 2007, Inspiral, merger and ring-down of equal-mass black-hole binaries, *Phys.Rev.D*, 75, 124018

Relativity in Fundamental Astronomy
Proceedings IAU Symposium No. 261, 2009
S. A. Klioner, P. K. Seidelman & M. H. Soffel, eds.

doi:10.1017/S1743921309990238

High-accuracy propagation of light rays

Pierre Teyssandier[1]

[1]Dépt SYstèmes de Référence Temps-Espace (SYRTE), CNRS-UMR 8630 & UPMC,
Observatoire de Paris, 61 avenue de l'Observatoire, 75014 Paris
email: Pierre.Teyssandier@obspm.fr

Abstract. We present a review of the different methods currently developed to determine the deflection of light rays due to gravity. The aim of these methods is primarily to calculate the angular distances with an accuracy of the order of microarcsecond.

Keywords. Gravitation, astrometry, reference systems.

1. Introduction

Space astrometric missions like Gaia or SIM are aimed to reach an accuracy of the order of microarcsecond (μas) in the measurements of the positions of celestial objects. We give a review of the main methods which are at our disposal to treat the high-accuracy propagation of light in the framework of metric theories of gravity.

2. Newtonian approach

The idea that light might be deflected by a gravitating body has appeared a long time before Einstein. A Newtonian deflection $\widehat{\delta}_N$ was calculated for an unbound orbit in the field of an isolated, spherically symmetric body of mass M around 1784 by Cavendish and then by Soldner in 1801 (see Will 1988). These old results are equivalent to the exact formula

$$\sin\frac{\widehat{\delta}_N}{2} = \frac{GM}{V^2 b}\left[1 + \left(\frac{GM}{V^2 b}\right)^2\right]^{-1/2}, \tag{2.1}$$

where G is the gravitational constant, V is the speed of light at infinity and b the impact parameter of the ray. For a ray grazing the surface of the Sun, Eq. (2.1) yields approximatively half the value predicted by general relativity, that is $\widehat{\delta}_N = 0.875$ arcsecond.

It may be concluded from Eq. (2.1) that the gravitational deflection of light should be taken into account in modern astrometry even if Newtonian theory was right.

3. Modern, relativistic approach: metric theories

Newton's theory is now replaced by the so-called relativistic metric theories. So we suppose henceforth that space-time $\mathcal{V}_4$ is endowed with a Lorentzian metric g describing gravity. Recall that this metric enables to define the scalar product $X.Y$ of two vectors X and Y at a given point: in any coordinate system (x^α), one has $X.Y = g_{\alpha\beta}X^\alpha Y^\beta = g^{\alpha\beta}X_\alpha Y_\beta$. We put $X^2 = X.X = g_{\alpha\beta}X^\alpha X^\beta$. The signature chosen for g is $(+,-,-,-)$.

At the 1 μas level of accuracy, the propagation of light may be analyzed within the geometric optics approximation. Then light rays are null geodesics of space-time (see Section 5). Recall that a curve is said to be null when its tangent vector l^α is a null vector, that is a vector such that

$$l^2 = g_{\alpha\beta} l^\alpha l^\beta = 0. \tag{3.1}$$

Null geodesics were determined in a closed-exact form for Schwarzschild metric in de Jans (1922). However, this exact solution is expressed in terms of elliptic functions which are difficult to handle, so that only perturbation solutions are used in practice. The total deflection of a light ray coming from infinity with an impact parameter $b \gg 2GM/c^2$ is given by

$$\widehat{\delta} = \frac{4GM}{c^2 b} + \frac{15\pi}{4}\left(\frac{GM}{c^2 b}\right)^2 + O(1/c^6). \tag{3.2}$$

It must be noted that the validity of the formula (3.2) is independent of the coordinate system because the impact parameter is a length defined for an (ideal) observer at rest at infinity, where space-time is Minkowskian.

For a ray grazing the Sun, the term of order c^{-4} in Eq. (3.2) amounts to 11 μas. Since the angle between the spacecraft rotation axis and the Sun direction is equal to 45 deg, the contribution of this term will remain completely unobservable by Gaia mission.

In what follows we examine how the gravitational deflection of light can be calculated for more general gravitational fields.

4. Angular distances in metric theories

Consider two light rays Γ and Γ' arriving at a point $x_o \in \mathcal{V}_4$. We assume that Γ and Γ' are emitted by point-like sources $\mathcal{E}$ and $\mathcal{E}'$, respectively. Let $\mathcal{O}(u)$ be an observer passing through x_o with a unit 4-velocity u, i.e. the vector of components $u^\mu = dx^\mu/ds$.

Space relative to $\mathcal{O}(u)$ at x_o is *defined* as the set $\Pi_{x_o}(u)$ of the tangent vectors orthogonal to the 4-velocity of $\mathcal{O}(u)$. As a consequence, the angular distance between $\mathcal{E}$ and $\mathcal{E}'$ as measured by $\mathcal{O}(u)$ is defined as the angle ϕ_u between the orthogonal projections of the rays Γ and Γ' on $\Pi_{x_o}(u)$ (see Soffel 1989 or Brumberg 1991). Let l and l' denote vectors tangent to Γ and Γ' at x_o, respectively. Taking into account that l and l' are null vectors, it may be seen that the angular distance ϕ_u is determined by the relation

$$\sin^2\frac{\phi_u}{2} = -\frac{(l'-l)^2}{4(u.l)(u.l')}, \quad 0 \leqslant \phi_u \leqslant \pi. \tag{4.1}$$

Two kinds of descriptions may be envisaged in modeling astrometric measurements.

1. One can introduce an orthonormal tetrad $e_{\underline{\alpha}}$ along the worldline of $O(u)$ such that

$$e_{\underline{0}} = u, \quad u.e_{\underline{i}} = 0, \quad e_{\underline{i}}.e_{\underline{j}} = -\delta_{ij}, \quad (i = 1, 2, 3).$$

The direction of the ray Γ is then defined by the spacelike components of the tangent vector l relative to the triad $e_{\underline{i}}$:

$$l^{\underline{i}} = -(l.e_{\underline{i}}).$$

Then ϕ_u may be obtained as a function of the direction of each ray by inserting $l = (l.u)u + l^{\underline{i}}e_{\underline{i}}$ and $l' = (l'.u)u + l'^{\underline{j}}e_{\underline{j}}$ into Eq. (4.1). This procedure may be very useful when it is necessary to take into account the attitude of the space station on which the measurements are performed, as in models RAMOD for Gaia (see, e.g., de Felice *et al.* 2006).

2. However, the angular distance may also be directly obtained by carrying out the calculation of $\sin^2\phi_u/2$ in the chosen reference frame (x^α) without introducing any tetrad. This point of view is close to the spirit of GREM model elaborated by the Dresden group working on Gaia (see Klioner 2003). It is the point of view chosen here.

Throughout this paper we assume that space-time may be covered by some global quasi-Galilean coordinate system $x^\alpha = (x^0, \boldsymbol{x})$ in which the metric is written as

$$g_{\mu\nu} = \eta_{\mu\nu} + h_{\mu\nu}, \qquad \eta_{\mu\nu} = \text{diag}\,(1, -1, -1, -1) \tag{4.2}$$

and we denote by $x_e = (x_e^0, \boldsymbol{x}_e)$ and $x_{e'} = (x_{e'}^0, \boldsymbol{x}_{e'})$ the points where Γ and Γ' are emitted, respectively. In what follows, the coordinate system may be identified with the BCRS recommended by IAU2000 resolutions.

We use the vector notations $\boldsymbol{a} = (a^1, a^2, a^3) = (a^i)$ and $\underline{\boldsymbol{c}} = (c_1, c_2, c_3) = (c_i)$. Then, given $\boldsymbol{a}, \boldsymbol{b}$ and $\underline{\boldsymbol{c}}$, we put $\boldsymbol{a}.\boldsymbol{b} = a^i b^i$, and $\boldsymbol{a}.\underline{\boldsymbol{c}} = a^i c_i$, the Einstein convention being systematically used for repeated indices.

The zeroth-order direction in which Γ is seen by an observer $\mathcal{O}(U)$ at rest relative to the coordinate system (x^α) passing through x_o is defined as the vector

$$\boldsymbol{N} = (N^i), \quad N^i = \frac{x_e^i - x_o^i}{|\boldsymbol{x}_e - \boldsymbol{x}_o|}. \tag{4.3}$$

Of course, an analogous vector $\boldsymbol{N}'$ may be defined for Γ'. We shall call $\phi_U^{(0)}$ the angle between $\boldsymbol{N}$ and $\boldsymbol{N}'$.

We put

$$\boldsymbol{\beta} = (\beta^i), \quad \beta^i = \frac{dx^i}{dx^0}, \quad \beta^2 = \delta_{ij}\beta^i\beta^i, \tag{4.4}$$

where the derivatives dx^i/dx^0 are taken along the worldline of $\mathcal{O}(u)$.

In the presence of gravity, the ratio $(l_i/l_0)_{x_o}$ is slightly different from N^i. As long as we may assume that there exists one and only one null geodesic between a source located at $\boldsymbol{x}_e$ and x_o, the difference $(l_i/l_0)_{x_o} - N^i$ is a function of $\boldsymbol{x}_e, t_o$ and $\boldsymbol{x}_o$. So we define the *(gravitational) deflection vector* as

$$\underline{\boldsymbol{\lambda}}(\boldsymbol{x}_e, t_o, \boldsymbol{x}_o) = (\lambda_i(\boldsymbol{x}_e, t_o, \boldsymbol{x}_o)), \quad \lambda_i(\boldsymbol{x}_e, t_o, \boldsymbol{x}_o) = \left(\frac{l_i}{l_0}\right)_{x_o} - N^i. \tag{4.5}$$

It is shown in Teyssandier & Le Poncin-Lafitte (2006) that the angular distance ϕ_u is determined by the equation

$$\sin^2\frac{\phi_u}{2} = K_{u/U}(\mathcal{E}, \mathcal{E}')\sin^2\frac{\phi_U}{2}, \tag{4.6}$$

where $K_{u/U}(\mathcal{E}, \mathcal{E}')$ is given by

$$K_{u/U}(\mathcal{E}, \mathcal{E}') = \frac{1}{1 + h_{00}}\,\frac{1 - \beta^2 + h_{00} + 2h_{0k}\beta^k + h_{kl}\beta^k\beta^l}{[1 + \boldsymbol{\beta}.(\boldsymbol{N} + \underline{\boldsymbol{\lambda}})][1 + \boldsymbol{\beta}.(\boldsymbol{N}' + \underline{\boldsymbol{\lambda}}')]} \tag{4.7}$$

and ϕ_U is the angular distance between $\mathcal{E}$ and $\mathcal{E}'$ as measured by $\mathcal{O}(U)$. It may be shown that

$$\sin^2\frac{\phi_U}{2} = (1 + h_{00})\Bigg[\sin^2\frac{\phi_U^{(0)}}{2} + \frac{1}{2}(\underline{\boldsymbol{\lambda}}' - \underline{\boldsymbol{\lambda}}).(\boldsymbol{N}' - \boldsymbol{N}) - \frac{1}{4}k^{ij}(N'^i - N^i)(N'^j - N^j)$$
$$+ \frac{1}{4}(\underline{\boldsymbol{\lambda}}' - \underline{\boldsymbol{\lambda}})^2 - \frac{1}{2}k^{ij}(\lambda_i' - \lambda_i)(N'^j - N^j) - \frac{1}{4}k^{ij}(\lambda_i' - \lambda_i)(\lambda_j' - \lambda_j)\Bigg], \tag{4.8}$$

with

$$k^{ij} = g^{ij} - \eta^{ij} = -h_{ij} + \eta^{\alpha\beta}h_{i\alpha}h_{j\beta} + O(h^3). \tag{4.9}$$

It is clear that the factor $K_{u/U}(\mathcal{E}, \mathcal{E}')$ describes the *aberration* due to the motion of the observer $\mathcal{O}(u)$ relative to the observer $\mathcal{O}(U)$.

Equations (4.6) and (4.8) show that the angular separation ϕ_u is theoretically determined when the deflection vector $\underline{\boldsymbol{\lambda}}$ is known. Let us emphasize that Eqs. (4.6)–(4.8) are rigorous.

Two kinds of procedures are currently available to determine the deflection vector $\underline{\boldsymbol{\lambda}}(\boldsymbol{x}_e, t_o, \boldsymbol{x}_o)$.

1. One can try to solve the differential equations satisfied by the null geodesics: this method works well when the rays are coming from infinity and can be extended to rays emitted at a finite distance by some adequate procedure (see, e.g., Brumberg 1991). This method is summarized in Sections 5 and 6 for rays emitted at infinity within the weak-field, linearized approximation.

2. One can also use procedures avoiding the integration of geodesic equations. These methods are particularly convenient when the sources of light rays and the observer are located at a finite distance. They are also more easy to extend to the higher orders of approximation. Some results are briefly presented in Sections 7 and 8.

At the level of accuracy required by Gaia or SIM missions, it is sufficient to know $\underline{\boldsymbol{\lambda}}(\boldsymbol{x}_e, t_o, \boldsymbol{x}_o)$ within the weak-field, linearized approximation in almost all the practical cases (see, e.g., Klioner 2003). The only known exception is the very special case of a ray grazing giant planets like Jupiter or Saturn (see Section 8). So we shall content ourselves with the linear approximation in the two next sections.

5. Equations of null geodesics

A geodesic of $(\mathcal{V}_4, g)$ is a parametrized curve $x^\rho(\zeta)$ satisfying the variational principle

$$\delta \int \frac{1}{2} g_{\alpha\beta}(x^\rho(\zeta)) l^\alpha l^\beta d\zeta = 0, \qquad l^\alpha = \frac{dx^\alpha(\zeta)}{d\zeta}. \tag{5.1}$$

Condition (5.1) means that any geodesic $x^\rho(\zeta)$ satisfies the following Euler-Lagrange equations

$$\frac{dl_\mu}{d\zeta} = \frac{1}{2}\frac{\partial h_{\alpha\beta}}{\partial x^\mu}(x^\rho(\zeta)) l^\alpha l^\beta, \qquad l_\mu = g_{\mu\nu} l^\nu. \tag{5.2}$$

Of course, any solution to Eqs. (5.2) representing a null geodesic has to fulfil the condition (3.1).

We assume henceforth that the gravitational perturbation $h_{\mu\nu}$ can be written as a power series in the gravitational constant G

$$h_{\mu\nu}(x, G) = \sum_{n=1}^{\infty} G^n g_{\mu\nu}^{(n)}(x). \tag{5.3}$$

So the solutions to Eqs. (5.2) and (3.1) may be written as

$$x^\rho(\zeta) = x^\rho_{(0)}(\zeta) + \sum_{n=1}^{\infty} G^n x^\rho_{(n)}(\zeta), \tag{5.4}$$

which implies for the covariant components of the tangent vector

$$l_\mu(\zeta) = l_\mu^{(0)}(\zeta) + \sum_{n=1}^{\infty} G^n l_\mu^{(n)}(\zeta) \tag{5.5}$$

and for the deflection vector

$$\underline{\boldsymbol{\lambda}}(\boldsymbol{x}_e, t_o, \boldsymbol{x}_o) = \sum_{n=1}^{\infty} G^n \underline{\boldsymbol{\lambda}}^{(n)}(\boldsymbol{x}_e, t_o, \boldsymbol{x}_o). \tag{5.6}$$

After substituting for l_μ from Eq. (5.5) into Eqs. (5.2) and (3.1), it is easy to see that $l_\mu^{(0)}$ is a constant:

$$l_\mu^{(0)} = K_\mu, \quad K_\mu = \text{const}, \tag{5.7}$$

where the K_μ have to fulfil the condition

$$\eta_{\alpha\beta} K^\alpha K^\beta = 0, \quad K^\alpha = \eta^{\alpha\mu} K_\mu. \tag{5.8}$$

For any null geodesic of 0th-order direction $-\boldsymbol{N}$ arriving at x_o, the constants K^i may be chosen so that $K^i = -N^i$. Hence $K^0 = 1$ as a consequence of (5.8). Denoting by ζ_o the (arbitrary) value of ζ at point x_o, the zeroth-order parametric equations of the null geodesic are then

$$x^0_{(0)}(\zeta) = \zeta - \zeta_o + x^0_o, \quad \boldsymbol{x}_{(0)}(\zeta) = -(\zeta - \zeta_o)\boldsymbol{N} + \boldsymbol{x}_o. \tag{5.9}$$

Taking (5.7) and (5.9) into account, and then adopting the notation

$$f_{,\mu}(x) = \frac{\partial f}{\partial x^\mu} \tag{5.10}$$

for the partial differentiation of any function of x, the differential equation satisfied by $l_\mu^{(1)}$ may be written as

$$\frac{dl_\mu^{(1)}}{d\zeta} = \frac{1}{2}\left[g^{(1)}_{00,\mu} - 2g^{(1)}_{0k,\mu}N^k + g^{(1)}_{kl,\mu}N^k N^l\right]_{x_{(0)}(\zeta)}. \tag{5.11}$$

A straightforward calculation shows that Eq. (3.1) reduces to a constraint equation which determines $l_0^{(1)}$:

$$l_0^{(1)} = \frac{1}{2}\left[g^{(1)}_{00} - 2g^{(1)}_{0k}N^k + g^{(1)}_{kl}N^k N^l + 2N^l l_l^{(1)}\right]_{x_{(0)}(\zeta)}. \tag{5.12}$$

As a consequence, the first-order term $\lambda_i^{(1)}$ is given by

$$\lambda_i^{(1)} = (\delta_i^j - N^i N^j) l_j^{(1)} - \frac{1}{2}N^i\left[g^{(1)}_{00} - 2g^{(1)}_{0k}N^k + g^{(1)}_{kl}N^k N^l\right]. \tag{5.13}$$

Thus the deflection vector at x_o is completely determined when the three spacelike covariant components $l_i^{(1)}$ are known. The next section is devoted to the expression of $l_i^{(1)}$ and $\lambda_i^{(1)}$ for a light ray emitted at infinity.

6. Light rays coming from infinity

In this section, we suppose that the gravitational potentials $h_{\mu\nu}$ and their first derivatives tend to zero when $|\boldsymbol{x}| \to \infty$. Assuming that the light ray is emitted at infinity in a direction $-\boldsymbol{N}$, an integration of Eqs. (5.11) yields for the value of $l_j^{(1)}$ at x_o

$$l_j^{(1)} = \frac{1}{2}\int_{-\infty}^{\zeta_o}\left[g^{(1)}_{00,j} - 2g^{(1)}_{0k,j}N^k + g^{(1)}_{kl,j}N^k N^l\right]_{x_{(0)}(\zeta)} d\zeta, \tag{6.1}$$

the integral being taken along the zeroth-order straight line defined by Eqs. (5.9).

We consider that the Solar System is an isolated system constituted by N slowly moving, self gravitating bodies A ($A = 1, 2, \ldots, N$). The integrals involved in Eqs. (6.1) have been studied using the technique of retarded potentials (see, e.g., Kopeikin & Schäfer 1999 and Kopeikin *et al.* 2006). However, it is generally estimated that the deflections due to the retarded effects are less than 1 μas and can therefore be currently neglected.

So we may content ourselves with using the parametrized post-Newtonian (1PPN) approximation developed in Klioner & Soffel (2004), which is a natural extension of the IAU2000 metric. Since a consistent treatment of light propagation requires to retain only the metric truncated at $1/c^3$, we take

$$g_{00}^{(1)} = -\frac{2w}{c^2} + O(4), \quad (g_{0k}^{(1)}) = (\gamma + 1)\frac{2\boldsymbol{w}}{c^3} + O(5), \quad g_{kl}^{(1)} = -\gamma\frac{2w}{c^2}\delta_{kl} + O(4), \tag{6.2}$$

where

$$w(x^0, \boldsymbol{x}) = \sum_{A=1}^{N} w_A(x^0, \boldsymbol{x}), \quad w_A(x^0, \boldsymbol{x}) = \int \frac{\rho_A(x^0, \boldsymbol{x}'_A)}{|\boldsymbol{x} - \boldsymbol{x}'_A|} d^3\boldsymbol{x}'_A, \tag{6.3}$$

$$\boldsymbol{w}(x^0, \boldsymbol{x}) = \sum_{A=1}^{N} \boldsymbol{w}_A(x^0, \boldsymbol{x}), \quad \boldsymbol{w}_A(x^0, \boldsymbol{x}) = \int \frac{\rho_A(x^0, \boldsymbol{x}'_A)\boldsymbol{v}'_A(x^0, \boldsymbol{x}'_A)}{|\boldsymbol{x} - \boldsymbol{x}'_A|} d^3\boldsymbol{x}'_A, \tag{6.4}$$

ρ_A and $\boldsymbol{v}'_A$ being the rest-mass density and the velocity-field of the matter constituting the body A, respectively.

A rough estimate shows that the Sun and the planets may be considered as axially symmetric bodies slowly spinning about their axis of symmetry. Moreover, the multipole expansion of the gravitomagnetic potentials $\boldsymbol{w}_A(x^0, \boldsymbol{x})$ may be completely neglected. We may content ourselves with the following expressions for $w_A(x^0, \boldsymbol{x})$ and $\boldsymbol{w}_A(x^0, \boldsymbol{x})$:

$$w_A(x^0, \boldsymbol{x}) = \frac{M_A}{|\boldsymbol{x} - \boldsymbol{x}_A|}\left\{1 - J_{1A}R_A\frac{\boldsymbol{k}_A.(\boldsymbol{x} - \boldsymbol{x}_A)}{|\boldsymbol{x} - \boldsymbol{x}_A|^2}\right.$$
$$\left. - J_{2A}R_A^2\frac{3[\boldsymbol{k}_A.(\boldsymbol{x} - \boldsymbol{x}_A)]^2 - (\boldsymbol{x} - \boldsymbol{x}_A)^2}{2|\boldsymbol{x} - \boldsymbol{x}_A|^4} + \cdots\right\} \tag{6.5}$$

$$\boldsymbol{w}_A(x^0, \boldsymbol{x}) = w_A(x^0, \boldsymbol{x})\boldsymbol{v}_A(x^0) + \frac{\boldsymbol{S}_A \times (\boldsymbol{x} - \boldsymbol{x}_A)}{2|\boldsymbol{x} - \boldsymbol{x}_A|^3} + \cdots, \tag{6.6}$$

where, for each body A, $\boldsymbol{x}_A$ is the point on the axis of symmetry supporting the multipole distribution, M_A the mass, $J_{1A}, J_{2A}, \ldots$ the mass-multipole moments, R_A the equatorial radius, $\boldsymbol{k}_A$ the unit vector on the axis of symmetry, $\boldsymbol{S}_A$ the intrinsic angular momentum ($\boldsymbol{S}_A = S_A\boldsymbol{k}_A$) and $\boldsymbol{v}_A = cd\boldsymbol{x}_A/dx^0$ ($\boldsymbol{x}_A$ is a function of x^0).

The insertion of Eqs. (6.5) and (6.6) in Eqs. (6.1) leads to expressions impossible to calculate analytically if the full variation in time of $\boldsymbol{x}_A$ is taken into account. However, one can obtain a correct estimate of integrals (6.1) assuming that each body A is fixed at its position $\boldsymbol{x}_{Ao}$ at the moment t_{Ao} of the closest approach of the body and the photon (Klioner & Kopeikin 1992 and Klioner 2003).

Let P_{Ao} be the projection of the position of point $\boldsymbol{x}_{Ao}$ on the straight line parallel to $\boldsymbol{N}$ passing through $\boldsymbol{x}_o$. Putting

$$\boldsymbol{b}_{Ao} = \boldsymbol{x}_{P_{Ao}} - \boldsymbol{x}_{Ao}, \quad \boldsymbol{p}_{Ao} = \frac{\boldsymbol{b}_{Ao}}{b_{Ao}}, \quad \boldsymbol{q}_{Ao} = \boldsymbol{p}_{Ao} \times \boldsymbol{N}, \quad \boldsymbol{n}_{Ao} = \frac{\boldsymbol{x}_o - \boldsymbol{x}_{Ao}}{|\boldsymbol{x}_o - \boldsymbol{x}_{Ao}|} \tag{6.7}$$

and denoting by α_{Ao} the angle between $-\boldsymbol{n}_{Ao}$ and $\boldsymbol{N}$, the deflection vector is given by

$$\underline{\boldsymbol{\lambda}}^{(1)}(\boldsymbol{N}, \boldsymbol{x}_o) = \sum_{A=1}^{N} \underline{\boldsymbol{\lambda}}_{M_A}^{(1)} + \underline{\boldsymbol{\lambda}}_{J_{1A}}^{(1)} + \underline{\boldsymbol{\lambda}}_{J_{2A}}^{(1)} + \underline{\boldsymbol{\lambda}}_{S_A}^{(1)} + \cdots, \tag{6.8}$$

where

$$\underline{\boldsymbol{\lambda}}^{(1)}_{M_A} = (\gamma+1)\frac{M_A}{c^2 b_{Ao}}\left[\sin\alpha_{Ao}\boldsymbol{N} + (1+\cos\alpha_{Ao})\boldsymbol{p}_{Ao}\right] + \cdots, \tag{6.9}$$

$$\underline{\boldsymbol{\lambda}}^{(1)}_{J_{1A}} = (\gamma+1)\frac{M_A}{c^2 b_{Ao}} J_{1A}\frac{R_A}{b_{Ao}}(1+\cos\alpha_{Ao})\Big[(\boldsymbol{k}_{Ao}.\boldsymbol{q}_{Ao})\boldsymbol{q}_{Ao} - (\boldsymbol{k}_{Ao}.\boldsymbol{p}_{Ao})\boldsymbol{p}_{Ao}\Big] + \cdots, \tag{6.10}$$

$$\underline{\boldsymbol{\lambda}}^{(1)}_{J_{2A}} = (\gamma+1)\frac{M_A}{c^2 b_{Ao}} J_{2A}\left(\frac{R_A}{b_{Ao}}\right)^2(1+\cos\alpha_{Ao})\Big\{\left[(\boldsymbol{k}_{Ao}.\boldsymbol{q}_{Ao})^2 - (\boldsymbol{k}_{Ao}.\boldsymbol{p}_{Ao})^2\right]\boldsymbol{p}_{Ao}$$
$$+2(\boldsymbol{k}_{Ao}.\boldsymbol{p}_{Ao})(\boldsymbol{k}_{Ao}.\boldsymbol{q}_{Ao})\boldsymbol{q}_{Ao}\Big\} + \cdots, \tag{6.11}$$

$$\underline{\boldsymbol{\lambda}}^{(1)}_{S_A} = (\gamma+1)\frac{S_A}{c^3 b^2_{Ao}}(1+\cos\alpha_{Ao})\left[(\boldsymbol{k}_{Ao}.\boldsymbol{q}_{Ao})\boldsymbol{p}_{Ao} + (\boldsymbol{k}_{Ao}.\boldsymbol{p}_{Ao})\boldsymbol{q}_{Ao}\right] + \cdots, \tag{6.12}$$

$\boldsymbol{k}_{Ao}$ being the vector $\boldsymbol{k}$ at instant t_{Ao} and $+\cdots$ standing for terms giving negligible contributions at the 1 μas level (case, e.g., of the translational gravitomagnetic terms).

The contributions of the mass-multipole moments J_1 and J_2 are analyzed in several works (see, e.g., Kopeikin & Makarov 2007 and Refs. therein). The contributions of the other moments J_n are thoroughly calculated in Le Poncin-Lafitte & Teyssandier (2008).

A test of the deflection due to the mass quadrupole moment of Jupiter, say $(J_2)_{Jup}$, was proposed in Crosta & Mignard (2006) in the context of the Gaia mission (project GAREX). This test seems feasible since the maximum quadrupolar deflection predicted for a ray grazing Jupiter amounts to $\left(\widehat{\delta}_{J_2}\right)_{Jup} = 240\ \mu$as.

A general account of the effects relevant to a 1μas accuracy is given in Klioner (2003). It must be noted that the deflection due to the gravitomagnetic term $\underline{\boldsymbol{\lambda}}^{(1)}_{S_A}$ is less than $1\,\mu$ as for the Sun and the giant planets.

7. Sources at a finite distance

Let us outline the new methods developed in Le Poncin-Lafitte *et al.* (2004) and Teyssandier & Le Poncin-Lafitte (2008). It follows from these works that the deflection vector is given by

$$\lambda_i = -\frac{c\partial\mathcal{T}_r}{\partial x^i_o}\left[1 - \frac{c\partial\mathcal{T}_r}{\partial x^0_o}\right]^{-1} - N^i, \tag{7.1}$$

where $\mathcal{T}_r$ is the so-called *reception time transfer function* giving the travel time of a photon as a function of the spatial position of the emitter $\boldsymbol{x}_e$, the instant of reception x^0_o and the spatial position $\boldsymbol{x}_o$ of the observer, so that $x^0_o - x^0_e$ can be written as

$$x^0_o - x^0_e = c\mathcal{T}_r(\boldsymbol{x}_e, x^0_o, \boldsymbol{x}_o). \tag{7.2}$$

Moreover, assuming that the reception time transfer function may be expanded as

$$c\mathcal{T}_r(\boldsymbol{x}_e, x^0_o, \boldsymbol{x}_o) = |\boldsymbol{x}_e - \boldsymbol{x}_o| + c\sum_{n=1}^{\infty} G^n\mathcal{T}_r^{(n)}(\boldsymbol{x}_e, x^0_o, \boldsymbol{x}_o), \tag{7.3}$$

it is shown that each perturbation term $\mathcal{T}_r^{(n)}(\boldsymbol{x}_e, x^0_o, \boldsymbol{x}_o)$ is given by an integral over the null geodesic of Minkowski space-time having $-\boldsymbol{N}$ as spatial direction and arriving at $(x^0_o, \boldsymbol{x}_o)$, that is the curve defined by the parametric equations

$$z_-(\zeta) = (x^0_o - \zeta|\boldsymbol{x}_e - \boldsymbol{x}_o|,\ \boldsymbol{x}_o + \zeta|\boldsymbol{x}_e - \boldsymbol{x}_o|\boldsymbol{N}). \tag{7.4}$$

One thus recovers the well-known expression for the first-order term $\mathcal{T}_r^{(1)}$:

$$c\mathcal{T}_r^{(1)}(\boldsymbol{x}_e, x_o^0, \boldsymbol{x}_o) = \frac{1}{2}|\boldsymbol{x}_e - \boldsymbol{x}_o| \int_0^1 \left[g_{(1)}^{00} + 2N^i g_{(1)}^{0i} + N^i N^j g_{(1)}^{ij}\right]_{z_-(\zeta)} d\zeta. \tag{7.5}$$

The higher-order terms $\mathcal{T}_r^{(n)}$ are very complicated. We shall content ourselves here with giving the expression of $c\mathcal{T}^{(2)}$ for a stationary gravitational field. In this case indeed, the quantities $\mathcal{T}_r^{(n)}$ do not depend on the instant of reception t_r. Then

$$\begin{aligned} c\mathcal{T}^{(2)}(\boldsymbol{x}_e, \boldsymbol{x}_o) = \frac{1}{2}|\boldsymbol{x}_e - \boldsymbol{x}_o| \int_0^1 \Bigg\{ & \left[g_{(2)}^{00} + 2N^i g_{(2)}^{0i} + N^i N^j g_{(2)}^{ij}\right]_{\boldsymbol{z}_-(\zeta)} \\ & + 2\left[g_{(1)}^{0i} + N^j g_{(1)}^{ij}\right]_{\boldsymbol{z}_-(\zeta)} \frac{c\partial\mathcal{T}^{(1)}}{\partial x^i}(\boldsymbol{z}_-(\zeta), \boldsymbol{x}_o) \\ & + \eta^{ij}\left[\frac{c\partial\mathcal{T}^{(1)}}{\partial x^i}\frac{c\partial\mathcal{T}^{(1)}}{\partial x^j}\right]_{(\boldsymbol{z}_-(\zeta), \boldsymbol{x}_o)} \Bigg\} d\zeta, \end{aligned} \tag{7.6}$$

where $\boldsymbol{z}_-(\zeta) = \boldsymbol{x}_o + \zeta|\boldsymbol{x}_e - \boldsymbol{x}_o|\boldsymbol{N}$.

8. Application to a static spherically symmetric metric

Let us consider the family of static, spherically symmetric metrics written in the form

$$ds^2 = \left(1 - \frac{2GM}{c^2 r} + 2\beta\frac{G^2M^2}{c^4 r^2} + \ldots\right)(dx^0)^2 - \left(1 + 2\gamma\frac{GM}{c^2 r} + \frac{3}{2}\delta\frac{G^2M^2}{c^4 r^2} + \cdots\right)\delta_{ij}dx^i dx^j, \tag{8.1}$$

where β and γ are the usual post-Newtonian parameters and δ is a supplementry post-post-Newtonian parameter ($\gamma = \beta = \delta = 1$ in general relativity). It follows from Eqs. (7.5) and (7.6) that (see Teyssandier & Le Poncin-Lafitte 2008)

$$\begin{aligned} \mathcal{T}(\boldsymbol{x}_e, \boldsymbol{x}_o) = & \frac{|\boldsymbol{x}_e - \boldsymbol{x}_o|}{c} + (\gamma+1)\frac{GM}{c^3}\ln\left[\frac{|\boldsymbol{x}_e| + |\boldsymbol{x}_o| + |\boldsymbol{x}_e - \boldsymbol{x}_o|}{|\boldsymbol{x}_e| + |\boldsymbol{x}_o| - |\boldsymbol{x}_e - \boldsymbol{x}_o|}\right] \\ & + \frac{G^2M^2}{c^5}\frac{|\boldsymbol{x}_e - \boldsymbol{x}_o|}{|\boldsymbol{x}_e||\boldsymbol{x}_o|}\left[\kappa\frac{\arccos(\boldsymbol{n}_e.\boldsymbol{n}_o)}{\sqrt{1-(\boldsymbol{n}_e.\boldsymbol{n}_o)^2}} - \frac{(1+\gamma)^2}{1+\boldsymbol{n}_e.\boldsymbol{n}_o}\right] + O(c^{-7}), \end{aligned} \tag{8.2}$$

where

$$\boldsymbol{n}_e = \frac{\boldsymbol{x}_e}{|\boldsymbol{x}_e|}, \quad \boldsymbol{n}_o = \frac{\boldsymbol{x}_o}{|\boldsymbol{x}_o|}, \quad \kappa = \frac{1}{4}(8 - 4\beta + 8\gamma + 3\delta). \tag{8.3}$$

Equation (8.2) generalizes a result given in Brumberg (1987) for general relativity ($\kappa = 15/4$).

Substituting for $\mathcal{T}$ from Eq. (8.2) into Eqs. (7.1) yields the deflection vector at point x_o. Let P be the foot of the perpendicular drawn from origin O to the straight line parallel to $\boldsymbol{N}$ passing through $\boldsymbol{x}_o$. Putting

$$b_{(0)} = |\boldsymbol{x}_P|, \quad \boldsymbol{p} = \frac{\boldsymbol{x}_P}{b_{(0)}}, \tag{8.4}$$

and then noting that the zeroth-order impact parameter $b_{(0)} = r_o \sin\alpha_o$, where α_o is the angle between $-\boldsymbol{n}_o$ and $\boldsymbol{N}$, we get in the case of a source located at infinity:

$$\begin{aligned} \underline{\boldsymbol{\lambda}} = \frac{GM}{c^2 b_{(0)}}\Bigg\{ & \left[(\gamma+1)\sin\alpha_o + \frac{GM}{c^2 b_{(0)}}(1+\cos\alpha_o)\left[2\kappa\sin^2\frac{\alpha_o}{2} - (\gamma+1)^2\right]\right]\boldsymbol{N} \\ & + \left[(\gamma+1)(1+\cos\alpha_o) + \frac{GM}{c^2 b_{(0)}}\left[\kappa\left(\pi - \alpha_o + \frac{1}{2}\sin 2\alpha_o\right) - \mathcal{C}^{(2)}\right]\right]\boldsymbol{p}\Bigg\}, \end{aligned} \tag{8.5}$$

where

$$\mathcal{C}^{(2)} = (\gamma+1)^2 \frac{(1+\cos\alpha_o)^2}{\sin\alpha_o}. \tag{8.6}$$

The contribution due to the second-order term $\mathcal{C}^{(2)}$ may compare with first-order contributions when α_o becomes sufficiently small. Indeed, this term yields a deflection of 16.1 μas for a light ray grazing Jupiter and observed at a distance of 6 AU (for $\gamma = 1$). It is exactly the result obtained in Zschocke & Klioner (2009) by a different calculation.

9. Some concluding remarks

The first-order approximation set out in Sections 5 and 6 is generally sufficient for space astrometric missions like Gaia or SIM. However, the special case of light rays grazing the surface of a giant planet and some highly precise tests of general relativity in foreseeable future will require calculations within the post-linear regime. The method presented in Sections 7 and 8 is able to vie with the integration of geodesic equations in the treatment of these problems. This procedure directly yields the deflection vector for a family of parametrized post-post-Newtonian spherically symmetric metrics.

A more radical improvement of the modelizations will require to treat the propagation of light in dynamical extensions of the IAU2000 metric including all the possible c^{-4} terms (see, e.g., Minazzoli & Chauvineau 2009 and Refs. therein). In the longer term, it will probably become necessary to take into account the cosmological background, as it is already outlined in Klioner & Soffel (2004).

References

Brumberg, V. A. 1987, *Kinematics Phys. Celest. Bodies*, 3, 6

Brumberg, V. A. 1991, *Essential Relativistic Celestial Mechanics* (Bristol: Adam Hilger), p. 200

Crosta, M. T. & Mignard, F. 2006, *Class. Quantum Grav.*, 23, 4853

de Felice, F., Vecchiato, A., Crosta, M. T., Bucciarelli, B., & Lattanzi, M. G. 2006, *ApJ*, 653, 1552

de Jans, C. 1922, *Mém. Acad. Roy. Belgique Cl. Sci.*, 6, pp. 1–26; 6, pp. 27–41

Klioner, S. A. & Kopeikin, S. M. 1992, *Astron. J.*, 104, 897

Klioner, S. A. & Soffel, M. H. 2000, *Phys. Rev. D*, 62, 024019

Klioner, S. A. 2003, *Astron. J.*, 125, 1580

Klioner, S. A & Soffel, M. H. 2004, in: C. Turon, K. S. O'Flaherty, & M. A. C. Perryman (eds.), *Proceedings of the Symposium "The Three-Dimensional Universe with Gaia"* (ESA SP-576, 2005), p. 305

Kopeikin, S. M. 1997, *J. Math. Phys.*, 38, 2587

Kopeikin, S. M. & Schäfer, G. 1999, *Phys. Rev. D*, 60, 124002

Kopeikin, S. M., Korobkov, P., & Polnarev, A. 2006, *Class. Quantum Gravity*, 23, 4299

Kopeikin, S. M. & Makarov, V. V. 2007, *Phys. Rev. D*, 75, 062002

Le Poncin-Lafitte C., Linet, B., & Teyssandier P. 2004, *Class. Quantum Gravity*, 21, 4463

Le Poncin-Lafitte C., & Teyssandier P. 2008, *Phys. Rev. D*, 77, 044029

Minazzoli, O., & Chauvineau, B. 2009, *Phys. Rev. D*, 79, 084027

Soffel, M. H. 1989, *Relativity in Astrometry, Celestial Mechanics and Geodesy* (Berlin: Springer-Verlag)

Teyssandier, P., & Le Poncin-Lafitte C. 2006, *arXiv:gr-qc/0611078*

Teyssandier, P., & Le Poncin-Lafitte C. 2008, *Class. Quantum Gravity*, 25, 145020

Will, C. M. 1988, *Am. J. Phys.*, 56, 413

Zschocke, S. & Klioner S. A. 2009, *arXiv:astro-ph/09043704*

Relativity in Fundamental Astronomy
Proceedings IAU Symposium No. 261, 2009
S. A. Klioner, P. K. Seidelman & M. H. Soffel, eds.

doi:10.1017/S174392130999024X

Relativistic aspects of rotational motion of celestial bodies

S. A. Klioner, E. Gerlach and M. H. Soffel

Lohrmann Observatory, Dresden Technical University, 01062 Dresden, Germany

Abstract. Relativistic modelling of rotational motion of extended bodies represents one of the most complicated problems of Applied Relativity. The relativistic reference systems of IAU (2000) give a suitable theoretical framework for such a modelling. Recent developments in the post-Newtonian theory of Earth rotation in the limit of rigidly rotating multipoles are reported below. All components of the theory are summarized and the results are demonstrated. The experience with the relativistic Earth rotation theory can be directly applied to model the rotational motion of other celestial bodies. The high-precision theories of rotation of the Moon, Mars and Mercury can be expected to be of interest in the near future.

---

1. Earth rotation and relativity

Earth rotation is the only astronomical phenomenon which is observed with very high accuracy, but is traditionally modelled in a Newtonian way. Although a number of attempts to estimate and calculate the relativistic effects in Earth rotation have been undertaken (e.g., Bizouard *et al.* (1992); Brumberg & Simon (2007) and references therein) no consistent theory has appeared until now. As a result the calculations of different authors substantially differ from each other. Even the way geodesic precession/nutation is usually taken into account is just a first-order approximation and is not fully consistent with relativity. On the other hand, the relativistic effects in Earth's rotation are relatively large. For example, the geodesic precession ($1.9''$ per century) is about 3×10^{-4} of general precession. The geodesic nutation (up to 200 μas) is 200 times larger than the accuracy goal of modern theories of Earth rotation. One more reason to carefully investigate relativistic effects in Earth rotation is the fact that the geodynamical observations yield important tests of general relativity (e.g., the best estimate of the PPN γ using a large range of angular distances from the Sun comes from geodesic VLBI data), and it is dangerous to risk that these tests are biased because of a relativistically flawed theory of Earth rotation.

Early attempts to model rotational motion of the Earth in a relativistic framework (see, for example, Brumberg 1972) made use of only one relativistic references system to describe both rotational and translational motions. That reference system was usually chosen to be quite similar to the BCRS. This resulted in a mathematically correct, but physically inadequate coordinate picture of rotational motion. For example, from that coordinate picture a prediction of seasonal LOD variations with an amplitude of about 75 microseconds has been put forward.

At the end of the 1980s a better reference system for modelling of Earth rotation has been constructed, that after a number of modifications and improvements has been adopted as GCRS in the IAU 2000 Resolutions. The GCRS implements Einstein's equivalence principle and represents a reference system in which the gravitational influence of external matter (the Moon, the Sun, planets, etc.) is reduced to tidal potentials. Thus, for physical phenomena occurring in the vicinity of the Earth the GCRS represents a reference system, the coordinates of which are, in a sense, as close as possible to measurable quantities. This substantially simplifies the interpretation of the coordinate description of physical phenomena localized in the vicinity of the Earth. One important application of the GCRS is modelling of Earth rotation. The price to pay when using GCRS is that one should deal not only with one relativistic reference system, but with several reference systems, the most important of which are the BCRS and the GCRS. This makes it necessary to clearly and carefully distinguish between parameters and quantities defined in the GCRS and those defined in the BCRS.

2. Relativistic equations of Earth rotation

The model which is used in this investigation was discussed and published by Klioner *et al.* (2001). Let us, however, repeat these equations again not going into physical details. The post-Newtonian equations of motion (omitting numerically negligible terms as explained in Klioner *et al.* 2001) read

$$\frac{d}{dT}\left(C^{ab}\,\omega^b\right)=F^a+L^a(\boldsymbol{C},\boldsymbol{\omega},\boldsymbol{\Omega}_{\rm iner}), \tag{2.1}$$

$$F^a=\sum_{l=1}^{\infty}\frac{1}{l!}\,\varepsilon_{abc}\,M_{bL}\,G_{cL}, \tag{2.2}$$

where $T=$ TCG, $\boldsymbol{C}=C^{ab}$ is the post-Newtonian tensor of inertia, $\boldsymbol{\omega}=\omega^a$ is the angular velocity of the post-Newtonian Tisserand axes (Klioner 1996), M_L are the multipole moments of the Earth's gravitational field defined in the GCRS, and G_L are the multipole moments of the external tidal gravito-electric field in the GCRS. In the simplest situation (a number of mass monopoles) G_L are explicitly given by Eqs. (19)–(23) of Klioner *et al.* (2001).

The additional torque L^a depends on $\boldsymbol{C}$, $\boldsymbol{\omega}$, as well as on the angular velocity $\boldsymbol{\Omega}_{\rm iner}$ describing the relativistic precessions (geodesic, Lense-Thirring and Thomas precessions). The definition of $\boldsymbol{\Omega}_{\rm iner}$ can be found, e.g., in Klioner *et al.* (2001). A detailed discussion of L^a, its structure and consequences will be published elsewhere (Klioner *et al.* 2009a).

The model of rigidly rotating multipoles (Klioner *et al.* 2001) represents a set of formal mathematical assumptions that make the general mathematical structure of Eq. (2.1) similar to that of the Newtonian equations of rotation of a rigid body:

$$C^{ab}=P^{ac}\,P^{bd}\,\overline{C}^{cd},\quad \overline{C}^{cd}=\text{const}, \tag{2.3}$$

$$M_{a_1a_2\ldots a_l}=P^{a_1b_1}\,P^{a_2b_2}\ldots P^{a_lb_l}\,\overline{M}_{b_1b_2\ldots b_l},\quad \overline{M}_{b_1b_2\ldots b_l}=\text{const},\quad l\geqslant 2, \tag{2.4}$$

where the orthogonal matrix $P^{ab}(T)$ is assumed to be related to the angular velocity ω^a used in (2.1) as

$$\omega^a=\frac{1}{2}\,\varepsilon_{abc}\,P^{db}(T)\,\frac{d}{dT}\,P^{dc}(T). \tag{2.5}$$

The meaning of these assumptions is that both the tensor of inertia C^{ab} and the multipole moments of the Earth's gravitational field M_L are "rotating rigidly" and that their rigid rotation is described by the same angular velocity ω^a that appears in the post-Newtonian equations of rotational motion. It means that in a reference system obtained from the GCRS by a time-dependent rotation of spatial axes, both the tensor of inertia and the multipole moments of the Earth's gravitational field are constant.

No acceptable definition of a physically rigid body exists in General Relativity. The model of rigidly rotating multipoles represents a minimal set of assumptions that allows one to develop the post-Newtonian theory of rotation in the same manner as one usually does within a Newtonian theory for rigid bodies. In the model of rigidly rotating multipoles, only those properties of Newtonian rigid bodies are saved which are indeed necessary for the theory of rotation. For example, no assumption on local physical properties ("local rigidity") is made. It has not been proved as a theorem, but it is rather probable that no physical body can satisfy assumptions (2.3)–(2.5). The assumptions of the model of rigidly rotating multipoles will be relaxed in a later stage of the work when non-rigid effects are discussed.

3. Post-Newtonian equations of rotational motions in numerical computations

Looking at the post-Newtonian equations of motion (2.1)–(2.5) one can formulate several problems to be solved before the equations can be used in numerical calculations:

A. How to parametrize the matrix P^{ab}?

B. How to compute M_L from the standard models of the Earth's gravity field?

C. How to compute G_L from a solar system ephemeris?

D. How to compute the torque $\varepsilon_{abc}\, M_{bL}\, G_{cL}$ out of M_L and G_L?

E. How to deal with different time scales (TCG, TCB, TT, TDB) appearing in the equations of motion, solar system ephemerides, used models of Earth gravity, etc.?

F. How to treat the relativistic scaling of various parameters when using TDB and/or TT instead of TCB and TCG?

G. How to find relativistically meaningful numerical values for the initial conditions and various parameters?

These questions are discussed below.

4. Relativistic definitions of the angles

One of the tricky points is the definition of the angles describing the Earth orientation in the relativistic framework. Exactly as in Bretagnon *et al.* (1997, 1998) we first define the rotated BCRS coordinates (x, y, z) by two constant rotations of the BCRS as realized by the JPL's DE403:

$$\begin{pmatrix} x \\ y \\ z \end{pmatrix} = R_x(23^\circ 26' 21.40928'')\, R_z(-0.05294'') \begin{pmatrix} x \\ y \\ z \end{pmatrix}_{\rm DE403} . \tag{4.1}$$

Then the IAU 2000 transformations between BCRS and GCRS are applied to the coordinates (t, x, y, z), t being TCB, to get the corresponding GCRS coordinates (T, X, Y, Z). The spatial coordinates (X, Y, Z) are then rotated by the time-dependent matrix P^{ij} to get the spatial coordinates of the terrestrial reference system (ξ, η, ζ). The matrix P^{ij} is represented as a product of three orthogonal matrices:

$$\begin{pmatrix} \xi \\ \eta \\ \zeta \end{pmatrix} = R_z(\phi)\, R_x(\omega)\, R_z(\psi) \begin{pmatrix} X \\ Y \\ Z \end{pmatrix} . \tag{4.2}$$

The angles ϕ, ψ and ω are used to parametrize the orthogonal matrix P^{ab} and therefore, to define the orientation of the Earth in the GCRS. The meaning of the terrestrial system (ξ, η, ζ) here is the same as in Bretagnon *et al.* (1997); this is the reference system in which we define the harmonic expansion of the gravitational field with the standard values of potential coefficients C_{lm} and S_{lm}.

5. STF model of the torque

The relativistic torque requires computations with symmetric and trace-free cartesian (STF) tensors M_L and G_L. For this project special numerical algorithms for numerical calculations have been developed. The detailed algorithms and their derivation will be published elsewhere (Klioner *et al.* 2009b). Let us give here only the most important formulas. For each l the component $D_a = \varepsilon_{abc}\, M_{bL-1}\, G_{cL-1}$ of the torque in the right-hand side of Eq. (2.1) can be computed as ($A_l = 4\, l\, \pi\, l!/(2l{+}1)!!$, $a^+_{lm} = \sqrt{l(l+1) - m(m+1)}$)

$$D_1 = \frac{1}{A_l}\left(\sum_{m=0}^{l-1} a^+_{lm}\,\left(-\mathcal{M}^R_{lm}\,\mathcal{G}^I_{l,m+1} + \mathcal{M}^I_{l,m+1}\,\mathcal{G}^R_{lm}\right) + \sum_{m=1}^{l-1} a^+_{lm}\,\left(\mathcal{M}^I_{lm}\,\mathcal{G}^R_{l,m+1} - \mathcal{M}^R_{l,m+1}\,\mathcal{G}^I_{lm}\right)\right), \tag{5.1}$$

$$D_2 = \frac{1}{A_l}\left(\sum_{m=0}^{l-1} a^+_{lm}\,\left(-\mathcal{M}^R_{lm}\,\mathcal{G}^R_{l,m+1} + \mathcal{M}^R_{l,m+1}\,\mathcal{G}^R_{lm}\right) + \sum_{m=1}^{l-1} a^+_{lm}\,\left(-\mathcal{M}^I_{lm}\,\mathcal{G}^I_{l,m+1} + \mathcal{M}^I_{l,m+1}\,\mathcal{G}^I_{lm}\right)\right), \tag{5.2}$$

$$D_3 = \frac{2}{A_l}\sum_{m=1}^{l} m\,\left(\mathcal{M}^I_{lm}\,\mathcal{G}^R_{lm} - \mathcal{M}^R_{lm}\,\mathcal{G}^I_{lm}\right). \tag{5.3}$$

The coefficients $\mathcal{G}^R_{lm}$ and $\mathcal{G}^I_{lm}$ characterizing the tidal field can be computed from Eqs. (19)–(23) of Klioner *et al.* (2001) as explicit functions of the parameters of the solar system bodies: their masses, positions, velocities and accelerations. A Fortran code to compute $\mathcal{G}^R_{lm}$ and $\mathcal{G}^I_{lm}$ for $l \leqslant 7$ and $0 \leqslant m \leqslant l$ has been generated automatically with a specially written software package for *Mathematica*. It is possible to develop a sort of recursive algorithm to compute $\mathcal{G}^R_{lm}$ and $\mathcal{G}^I_{lm}$ for any l similar to the corresponding algorithms for, e.g., Legendre polynomials.

The coefficients $\mathcal{M}^R_{lm}$ and $\mathcal{M}^I_{lm}$ characterizing the gravitational field of the Earth can be computed as

$$\mathcal{M}^R_{l0} = \frac{l!}{(2l-1)!!}\left(\frac{4\pi}{2l+1}\right)^{1/2} M_E\, R^l_E\, C_{l0}, \tag{5.4}$$

$$\mathcal{M}^R_{lm} = (-1)^m\,\frac{1}{2}\,\frac{l!}{(2l-1)!!}\left(\frac{4\,\pi}{2l+1}\,\frac{(l+m)!}{(l-m)!}\right)^{1/2} M_E\, R^l_E\, C_{lm},\ \ 1 \leqslant m \leqslant l, \tag{5.5}$$

$$\mathcal{M}^I_{lm} = (-1)^{m+1}\,\frac{1}{2}\,\frac{l!}{(2l-1)!!}\left(\frac{4\,\pi}{2l+1}\,\frac{(l+m)!}{(l-m)!}\right)^{1/2} M_E\, R^l_E\, S_{lm},\ \ 1 \leqslant m \leqslant l, \tag{5.6}$$

where M_E is the mass of the Earth, R_E its radius, C_{lm} and S_{lm} are the usual potential coefficients of the Earth's gravitational field. If only Newtonian terms are considered in the torque, this formulation with STF tensors is fully equivalent to the classical formulation with Legendre polynomials (e.g., Bretagnon *et al.* 1997, 1998). If the relativistic terms are taken in account, the only known way to express the torque is that with STF tensors.

6. Time transformations

An important aspect of relativistic Earth rotation theory is the treatment of different relativistic time scales. The transformation between TDB and TT *at the geocenter* (all the transformations in this Section are meant to be "evaluated at the geocenter") are computed along the lines of Section 3 of Klioner (2008b). Namely,

$$\mathrm{TT} = \mathrm{TDB} + \Delta\mathrm{TDB}(\mathrm{TDB}), \tag{6.1}$$

$$\mathrm{TDB} = \mathrm{TT} - \Delta\mathrm{TT}(\mathrm{TT}), \tag{6.2}$$

$$\mathrm{TCG} = \mathrm{TCB} + \Delta\mathrm{TCB}(\mathrm{TCB}), \tag{6.3}$$

$$\mathrm{TCB} = \mathrm{TCG} - \Delta\mathrm{TCG}(\mathrm{TCG}), \tag{6.4}$$

so that

$$\frac{d\Delta\mathrm{TDB}}{d\mathrm{TDB}} = A_{\mathrm{TDB}} + B_{\mathrm{TDB}}\,\frac{d\Delta\mathrm{TCB}}{d\mathrm{TCB}}, \tag{6.5}$$

$$A_{\mathrm{TDB}} = \frac{L_B - L_G}{1 - L_B}, \tag{6.6}$$

$$B_{\mathrm{TDB}} = \frac{1 - L_G}{1 - L_B} = A_{\mathrm{TDB}} + 1, \tag{6.7}$$

$$\frac{d\Delta\mathrm{TT}}{d\mathrm{TT}} = A_{\mathrm{TT}} + B_{\mathrm{TT}}\,\frac{d\Delta\mathrm{TCG}}{d\mathrm{TCG}}, \tag{6.8}$$

$$A_{\mathrm{TT}} = \frac{L_B - L_G}{1 - L_G}, \tag{6.9}$$

$$B_{\mathrm{TT}} = \frac{1 - L_B}{1 - L_G} = 1 - A_{\mathrm{TT}}, \tag{6.10}$$

$$\frac{d\Delta\mathrm{TCB}}{d\mathrm{TCB}} = F(\mathrm{TCB}) = \frac{1}{c^2}\,\alpha(\mathrm{TCB}) + \frac{1}{c^4}\,\beta(\mathrm{TCB}), \tag{6.11}$$

$$\frac{d\Delta\mathrm{TCG}}{d\mathrm{TCG}} = \frac{F(\mathrm{TCG} - \Delta\mathrm{TCG})}{1 + F(\mathrm{TCG} - \Delta\mathrm{TCG})}, \tag{6.12}$$

where the functions α and β are given by Eqs. (3.3)–(3.4) of Klioner (2008b) and Eq. (6.5) represents a computational improvement of Eq. (3.8) of Klioner (2008b). Clearly, the derivatives $d\Delta\mathrm{TCB}/d\mathrm{TCB}$ and $d\Delta\mathrm{TCG}/d\mathrm{TCG}$ must be expressed as functions of TDB and TT, respectively, when used in (6.5)–(6.8).

The differential equations for ΔTDB and ΔTT are first integrated numerically for the whole range of the solar system ephemeris (any ephemeris with DE-like interface can be used with the code). The initial conditions for ΔTDB and ΔTT are chosen according to the IAU 2006 Resolution defining TDB: for $JD_{\mathrm{TT}} = 2443144.5003725$ one has $JD_{\mathrm{TDB}} = 2443144.5003725 - 6.55 \times 10^{-5}/86400$ and vice versa. The results of the integrations for the pairs ΔTDB and $d\Delta\mathrm{TDB}/d\mathrm{TDB}$, and ΔTT and $d\Delta\mathrm{TT}/dTT$ are stored with a selected step in the corresponding time variable (TDB for ΔTDB and its derivative, and TT for ΔTT and its derivative). A cubic spline on the equidistant grid is then constructed for each of these 4 quantities. The accuracy of the spline representation is automatically estimated using additional data points computed during the numerical integration. These additional data points lie between the grid points used for the spline and are only used to control the accuracy of the spline. The splines, precomputed and validated in this way, are stored in files and read in by the main code upon request. These splines are directly used for time transformation during the numerical integrations of Earth rotation. Although this spline representation requires significantly more stored coefficients than, for example, a representation with Chebyshev polynomials with the same accuracy, the spline representation has been chosen because of its extremely high computational efficiency. More sophisticated representations may be implemented in future versions of the code.

7. Relativistic scaling of parameters

Obviously, there are two classes of quantities entering Eqs. (2.1)–(2.5) that are defined in the BCRS and GCRS and, therefore, naturally parametrized by TCB and TCG, respectively. The relevant quantities defined in the GCRS and parametrized by TCG are: (1) the orthogonal matrix P^{ab} and quantities related to that matrix: angular velocity ω^a and corresponding Euler angles φ, ψ and ω; (2) the tensor of inertia C^{ab}; (3) the multipole moment of Earth's gravitational field M_L. In principle, (a) G_L and (b) $\Omega^a_{\rm iner}$ are also defined in the GCRS and parametrized by TCG, but these quantities are computed using positions $\boldsymbol{x}_A$, velocities $\boldsymbol{v}_A$ and accelerations $\boldsymbol{a}_A$ of solar system bodies. The orbital motion of solar system bodies is modelled in BCRS and parametrized by TCB. The definition of G_L is conceived in such a way that positions, velocities and accelerations of solar system bodies in the BCRS should be taken at the moment of TCB corresponding to the required moment of TCG with spatial location taken at the geocenter. Let us recall that the transformation between TCB and TCG is a 4-dimensional one and requires the spatial location of an event to be known.

7.1. *Change of the independent variable of the equations*

It is important to realize that the post-Newtonian equations of motion are only valid if non-scaled time scales TCG and TCB are used. If TT and/or TDB are needed, the equations should be changed correspondingly. In order to use TT instead of TCG, simple rescaling of the first and second derivatives of the angles entering the equations of rotational motion should be applied :

$$\frac{d\theta}{d\mathrm{TCG}} = (1 - L_G)\,\frac{d\theta}{d\mathrm{TT}}, \tag{7.1}$$

$$\frac{d^2\theta}{d\mathrm{TCG}^2} = (1 - L_G)^2\,\frac{d^2\theta}{d\mathrm{TT}^2}, \tag{7.2}$$

where θ is any of the angles φ, ψ and ω used in the equations of motion to parametrize the orientation of the Earth. If TDB is used as the independent variable, the corresponding formulas are more complicated:

$$\frac{d\theta}{d\mathrm{TCG}} = (1 - L_G)\left(\left.\frac{d\mathrm{TT}}{d\mathrm{TDB}}\right|_{\boldsymbol{x}_E}\right)^{-1}\frac{d\theta}{d\mathrm{TDB}}, \tag{7.3}$$

$$\begin{aligned}\frac{d^2\theta}{d\mathrm{TCG}^2} = {} & (1 - L_G)^2\left(\left.\frac{d\mathrm{TT}}{d\mathrm{TDB}}\right|_{\boldsymbol{x}_E}\right)^{-2}\frac{d^2\theta}{d\mathrm{TDB}^2} \\ & -(1 - L_G)^2\left(\left.\frac{d\mathrm{TT}}{d\mathrm{TDB}}\right|_{\boldsymbol{x}_E}\right)^{-3}\left.\frac{d^2\mathrm{TT}}{d\mathrm{TDB}^2}\right|_{\boldsymbol{x}_E}\frac{d\theta}{d\mathrm{TDB}},\end{aligned} \tag{7.4}$$

where the derivatives of TT with respect to TDB should be evaluated at the geocenter (i.e., for $\boldsymbol{x} = \boldsymbol{x}_E$). These relations must be substituted into the equations of rotational motion to replace the derivatives of the angles φ, ψ and ω with respect to TCG as appear, e.g., in Eqs. (7)–(9) of Bretagnon *et al.* (1998). It is clear that the parametrization with TDB makes the equations more complicated.

7.2. *Origin of the numerical parameters*

The values of the parameters naturally entering the equations of rotational motion must be interpreted as unscaled (TCB-compatible or TCG-compatible) values. If scaled (TT-compatible or TDB-compatible) values are used, the scaling must be explicitly taken into account. The relativistic scaling of parameters read (see, e.g., Klioner 2008a):

$$GM_A^{\rm TT} = (1 - L_G)\, GM_A^{\rm TCG}, \; GM_A^{\rm TCG} = GM_A^{\rm TCB}, \qquad GM_A^{\rm TDB} = (1 - L_B)\, GM_A^{\rm TCB}, \tag{7.5}$$

$$X^{\rm TT} = (1 - L_G)\, X^{\rm TCG}, \; x^{\rm TDB} = (1 - L_B)\, x^{\rm TCB}, \tag{7.6}$$

$$V^{\rm TT} = V^{\rm TCG}, \; v^{\rm TDB} = v^{\rm TCB}, \tag{7.7}$$

$$A^{\rm TT} = (1 - L_G)^{-1}\, A^{\rm TCG}, \; a^{\rm TDB} = (1 - L_B)^{-1}\, a^{\rm TCB}, \tag{7.8}$$

where GM_A is the mass parameter of a body, x, v, and a are parameters representing spatial coordinates (distances), velocities and accelerations in the BCRS, respectively, while X, V, and A are similar quantities in the GCRS.

Now, considering the source of the numerical values of the parameters used in the equations of Earth rotation, we can see the following.

a. The position $\boldsymbol{x}_A$, velocities $\boldsymbol{v}_A$, accelerations $\boldsymbol{a}_A$ and mass parameters GM_A of the massive solar system bodies are taken from standard JPL ephemerides and are TDB-compatible.

b. The radius of the Earth comes together with the potential coefficients C_{lm} and S_{lm} from a model of the Earth's gravity field (e.g., GEMT3 was used in SMART). These values come from SLR and dedicated techniques like GRACE. GCRS and TT-compatible quantities are used to process these data. Therefore, the value of the radius of the Earth is TT-compatible. Obviously, C_{lm} and S_{lm} have the same values when used with any time scale. The mass parameter GM_E of the Earth, coming with the Earth gravity models, is also TT-compatible.

c. From the definitions of $\mathcal{M}^R_{lm}$ and $\mathcal{M}^I_{lm}$ given above and formulas for G_L given by Eqs. (19)–(23) of Klioner *et al.* (2001), it is easy to see that the TCG-compatible torque given by Eq. (2.2) can be computed using TDB-compatible values of mass parameters $GM_A^{\rm TDB}$, positions $\boldsymbol{x}_A^{\rm TDB}$, velocities $\boldsymbol{v}_A^{\rm TDB}$ and accelerations $\boldsymbol{a}_A^{\rm TDB}$ of all external bodies, TDB-compatible value of the mass parameter of the Earth $GM_E^{\rm TDB}$, and the value of the Earth's radius formally rescaled from TT to TDB as $R_E^{\rm TDB} = (1 - L_B)\,(1 - L_G)^{-1}\, R_E^{\rm TT}$. Denoting the resulting torque by $F^a_{\rm TDB}$, it can be seen that the TCG-compatible value is $F^a_{\rm TCG} = (1 - L_B)^{-1}\, F^a_{\rm TDB}$.

d. The values of the Earth's moments of inertia $\mathcal{A}_i$, $i = 1, 2, 3$ can be represented as $G\mathcal{A}_i = GM_E R_E^2 k_i$, where k_i is a factor characterizing the distribution of matter inside the Earth. Clearly, the factors k_i do not depend on the scaling. Therefore, the moments of inertia can be scaled as

$$\mathcal{A}_i^{\rm TT} = (1 - L_G)^3\, \mathcal{A}_i^{\rm TCG}. \tag{7.9}$$

The last question is how to interpret the values of the moments of inertia $\mathcal{A}_i = (A, B, C)$ and the initial conditions for the angles φ, ψ and ω and their derivatives given in Bretagnon *et al.* (1998). Obviously, the initial angles at J2000 are independent of the scaling. For the other parameters in question it is not possible to clearly claim if the given values are TDB-compatible or TT-compatible. Arguments in favor of both interpretations can be given. A rigorous solution here is only possible when all calculations leading to these values are repeated in the framework of General Relativity. In this paper we prefer to interpret the SMART values of $\mathcal{A}_i$, $\dot\varphi$, $\dot\psi$ and $\dot\omega$ as being TT-compatible. Therefore, if TDB is used as independent variable, the values of the derivatives should be changed accordingly. For any of the angles one has

$$\frac{d\theta}{d\mathrm{TDB}} = \left(\left. \frac{d\mathrm{TT}}{d\mathrm{TDB}} \right|_{\boldsymbol{x}_E} \right) \frac{d\theta}{d\mathrm{TT}} \,. \tag{7.10}$$

8. Geodesic precession and nutation

In the framework of our model geodesic precession and nutation are taken into account in a natural way by including the additional torque that depends on $\Omega^a_{\rm iner}$ in the equations of rotational motion:

$$L^a = \varepsilon_{abc}\, C^{bd}\, \omega^d\, \Omega^c_{\rm iner} - \frac{d}{dT}\left(C^{ab}\, \Omega^b_{\rm iner}\right). \tag{8.1}$$

The first term of the additional torque reflects the fact that the GCRS of the IAU is defined to be kinematically non-rotating (see Soffel *et al.* 2003). The second term has been usually hidden by the corresponding re-definition of the post-Newtonian spin (Damour, Soffel & Xu 1993; Klioner & Soffel 2000). It can be demonstrated that this second term must be explicitly taken into account to maintain the consistency between dynamically and kinematically non-rotating solutions. Further details will be published elsewhere (Klioner *et al.* 2009a). Using the additional torque L^a in Eq. (2.1) is a rigorous way to take geodesic precession/nutation into account.

The standard way to account for geodesic precession/nutation that was used up to now by a number of authors can be described as follows: (1) solve the purely Newtonian equations of rotational motion and consider this solution as a relativistic one in a dynamically non-rotating version of the GCRS and (2) add the precomputed geodesic precession/nutation to it. The second step is fully correct since the geodesic precession/nutation is by definition the rotation between the kinematically and dynamically non-rotating versions of the GCRS and it can be precomputed, because it is fully independent of the Earth rotation. The inconsistency of the first step comes from the fact that in the computation of the Newtonian torque the coordinates of the solar system bodies are taken from an ephemeris constructed in the BCRS. However, the dynamically non-rotating version of the GCRS *rotates* relative to the BCRS with angular velocity $\Omega^a_{\rm iner}(T)$. This means that the BCRS coordinates of solar system bodies should be first rotated into "dynamically non-rotating coordinates", and only after that rotation those coordinates can be used to compute the Newtonian torque. For this reason this procedure does not lead to a correct solution in the kinematically non-rotating GCRS (see Fig. 1). We will call such solutions in this paper "SMART-like kinematical solutions".

On the other hand, there are two ways to obtain a correct kinematically non-rotating solution: (1) use the torque given by Eq. (8.1) in the equations of motion, (2) compute the geodesic precession/nutation matrix, apply the geodesic precession/nutation to the solar system ephemeris, integrate (2.1) without L^a with the obtained rotated ephemeris (the correct solution in a dynamically non-rotating version of the GCRS is obtained in this step), apply the geodesic precession/nutation matrix to the solution. We have implemented both ways in our code and checked explicitly that they give the same solution (to within about 0.001 μas over 150 years). It is interesting to note that the rotational matrix of geodesic precession/nutation (that is, the matrix defining a rotation with the angular velocity $\Omega^a_{\rm iner}$) cannot be parametrized by normal Euler angles. We have used, therefore, the quaternion representation for that matrix.

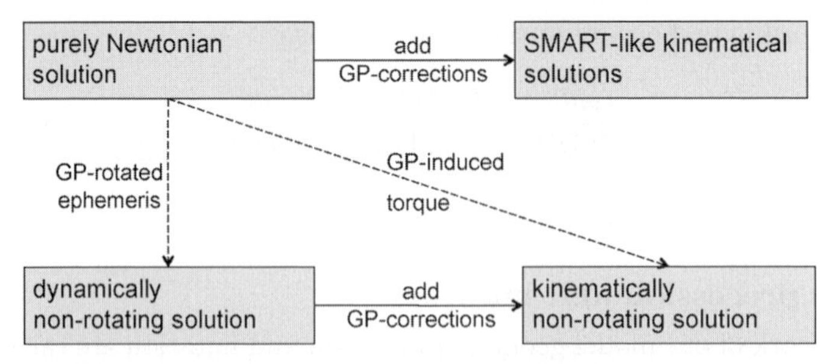

Figure 1. Scheme of the two ways to obtain a kinematically non-rotating solution from a purely Newtonian one, and an illustration of the relation between the correct kinematically non-rotating solution and SMART-like kinematical solutions. "GP" stands for geodesic precession/nutation. Each gray block represents a solution. A solid line means: add precomputed geodesic precession/nutation into a solution to get a new one. A dashed line means: recompute a solution with indicated change in the torque model.

9. Overview of the numerical code

A code in Fortran 95 has been written to integrate the post-Newtonian equations of rotational motion numerically. The software is carefully coded to avoid numerical instabilities and excessive round-off errors. Two numerical integrators with dense output – ODEX and Adams-Bashforth-Moulton multistep integrator – can be used for numerical integrations. These two integrators can be used to crosscheck each other. The integrations are automatically performed in two directions – forwards and backwards – that allows one to directly estimate the accuracy of the integration. The code is able to use any type of arithmetic available with a given current hardware and compiler. For a number of operations, which have been identified as precision-critical, one has the possibility to use either the library FMLIB Smith (2001) for arbitrary-precision arithmetic or the package DDFUN that uses two double-precision numbers to implement quadrupole-precision arithmetic (Bailey 2005). Our current baseline is to use ODEX with 80 bit arithmetic. The estimated errors of numerical integrations after 150 years of integration are below 0.001 μas.

Several relativistic features have been incorporated into the code: (1) the full post-Newtonian torque using the STF tensor machinery, (2) rigorous treatment of geodesic precession/nutation as an additional torque in the equations of motion, (3) rigorous treatment of time scales (any of the four time scales – TT, TDB, TCB or TCG – evaluated at the geocenter can be used as the independent variable of the equations of motion (TCG being physically preferable for this role), (4) correct relativistic scaling of constants and parameters. All these "sources of relativistic effects" can be switched on and off independently of each other.

In order to test our code and the STF-tensor formulation of the torque, we have coded also the classical Newtonian torque with Legendre polynomials as described by Bretagnon *et al.* (1997, 1998), and integrated our equations for 150 years with these two torque algorithms. Maximal deviations between these two integrations were 0.0004 μas for ϕ, 0.0001 μas for ψ, and 0.0002 μas for ω. This demonstrates both the equivalence of the two formulations and the correctness of our code.

We have also repeated the Newtonian dynamical solution of SMART97 using the Newtonian torque, the JPL ephemeris DE403, and the same initial values as in Bretagnon *et al.* (1998). Jean-Louis Simon (2007) has provided us with the unpublished full version of SMART97 (involving about 70000 Poisson terms for each of the three angles). We have calculated the differences between that full SMART97 series and our numerical

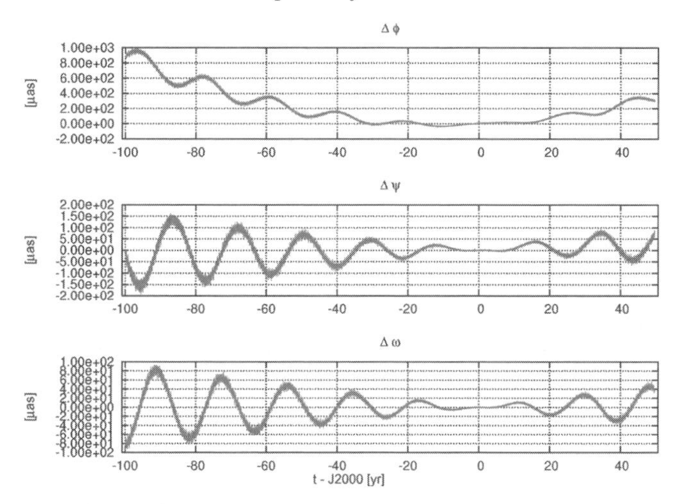

Figure 2. Differences (in μas) between the published kinematical SMART97 solution and the correct kinematically non-rotating solution (with post-Newtonian torques, relativistic scaling and time scaled neglected).

integration over 150 years. Analysis of the results and a comparison to Bretagnon *et al.* (1998) have demonstrated that our integrations reproduce SMART97 within the full accuracy of the latter.

10. Relativistic vs. Newtonian integrations

We have performed a series of numerical calculations comparing purely Newtonian integrations with integrations where relativistic effects are taken into account. The same initial conditions and parameters that we used to reconstruct the SMART97 solution were used for all integrations (see below). The results are illustrated on Figs. 2–5. The difference between the kinematical SMART97 solution and the consistent kinematically-non-rotating solution obtained as described in Section 8 is shown in Fig. 2. Fig. 3 shows the effects of the post-Newtonian torque. The effects of the relativistic scaling and time scales are depicted in Fig. 4. Finally, Fig. 5 demonstrates the differences between a SMART-like kinematical solution and our full post-Newtonian integration. A detailed analysis of these results will be done elsewhere (Klioner *et al.* 2009b).

To complete the consistent post-Newtonian theory of Earth rotation the parameters (first of all, the moments of inertia of the Earth) should be fitted to be consistent with the observed precession rate. This task will be discussed and treated in the near future.

11. Relativistic effects in rotational motion of other bodies

The same numerical code can be applied to model the rotational motion of other bodies. Especially, high-accuracy models of rotational motion of the Moon, Mercury and Mars are of interest because of the planned space missions to Mercury and Mars, and the expected improvements of the accuracy of LLR. Most of the changes in the code are trivial and concern the numerical values of the constants. One important improvement of the code is necessary for the Moon: the figure-figure interaction with the Earth must be taken into account. Using the STF approach to compute the torque, this task is not difficult.

The relativistic effects in the rotation of Moon, Mars and Mercury may be significantly larger than in the rotation of Earth. In Table 1 the amplitudes of geodesic precession

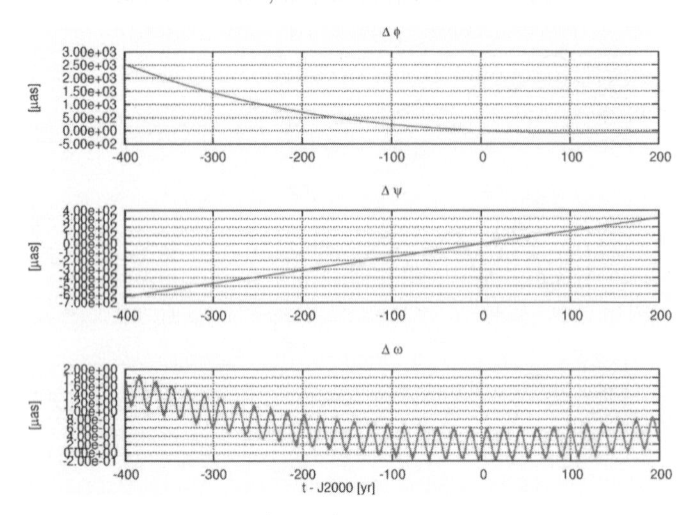

Figure 3. The effect (in μas) of the post-Newtonian torque.

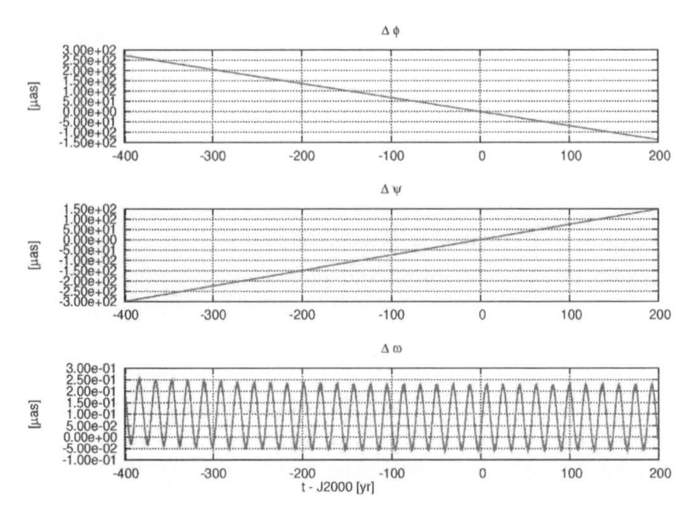

Figure 4. The effect (in μas) of the relativistic scaling and time scales.

body	geodesic precession [$''$ per century]	geodesic nutation [μas]
Earth	1.92	153
Moon	1.95	154
Mercury	21.43	5080
Venus	4.32	85
Mars	0.68	567

Table 1. Magnitude of geodesic precession/nutation for various bodies

and nutation are given for several solar system bodies. One can see the large effects for Mercury and Mars. Besides an early investigation of Bois & Vokroulicky (1995) suggests that the effects of the relativistic torque for the Moon may attain 1 mas. Our approach allows one to investigate the rotational motion of the Moon, Mars and Mercury in a rigorous relativistic framework.

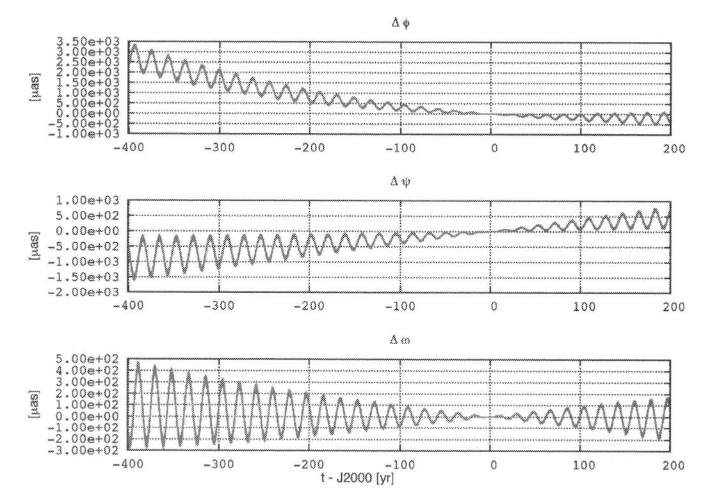

Figure 5. Difference between a purely Newtonian integration rotated for geodesic precession/nutation in a SMART-like way (see Section 8) and our solution that includes all relativistic effects discussed here.

References

Bailey, D. H., 2005, Computing in Science and Engineering, 7(3), 54; see also `http://crd.lbl.gov/~dhbailey/mpdist/`

Bizouard, C., Schastok, J., Soffel, M. H., & Souchay, J., (1992) In: *Journées 1992*, N. Capitaine (ed.), Observatoire de Paris, 76

Bois, E. & Vokrouhlicky, D., 1995, *A&A*, 300, 559

Bretagnon, P., Francou, G., Rocher, P., & Simon, J. L., 1997, *A&A*, 319, 305

Bretagnon, P., Francou, G., Rocher, P., & Simon, J. L., 1998, *A&A*, 329, 329

Brumberg, V.A., 1972, Relativistic Celestial Mechanics, Nauka: Moscow, in Russian

Brumberg, V. A., Simon, J.-L., 2007, Notes scientifique et techniques de l'insitut de méchanique céleste, S088

Damour, T., Soffel, M., & Xu, Ch., 1993, Phys. Rev. D 47, 3124

Klioner, S. A., 1996, In: "Dynamics, Ephemerides, and Astrometry of the Solar System", Ferraz-Mello, S. and Morando, B., Arlot, J.-E. (eds.), Springer, New York, 309

Klioner, S. A., 2008a, *A&A*, 478, 951

Klioner, S. A., 2008b, In: A Giant Step: from Milli- to Micro-arcsecond Astrometry, W. Jin, I. Platais, M. Perryman (eds.), Cambridge University Press, Cambridge, 356

Klioner, S. A. & Soffel, M., 2000, Phys. Rev. D 62, 024019

Klioner, S. A., Soffel, M., Xu. C., & Wu, X., 2001, In: Influence of geophysics, time and space reference frames on Earth rotation studies, N. Capitaine (ed.), Paris Observatory, Paris, 232

Klioner, S. A., Gerlach, E., & Soffel, M., 2009a, "Rigorous treatment of geodesic precession in the theory of Earth rotation", in preparation

Klioner, S. A., Gerlach, E., & Soffel, M., 2009b, "Post-Newtonian theory of Earth rotation", in preparation

Simon, J.-L., 2007, private communication

Smith, D., 2001, "FMLIB", `http://myweb.lmu.edu/dmsmith/FMLIB.html`

Soffel, M., Klioner, S. A., Petit, G. *et al.*, 2003, *Astron.J.*, 126, 2687

Relativity in Fundamental Astronomy
Proceedings IAU Symposium No. 261, 2009
S. A. Klioner, P. K. Seidelman & M. H. Soffel, eds.

doi:10.1017/S1743921309990251

A relativistic orbit model for the LISA mission to be used in LISA TDI simulators

Sophie Pireaux[1] **and Bertrand Chauvineau**[2]

[1]Department 1, Royal Observatory of Belgium,
3 avenue Circulaire, 1180 Brussels, Belgium
email: sophie.pireaux@oma.be

[2]ARTEMIS Department, Observatoire de la Côte d'Azur,
Avenue de Copernic, Grasse, France
email: bertrand.chauvineau@oca.eu

Abstract. The LISA mission is an interferometer, formed by three spacecraft, that aims at the detection of gravitational waves in the $[10^{-4}, 10^{-1}]$ Hz frequency band. Present LISA TDI simulators, aimed at validating the novel Time Delay Interferometry method, use a classical Keplerian orbit model at first order in eccentricity in the gravitational field of a spherical non-rotating Sun, without planets. We propose to use the same model but described in the framework of relativistic gravity, and we focus here on quantifying the differences between classical and relativistic orbits for the LISA spacecraft, under the same assumptions.

Keywords. gravitation, relativity, methods: analytical, numericical

1. Introduction

The Laser Interferometer Space Antenna (LISA) is a joint ESA-NASA mission to be launched in 2018 (at the earliest) [LISA Pre-Phase A Report (1998)]. From the point of view of France, it involves the CNES and the LISAFrance group [LISA-France (2009)]. LISA consists in 3 spacecraft, each following a free-falling test-mass, and of double laser-links connecting the free-falling test-masses.

The aim of LISA is the detection of gravitational waves in a frequency band complementary to that of Earth based detectors. Indeed, a gravitational wave is a propagating spacetime deformation. As it passes by, a free-falling mass moves like a cork on an oscillating water surface. So, the interferometric armlength between a pair of free-falling test-masses varies. To detect gravitational waves implies to monitor very precisely the interferometric armlength via phase shifts (or fractional frequency fluctuations): a $\Delta L/L$ of the order of 10^{-23} with $L = 5 \cdot 10^9$ m, the nominal interdistance between LISA spacecraft.

However, a huge challenge faces LISA. Laser frequency and optical bench noises are well above ($\sim 10^{-13}$ in fractional frequency units), by orders of magnitude, the gravitational wave threshold ($\sim 10^{-21}$ in fractional frequency units). To solve this problem, a new metrology technique, the so-called Time-Delay Interferometry (TDI) was developed [Shaddock *et al.* (2003)]. It is based on the precise knowledge of photon time transfer t_{ij} between a pair of LISA spacecraft i and j. That is the delay taken by a photon to travel the interdistance L_{ij}. At lowest order, that is the L_{ij} distance divided by the speed of light. Moreover, the LISA detector is complex. Its sensitivity depends to a large extent on the different noise contributions and on the efficiency of the TDI novel technique. Therefore, and because a laboratory replica of the system is not totally achievable, the performance of LISA (and of TDI in particular) can only be studied with computer simulations of the different processes involved. Such is the aim of the LISACode software [Petiteau *et al.* (2008)] developed by the LISAFrance group, or of other simulators in the USA, namely Synthetic LISA [Vallisneri (2005)] and LISA Simulator [Cornish *et al.* (2004)]. Among the processes to be implemented in a LISA simulator, the orbit model of the spacecraft, providing positions, velocities and interdistances of spacecraft needed for TDI, is the subject of the present paper.

2. The three LISA orbit models used

The orbits of LISA spacecraft have the following characteristics. The three spacecraft should have a drag-free motion at an average interdistance of L, forming a triangle that rotates around its center of mass and follows the Earth in its orbit around the Sun at a distance $a = 1$ Astronomical Unit.

From the point of view of orbit models, up to now, simulators for TDI LISA [LISACode: Petiteau *et al.* (2008), Synthetic LISA: Vallisneri (2005) and LISA Simulator: Cornish *et al.* (2004)] being used by the Mock LISA Data Challenge (MLDC) task force make the following additional simplifying assumptions. Each spacecraft follows perfectly a test mass that is itself perfectly shielded from non-gravitational forces and feels no constraints (for simplicity, one test mass per spacecraft is modeled). As the gravitational field is concerned, solely a spherical non-rotating Sun is considered. Departures from the above assumptions on orbits are presently considered as part of the noise budget in TDI: residual laser frequency and optical bench noises, detector shot noise, ultra-stable oscillator noise, scattered-light noise, laser-beam pointing instability, acceleration noise, inertial mass noise and others (as specified in Table 1 of reference [Petiteau *et al.* (2008)]).

In the present study, we considered three different orbit models described in the next subsections.

2.1. *Classical orbits*

The first model follows what is used presently in LISA TDI simulators (LISACode, Synthetic LISA, LISA Simulator): classical orbits. Usually, in orbit determination, Newton's second law of motion around a central body, with additional 1PN relativistic corrections, are numerically integrated. However, for LISA, up to now, neither relativistic corrections nor planetary perturbations to orbits are considered by the MLDC task force [Petiteau *et al.* (2008), Vallisneri (2005), Cornish *et al.* (2004) and Arnaud *et al.* (2007)]. This means Keplerian orbits for each LISA spacecraft. The Keplerian orbits are furthermore developped up to first order in eccentricity in present MLDCs [Arnaud *et al.* (2007)]. A special angle, $\nu = \frac{\pi}{3} + \frac{5}{8}\frac{L}{2a}$, is selected for the plane of the LISA triangle with the ecliptic. Indeed, it was shown to minimise LISA natural armlength variations and make it easier to detect gravitational waves. The 3 spacecraft orbits share the same very small eccentricity, $e = \sqrt{1 + \frac{4}{\sqrt{3}}\frac{L}{2a}\cos\nu + \frac{4}{3}\left(\frac{L}{2a}\right)^2} - 1 \simeq 0.0096$. They have also the same orbit inclination, $i = arctg\,(\frac{\frac{L}{2a}\sin\nu}{\sqrt{3}/2+\frac{L}{2a}\cos\nu})$, and orbital period, $T_{cl} \sim 1$ year. But each spacecraft orbit $k = 1, 2, 3$ is out of phase with respect to number 1 by $\vartheta_k \equiv -2\,(k-1)\,\frac{\pi}{3}$ [Nayak & Vinet (1999)].

2.2. *Relativistic Motion Integrator (RMI) method*

Since our aim is to estimate relativistic effects in LISA orbits, we used the Relativistic Motion Integrator (RMI) method [Pireaux *et al.* (2006), Pireaux & Chauvineau (2008)]. Instead of integrating Newton plus relativistic corrections, RMI integrates numerically straightaway the relativistic equations of motion, i.e. geodesic equations for a given metric, $g_{\beta\gamma}$, and possibly an additional term for non-gravitational forces. Doing so, both *classical* and *relativistic* gravitational effects (up to the corresponding order of the chosen metric) are taken into account natively.

When applying this RMI method to LISA, following LISA's simplifying assumptions, there is no non-gravitational force term and thus we integrate

$$\frac{d^2x^l}{dt^2} = \left[-\Gamma^l_{\beta\gamma}(x^\mu) + \frac{1}{c}\Gamma^0_{\beta\gamma}(x^\mu)\cdot\frac{dx^l}{dt}\right]\cdot\frac{dx^\beta}{dt}\cdot\frac{dx^\gamma}{dt} \tag{2.1}$$

where $l = 1, 2, 3$, $x^{\alpha=0,1,2,3} \equiv (ct, x, y, z)$, $\Gamma^l_{\beta\gamma}$ are Christoffel symbols associated with the metric and c is the speed of light in vacuum. We selected the Barycentric Coordinate Reference System metric recommended by the IAU 2000 resolutions; and in order to compare the RMI relativistic orbits for LISA with the standard classical orbits used for the MLDCs, we used it without planets, for a spherical non-rotating Sun (in which case $\Gamma^l_{\beta\gamma}$ does not explicitly depend on t).

2.3. *Relativistic analytical development*

Our third model for LISA orbits is an analytical development up to first order in the eccentricity e of the orbit and up to first order in GM/c^2, where G is Newton's constant and M, the solar mass [Pireaux & Chauvineau (2008)].

Why a development at first order in eccentricity? Because LISA's eccentricity is small and because TDI and classical orbit models for LISA used by the MLDC task force have been developed using first order in eccentricity approximations [Arnaud *et al.* (2007): the pseudo-LISA set of conventions].

Hence, the geodesic equations are developed to obtain the purely relativistic radial, $\Delta r \equiv r_{rel} - r_{cl}$ ($_{rel}$ and $_{cl}$ for relativistic and classical orbit models, respectively), and along track, $\Delta l \equiv a \cdot \Delta\theta \equiv a \cdot (\theta_{rel} - \theta_{cl})$, effects at first Post-Newtonian (PN) order as the sum of a zeroth order, $^{[0]}$, and first order, $^{[1]}$, in eccentricity for each satellite $k = 1$ to 3:

$$\begin{aligned}
\overset{[0]}{\Delta\theta_k} &= -6\frac{GM}{a\,c^2}\left\{\begin{array}{l} +n_{cl}t - \cos(n_{cl}t_{kp})\sin(n_{cl}(t-t_{kp})) \\ -\sin(n_{cl}t_{kp})\cos(n_{cl}(t-t_{kp})) \end{array}\right\} \\
\overset{[1]}{\Delta\theta_k} &= +e\frac{GM}{a\,c^2}\left\{\begin{array}{l} +2\sin(n_{cl}t_{kp}) - 21\cos(n_{cl}t_{kp})\,n_{cl}t \\ -18\;n_{cl}t\cos(n_{cl}(t-t_{kp})) \\ +22\cos(n_{cl}t_{kp})\sin(n_{cl}t_{kp})\cos(n_{cl}(t-t_{kp})) \\ +\left\{2+22\cos^2(n_{cl}t_{kp})\right\}\sin(n_{cl}(t-t_{kp})) \\ +15\sin(n_{cl}t_{kp})\cos^2(n_{cl}(t-t_{kp})) \\ +15\cos(n_{cl}t_{kp})\sin(n_{cl}(t-t_{kp}))\cos(n_{cl}(t-t_{kp})) \end{array}\right\} \\
\overset{[0]}{\Delta r_k} &= +3\frac{GM}{c^2}\left\{\begin{array}{l} +1 - \cos(n_{cl}t_{kp})\cos(n_{cl}(t-t_{kp})) \\ +\sin(n_{cl}t_{kp})\sin(n_{cl}(t-t_{kp})) \end{array}\right\} \\
\overset{[1]}{\Delta r_k} &= +e\frac{GM}{c^2}\left\{\begin{array}{l} +20\cos(n_{cl}t_{kp}) - 9\;n_{cl}t\sin(n_{cl}(t-t_{kp})) \\ -\left\{3+11\cos^2(n_{cl}t_{kp})\right\}\cos(n_{cl}(t-t_{kp})) \\ +11\cos(n_{cl}t_{kp})\sin(n_{cl}t_{kp})\sin(n_{cl}(t-t_{kp})) \\ -6\cos(n_{cl}t_{kp})\cos^2(n_{cl}(t-t_{kp})) \\ +6\sin(n_{cl}t_{kp})\sin(n_{cl}(t-t_{kp}))\cos(n_{cl}(t-t_{kp})) \end{array}\right\}
\end{aligned} \tag{2.2}$$

We see that these expressions, in polar coordinates ($x = r\cos\theta$, $y = r\sin\theta$), as functions of the derivative of the classical angle, $n_{cl} \equiv d\theta_{cl}/dt$, and of the time of passage of spacecraft k at perihelion, t_{kp}, already contain quite a few terms.

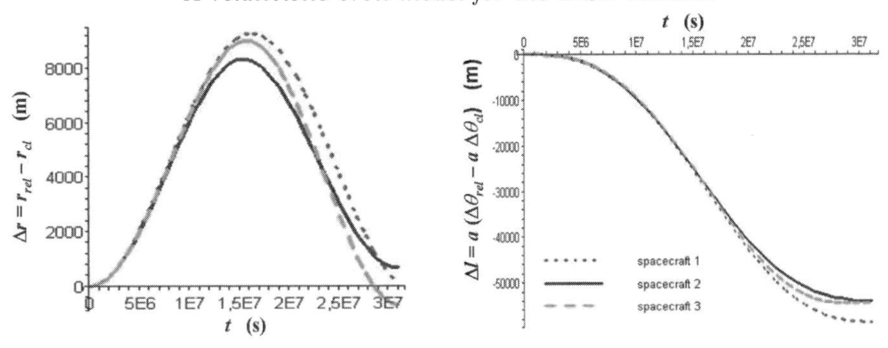

Figure 1. Purely general relativistic effect in [a, left]: radial distance (Δr); [b, right]: along track distance (Δl).

3. Orbit model comparison

We now compare the different orbit models used. First, we check that the spacecraft ephemerides produced by RMI agree with the analytical development at first order in GM/c^2 and in eccentricity, up to $e^2 \cdot GM/c^2$ terms.

Second, we estimate the purely relativistic effects (Δr and Δl) in LISA orbits by computing the differences between relativistic spacecraft ephemerides obtained with RMI and classical ones for the same initial coordinate conditions. Figure 1a shows Δr for the three LISA spacecraft for one classical period, as a function of the barycentric coordinate time. We see that this radial effect can reach about 9 km.
Figure 1b shows Δl for one classical period: it can reach about 60 km.

These relativistic effects impact on the satellite interdistance L_{ij} which is crucial in a TDI analysis. We first recall that even classically, the LISA triangle breathes around the nominal interferometric armlength of 5 million km. The armlength flexing amplitude is about 48 000km. Now, relativistic effects add a small contribution to the flexing: up to 3 km after a period, as is shown in Figure 2. This additional relativistic contribution to the arm flexing is interesting for the TDI method since it impacts on the time taken by photons to travel along those interferometric arms. Indeed, at zeroth order in $1/c^2$, the nominal 5 million km armlength means 16.7 s of travel time; the flexing amplitude of 48000 km means 0.16 s fluctuation. At half order, the Sagnac and parallax effects amount to 960 km, that is $3 \cdot 10^{-3}$ s. While at first order, the Shapiro contribution, thanks to the supressing effect of LISA configuration, is less than 30 m, that is 10^{-7} s, as we computed in [Chauvineau *et al.* (2005)], with the classical orbit model. Now, we have shown that purely relativistic effects in orbits cause a correction up to 3 km in a period, meaning an effect of several milliseconds in the photon flight time at zeroth order in $1/c^2$.

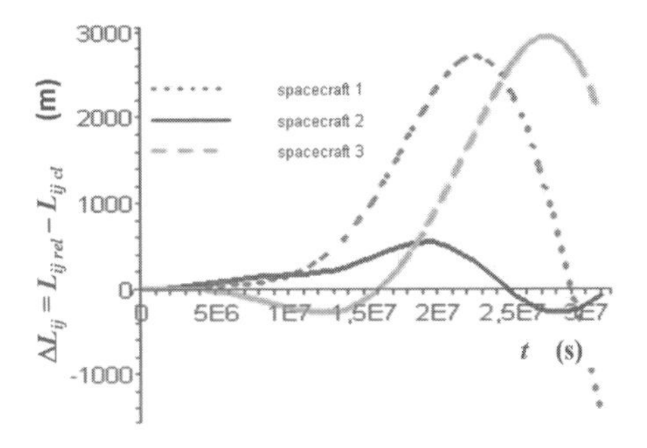

Figure 2. Purely general relativistic effect in armlength.

4. Conclusions

The big relativistic picture in LISA. Time Delay Interferometry (TDI) is a novel technique for spacecraft constellations interconnected by laser links. TDI data-preprocessing [Shaddock *et al.* (2003)] should help mitigate laser frequency and optical bench noises in interferometry, and bring them down to LISA specifications. TDI ranging [Tinto *et al.* (2005)] will allow to measure the photon time transfer between two spacecraft. Using an appropriate (with respect to requested precision) laser link model, one can then compute the spacecraft relative interdistances. The LISA mission is a good example to stress that a coherent general relativistic approach is needed. In particular since LISA is a very complex mission and the TDI method must be validated. Hence, the need for LISA TDI simulators. We stress that coherence is needed, in those, between the photon time transfer model [Chauvineau *et al.* (2005)], the orbit model, coordinates and timescales transformations [Pireaux (2007)]. We studied here the second point.

Relativistic orbit model for LISA TDI simulators. Since, in the simulator named LISACode, as well as in other LISA TDI simulators used by the LISA Mock Data Challenge (LMDC) task force, the orbit model used so far is classical while the laser link is relativistic, we needed to quantify the impact of relativistic effects in LISA orbits on TDI. We have shown that a numerical classical model for LISA orbits in the gravitational field of a non-rotating spherical Sun without planets can be wrong, with respect to the relativistic version of the same model, by as much as about ten kilometers in radial distance during a year and up to about 60 kilometers in along track distance or 3 km in terms of spacecraft interdistance after a year... with consequences on estimated photon flight times between spacecraft used in TDI.

Note that relativistic orbits for the LISA spacecraft, taking into account planetary and the Moon perturbations, have been obtained through the integration of the Einstein-Infeld-Hoffmann (EIH) equations [Folkner *et al.* (1997)] and are used for the sake of orbit selection/analysis/formation control studies.

Strength of the Relativistic Motion Integrator (RMI) method. RMI can be used to compute relativistic orbits for different missions (whether barycentric or planetocentric): only the central body parameters and initial conditions, mission parameters in the corresponding RMI modules change. When updating or changing the metric, only the metric module in RMI needs to be updated. No need to recompute additional analytical developments. Indeed, RMI includes any gravitational contribution at the corresponding order of the metric (whether 1PN or higher). RMI has been validated using a 1PN (Post-Newtonian) development [Hees & Pireaux (2009)] and is a coherent native relativistic approach. It should be preferred to "Newton plus relativistic correction" methods, since analytical developments might become cumbersome as the precision of measurements increases and more than 1PN terms are required.

References

Arnaud, K. A., Babak, S., Baker, J. G., Benacquista, M. J., Cornish, N. J., Cutler, C., Finn, L. S., Larson, S. L., Linttenberg, T., Porter, E. K., Vallisneri, V., Vecchio, A., & Vinet, J-Y. (the Mock LISA Challenge Task Force) 2007, *arXiv:0701170v4(gr-qc)*

Chauvineau, B., Pireaux, S., Regimbau, T., & Vinet, J-Y. 2005, *Phys. Rev. D*, 72, 122003

Cornish, N. J., Rubbo, L. J., & Poujade, O. 2004, *Phys. Rev. D*, 69, 082003

Folkner, W. M., Hechler, F., Sweetser, T. H., Vincent, M. A., & Bender, P. L., 1997, *CQG*, 14, 1405-1410; and private communication from Dr. Folkner on the 30th July 2009.

Hees, A. & Pireaux, S. 2009, *this proceedings*, 144

LISA-France web site at http://www.apc.univ-paris7.fr/LISA-France/analyse.phtml

LISA Pre-Phase A Report 1998, *2nd Ed.*, http://www.srl.caltech.edu/lisa/documents/PrePhaseA.pdf

Nayak, K. R., & Vinet, J-Y. 2005, *CQG*, 22, S437–S443

Petiteau, A., Auger, G., Halloin, H., Jeannin, O., Pireaux, S., Plagnol, E., Regimbau, T., & Vinet, J-Y. 2008, *Phys. Rev. D*, 77023002

Pireaux, S., Barriot, J-P., & Rosenblatt, P. 2006, *Acta Astronautica*, 59, 517–523; gr-qc/06022008

Pireaux, S. 2007, *CQG*, 24, 2271-2281

Pireaux, S. & Chauvineau, B. 2008, *arXiv:0801.3627v1(gr-qc)*

Shaddock, D. A., Tinto, M., Estabrook, F. B., & Armstrong, J. W. 2003, *Phys. Rev. D*, 68, 061303(R)

Tinto, M., Vallisneri, M., & Armstrong, J. W. 2005, *Phys. Rev. D*, 71, 041101

Vallisneri, M. 2005, *Phys. Rev. D*, 71, 022001

Relativity in Fundamental Astronomy
Proceedings IAU Symposium No. 261, 2009
S. A. Klioner, P. K. Seidelman & M. H. Soffel, eds.

doi:10.1017/S1743921309990263

Proper stellar directions and astronomical aberration

Mariateresa Crosta[1] **and Alberto Vecchiato**[1]

[1]Osservatorio Astronomico di Torino-INAF
email: crosta@oato.inaf.it, vecchiato@oato.inaf.it

Abstract. The general relativistic definition of astrometric measurement needs an appropriate use of the concept of reference frame, which should then be linked to the conventions of the IAU Resolutions (Soffel *et al.*, 2003), which fix the celestial coordinate system. A consistent definition of the astrometric observables in the context of General Relativity is also essential to find uniquely the stellar coordinates and proper motion, this being the main physical task of the inverse ray tracing problem. Aim of this work is to set the level of reciprocal consistency of two relativistic models, GREM and RAMOD (Gaia, ESA mission), in order to guarantee a physically correct definition of light direction to a star, an essential item for deducing the star coordinates and proper motion within the same level of measurement accuracy.

Keywords. space astrometry, relativity, reference systems, catalogs, solar system, gravitation

1. Introduction

Modern space mission like Gaia (Turon *et al.*, 2005) requires that any astrometric measurement has to be modelled in a way that light propagation and detection are both conceived in a general relativistic framework. One needs, in fact, to solve the relativistic equations of the null geodesic which describes the trajectory of a photon emitted by a star and detected by an observer with an assigned state of motion. The whole process takes place in a geometrical environment generated by an N-body distribution as could be that of our Solar System.

The astrometric problem consists in the determination, from a prescribed set of observational data (hereafter *observables*) of the astrometric parameters of a star, namely its coordinates, parallax, and proper motion. Essential to the solution of the inverse ray tracing is the identification, as boundary conditions, of the local observer's line-of-sight defined in a suitable reference frame. However, while in classical (non relativistic) astrometry these quantities are well defined, in General Relativity (GR) they must be interpreted consistently with the relativistic framework of the model. Similarly, the parameters describing the attitude and the center-of-mass motion of the satellite need to be defined consistently with the chosen relativistic model.

At present, two conceptual frameworks are foreseen for Gaia. The first model, named GREM (Gaia Relativistic Model) and described in Klioner, 2003, has been formulated according to a Parametrized Post Newtonian (PPN) scheme accurate to 1 micro-arcsecond. GREM is considered as baseline for the Gaia data reduction. The second model, RAMOD (Relativistic Astrometric MODel), is conceived to solve the inverse ray-tracing problem in a general relativistic framework not constrained a priori by any target accuracy. At present, the RAMOD full solution requires the integration of a set of coupled non linear differential equations which allows to trace back the light trajectory to the initial position of the star and which naturally entangles, all together, the contributions by the aberration and all those due to the curvature of the background geometry.

Since both models are used for the Gaia data reduction with the purpose to create a catalog of positions and proper motions based on measurements of *absolute* astrometry, any inconsistency in the relativistic model(s) would invalidate the quality and reliability of the estimates, indeed all the related scientific outputs. This alone is sufficient reason for making a theoretical comparison of the existing approaches for Gaia a necessity.

In this work we present the first step of the theoretical comparison, showing how it is possible to isolate the aberration terms from the global RAMOD construct and recasting them in a GREM-like formula.

2. The RAMOD frames

The set-up of any astrometric model implies, primarily, the identification of the gravitational sources and of the background geometry. Then one needs to label the space-time points with a coordinate system. The above steps fix a reference frame with respect to which one describes the light trajectory, the motion of the stars and that of the observer.

The RAMOD framework is based on the weak-field requirements for the background geometry, which in turn has to be specialized to the particular case one wants to model. For a Gaia-like mission, we can assume the Solar System as the only source of gravity, i.e. a physical system gravitationally bound, in the weak field and small velocities regime. With these assumptions the background geometry is given by the following line element $\mathrm{d}s^2 \equiv g_{\alpha\beta}\mathrm{d}x^\alpha \mathrm{d}x^\beta = (\eta_{\alpha\beta} + h_{\alpha\beta} + \mathcal{O}(h^2))\,\mathrm{d}x^\alpha \mathrm{d}x^\beta$, where $\mathcal{O}(h^2)$ collects all non linear terms in h, the coordinates are $x^0 = ct, x^1 = x, x^2 = y, x^3 = z$, the origin being fixed at the barycenter of the Solar System, and $\eta_{\alpha\beta}$ is the Minkowskian metric. Then, only first order terms in the metric perturbation h are retained. These terms already include all of the possible $(v/c)^n$-order expansions of post-Newtonian approach (v is the velocity of the source and c the velocity of light), but just those up to $(v/c)^3$ are needed to reach the micro-arcsecond accuracy required for Gaia.

The global and the local BCRS. In RAMOD a Barycentric Celestial Reference System (BCRS) is identified requiring that a smooth family of space-like hypersurfaces exists with equation $t(x, y, z) = \text{constant}$ (see de Felice *et al.*, 2004). The function t can be taken as a time coordinate. On each of these $t(x, y, z) = \text{constant}$ hypersurfaces one can choose a set of Cartesian-like coordinates centered at the barycenter of the Solar System (B) and running smoothly as parameters along space-like curves which point to distant cosmic sources. The parameters x, y, z, together with the time coordinate t, provides a basic coordinate representation of the space-time according to the IAU resolutions. Any tensorial quantity will be expressed in terms of coordinate components relative to coordinate bases induced by the BCRS.

Moreover, the RAMOD geometrical construct allows the existence, at any space-time point, of a unitary four-vector $\mathbf{u}$ which is tangent to the world line of a physical observer at rest with respect to the spatial grid of the BCRS defined as:

$$u^\alpha = (-g_{00})^{-1/2}\,\delta_0^\alpha = \left(1 + \frac{h_{00}}{2}\right)\delta_0^\alpha + \mathcal{O}\left(h^2\right). \tag{2.1}$$

The totality of these four-vectors over the space-time forms a vector field which is proportional to a time-like and asymptotically Killing vector field (de Felice *et al.*, 2004). To the order of accuracy required for Gaia, the rest space of $\mathbf{u}$ can be locally identified by a spatial triad of unitary and orthogonal vectors lying on a surface which differs from the $t = \mathit{constant}$ one, but chosen in such a way that their spatial components point to the local coordinate directions as chosen by the BCRS. This frame will be called *local BCRS*, represented by a tetrad whose spatial axes (the triad) coincide with the local coordinate axes, but whose origin is the barycenter of the satellite. This triad is (Bini *et al.*, 2003)

$$\lambda_{\hat{a}}^{\alpha} = h_{0a}\,\delta_0^{\alpha} + \left(1 - \frac{h_{00}}{2}\right)\delta_a^{\alpha} + \mathcal{O}\left(h^2\right) \tag{2.2}$$

for $a = 1, 2, 3$. In RAMOD any physical measurement refers to the local BCRS.

The proper reference frame for the satellite. The tensorial quantity which expresses a proper reference frame of a given observer is a *tetrad adapted to that observer*, namely a set of four unitary mutually orthogonal four-vectors $\{\lambda_{\hat{\alpha}}\}$ one of which, *i.e.* $\lambda_{\hat{0}}$, is the observer's four-velocity while the other $\lambda_{\hat{a}}$s form a spatial triad of space-like four-vectors. The physical measurements made by the observer (satellite) represented by such a tetrad are obtained by projecting the appropriate tensorial quantities on the tetrad axes. As far as RAMOD is concerned, in the case of a Gaia-like mission, an explicit analytic expression for a tetrad adapted to the satellite four-velocity exists and can be found in (Bini *et al.*, 2003). The spatial axes of this tetrad, $E_{\hat{a}}$, are used to model the attitude of the satellite.

3. RAMOD and GREM first comparison

Any comparison between RAMOD and GREM requires that both use the same metric. In the Gaia context the metric perturbation terms can be chosen as $h_{00} = 2w/c^2$, $h_{0i} = w^i/c^3$, where w and w^i are, respectively, the gravitational potential and the vector potential generated by all the sources inside the Solar System according the IAU resolution B1.3 (Soffel *et al.*, 2003).

GREM reproduces in a relativistic framework the classical approach of astrometry, where the quantities which ultimately enter the catalogue are referred to a global inertial reference system, taking into account, one by one and independently from each other, effects such as aberration and parallax. The BCRS is, for this model, the equivalent of the inertial reference system of the classical approach, while the final expression of the star direction in the BCRS is obtained after converting the observed direction into coordinate ones in several steps which divide the effects of the aberration, the gravitational deflection, the parallax, and proper motion (Klioner, 2003). In GREM the observed direction to the source ($\mathbf{s}$) with respect to the local inertial frame ($\mathcal{X}^{\alpha}$) of the observer is

$$s^i = -\frac{\mathrm{d}\mathcal{X}^i}{\mathrm{d}\mathcal{X}^0}. \tag{3.1}$$

The coordinate direction to the light source at the satellite location $\mathbf{x}_s$ is defined by the four-vector $p^{\alpha} = (1, p^i)$, where $p^i = c^{-1}dx^i/dt$, x^i and t being the BCRS coordinates.

The observables in RAMOD, instead, are the three direction cosines which identify the local line-of-sight to the observed object, relative to a spatial triad $E_{\hat{a}}$ associated to a given observer $\mathbf{u}'$; namely:

$$\cos\psi_{\hat{a}} = \frac{P(u')_{\alpha\beta}k^{\alpha}E_{\hat{a}}^{\beta}}{\left(P(u')_{\alpha\beta}k^{\alpha}k^{\beta}\right)^{1/2}} \equiv \mathbf{e}_{\hat{a}}, \tag{3.2}$$

where the final $\mathbf{e}_{\hat{a}}$ has to be intended a shorthand notation for $\cos\psi_{\hat{a}}$, k^{α} is the four-vector tangent to the null geodesic connecting the star to the observer, and $P(u')$ is the operator which projects on the rest-space of $\mathbf{u}'$; all the quantities are obviously computed at the event of the observation. The RAMOD formulation naturally entangles in the previous formula every GR "effect" when (as in Bini *et al.*, 2003) the attitude frame $E_{\hat{a}}$ is that of a Gaia-like observer. Therefore, to retrieve the aberration in RAMOD, one needs to specialize Eq. 3.2 to the case of a tetrad $\{\tilde{\lambda}_{\hat{\alpha}}\}$ adapted to the center of mass of the satellite assumed with no attitude parameters. In this case the observation equation will give a relation between the "aberrated" direction represented by the cosines as measured by the satellite and the "aberration-free" direction. The latter is given by quantity $\bar{l}^\alpha = P^\alpha_\beta(u)k^\beta$ referred to the local BCRS frame $\{\lambda_{\hat{a}}\}$, where $\bar{l}^\alpha$ was introduced in RAMOD (de Felice *et al.*, 2004) as the unitary four-vector which represents the *local line-of-sight* of the photon as seen by $\mathbf{u}$. The tetrad $\{\tilde{\lambda}_{\hat{a}}\}$ differ from the local BCRS's $\{\lambda_{\hat{a}}\}$ for a boost transformation with four-velocity u^α_s of the satellite (Jantzen *et al.*, 1992, Bini *et al.*, 2003) and plays the same role of the CoMRS (Center-of-Mass Reference System) defined for Gaia (Klioner, 2004). Considering the IAU metric, we obtain (v^i stands for the spatial coordinate velocity of the satellite)

$$\begin{aligned}\tilde{\mathbf{e}}_{\hat{a}} \approx & \bar{l}^a + \frac{1}{c}\left[-v^a + \left(\delta_{ij}v^i\bar{l}^j\right)\bar{l}^a\right] + \frac{1}{c^2}\left\{w\bar{l}^a - \frac{1}{2}\left(\delta_{ij}v^i\bar{l}^j\right)v^a + \left[\left(\delta_{ij}v^i\bar{l}^j\right)^2 - \frac{1}{2}v^2\right]\bar{l}^a\right\} \\ & + \frac{1}{c^3}\left\{-2wv^a - \frac{1}{2}\left(\delta_{ij}v^i\bar{l}^j\right)^2 v^a + \bar{l}^a\left[3w\left(\delta_{ij}v^i\bar{l}^j\right)\right.\right. \\ & \left.\left. + \left(\delta_{ij}v^i\bar{l}^j\right)^3 - \frac{1}{2}v^2\left(\delta_{ij}v^i\bar{l}^j\right) + w\left(\delta_{ij}v^i\bar{l}^j\right)\right]\right\} + \mathcal{O}\left(v^4/c^4\right)\end{aligned} \tag{3.3}$$

where $\tilde{\mathbf{e}}_{\hat{a}}$ are the cosines related to the tetrad without the attitude parameters.

At a first glance, the last expression shows differences in terms up to the $(v/c)^2$ order (note in particular the term $w\bar{l}^a$) and of the $(v/c)^3$ order which cannot allow to straightforwardly compare, as expected, the above expression to the GREM vectorial one (see Klioner, 2003), where the aberration is expressed in terms of a vector $\mathbf{n} = p^i/p$, which represents the "aberration-free" coordinate line of sight of the observed star at the position of the satellite momentarily at rest. To compare formula 3.3 with GREM's formula, $\mathbf{n}$ and $\bar{l}^\alpha$ we need to reduce $\bar{l}^\alpha$ to its coordinate Euclidean expression. In RAMOD, as said, $\bar{l}^\alpha$ represents the normalized *local line-of-sight* of the observed star *as seen* by the local barycentric observer $\mathbf{u}$. From the physical point of view $\mathbf{n}$ and $\bar{l}^\alpha$ have the same meaning, as the *observed* "aberration free" direction to the star. From the definition of n^i, we recover $n^i = p^i(1 + h_{00} + h_{0i}p^i) + \mathcal{O}(h^2)$. On the other hand, using the definition of $\bar{l}^\alpha$ in de Felice *et al.*, 2004, and from $u^0 = (-g_{00})^{-1/2}$ and $k^i/k^0 = c^{-1}\mathrm{d}x^i/\mathrm{d}t \equiv p^i$, it can be easily shown that its spatial components can be approximated as

$$\bar{l}^i = n^i\left(1 - \frac{h_{00}}{2}\right) + \mathcal{O}\left(\frac{v^4}{c^4}\right). \tag{3.4}$$

Combining Eq. 3.3 with 3.4 and setting $(\delta_{ij}v^i n^j) \equiv \mathbf{v}\cdot\mathbf{n}$ to ease the notation, we obtained

$$\begin{aligned}\tilde{\mathbf{e}}_{\hat{a}} = & n^a + \frac{1}{c}\left[-v^a + (\mathbf{v}\cdot\mathbf{n})\,n^a\right] + \frac{1}{c^2}\left\{-\frac{1}{2}(\mathbf{v}\cdot\mathbf{n})\,v^a + \left[(\mathbf{v}\cdot\mathbf{n})^2 - \frac{1}{2}v^2\right]n^a\right\} + \\ & \frac{1}{c^3}\left\{-2wv^a - \frac{1}{2}(\mathbf{v}\cdot\mathbf{n})^2 v^a + (\mathbf{v}\cdot\mathbf{n})\,n^a\left[2w + (\mathbf{v}\cdot\mathbf{n})^2 - \frac{1}{2}v^2\right]\right\} + \mathcal{O}\left(\frac{v^4}{c^4}\right).\end{aligned} \tag{3.5}$$

In this way the right-hand side of the aberration expression of RAMOD is rewritten with the GREM quantities at the $(v/c)^3$ order. The final step is to find a relation between $\tilde{\mathbf{e}}_{\hat{a}}$ and s^i. Using the definition of the projection operator and the tetrad property $(\lambda^{\hat{\mu}}_\alpha\lambda_{\hat{\mu}\beta} = g_{\alpha\beta})$, we have:

$$\tilde{\mathbf{e}}_{\hat{a}} \equiv \frac{P(u)_{\alpha\beta}k^\alpha\tilde{\lambda}^\beta_{\hat{a}}}{\left(P(u)_{\alpha\beta}k^\alpha k^\beta\right)^{1/2}} = \frac{k^\alpha\tilde{\lambda}^{\hat{a}}_\alpha}{|g_{\alpha\beta}u^\alpha k^\beta|} = -\frac{k^\alpha\tilde{\lambda}^{\hat{a}}_\alpha}{g_{\alpha\beta}\tilde{\lambda}^\alpha_{\hat{0}}k^\beta} = \frac{k^\alpha\tilde{\lambda}^{\hat{a}}_\alpha}{k^\beta\tilde{\lambda}^{\hat{0}}_\beta} = \frac{\mathrm{d}\tilde{x}^{\hat{a}}}{\mathrm{d}\tilde{x}^{\hat{0}}}. \tag{3.6}$$

The tetrad components of the light ray can be directly associated to CoMRS coordinates, i.e. to a *coordinate-induced tetrad* (as done in Klioner, 2004), if the boosted local BCRS tetrad coordinates $\tilde{x}^{\hat{\alpha}}$ are equivalent to the CoMRS ones $\mathcal{X}^{\alpha}$. This is true if the origins of the two reference systems coincide and only locally, since the tetrad are not in general holonomic. So, in the case of Gaia, from 3.1 it is $\tilde{\mathbf{e}}_{\hat{a}} = -s^a$. Therefore one can finally write relation 3.5 (where $\mathbf{n}\cdot\mathbf{n} = 1$ and $v^2 = \delta_{ij}v^i v^j \equiv \mathbf{v}\cdot\mathbf{v}$)

$$
\begin{aligned}
s^a = & -n^a + \frac{1}{c}\left[\mathbf{n}\times(\mathbf{v}\times\mathbf{n})\right]^a + \frac{1}{c^2}\left\{(\mathbf{v}\cdot\mathbf{n})\left[\mathbf{n}\times(\mathbf{v}\times\mathbf{n})\right]^a + \frac{1}{2}\left[\mathbf{v}\times(\mathbf{n}\times\mathbf{v})\right]^a\right\} \\
& + \frac{1}{c^3}\left\{\left[(\mathbf{v}\cdot\mathbf{n})^2 + 2w\right]\left[\mathbf{n}\times(\mathbf{v}\times\mathbf{n})\right]^a + \frac{1}{2}(\mathbf{v}\cdot\mathbf{n})\left[\mathbf{v}\times(\mathbf{n}\times\mathbf{v})\right]^a\right\} + \mathcal{O}\left(\frac{v^4}{c^4}\right) \qquad (3.7)
\end{aligned}
$$

which is the formula for the aberration in GREM.

4. Conclusions

The direction cosines being physical quantities not depending on the coordinates, are a powerful tool to compare the astrometric relativistic models: their physical meaning allow us to correctly interpret the astrometric parameters in terms of coordinate quantities. This justified the conversion of the physical stellar proper direction of RAMOD into its analogous Euclidean coordinate counterpart, which ultimately leads to the derivation of a GREM-style aberration formula 3.7. Note, also, that the observables of RAMOD can be matched with components of the observed s^i of GREM only if the origins of the boosted local BCRS tetrad in RAMOD and of the CoMRS in GREM coincide.

Moreover, the direction cosines (i.e. the astrometric measurements strictly dependent on the mathematical characterization of the attitude) taken as a function of the local line-of-sight (the *physical one*), at the time of observation, allow to fix the boundary conditions needed to solve the master equations and to determine *uniquely* the star coordinates. The vector $\mathbf{n}$, i.e. the "aberration-free" counterpart of l^{α} in GREM, is instead used to derive the aberration effect (in a coordinate language) and there is no need to connect it with a RAMOD-like boundary value problem.

Acknowledgements

This presentation was supported by the IAU grants.

References

Soffel M., Klioner S. A., Petit G., Wolf P., Kopeikin S. M., Bretagnon P., Brumberg V. A., Capitaine N., Damour T., Fukushima T. *et al.*, 2003, *Astron. J.*, 126, 2687

Eds. Turon C., O'Flaherty K. S., & Perryman M. A. C., 2005, *The Three-Dimensional Universe with Gaia, Publ. Astron. Soc. Pac.*, 120, 38

Bini D., Crosta M., & de Felice F. 2003, *Class. Quantum Grav.*, 20, 4695

de Felice F., crosta M., Vecchiato A., Lattanzi M. G., & Bucciarelli B., 2004, *Astrophys. J.*, 607, 580.

de Felice F., Vecchiato A., Crosta M., Lattanzi M. G., & Bucciarelli B., 2006 *Astrophys. J.*, 653, 1552

de Felice F & Preti G., 2006, *Class. Quantum Grav.*, 23, 5467

Klioner S. A., 2003 *Astron. J.*, 125, 1580

Gravitation, Misner C. W., Thorne, K. S., & Wheeler J. A., 1973 *San Francisco: W.H. Freeman and Co.*

Jantzen R. T., Carini P., Bini D., 1992 *Ann. Phys.*, 215, 1

Klioner S. A., 2004 *Phys. Rev. D*, 69, 124001

Relativity in Fundamental Astronomy
Proceedings IAU Symposium No. 261, 2009
S. A. Klioner, P. K. Seidelman & M. H. Soffel, eds.

doi:10.1017/S1743921309990275

Spectroscopic binary mass determination using relativity

Shay Zucker[1] **and Tal Alexander**[2]

[1]Dept. of Geophysics & Planetary Sciences, Raymond and Beverly Sackler Faculty of Exact Sciences, Tel Aviv University, Tel Aviv 69978, Israel
email: `shayz@post.tau.ac.il`

[2]Faculty of Physics, Weizmann Institute of Science PO Box 26, Rehovot 76100, Israel
email: `Tal.Alexander@weizmann.ac.il`

Abstract. High-precision radial-velocity techniques, which enabled the detection of extra-solar planets, are now sensitive to the lowest-order relativistic effects in the data of spectroscopic binary stars (SBs). We show how these effects can be used to derive the absolute masses of the components of eclipsing single-lined SBs and double-lined SBs from Doppler measurements alone. High-precision stellar spectroscopy can thus substantially increase the number of measured stellar masses, thereby improving the mass-radius and mass-luminosity calibrations.

Keywords. binaries: close, binaries: spectroscopic, celestial mechanics, methods: data analysis, relativity, techniques: radial velocities

---

1. Introduction

During the current IAU Symposium, optical Doppler measurements of binary stars were hardly mentioned. Usually, one doesn't expect to see any relativistic effects in the optical Doppler data, apart from long term (secular) effects, with an accumulating nature, such as the relativistic precession of the periastron. However, the recent two decades have witnessed the emergence of precise radial-velocity (RV) measurements, with long-term precisions of a few meters per second, that are now routinely obtained by several telescopes around the world. The most notable scientific achievement of precise RV measurements has been the detection of planets orbiting solar-type stars (Mayor & Queloz 1995, Marcy & Butler 1996). In this paper we show that these new techniques are already sensitive to the lowest-order (Einstein delay) terms, and that they can be used to add more data about the binary star, which is not attainable in any other way. Kopeikin & Ozernoy (1999) have already detailed the relativistic effects one expects to find in the Doppler measurements of binary stars. Here we focus on the effects that we expect to practically measure in SBs, and identify the additional information they may provide. We have already introduced our main arguments in an ApJ Letter (Zucker & Alexander 2007).

The typical velocities of components of close binary stars can be as high as $150\,\mathrm{km\,s^{-1}}$, $\beta \equiv v/c \sim \mathcal{O}(10^{-4})$. The classical Doppler shift formula predicts a relative wavelength shift $\Delta\lambda/\lambda$ of order β. The next order corrections are $\beta^2 \sim \mathcal{O}(10^{-8})$, which translate to $\mathcal{O}(1\,\mathrm{m\,s^{-1}})$. Terms of order β^3 are beyond foreseen technical capabilities. We thus limit our analysis to $\mathcal{O}(\beta^2)$ effects — the transverse Doppler shift (time dilation) and the gravitational redshift. We focus on effects we can detect during relatively short observing runs, so we ignore the higher-order secular terms.

2. Single-lined spectroscopic binary

The Keplerian RV curve of a single-lined spectroscopic binary (SB1) can be presented as:

$$V_{R1} = K_1 \cos\omega \cos\nu - K_1 \sin\omega \sin\nu + eK_1 \cos\omega + V_{R0} \ , \tag{2.1}$$

where K_1 is the primary star RV semi-amplitude, ω is the argument of periastron, e is the eccentricity, V_{R0} is the center-of-mass RV, and ν is the time-dependent true anomaly. The customary procedure to solve an SB1 is to fit this orbital model to the observed RV data. This fit is achieved through some optimization algorithm that scans the (P, T, e) space (period, periastron time and eccentricity). For each trial set of values for these three parameters, the algorithm produces the corresponding $\nu(t)$ and then solves analytically for $K_{C1}(= K_1 \cos\omega)$, $K_{S1}(= -K_1 \sin\omega)$, and V_{R0}, which appear linearly in the expression for V_{R1}.

We have recently shown (Zucker & Alexander 2007) that including the β^2 effects changes Equation 2.1 by simply modifying the linear elements K_{S1}, K_{C1} and V_{R0}. The modified elements are given by:

$$K'_{S1} = -K_1 \left(1 + \frac{V_{R0}}{c}\right) \sin\omega \tag{2.2a}$$

$$K'_{C1} = K_1 \left[\left(1 + \frac{V_{R0}}{c}\right) \cos\omega + \frac{e}{\sin^2 i} \frac{2K_1 + K_2}{c}\right] \tag{2.2b}$$

$$V'_{R0} = V_{R0} + \frac{1 - e^2}{\sin^2 i} \frac{K_1}{c} \left(\frac{3}{2} K_1 + K_2\right) + \frac{V_0^2}{2c} \ . \tag{2.2c}$$

The modified linear elements are more difficult to interpret now. The three quantities K'_{S1}, K'_{C1}, and V'_{R0} depend on the six elements K_1, K_2, ω, $\sin i$, V_{R0}, and V_0. Thus, Equations (2.2) are under-determined and we cannot completely infer the six elements above, unless some additional independent information is available, or further assumptions are introduced.

Such independent information may be available through precise photometry of eccentric eclipsing binaries. There, the shapes and widths of the eclipses as well as the phase differences between primary and secondary eclipses can be used to estimate ω and $\sin i$. In this case, we may derive K_2 – the RV amplitude of the secondary star. By obtaining K_2 we effectively turn the binary into a double-lined spectroscopic binary (SB2), in which both K_1 and K_2 are measured. Together with the known inclination, we then obtain full knowledge of the component masses.

Curiously, another result of including the relativistic terms is that an eccentric binary should always display an apparent RV signature, even in the extreme case where the inclination is exactly zero and the orbit is observed face on. Nevertheless, such small values of the inclination are extremely rare and this possibility is not realistic.

3. Double-lined spectroscopic binary

In the case of a Keplerian SB2, there are two sets of measured RVs. The two sets of RVs share the same fundamental orbital elements (P, T, e, ω, and V_{R0}), and the only difference between them is their amplitudes K_1 and $-K_2$. The common procedure is to scan the space of the four parameters (P, T, e, ω), and then solve analytically for the three linear elements K_1, K_2, and V_{R0}. However, when we incorporate the relativistic corrections above, we see that we now have two RV curves with two different sets of derived amplitudes, arguments of periastra, and center-of-mass velocities. The two RV curves still share the same period, periastron time and eccentricity.

The four equations corresponding to the equations 2.2*a* and 2.2*b* have four unknowns: K_1, K_2, ω, and $\sin i$. V_{R0} appears in these four equations always divided by the speed of light. Thus, we can safely use its approximate derived value, from either set of measured velocities, since the discrepancy will be only $\mathcal{O}(\beta^3)$. Solution of this set of equations will yield a more accurate estimate of the first three unknowns, but more importantly, it will yield an estimate of $\sin i$. Retaining only leading order terms, we can arrive at the following solution:

$$\sin^2 i = \frac{3e}{\omega_2' - \omega_1'} \frac{K_{S2}' + K_{S1}'}{c} , \tag{3.1}$$

where ω_1' and ω_2' are the *apparent* arguments of periastra of the two components. The fact that the two arguments of periastra differ is a pure relativistic effect. Note that classically, there is no way to estimate $\sin i$ from pure RV data alone. The value of $\sin i$ is usually obtained only from the analysis of SBs that are eclipsing or astrometric binaries.

4. Discussion

We present here an approach to extract more information from RV data of a spectroscopic binary, based on the inclusion of relativistic effects. The practical potential lies in the time-dependent parts of the relativistic terms which are closely linked to the variation of the distance between the binary components. Thus, these terms are especially useful for orbits that are eccentric enough.

The approach is mainly useful in the case of SB2s, but precise photometry or astrometry may add the required information to utilize relativity in SB1s as well. Precise photometry space missions, like MOST (Walker *et al.* 2003), CoRoT (Baglin *et al.* 2003), and Kepler (Basri *et al.* 2005), may be able to provide precise enough measurements of ω and the inclination of eclipsing binaries from the analysis of their light curves.

In any case, it is essential that the data quality be high enough to be sensitive to variations of the required order, namely one meter per second or less. Currently, the best precision is obtained by the ESO HARPS fiber-fed echelle spectrograph, where the RV error can be as low as $1\,\mathrm{m\,s^{-1}}$ in certain cases and maybe even less (Lovis *et al.* 2005). In the future, much better precisions can be hoped for, on instruments designed for the "extremely large telescopes" (e.g., Pasquini *et al.* 2006).

Currently, precisions of a few meters per second are still difficult to obtain, and besides using the best instruments available, there are also several limitations imposed by the star itself. Thus, early-type stars or rapid rotators, where the spectral lines are significantly broadened, do not lend themselves easily to high-precision RV measurements. Stellar oscillations and star spots are also a concern as they can cause apparent RV modulation. In addition, analyzing SB2s with the same level of precision as SB1s has not been easy until recently, when TODCOR (Zucker & Mazeh 1994) was applied successfully to high-precision spectra by several teams (Zucker *et al.* 2004,Udry *et al.* 2004,Konacki 2005).

The effects we have examined are obviously most useful when the orbits are eccentric and the RV amplitudes are large enough. Large RV amplitudes are typical for close binaries, which are expected to have undergone orbital circularization and usually have vanishing eccentricities. However, eccentricity somewhat increases the RV amplitude, and even relatively wide binaries, with high enough eccentricities, can display quite large RV amplitudes, as the next two paragraphs demonstrate.

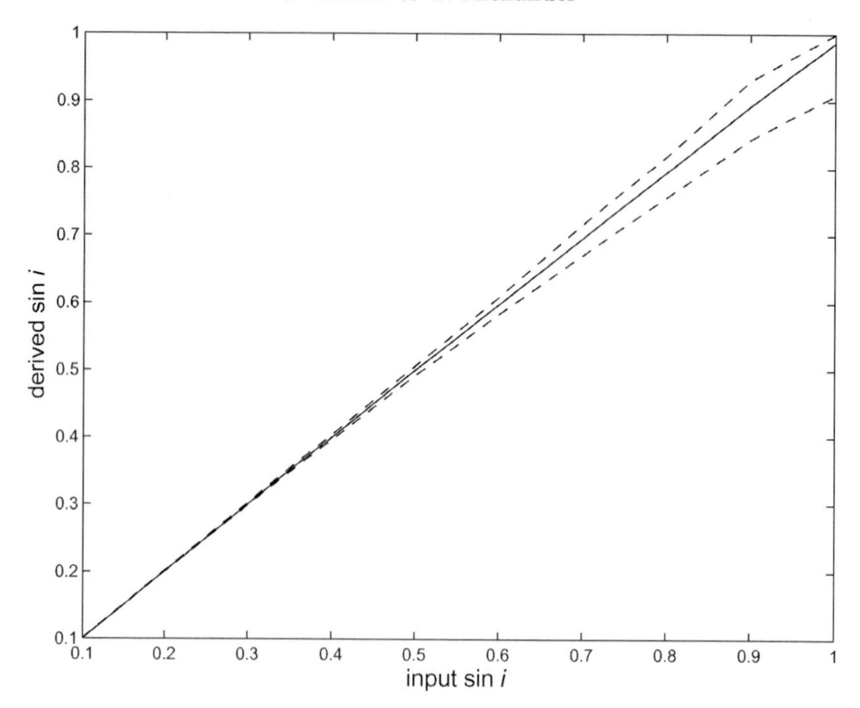

Figure 1. The results of simulations of 12 Boo including relativistic effects, assuming RV precision of $1\,\mathrm{m\,s^{-1}}$. The plot shows percentiles of the derived $\sin i$ against the input simulated value. The dashed lines represent the first and third quartile, and the solid line is the median.

To demonstrate the relevance of the suggested approach in real-life cases we chose to examine the SB2 12 Boo. Recently, Tomkin & Fekel (2006) published a precise solution of 12 Boo based on RVs obtained at the 2.1-m telescope at the McDonald Observatory and at the Coudé feed telescope at Kitt Peak National Observatory, with RV precisions of 0.1–$0.2\,\mathrm{km\,s^{-1}}$. The system has a period of 9.6 days, eccentricity 0.2, and both RV semi-amplitudes are close to $70\,\mathrm{km\,s^{-1}}$. These orbital parameters translate to an expected relativistic amplitude variation of about $10\,\mathrm{m\,s^{-1}}$. Furthermore, with a declination of $+25°$, the star is observable by HARPS. Its brightness (5th magnitude) and spectral type (F9IV) make it fairly reasonable to expect a precision of $1\,\mathrm{m\,s^{-1}}$ with HARPS. We used the available 24 RVs from Tomkin & Fekel (2006) and augmented them with only 3 simulated HARPS measurements (assuming errors of $1\,\mathrm{m\,s^{-1}}$) , including the relativistic effects. For each assumed value of $\sin i$ we produced 1000 sets of simulated measurements and solved for the orbital elements, using Equation 3.1 to estimate $\sin i$. Figure 1 shows the median of the derived values in solid line, and the 25% and 75% percentiles in dashed lines.

The figure demonstrates that with reasonable efforts, $\sin i$ can be measured satisfactorily. In the worst case where $\sin i = 1$, the standard deviation of the derived inclination is 0.14 and a few more precise measurements can reduce this value significantly. An additional advantage of this test case is that the inclination of 12 Boo has already been measured by interferometry and is known to be 108° (Boden et al. 2005). In real observations, more than three precise measurements may be needed in order to account for differences in zero points between instruments. In fact we have already secured several HARPS exposures of 12 Boo and after some technical difficulties, we are now at the process of analysing it.

Care must be taken to model correctly any other effects of order β^2 that might contaminate the data. One such effect is the light-travel-time effect. This effect can be easily approximated to the relevant order by adding the following term to V_{R1} (and a corresponding one to V_{R2}):

$$\Delta V_{\rm LT} = \frac{K_1^2}{c} \sin^2(\nu + \omega)\,(1 + e\cos\nu) \ . \tag{4.1}$$

Another effect which should be analyzed carefully is the tidal distortion of the stellar components, in particular close to periastron. This distortion may affect the spectral lines, introducing line asymmetry, which can bias the estimated Doppler shift. RV extrasolar planet surveys use the line-bisector analysis (e.g., Queloz *et al.* 2001) to quantify such time-dependent asymmetries. Further development of this technique may be the key to disentangle the tidal distortion and the relativistic effects.

One important application of the proposed method is to calibrate the low-mass end of the mass-luminosity relation, to better understand the stellar-substellar borderline. This mass regime is still poorly constrained, since low mass SB2s are quite rare due to the special photometric, spectroscopic and geometric requirements (Ribas 2006). Large efforts are in progress to obtain accurate stellar masses in this regime, including adaptive optics, interferometry and in the future space interferometry (e.g., Henry *et al.* 2005). We propose a new, relatively accessible tool to accomplish this goal, where the only requirements are spectroscopic. Precise RVs for low-mass SB2s were already measured by Delfosse *et al.* (1999). Using the method presented here, their absolute masses may be derived with a relatively small observational effort. No other method exists yet to derive this information purely from RV measurements.

Acknowledgment

S.Z. wishes to express his gratitude to the organizing committees for their support in the form of a travel grant.

References

Baglin, A. 2003, *Adv. Sp. Res.*, 31, 345
Basri, G., Borucki, W. J., & Koch, D. 2005, *New Astron. Revs*, 49, 478
Boden, A. F., Torres, G., & Hummel, C. A. 2005, *ApJ*, 627, 464
Delfosse, X., Forveille, T., Beuzit, J.-L., Udry, S., Mayor, M., & Perrier, C. 1999, *A&A*, 344, 897
Henry, T. J., *et al.* 2005, *BAAS*, 37, 1356
Konacki, M. 2005, *ApJ*, 626, 431
Kopeikin, S. M. & Ozernoy, L. M. 1999, *ApJ*, 523, 771
Lovis, C., *et al.* 2005, *A&A*, 437, 1121
Marcy, G. W. & Butler, R. P. 1996, *ApJ*, 464, L147
Mayor, M. & Queloz, D. 1995, *Nature*, 378, 355
Pasquini, L., *et al.* 2006, in IAU Symp. 232, The Scientific Requirements for Extremely Large Telescopes, ed. P.A. Whitelock, M. Dennefeld & B. Leibundgut (Cambridge: Cambridge Univ. Press), p. 193
Queloz, D., *et al.* 2001, *A&A*, 379, 279
Ribas, I. 2006, *Ap&SS*, 304, 89
Tomkin, J. & Fekel, F. C. 2006, *AJ*, 131, 2652
Udry, S., Eggenberger, A., Mayor, M., Mazeh, T., & Zucker, S. 2004, *Rev. Mexicana AyA*, 21, 207
Walker, G., *et al.* 2003, *PASP*, 115, 1023
Zucker, S. & Alexander, T. 2007, *ApJ*, 645, L83
Zucker, S. & Mazeh, T. 1994, *ApJ*, 420, 806
Zucker, S., Mazeh, T., Santos, N. C., Udry, S., & Mayor, M. 2004, *A&A*, 426, 695

Relativity in Fundamental Astronomy
Proceedings IAU Symposium No. 261, 2009
S. A. Klioner, P. K. Seidelman & M. H. Soffel, eds.

doi:10.1017/S1743921309990287

Gravitational light deflection, time delay and frequency shift in Einstein-Aether theory

Kai Tang[1,2], Tian-Yi Huang[3], Zheng-Hong Tang[1]

[1]Shanghai Astronomical Observatory,Chinese Academy of Sciences;tangkai@shao.ac.cn;zhtang@shao.ac.cn
[2]Graduate School of Chinese Academy of Sciences
[3]Department of Astronomy, Nanjing University;tyhuang@nju.edu.cn

Abstract. Einstein-Aether gravity theory has been proven successful in passing experiments of different scales. Especially its Eddington parameters β and γ have the same numerical values as those in general relativity. Recently Xie and Huang (2008) have advanced this theory to a second post-Newtonian approximation for an N-body model and obtained an explicit metric when the bodies are point-like masses. This research considers light propagation in the above gravitational field, and explores the light deflection, time delay, frequency shift etc. The results will provide for future experiments in testing gravity theories.

Keywords. Einstein-Aether, light deflection, time delay, frequency shift

1. Introduction

Einstein-Aether theory is a vector-tensor theory. It postulates a dynamical unit time-like vector field besides the metric [1]. This vector can be viewed as the minima structure required to determine a local preferred rest frame. We call it the "aether" as it is ubiquitous and determines a locally preferred frame at every point of space-time [2].

Recently, Xie and Huang (2008) have advanced this theory to a second post-newtonian approximation for a N-body model and obtained an explicit metric when the bodies are point-like masses [3]. In this research, we only consider the case that light closely passes the limb of the sun and neglect the contributions from planets. And the metric in isotropic gauge for such a practical 2PN light deflection experiments can be simplified as

$$\begin{aligned} ds^2 = & \left(-1 + \frac{2GM}{c^2 r} - \frac{2G^2M^2}{c^4 r^2}\right) c^2 dt^2 \\ & + \left(1 + \frac{2GM}{c^2 r} + \frac{3}{2}\frac{G^2M^2}{c^4 r^2} + \frac{1}{4}c_{14}\frac{G^2M^2}{c^4 r^2}\right)\left(dr^2 + r^2 d\theta^2 + r^2 \sin^2\theta d\phi^2\right), \end{aligned} \tag{1.1}$$

which is quite close to that adopted by Brumberg (1991) [4], Teyssandier and Le Poncin-Lafitte (2008) [5] and Klioner and Zschocke (2009) [6].

There is a 2PN parameter c_{14} which is the only non-GR parameter remaining as the deviation from general relativity. It shows the difference between these two gravity theories. The measurement of c_{14} can be achieved in future experiment with light propagating in the field of the Sun.

Since the field is isotropic, we may consider the space track of the light to be confined to the equatorial plane, that $\theta = \frac{\pi}{2}$. And we define the coordinate system as shown in Figure 1. The origin O is the mass center of the Sun. D is the nearest point to the Sun in the light propagation, **OD** is the pole axis. And the light ray propagates between two

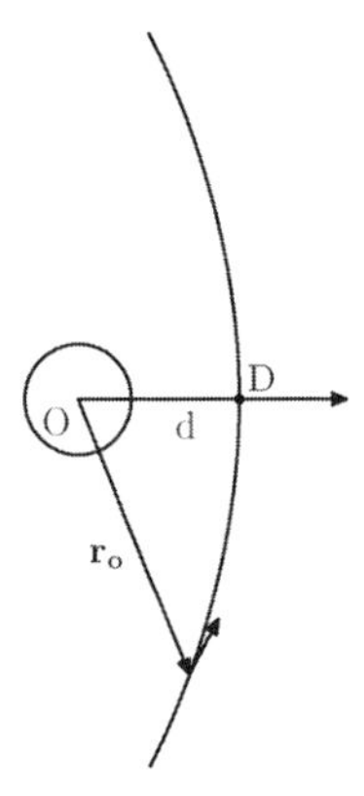

Figure 1. Light propagation in the gravitational field of the Sun

points, being emitted at a position $\mathbf{r_s}(r_s, \varphi_s)$ at the time moment t_s and received at a position $\mathbf{r_o}(r_o, \varphi_o)$ at the time moment t_o.

2. Light propagation

Metric(1.1) does not depend explicitly on t and ϕ, so we can make use of two invariants to simplify our derivation. And the space track of light in our isotropic spherical coordinates is as follows,

$$\varphi = \arccos\frac{d}{r} + \frac{2GM}{c^2 d}\sqrt{\frac{r-d}{r+d}} + \frac{G^2M^2}{c^4}\left[-\frac{2(3d+2r)\sqrt{r^2-d^2}}{(d+r)^2 d^2} + \frac{15}{4d^2}\arccos\frac{d}{r} + \frac{c_{14}}{8d^2}\arccos\frac{d}{r}\right], \tag{2.1}$$

in which d is the closest distance of the light path from the Sun.

For testing these two gravity theories, we obtain the post-post-Newtonian equation of light deflection, time delay and frequency shift in a static gravitational field from the metric(1.1).

1. Light Deflection

When the source is at infinity, we found the angle between the observed and the coordinate direction of the source, which is the light deflection, as the following,

$$\frac{2GM}{c^2 d}\left(1+\frac{\sqrt{r_o^2-d^2}}{r_o}\right) + \frac{G^2M^2}{c^4}\left[\frac{15\pi}{8d^2} - \frac{4}{d^2} - \frac{\sqrt{r_o^2-d^2}(d^2+dr_o+16r_o^2)}{4r_o^2(d+r_o)d^2} + \frac{15}{4d^2}\arccos\frac{d}{r_o}\right] + \frac{G^2M^2}{c^4}\left[\frac{c_{14}}{8d^2}\arccos\frac{d}{r_o} + \frac{c_{14}\pi}{16d^2} + \frac{c_{14}\sqrt{r_o^2-d^2}}{8dr_o^2}\right], \tag{2.2}$$

in which

$$d = r_0 \sin\beta - \frac{2GM}{c^2}(1-\sin\beta), \tag{2.3}$$

where β is the angle between the observed direction of the source and $\mathbf{r_o}$.

2. Time Delay

We denote $\mathbf{R} = \mathbf{r_s} - \mathbf{r_o}, R = |\mathbf{R}|$. The formula for the time of light propagation is

$$R + \frac{2GM}{c^2}\ln\frac{r_s + r_o + R}{r_s + r_o - R} + \frac{G^2M^2}{c^4}\frac{2R}{|\mathbf{r}_o \times \mathbf{r}_s|^2}\left[(r_o - r_s)^2 - R^2\right]$$
$$+ \frac{G^2M^2}{c^4}\frac{R}{|\mathbf{r}_o \times \mathbf{r}_s|}\left(\frac{15}{4}\arccos\frac{r_o^2 + r_s^2 - R^2}{2r_s r_o} + \frac{c_{14}}{8}\arccos\frac{r_o^2 + r_s^2 - R^2}{2r_s r_o}\right). \tag{2.4}$$

When $c_{14} = 0$, it's the same as the result previously derived by different approaches [4] [5].

3. Frequency Shift

The observed gravitational shift of frequency is

$$1 + z = \frac{d\tau_s}{d\tau_o} = \frac{d\tau_s}{dt_s}\frac{dt_s}{dt_o}\frac{dt_o}{d\tau_o}, \tag{2.5}$$

where

$$\frac{d\tau_s}{dt_s} = 1 - \frac{GM}{c^2 r_s} - \frac{\dot{\mathbf{r}}_s^2}{2c^2} + \frac{G^2M^2}{2c^4 r_s^2} - \frac{3GM}{2c^2 r_s}\frac{\dot{\mathbf{r}}_s^2}{c^2} - \frac{1}{8}\frac{\dot{\mathbf{r}}_s^4}{c^4}, \tag{2.6}$$

$$\frac{dt_o}{d\tau_o} = 1 + \frac{GM}{c^2 r_o} + \frac{\dot{\mathbf{r}}_o^2}{2c^2} + \frac{G^2M^2}{2c^4 r_o^2} + \frac{5GM}{2c^2 r_o}\frac{\dot{\mathbf{r}}_o^2}{c^2} + \frac{3}{8}\frac{\dot{\mathbf{r}}_o^4}{c^4}, \tag{2.7}$$

$$\frac{dt_s}{dt_o} = \frac{1 - A\frac{\mathbf{R}\cdot\dot{\mathbf{r}}_o}{cR} + B\frac{\mathbf{r}_o\cdot\dot{\mathbf{r}}_o}{cr_o} + D\frac{(\mathbf{r}_o\times\mathbf{r}_s)\cdot(\dot{\mathbf{r}}_o\times\mathbf{r}_s)}{c|\mathbf{r}_o\times\mathbf{r}_s|}}{1 - A\frac{\mathbf{R}\cdot\dot{\mathbf{r}}_s}{cR} - C\frac{\mathbf{r}_s\cdot\dot{\mathbf{r}}_s}{cr_s} - D\frac{(\mathbf{r}_o\times\mathbf{r}_s)\cdot(\mathbf{r}_o\times\dot{\mathbf{r}}_s)}{c|\mathbf{r}_o\times\mathbf{r}_s|}}, \tag{2.8}$$

with $\dot{\mathbf{r}}_o = d\mathbf{r}_o/dt_o$ and $\dot{\mathbf{r}}_s = d\mathbf{r}_s/dt_s$,

$$A = 1 + \frac{4GM}{c^2}\frac{r_o + r_s}{(r_o + r_s)^2 - R^2} + \frac{2G^2M^2}{c^4}\frac{(r_o - r_s)^2 - 3R^2}{|\mathbf{r}_o \times \mathbf{r}_s|^2}$$
$$+ \frac{(\frac{15}{4} + \frac{c_{14}}{8})}{|\mathbf{r}_o \times \mathbf{r}_s|}\frac{G^2M^2}{c^4}\left[\arccos\frac{r_o^2 + r_s^2 - R^2}{2r_s r_o} + \frac{2R^2}{\sqrt{(r_o + r_s)^2 - R^2}\sqrt{R^2 - (r_o - r_s)^2}}\right], \tag{2.9}$$

$$B = -\frac{4GM}{c^2}\frac{R}{(r_o + r_s)^2 - R^2} + \frac{4G^2M^2}{c^4}\frac{R(r_o - r_s)}{|\mathbf{r}_o \times \mathbf{r}_s|^2}$$
$$- (\frac{15}{4} + \frac{c_{14}}{8})\frac{G^2M^2}{c^4}\frac{R}{|\mathbf{r}_o \times \mathbf{r}_s|}\frac{r_o^2 + R^2 - r_s^2}{r_o\sqrt{(r_o + r_s)^2 - R^2}\sqrt{R^2 - (r_o - r_s)^2}}, \tag{2.10}$$

$$C = -\frac{4GM}{c^2}\frac{R}{(r_o + r_s)^2 - R^2} + \frac{4G^2M^2}{c^4}\frac{R(r_s - r_o)}{|\mathbf{r}_o \times \mathbf{r}_s|^2}$$
$$- (\frac{15}{4} + \frac{c_{14}}{8})\frac{G^2M^2}{c^4}\frac{R}{|\mathbf{r}_o \times \mathbf{r}_s|}\frac{r_s^2 + R^2 - r_o^2}{r_s\sqrt{(r_o + r_s)^2 - R^2}\sqrt{R^2 - (r_o - r_s)^2}}, \tag{2.11}$$

$$D = -\frac{4G^2M^2}{c^4}\frac{R\left[(r_o - r_s)^2 - R^2\right]}{|\mathbf{r}_o \times \mathbf{r}_s|^3} - (\frac{15}{4} + \frac{c_{14}}{8})\frac{G^2M^2}{c^4}\frac{R\arccos\frac{r_o^2 + r_s^2 - R^2}{2r_s r_o}}{|\mathbf{r}_o \times \mathbf{r}_s|^2}. \tag{2.12}$$

3. Discussion

The analytical solution for light deflection, time delay and frequency shift is derived in this note. The terms containing c_{14} are the main difference between general relativity and Einstein-Aether theory. This result can be provided for future experiments in testing gravity theories. Klioner and Zschocke have gotten the time delay in harmonic gauge [6], and the result we got is in isotropic condition. The comparison of the two works can be achieved through a coordinate transformation. The transformation is the Formula 129 given in [3]. Detailed procedure of derivation will be given elsewhere.

4. Acknowledgement

The authors thank the editor and referee for their kind comments. This work is supported by the Natural Science Foundation of China under the Grant Nos. 10673026 and 10873014.

References

[1] Jacobson, T. & Mattingly, D. 2001, Phys.Rev.D, 64, 024028
[2] Jacobson, T. & Mattingly, D. 2004, Phys.Rev.D, 70, 024003
[3] Xie, Y. & Huang, T.-Y. 2008, Phys.Rev.D, 77, 124049
[4] Brumberg, V. A. 1991, Bristol, England and New York, Adam Hilger, 1991, 271 p.,
[5] Teyssandier, P. & Le Poncin-Lafitte, C. 2008, Classical and Quantum Gravity, 25, 145020
[6] Klioner, S. A. & Zschocke, S. 2009, arXiv:0902.4206

Relativity in Fundamental Astronomy
Proceedings IAU Symposium No. 261, 2009
S. A. Klioner, P. K. Seidelman & M. H. Soffel, eds.

doi:10.1017/S1743921309990299

A relativistic motion integrator: numerical accuracy and illustration with BepiColombo and Mars-NEXT

A. Hees[1] **and S. Pireaux**[2]

[1,2]Royal Observatory of Belgium (ROB)
Avenue Circulaire 3, 1180 Bruxelles, Belgium

[1]aurelien.hees@oma.be [2]sophie.pireaux@oma.be

Abstract. Today, the motion of spacecraft is still described by the classical Newtonian equations of motion plus some relativistic corrections. This approach might become cumbersome due to the increasing precision required. We use the Relativistic Motion Integrator (RMI) approach to numerically integrate the native relativistic equations of motion for a spacecraft. The principle of RMI is presented. We compare the results obtained with the RMI method with those from the usual Newton plus correction approach for the orbit of the BepiColombo (around Mercury) and Mars-NEXT (around Mars) orbiters. Finally, we present a numerical study of RMI and we show that the RMI approach is relevant to study the orbit of spacecraft.

Keywords. gravitation, relativity, methods: numerical

1. The Relativistic Motion Integrator (RMI) and an analytical development

The software RMI presented in Pireaux *et al.* (2006), Pireaux *et al.* (2008) numerically integrates the relativistic equations of motion

$$\frac{d^2X^\alpha}{d\tau^2} = -\Gamma^\alpha_{\mu\nu}\frac{dX^\mu}{d\tau}\frac{dX^\nu}{d\tau} \tag{1.1}$$

for a given metric $G_{\mu\nu}$ where $X^\mu = (cT, X, Y, Z)$ are the coordinates, τ is the proper time and $\Gamma^\alpha_{\mu\nu}$ are the Christoffel symbols of the metric considered, derived numerically. As an example, we use this integrator with the planetocentric metric advised by the IAU 2000 resolutions, described in Soffel *et al.* (2003) and characterised by a scalar potential W and a vector potential W^i. Until now, we considered only the central body of mass M, so that we have:

$$\begin{cases} W(X^\alpha) &= \frac{GM}{R}\left[1+\sum_{l=2}^{\infty}\sum_{m=0}^{l}\left(\frac{R_e}{R}\right)^l P_{lm}(\cos\theta)\left(C_{lm}\cos m\phi + S_{lm}\sin m\phi\right)\right] \\ W^i(X^\alpha) &= -\frac{G}{2}\frac{(\mathbf{R}\times\mathbf{S})^i}{R^3} \end{cases} \tag{1.2}$$

where G is Newton's gravitational constant, c the speed of light, C_{lm} and S_{lm} are related to the central gravity field, R_e is the equatorial radius of the central body, while the vector $\mathbf{S}$ is its spin moment and $R = \sqrt{X^2+Y^2+Z^2}$.

It is possible to develop analytically the equations of motion (1.1) at first Post-Newtonian (1PN) order. Doing so, one gets $\frac{d^2\mathbf{R}}{dt^2} = -\frac{GM}{R^3}\mathbf{R} + \text{corr}$, where the corrections are composed of different types of forces: a Newtonian correction coming from the harmonics (proportional to C_{lm} or S_{lm}), a relativistic Schwarzschild acceleration (proportional to $1/c^2$), a relativistic correction coming from the harmonics (proportional to C_{lm}/c^2 or S_{lm}/c^2), a relativistic coupling between harmonics (proportional

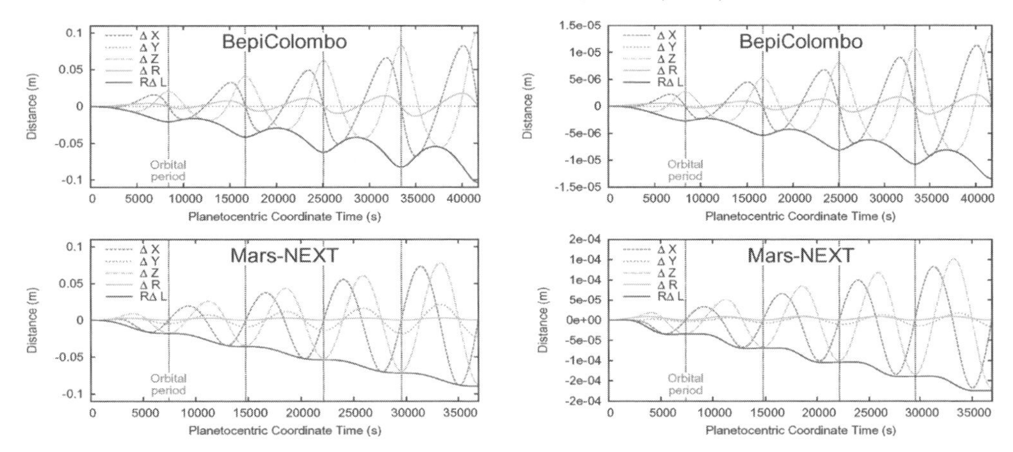

Figure 1. Corrections due to the relativistic Schwarzschild acceleration.

Figure 2. Corrections due to the relativistic contribution from the first harmonics (C_{20}, C_{22} and S_{22}).

to $C_{lm}C_{lm}/c^2$, $C_{lm}S_{lm}/c^2$ or $S_{lm}S_{lm}/c^2$) and finally a relativistic Lense-Thirring acceleration (proportional to the spin momentum over c^2).

2. Results for the BepiColombo and Mars-NEXT missions

The analytical development described above is used to assess the order of magnitude of each separate effect and to validate the RMI method. Figures 1, 2 and 3 show the separate impact of the relativistic effects in terms of cartesian coordinates (X, Y, Z), radial distance R and $L = \omega + w$, where ω is the argument of pericenter and w is the true anomaly. The orbital parameters of the BepiColombo mission can be found in Balog *et al.* (2000): $a = 3389$ km, $e = 0.162$, $i = 90^\circ$. The orbital parameters of the Mars-NEXT mission (see Chicarro *et al.* (2008)) are: $a = 3896$ km, $e = 0$ and $i = 75^\circ$. The numerical integration has been performed over 5 orbital periods.

3. Numerical precision

The numerical derivative has to be treated carefully. In the implementation of RMI, we used a fourth-order numerical derivative

$$f'(x) \approx D_h + \mathcal{O}(h^4) = \frac{f(x-2h) - 8f(x-h) + 8f(x+h) - f(x+2h)}{12h} + \mathcal{O}(h^4). \quad (3.1)$$

For large h, the discretization error is important, while for small h, the roundoff error increases. We use an optimal derivation step computed analytically for a function $\frac{1}{r}$ (see Figure 4), given by $h_{\rm opt} = (45\epsilon a^6 c^2/(960GM))^{1/5}$ (Kincaid and Cheney (2002)) where ϵ is the machine precision and a is the semi-major axis. As can be seen on Figure 4, we derive $H_{\mu\nu} = G_{\mu\nu} - \eta_{\mu\nu}$, with $\eta_{\mu\nu}$ the Minkowski metric, instead of $G_{\mu\nu}$ since it is more stable numerically. Moreover, we use a Richardson extrapolation in order to increase the precision on the derivative (Richardson (1927)). This extrapolation uses two estimations of the derivative of order 4 with different step size (D_h and $D_{h/k}$ with k a real factor) to construct an estimation of order 8

$$f'(x) \approx \frac{k^4 D_{h/k} - D_h}{k^4 - 1} + \mathcal{O}(h^8). \quad (3.2)$$

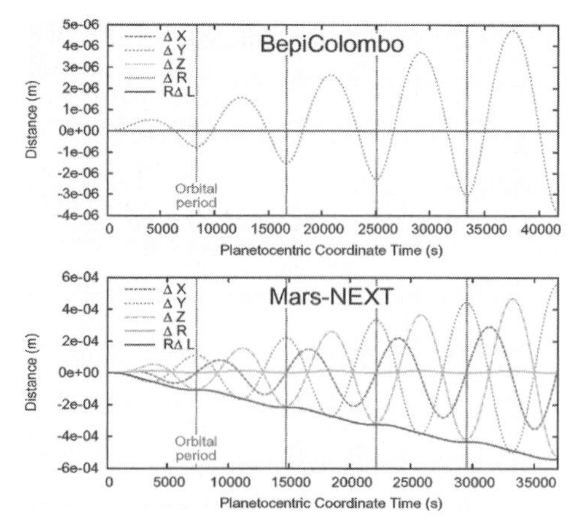

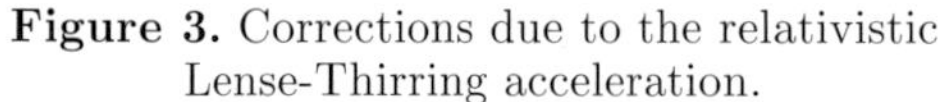

Figure 3. Corrections due to the relativistic Lense-Thirring acceleration.

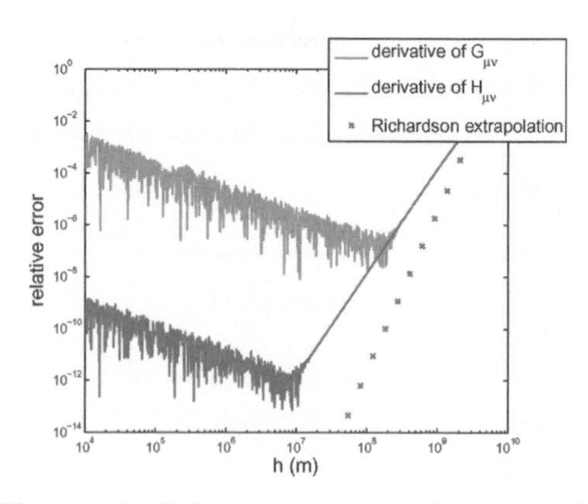

Figure 4. Relative precision of the metric derivative as a function of the discretization step size h.

Typically, the value of k is often chosen as 2 or 1.5. We can see on Figure 4 that this procedure can increase the derivative precision. It is furthermore less sensitive to the choice of h_{opt}. With such an implementation, it is possible to show that the relative precision of RMI is of the order of 10^{-12} in double precision and 10^{-22} in quadruple precision.

4. Conclusion

We have shown that RMI is useful to compute relativistic orbits for different missions. It includes all the relativistic effects (up to the corresponding order of the metric). It is quite easy to use, since the user only has to change the metric module if he wants to change the metric. Until now, RMI is only a prototype that is more time consuming than the usual 1PN approach. Nevertheless, it is possible to reduce drastically the integration time required by the RMI method via proper coding and using parallelization (the computation of the Christoffel symbols can easily be parallelized).

Acknowledgments

A. Hees is a research fellow from the FRS-FNRS (Belgian Fund for Scientific Research) for his PhD thesis at ROB-UCL (Université Catholique de Louvain, Belgium) and both authors acknowledge a FNRS and a LOC grant to attend to the IAU 261 symposium.

References

Pireaux, S., Barriot, J.-P., & Rosenblatt, P. 2006, *Acta Astronautica*, 59-517
Pireaux, S., Chauvineau, B., & Hees, A. 2009, arXiv: 0801.3637v2 (gr-qc)
Balog, A. *et al.* 2000, *ESA-SCI(2000)1*
Chicarro, A., ESA. 2008, *Lunar and Planetary Science XXXIX*
Soffel, M., Klioner, S. *et al.* 2003, *AJ* 126-2687
Kincaid, D. & Cheney, W. 2002, Numerical analysis: Mathematics of Scientific Computing, *American Mathematical Society*
Richardson, L. S. 1927, *Phil. Trans. of the Royal Society of London*, A226-299

Relativity in Fundamental Astronomy
Proceedings IAU Symposium No. 261, 2009
S. A. Klioner, P. K. Seidelman & M. H. Soffel, eds.

doi:10.1017/S1743921309990305

The motion of vibrating systems in Schwarzchild spacetime

A. Hees[1], L. Bergamin[2] and P. Delva[3]

[1,2,3]European Space Agency, The Advanced Concepts Team
Keplerlaan 1, 2201 AZ Noordwijk, The Netherlands

[1]Royal Observatory of Belgium (ROB)
Avenue Circulaire 3, 1180 Bruxelles, Belgium

[1]aurelien.hees@oma.be [2]bergamin@tph.tuwien.ac.at [3]Pacome.Delva@esa.int

Abstract. In this communication, the effects of vibrations at high frequencies onto a freely falling two-body system in Schwarzschild spacetime are investigated. We present these effects for different kinds of reference motions, all of which are placed in regions of weak gravitation: circular orbits, radial free fall (with different initial velocities) and radial free fall with a small tangential velocity. The vibrations induce a change in the motion of the vibrating system, which is characterized by a radial deviation between the vibrating system and the reference motion of the non-vibrating system. For a circular orbit, we show that the maximal radial deviation increases linearly with the initial radius. For a radial free fall, the radial deviation after one oscillation decreases quadratically with respect to the initial radius.

Keywords. gravitation, general relativity, Schwarzchild geometry, vibrating systems

---

1. Introduction

Changing the shape of a free-falling body in a gravitational field can induce a change in its motion. In Newtonian physics, such effects are well known and could be used in space technology for propellentless propulsion as presented in Martinez *et al.* (1987). These proposals are based on resonant effects, where the change of shape is linked to the motion of the body in space. More recently, similar situations have been studied in the framework of general relativity (see for example Wisdom (2003), Guéron *et al.* (2006), Avron and Kenneth (2006) , Guéron and Mosna (2007)) where non-resonant effects can become large.

In the present communication, we consider the free fall of an oscillating two-body system around a central body, where the deviation of the motion of a vibrating system with respect to the motion of the same system without vibration is investigated. Three different classes of initial conditions (or reference motions) will be presented: a circular orbit (described in Bergamin *et al.* (2009b)), a radial free fall (described in Guéron and Mosna (2007), Bergamin *et al.* (2009a)) and a radial free-fall with a tangential initial velocity. We will show that in general relativity new interesting effects appear even in regions of weak gravitation, which are not present in Newtonian physics, and we will describe how these effects depend on the initial conditions.

2. The model

A vibrating or oscillating system is implemented as a collection of point masses whose relative positions are related by time dependant constraints. The model used here is described in Guéron and Mosna (2007), Bergamin *et al.* (2009a), Bergamin *et al.* (2009b):

two point masses connected by a massless tether, whose length $l(t)$ is imposed by an oscillating constraint. The constraint is described by four parameters: its frequency ω, its amplitude δ_l, its minimum length l_0 and an asymmetry parameter α (it describes how much the oscillation fails to be symmetric in time). The vibrating system evolves in the surrounding of a central body described by a Schwarzschild spacetime. We implement the oscillating constraint in Schwarzschild coordinates. We do not describe a particular system, such as a molecule or a crystal. To do so, one would have to describe this particular system in the Fermi-Walker coordinates of the reference motion, which are in good approximation comoving coordinates of the vibrating system. The system dynamics is studied in the orbital plane: each body is described in terms of two variables (r_i, ϕ_i) (where $i = 1, 2$), and we define the geometrical center of the system whose radial coordinates is $r = (r_1 + r_2)/2$.

The equations of motion of the vibrating system are derived from the action

$$S = -\int \mathrm{d}t \left[\sqrt{L_1} + \sqrt{L_2} + \lambda\left(r_2 - r_1 - \sqrt{l^2(t) - r_1^2\theta^2}\right)\right], \tag{2.1}$$

where λ is a Lagrange multiplier, t is the Schwarzschild time coordinate, $l(t)$ is the length of the constraint, $\theta = \phi_2 - \phi_1$, and

$$L_i = \left(1 - \frac{r_s}{r_i}\right) - \left(1 - \frac{r_s}{r_i}\right)^{-1} \dot{r}_i^2 - r_i^2\dot{\varphi}_i^2 \tag{2.2}$$

where $\dot{( )} \equiv \mathrm{d}/dt$ and r_s is the Schwarzschild radius. In addition to these equations of motion, one has to choose initial conditions. Different initial conditions will be presented here: an initial condition leading to a circular orbit, different initial conditions for a radial free fall and some intermediate cases.

3. Circular orbit

For the first set of initial conditions we choose vanishing radial velocity $\dot{r}(t = 0) = 0$ with $r(t = 0) = r_0$, while $\phi_i(t = 0) = 0$ and $\dot{\phi}_i(t = 0)$ is determined in such a way that the reference motion (for the non-vibrating system) is a circular orbit. This situation is thoroughly discussed in Bergamin *et al.* (2009b). Here, the deviation of the motion of the vibrating system with respect to the motion of the non-vibrating one is characterized by two effects: a radial deviation and an oscillation of the system around the radial axis. The radial deviation implies that the trajectory becomes a quasi-elliptical orbit. It is possible to quantify the maximal radial deviation Δr between the vibrating system and the non-vibrating one. The analytical expansion in Bergamin *et al.* (2009b) leads to

$$\begin{aligned}\frac{\Delta r}{r_0} &= \frac{\delta_l^2\omega^2}{c^2}\frac{S_1(\alpha)}{\left(1-\frac{r_s}{r_0}\right)\left(1-\frac{3r_s}{r_0}\right)} + \frac{5}{2}\frac{\delta_l}{r_0}\frac{1-\frac{9r_s}{r_0}}{1-\frac{3r_s}{r_0}}\left(2\frac{l_0}{r_0}S_2(\alpha) + \frac{\delta_l}{r_0}S_3(\alpha)\right) \\ &\quad + \mathcal{O}\left(\frac{\dot{l}^2}{c^2}\frac{l^2}{r_0^2}\right) + \mathcal{O}\left(\frac{\dot{l}^4}{c^4}\right) + \mathcal{O}\left(\frac{l^4}{r_0^4}\right)\end{aligned} \tag{3.1}$$

where c is the speed of light and $S_i(\alpha)$ are some functions in α of the order one. The first term of this expansion is a purely relativistic effect not present in Newtonian theory and it grows linearly with the initial radius r_0. The second term is present in Newtonian theory (with $r_s/r_0 \to 0$) and it decreases as $1/r_0$. Then the difference between general relativity and the Newtonian theory of gravitation increases linearly with the initial radius. This analytical result is confirmed by numerical simulations as can be seen in Figure 1 (a).

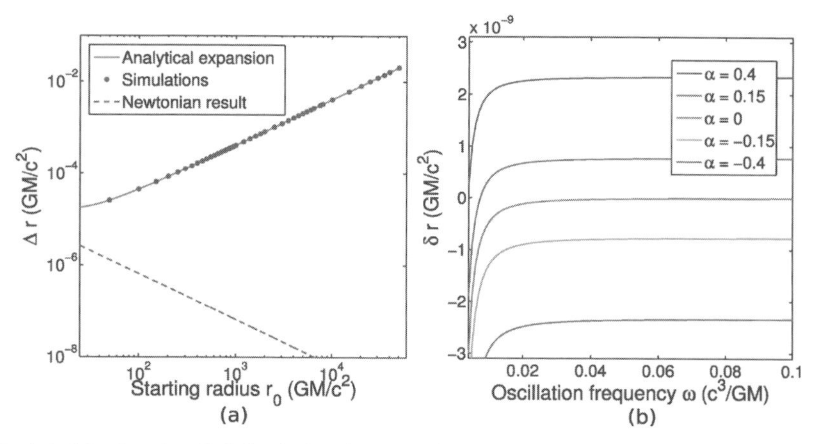

Figure 1. (a): Maximal radial deviation in case of a circular orbit as a function of the starting radius r_0 for $\omega = 0.05$: comparison of numerical simulations and an analytical expansion. Additionally, the leading classical gravitational term has been included in the figure to illustrate the difference between general relativity and Newtonian gravity. (b): Radial deviation after one oscillation period, in case of a radial free-fall with a vanishing initial velocity as a function of the oscillation frequency, for different values of the asymmetry parameter α and for $r_0 = 120\, GM/c^2$.

In summary, for a circular orbit the relativistic radial deviation increases linearly with the radius, increases quadratically with the vibration frequency ω and the oscillation amplitude δ_l, while the effect of the asymmetry parameter α is quite small.

4. Radial free fall

Now, we consider the same system in a radial free fall towards a central body. This situation is studied in Guéron and Mosna (2007), Bergamin *et al.* (2009a). The initial conditions are $\phi_i(t=0) = \dot{\phi}_i(t=0) = 0$, $r(t=0) = r_0$ and $\dot{r}(t=0) = v_0$. The case with $v_0 = 0$ was discussed by Guéron and Mosna (2007). In that case, it is possible to slow down or to accelerate the free fall of a system depending on the parameter α. Guéron showed that after one oscillation period, the radial deviation is given by

$$\delta r \approx \Gamma_\alpha \frac{\delta_l^2}{r_0^2} \frac{GM}{c^2}, \tag{4.1}$$

where M is the mass of the central body, G the gravitational constant and Γ_α is a factor associated with the asymmetry parameter. This radial deviation again is a purely relativistic effect and it decreases quadratically with the initial radius. Here the asymmetry parameter is very important since δr vanishes for $\alpha = 0$, is positive (it corresponds to a deceleration) for positive values of α and negative (it corresponds to an acceleration) for negative values of α, as can be seen in Figure 1 (b).

Now, we extend the particular initial condition $v_0 = 0$ studied by Guéron and Mosna (2007) to non-vanishing initial velocities. In Figure 2 (a), the radial deviations for different initial velocities $\dot{r}(t=0) = v_0$ for a positive value of α are presented. As can be seen, the effect increases if the initial velocity is negative (the system already falls down at $t = 0$); while in the opposite case (the system is thrown up at $t = 0$), the effect decreases to become negative at higher velocity. This behavior is certainly not surprising, since the work performed by the oscillations is expected to change sign when the system moves upwards.

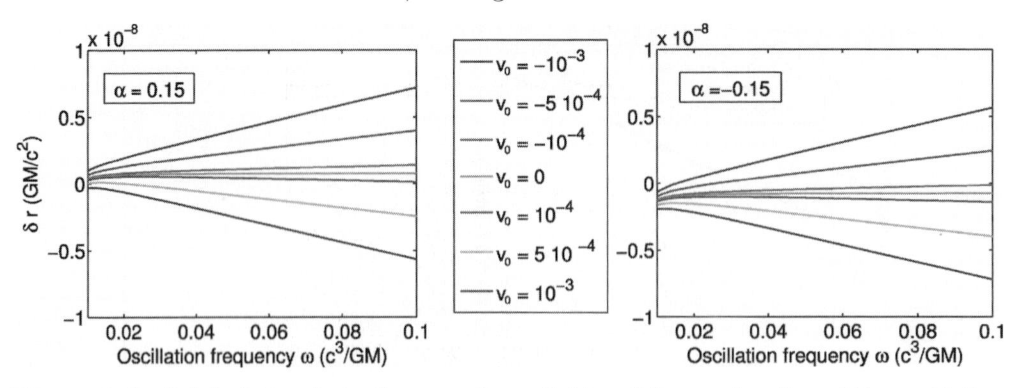

Figure 2. Radial deviation δr in the case of a radial free fall, as a function of the oscillation frequency (ω), for different initial radial velocities $\dot{r}(t=0) = v_0$ (expressed as a fraction of c). In these examples, $r_0 = 120\ GM/c^2$ and (a): asymmetry parameter $\alpha = 0.15$. (b): asymmetry parameter $\alpha = -0.15$.

Still, changing the sign of the asymmetry parameter for $v_0 \geqslant 0$ is not an option to obtain a positive displacement (to slow down the free fall). This situation is depicted in Figure 2 (b), where different initial velocities with negative values of α are presented. The change in the sign of α induces a vertical shift of all curves. Thus, a positive effect is only achievable for a system falling down at high velocities ($v_0 \gg 0$), while for initial conditions $v_0 = 0$ a negative value results over the whole frequency range as seen in Figure 1 (b).

In summary, for a radial free fall, the relativistic radial deviation decreases quadratically with the initial radius, has a plateau in the frequency and the effect of the asymmetry parameter α is important (no effect is found for a symmetric constraint). The radial deviation δr is increased if the initial radial velocity is negative (the system is already falling down $v_0 < 0$) and is decreased in the opposite case.

5. Radial free fall with angular velocity

We have presented the effects of vibrations on a circular orbit and on a radial free fall with different initial conditions. Now, we will study the intermediate case: the deviation of a vibrating system falling down towards a central body with a small angular velocity. The initial angular velocity is expressed as a fraction of the angular velocity of the reference circular orbit, $\dot{\phi}_i(t=0) = \beta\dot{\phi}_c$, where $\beta = 0$ implies a radial free fall and $\beta = 1$ a circular orbit. The remaining initial conditions are $\phi_i(t=0) = 0$, $r(t=0) = r_0$ and $\dot{r}(t=0) = 0$. Figure 3 shows the behavior of the radial deviation δr for different initial angular velocities. It can be seen that the effect (δr) decreases if the angular velocity is increased and becomes negative at high β values. Still, the radial relativistic displacement δr is not very sensible to the initial β conditions as long as the angular velocity remains small. This shows that the vibrations indeed can be used for "gliding".

Similar conclusions apply for negative values of α but since for these values, $\delta r \leqslant 0$ at $\beta = 0$, we do not reproduce these cases explicitly.

In summary, a small angular velocity has a very small influence on the relativistic radial deviation δr and for high angular velocity, the system will orbit around the central mass.

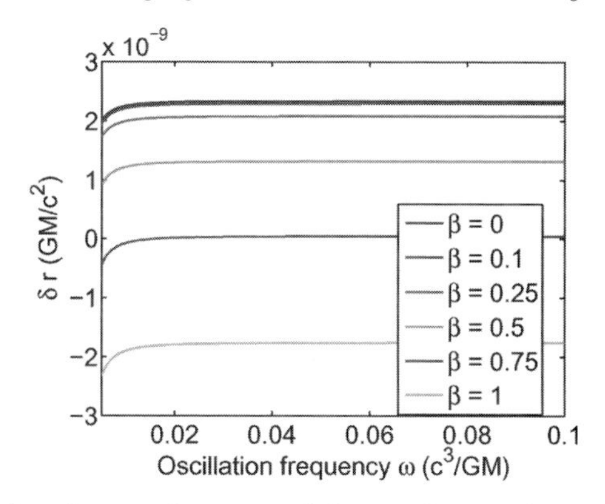

Figure 3. Radial deviation δr as a function of the oscillation frequency ω in case of a radial free fall, with different value of initial angular velocity, expressed as a fraction β of the circular angular velocity. In this example, $r_0 = 120\ GM/c^2$ and $\alpha = 0.4$

6. Conclusion

In this communication we have discussed the influence of oscillations onto the trajectory of a small system in Schwarzschild spacetime in weak gravity field. We have shown that a systematic expansion of these effects in Schwarzschild coordinates can be dominated by a term not present in Newtonian theory of gravitation even in weak gravity field. These effects manifest themselves as a radial deviation from the reference motion and depend on the initial conditions of the system. In particular, the differences between a radial free fall and a circular orbit have been worked out. For the circular case, the maximal radial deviation is increasing linearly with the starting radius, whereas, for the radial free fall, the radial deviation after one oscillation decreases quadratically. Hence, the circular case could be more interesting for experimental verifications. The second difference is the role played by the asymmetry parameter of the vibration. For the circular case, it has a minor impact on the result, while in the radial free fall, it is crucial in order to get a relativistic effect.

Acknowledgments

The authors would like to thank D. Izzo for important discussions on the topic. A. Hees is a research fellow from FRS-FNRS (Belgian Fund for Scientific Research) for his PhD thesis at ROB-UCL (Université Catholique de Louvain, Belgium) and he acknowledges a FNRS and a NSF grant to attend to the IAU 261 symposium.

References

Martinez-Sanchez, M. & Gavit., S. A., 1987, *J. Guid. Control Dyn.* 10-233
Wisdom, J. 2003, *Science* 299-2865.
Gueron, E., Maia, C. A. S., & Matsas, G. E. A. 2006, *Phys. Rev. D* 73/024020
Avron, J. E. & Kenneth., O. 2006, *New J. Phys.*, 8-68
Guéron, E. & Mosna, R. A. 2007, *Phys. Rev. D*, 75/081501
Bergamin, L., Delva, P., & Hees, A. 2009 (a), arXiv: gr-qc/0901.2298
Bergamin, L., Delva, P., & Hees, A. 2009 (b), *Class. Quantum Grav.*, 26/185006

Relativity in Fundamental Astronomy
Proceedings IAU Symposium No. 261, 2009
S. A. Klioner, P. K. Seidelman & M. H. Soffel, eds.

doi:10.1017/S1743921309990317

Gravitomagnetic effects of a massive and slowly rotating sphere with an equatorial mass current on orbiting test particles

Leonardo Castañeda[1], Fernando Fandiño[1,2], William Almonacid[1], Edilberto Suárez[2] and Giovanni Pinzón[1]†

[1]Observatorio Astronómico Nacional, Universidad Nacional de Colombia, Bogotá, Colombia
[2]Universidad Distrital "Francisco Jose de Caldas", Bogotá, Colombia

Abstract. Within the framework of linearized Einstein field equations we compute the gravitomagnetic effects on a test particle orbiting a slowly rotating, spherical body with a rotating matter ring fixed to the equatorial plane. Our results show that the effect on the precession of particle orbits is increased by the presence of the ring.

Keywords. Gravitation, Solar System

The analogy between classical electrodynamics and the linearized Einstein field equations has been largely studied (Misner, Thorne & Wheeler 1973). This analogy was recently revised by Franklin & Baker (2007) in the frame of a test particle orbiting a slowly rotating sphere. In this work we have introduced a massive rotating equatorial ring surrounding the sphere in order to simulate the effect of an axisymmetric mass distribution.

The gravito-electric and gravito-magnetic potentials obey a Maxwell-like set of equations (Franklin & Baker 2007) directly derived from linearized General Relativity (Soffel 1989). For a sphere with Newtonian angular momentum $\ell_S = \frac{2}{5}MR^2\omega_S$, the corresponding potentials are given by:

$$V_S(r) = -\frac{GM}{r}, \qquad \mathbf{A}_S(\mathbf{r}) = -\frac{G}{c^2}\frac{\ell_S}{2}\frac{\sin\theta}{r^2}\hat{\varphi}. \tag{1}$$

On the other hand, for the ring with Newtonian angular momentum $\ell_R = ma^2\omega_R$, we have for the potentials:

$$V_R(\mathbf{r}) = -Gm\sum_{n=0}^{\infty}\frac{(-1)^n(2n)!}{2^{2n}(n!)^2}P_{2n}(\cos\theta)\frac{a^{2n}}{r^{2n+1}},$$

$$\mathbf{A}_R(\mathbf{r}) = -\frac{G}{c^2}\ell_R\sum_{n=0}^{\infty}(-1)^n\frac{(2n)!}{(2n+2)!}\frac{(2n+1)!}{(2^n n!)^2}P^1_{2n+1}(\cos\theta)\frac{a^{2n}}{r^{2n+2}}\hat{\varphi}. \tag{2}$$

Restricting the discussion to the mass-monopole and spin dipole, the gravito-electric and gravito-magnetic potentials for the composite system take the form:

$$V(\mathbf{r}) = -G\frac{(M+m)}{r},$$

$$\mathbf{A}(\mathbf{r}) = -\frac{G}{2c^2}(\ell_S+\ell_R)\frac{\sin\theta}{r^2}\hat{\varphi}. \tag{3}$$

† email: lcastanedac@unal.edu.co, jffandilloc@unal.edu.co

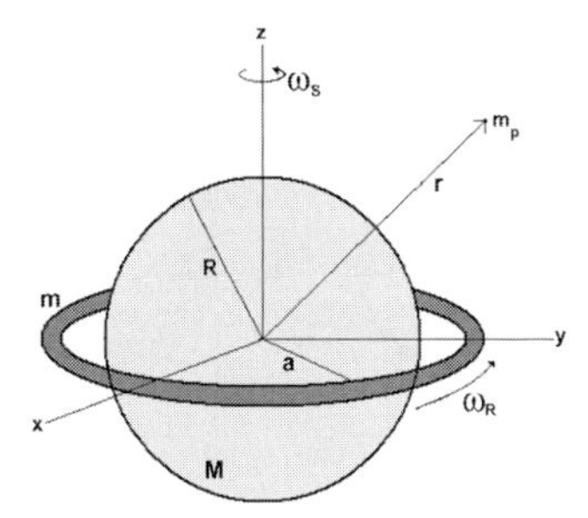

Figure 1. Spinning sphere with radius R, angular frequency ω_S and equatorial mass current

The Lagrangian for a test particle in the field of these potentials (neglecting c^{-2} terms from the Schwarzschild problem) reads:

$$L(\mathbf{r},\mathbf{v},t) = \frac{m_p}{2}\left(\dot{r}^2 + r^2\dot{\theta}^2 + r^2\sin^2\theta\dot{\varphi}^2\right) + Gm_p\frac{(M+m)}{r} - \frac{Gm_p}{2c^2}\frac{(\ell_S+\ell_R)}{r}\sin^2\theta\dot{\varphi}. \tag{4}$$

Therefore, the conserved quantities are: the z-component of the angular momentum and the energy of the particle. Both quantities constraint the initial conditions for the test particle.

$$\begin{aligned} J_z &= \frac{\partial L}{\partial\dot{\varphi}} = m_p\left[r^2\dot{\varphi} - \frac{G}{2c^2}\frac{(\ell_S+\ell_R)}{r}\right]\sin^2\theta\ , \\ E &= \frac{m_p}{2}\left(\dot{r}^2 + r^2\dot{\theta}^2 + r^2\sin^2\theta\dot{\varphi}^2\right) - Gm_p\frac{(M+m)}{r}\ . \end{aligned} \tag{5}$$

From equation (4) we obtain the equations of motion for the test particle:

$$\begin{aligned} \ddot{r} &= r\dot{\theta}^2 + r\sin^2\theta\dot{\varphi}^2 - G\frac{(M+m)}{r^2} + \frac{G}{2c^2}\frac{(\ell_S+\ell_R)}{r^2}\sin^2\theta\dot{\varphi}\ , \\ \ddot{\theta} &= \frac{\sin 2\theta}{2}\dot{\varphi}^2 - \frac{2}{r}\dot{r}\dot{\theta} - \frac{G}{2c^2}\frac{(\ell_S+\ell_R)}{r^3}\sin 2\theta\dot{\varphi}\ , \\ \ddot{\varphi} &= -\frac{2}{r}\dot{r}\dot{\varphi} - 2\cot\theta\dot{\theta}\dot{\varphi} + \frac{G}{c^2}\frac{(\ell_S+\ell_R)}{r^3}\cot\theta\dot{\theta} - \frac{G}{2c^2}\frac{(\ell_S+\ell_R)}{r^4}\dot{r}. \end{aligned} \tag{6}$$

In order to get the trajectory of the test particle we conducted numerical integrations of equations (6). It is important to note that the term related with the matter current, originating in the gravito-magnetic field, is a perturbation of the gravito-electric field. In fact, the gravito-magnetic terms are of order $(v/c)^2$, i.e., the effect is important for large timescales. Gravito-magnetic effects onto the system are "amplified" by using $c = 1$. We constraint our simulations to planar and closed orbits setting $\theta = \pi/2$ and $\dot{\theta} = 0$. 3D simulations will be shown in a forthcoming paper (Castañeda *et al.*, 2009). For the orbits in the plane, following the Bertrand's Theorem (Goldstein, Polle & Safko 2000), we impose an initial angular velocity for the test particle to obtain elliptic orbits. Initial conditions are therefore those corresponding to the perihelion, r_p, where $\varphi = 0$, $\dot{r} = 0$ and $\dot{\varphi}$ are obtained from the first integrals of motion given by equations (5).

Numerical simulations were conducted using a Runge-Kutta method implemented in Matlab 7.0 and compared with a 3D code written in C. From Figure (2) the ring's gravito-magnetic contribution to the orbital precession is obvious, which is the central issue of this paper. Our main goal has been to study a system of astrophysical interest; our results are a first step to understand the main features from gravito-magnetic contributions. Gravito-magnetic effects, such as the Lense-Thirring one, follow directly from our simulations, as

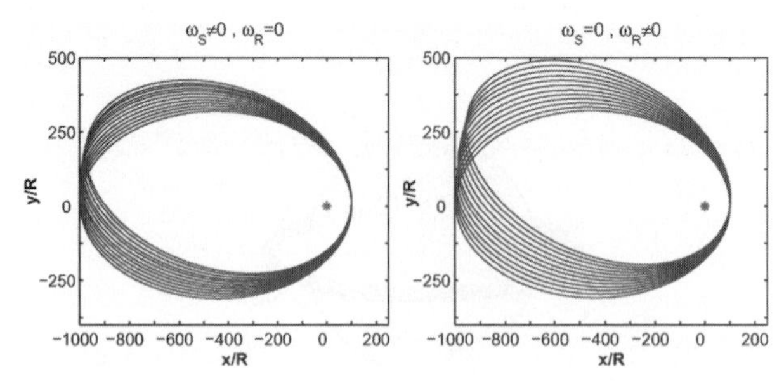

Figure 2. Left panel: Orbits of the test particle in the x-y plane for the test particle with the ring at rest. Right panel: Sphere at rest. Parameters used in the simulation: $c = 1$, $M = 10^{24}kg$, $R = 2 \times 10^4 m$, $m = M/4$, $a = 1.5R$, $r_p = 100R$, $ra = 1000R$ y $m_p = 1kg$.

will be shown in Castañeda *et al.*, 2009. Finally, Figure (3) shows a family of trajectories for different ring velocities for a fixed sphere rotation velocity. The upper panel gives the effect of corotation between the sphere and the ring; the lower one shows the effect of counter-rotating bodies. The result is the precession of the orbit as was shown in a special case by Franklin & Baker (2007). However, we note that the effect of the equatorial mass current is to increase the precession velocity. In fact, as the velocity of the ring increases, the effect on the precession is more noticeable. These results motivate us to extend the present problem to further studies.

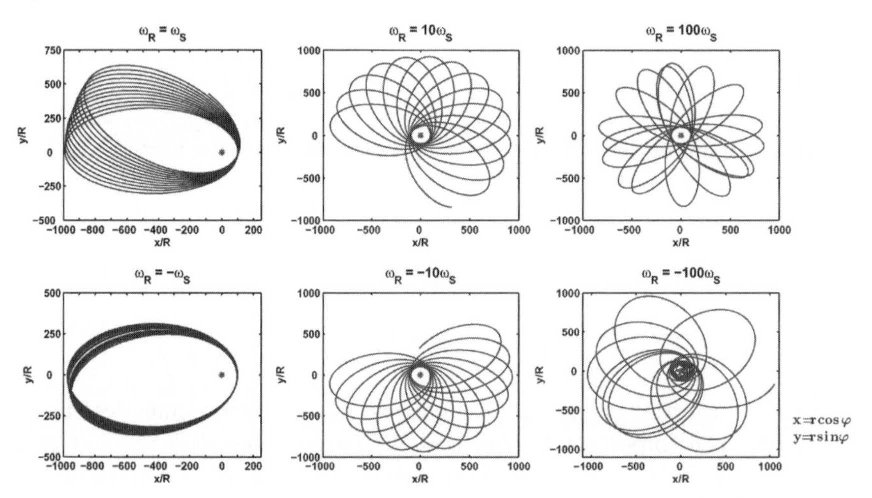

Figure 3. Orbits of the test particle in the x-y plane for $\omega_S = 2 \times 10^{-8} s^{-1}$ and $\omega_R = k\omega_S$ with $k = -100, -10, -1, 1, 10, 100$.

References

Castañeda, L., Fandiño, F., Almonacid, W., Izquierdo, D., & Pinzón G. 2009, in preparation
Franklin, J. & Baker, T. 2007, *Am. J. Phys.* 75, 4
Goldstein, H., Polle, C., & Safko, J. 2000, *Classical Mechanics* NY: Addison-Wesley
Jackson, J. D. 1998, *Classical Electrodynamics* NY: John Wiley & Sons
Misner, C. W., Thorne, K. S., & Wheeler, J. A. 1973, *Gravitation* NY: Freeman
Soffel, M. H. 1989, *Relativity in Astrometry, Celestial Mechanics and Geodesy*, Springer

Relativity in Fundamental Astronomy
Proceedings IAU Symposium No. 261, 2009
S. A. Klioner, P. K. Seidelman & M. H. Soffel, eds.

doi:10.1017/S1743921309990329

Relativistic aspects of the JPL planetary ephemeris

W. M. Folkner[1]

[1]Jet Propulsion Laboratory, California Institue of Technology,
4800 Oak Grove Dr., Pasadena, CA, 91009, USA
email: william.m.folkner@jpl.nasa.gov

Abstract. The orbits of the planets as represented by the JPL planetary ephemerides are now primarily determined by radio tracking of spacecraft. Analysis of the data and propagation of the orbits relies on an internally consistent set of equations of motion and propagation of radio signals including relativistic effects at the centimeter level. The planetary ephemeris data set can be used to test some aspects of the underlying theory such as estimates of PPN parameters γ and β, time variation in the gravitational constant G, rotation of the solar system relative to distant objects (Mach's principle), and place stringent limits on the possible violation of the inverse-square law.

Keywords. ephemerides, relativity, gravitation

---

1. Introduction

Fitting the orbits of the planets, Sun, and Moon as done in the planetary ephemerides developed at JPL and other institutions must take relativistic effects into account in order to fit the data to the current measurement accuracy. The orbits of the Earth-Moon barycenter and Mars are constrained by copious amounts of radio range measurements to Mars landers and orbiters along with very-long baseline interferometer (VLBI) measurements of Mars orbiters that determine the orbital orientations relative to the IAU celestial reference frame (ICRF). Of the outer planets, Saturn is now best determined through the use of radio range and VLBI measurements of the Cassini spacecraft. Besides determining the planetary orbits, these data can be used to estimate possible corrections to the nominal relativistic models. Preliminary estimates of gravity parameters of interest from the Mars and Saturn data sets are given below. More thorough analyses are in development.

Data for the orbits of Mercury, Venus and the Moon are also useful for testing relativity on the scale of the solar system. However these data are not discussed here, though they are used to constrain the orbits of those objects.

2. Solar system models

The basic equations of motion for the solar system are from Einstein et al. (1938) giving the acceleration of the planets, Sun, and Moon as point masses, which are then numerically integrated. In addition the gravitational effects of the solar oblateness, the effects of more than 300 asteroids, and the tidal interactions of the Earth and Moon are taken into account. The initial conditions (orbital elements) of the planets and Moon are estimated as part of the fitting process. The orbit of the Sun is constrained by the conservation of linear and angular momentum. The mass parameters of the planets are derived from the analysis of spacecraft tracking data taken in orbit or during planetary encounters, except for the dwarf planet Pluto for which the mass parameter is estimated from astrometric observations of the Pluto system. The mass parameter of the Sun and the solar oblateness parameter J_2 are estimated from the ephemeris data.

Because the orbit of Mars is significantly affected by asteroids, the treatment of the asteroid mass parameters has a large effect (up to two orders of magnitude) on the uncertainties in the estimated parameters. For the preliminary estimates given below, we estimate the mass parameters for the 67 asteroids with the largest effect on the Mars orbit as identified in Konopliv *et al.* (2006). The mass parameters for an additional 276 asteroids are modeled in three taxonomic classes and the mean density for each class estimated, based on a nominal radius for each asteroid determined by infrared photometry (e.g. Matson *et al.* 1986). The Mars spacecraft data set is not sufficient to produce accurate estimates for all 67 individual asteroids, due to large correlations through the effects of the asteroids on the orbit of Mars. This approach results in significantly larger uncertainties in gravity-related parameters than other works, such as Pitjeva (2004), in which only a few asteroid mass parameters are individually estimated. We feel that the estimation of 67 individual asteroid mass parameters gives a more realistic indication of the uncertainties. With a significant number of asteroid mass parameters now being accurately determined through the analysis of astrometric measurements of asteroids that have undergone deflections from asteroid-asteroid interactions (e.g. Baer & Chesley 2008), improved results may be expected in the near future.

Our modeling of the propagation of radio signals between Earth tracking stations and spacecraft uses the formalism by Moyer (2000). The orbits of spacecraft about Mars are estimated from radio Doppler measurements with an accuracy better than one meter (e.g. Konopliv *et al.* 2006). After fitting, the radio range measurements to Mars landers and orbiters have a scatter with root-mean-square residuals of about 2 meters in Earth-Mars distance. The range measurement scatter is dominated by Earth tracking station calibrations done before each tracking pass. Because this calibration error is common to each range measurement, we use only the average range measurement for each tracking pass. Since there are an average of 70 range measurements per tracking pass, an estimation which includes each range measurement, ignoring the correlated effect due to the calibration process, would result in an estimate too optimistic by almost an order of magnitude.

3. PPN Parameter estimation

The parameterized post-Newtonian (PPN) parameters γ and β characterize a range of possible theories for gravity, with values of unity under Einstein's theory of general relativity.

The PPN parameter γ mainly appears in measurements of radio range delay during solar conjunction (e.g. Shapiro 1964). An estimate of γ from the effective delay on radio range measurements to the Viking landers during the solar conjunction of 1976 was given with an uncertainty of 0.002 by Reasenberg *et al.* (1979). From the radio range measurements from the Viking landers during the solar conjunctions of 1976 and 1979, combined with radio range and VLBI measurements from later orbiting spacecraft not during conjunction, we find that $(\gamma - 1) = 0.00009 \pm 0.00070$. This uncertainty is larger than that given by Pitjeva (2004) mainly due to our treatment of the uncertainties in the asteroid mass parameters. In normal ephemeris development we have not previously included radio range measurements from recent Mars orbiting spacecraft near solar conjunction due to concerns about possible correlations of the solar plasma delay with other parameters such as PPN γ. In a preliminary look at radio range measurements during the

Mars conjunction of 2009 we find that the solar plasma and PPN γ effects are reasonably de-correlated and get an improved estimate of $(\gamma - 1) = 0.00030 \pm 0.00050$. A better estimate has been given from an experiment with Cassini using a multi-band radio experiment to cancel the solar plasma with an uncertainty a factor of 25 lower by Bertotti *et al.* (2003).

The PPN parameter β mainly appears in the Mars spacecraft data set through the precession of the perihelion of Mars, an effect predicted by general relativity to be $\delta\phi/\mathrm{orbit} = (2 - \beta - 2\gamma)6\pi \mathrm{M}_\odot/[3\mathrm{a}(1 - \mathrm{e}^2)]$ with a nominal value of 1.35"/century. With the current Mars spacecraft data set we find $(2\gamma - \beta - 1) = +0.0004 \pm 0.0012$. The uncertainty is larger than some other published values due to our treatment of uncertainties in the asteroid mass parameters. An analysis of radio range measurement to the Venus Express orbiter may be expected to give a better constraint due to less perturbations by the asteroids. A tighter constraint on β has been derived from lunar laser ranging analysis (e.g. Williams *et al.* 2004).

4. Inverse-square law

Analysis of radio tracking data of some spacecraft appear to give a small anomalous acceleration of the spacecraft radially toward from the Sun (e.g. Anderson *et al.* 1998). It has been recognized that radio range measurements to Mars from the Viking landers show no such anomalous acceleration at a much lower level. From the current Mars radio range data set, we find upper bounds on a radial acceleration of Earth and Mars to be less than 3×10^{-14} m/s^2 and 8×10^{-14} m/s^2 respectively. A tighter constraint on such an effect can be derived from radio tracking data of the Cassini spacecraft in orbit about Saturn, giving a radial acceleration of Saturn as $< 1 \times 10^{-14}$ m/s^2.

Another common test is to look for a time variation in the gravitation constant G. From the Mars spacecraft data set we can estimate a time variation in the mass parameter of the Sun. We find that $1/(GM_\odot) \times d(GM_\odot)/dt < 2 \times 10^{-13}$/year, comparable to the uncertainty given from lunar laser ranging by Williams *et al.* (2004). By comparison, the mass loss of the Sun due to emission of photons is about 7×10^{-14}/year. Assuming the mass loss from the Sun is comparable with that from photon emission, the estimate of the rate of change of the Sun's mass parameter gives an upper bound on the rate of change in the gravitation constant G. We have looked at the Mars spacecraft data set to see if it can be used to separately distinguish a rate of change of the Sun's mass and the gravitation constant G and we find that the two are almost completely correlated and hence not distinguishable.

5. Mach's principle

The orbits of the planets are integrated in a locally inertial (non-rotating) reference frame through use of the Einstein-Infeld-Hofmann equations of motion. Lunar laser ranging and radio ranging to Mars-orbiting spacecraft determine the internal dynamics of the solar system to very high accuracy. For example, the mean motion of Mars relative to Earth can be determined with an accuracy of about 0.001"/century without use of any observations relative to objects outside the solar system. The dynamical ephemeris is usually aligned to an external coordinate system through use of angular measurements of planets or spacecraft in orbit about planets relative to stars or extra-galactic radio sources. We normally assume that the locally inertial reference frame is not rotating with respect to the rest of the universe. This assumption (Mach's principle) can be tested by comparing the mean motion of Mars relative to Earth determined from ranging measurements with the mean motion determined from VLBI observations of Mars-orbiting spacecraft. In a preliminary assessment of this effect we find that the dynamical rotation rate of the solar system relative to extra-galactic radio sources is less than 0.004"/century. This upper bound is larger than the expected rotation rate of the solar system about the galactic center. The estimate could be affected by systematic errors in the catalog of extra-galactic radio sources as well as by limits on modeling the motions of Earth and Mars. The uncertainty in the rate is expected to improve in the near future with updates in the catalogs of radio sources and subsequent reprocessing of Mars spacecraft VLBI measurements in the coming year.

6. Acknowledgements

The results presented here are based on the enormous efforts of reducing spacecraft orbits relative to the planets and other aspects of radio measurement processing, that were provided by Alex Konopliv, Robert Jacobson, and James Border of JPL and Trevor Morely and colleagues at ESOC. This research was carried out in part at the Jet Propulsion Laboratory, California Institute of Technology, under contract with the National Aeronautics and Space Administration.

References

Anderson, J. D., Laing, P. A., Lau, E. L., Liu, A. S., Nieto, M. M., & Turyshev, S. G. 1998, *Phys. Rev. Lett.*, 81, 2858

Baer, J. & Chesley, S. R. 2008, *Celest. Mech. Dyn. Astr.*, 100, 27

Bertotti, B., Iess, L., & Tortora, P. 2003, *Nature*, 425, 374

Einstein, A., Infeld, L., & Hoffmann, B. 1938, *Ann. Math*, 39, 65

Konopliv, A. S., Yoder, C. F., Standish, E. M.,Yuan, D. N., & Sjogren, W. L. 2006, *Icarus*, 182, 23

Matson, D. L., Veeder, G. J., Tedesco, E. F., Lebronsky, L. A. & Walker, R. G. 1986, *Adv. Space Res.* 6(7), 47

Moyer, T. D. 2000, *Formulation for Observed and Computed Values of Deep Space Network Data Types for Navigation* JPL Publication 00-7 (Pasadena: Jet Propulsion Laboratory, California Institute of Technology)

Pitjeva, E. V. 2004, *Astron. Lett.*, 31, 340

Reasenberg, R. D., Shapiro, I. I., MacNeil, P. E., Goldstein, R. B., Briedenthal, J. C., Brinkle, J. P., Cain, D. I., Kaufman, T. M., Komarek, T. A., & Zygielbaum, A. I. 1979, *ApJ*, 234, L219

Shapiro, I. I. 1964, *Phys. Rev. Lett.*, 13, 789

Williams, J. G., Turyshev, S. G., & Boggs, D. H. 2004, *Phys. Rev. Lett.*, 93, 261101

Relativity in Fundamental Astronomy
Proceedings IAU Symposium No. 261, 2009
S. A. Klioner, P. K. Seidelman & M. H. Soffel, eds.

doi:10.1017/S1743921309990330

Gravity tests with INPOP planetary ephemerides

A. Fienga[1,2], J. Laskar[1], P. Kuchynka[1], C. Leponcin-Lafitte[3], H. Manche[1] and M. Gastineau[1]

[1]Astronomie et Systèmes Dynamiques, IMCCE-CNRS UMR8028, 77 Av. Denfert-Rochereau, 75014 Paris, France and [2]Observatoire de Besançon- CNRS UMR6213, 41bis Av. de l'Observatoire, 25000 Besançon and [3] SYRTE-CNRS UMR8630, Observatoire de Paris, 77 Av. Denfert-Rochereau, France

Abstract. We present here several gravity tests made with the latest INPOP08 planetary ephemerides. We first propose two methods to estimate the PPN parameter β and its correlated value, the Sun J_2, and we discuss the correlation between the Sun J_2 and the mass of the asteroid ring. We estimate a possible advance in the planet perihelia. We also show that no constant acceleration larger than 1/4 of the Pioneer anomaly is compatible with the observed motion of the planets in our Solar System.

1. Introduction

Since 1981 and the use in the JPL NASA orbit determination software of the Einstein-Hoffman equations of motion in the frame of the PPN formulism (Will 1971, Moyer 1981), deviation of the gravity from the General Relativity (GR) theory can be measured by the estimation of parameters scaling the gravity to the GR, essentially γ, β. The most precise measurement of γ can be obtained in experiments of light deflection by the Sun or a major planet and gravitational time delay experiments (Kopeikin and Makarov 2008, Bertotti *et al.* 2003). However β can only be estimated in using the advance of the planet perihelion in association with a γ determination (Will 2006, Williams *et al.* 2009).

At the same time, the planetary ephemerides have made a huge improvement in modeling of solar system bodies, and the analysis of tracking observations of spacecraft became a powerful tool to constrain the dynamics of these bodies and to determine the PPN parameters such as γ and β (Pitjeva 1986).

Besides the estimations of parameters such as PPN β, the planetary ephemerides are an interesting tool to study the impact of the Pioneer anomaly (PA) in the solar system. Since the unexplained acceleration exhibited by the two Pioneer spacecrafts, sent more than 30 years ago to the limit of our solar system, was confirmed by Anderson *et al.* (2002), many possible explanations were proposed and investigated. Some of the proposed explanations can be tested thanks to planetary ephemerides. Finally as GR was the explanation of the secular advance of the perihelion of Mercury detected by the old optical observations, one can be interested in using new accurate radar tracking data to detect supplementary advances unexplained by GR (Pitjeva 1986).

In this paper, we give several estimations of PPN β, as well as the Sun oblateness coefficient J_2. We show how the planetary ephemerides can help to solve the Pioneer anomaly, and we estimate the limit of detection of possible extra advances in perihelia based on modern observations of planets.

Table 1. The first 2 columns give the a-priori INPOP uncertainties in geocentric angles and distances limited by the observation accuracy. In the third column are the estimates of the general relativity and Sun oblateness contribution to the perihelion rate $\dot{\omega}$, of Mercury, Venus and Mars. Column 4 gives the S/N ratio estimated over the time period of column 5.

Planets	INPOP accuracy angle	distance	$\dot{\omega}$ "/yr	S/N	period years	Planets	INPOP accuracy angle	distance	$\dot{\omega}$ "/yr	S/N	period year
Venus	0.001"	4m	0.086	172	2	Mars	0.001"	2m	0.013	130	10
				344	4					390	30
Mercury	0.050"	1km	0.43	300	35						

2. Determination of PPN β and the Sun oblateness J_2

2.1. *Planetary ephemerides accuracy*

Thanks to the high precision achieved with the observations deduced from spacecraft tracking, it becomes possible to estimate relativistic parameters γ and β of the Parametrized Post-Newtonian formalism of General Relativity (Will, 1993). Nevertheless, if γ plays a role in the equations of motion, it is worth noting that light propagation is only sensitive to that parameter. PPN γ can then be estimated with high accuracy by light deflection measurements by VLBI (Shapiro *et al.* 2004, Lambert & Le Poncin-Lafitte, 2009), by time delay during an interplanetary roundtrip, and by Doppler tracking data of a space mission (see for instance the Cassini experiment, Bertotti *et al.* 2003). This is also why, in the following, we assume $\gamma = 1$ in order to test only the sensitivity of PPN β on the perihelion's advance of planets. However, the Sun oblateness J_2 plays also a key role in this phenomena. Indeed, the usual expression of the advance of perihelion is given by (Will 2006)

$$\Delta\omega = \frac{2\varpi(2\gamma - \beta + 2)GM_{\text{sun}}}{a(1-e^2)c^2} + \frac{3\varpi J_2 R_{\text{sun}}^2}{a^2(1-e^2)^2} \tag{2.1}$$

where G and c are the newtonian gravitational constant and the speed of light in vacuum, respectively. J_2, M_{sun} and R_{sun} are the Sun oblateness, mass and equatorial radius, respectively, while a and e are the semi-major axis and the eccentricity of the precessing planet. The PPN β is, thus, correlated with the Sun oblateness J_2 through this linear relation. But, the β coefficient varies as $1/a$, while the J_2 coefficient is proportional to $1/a^2$. Using data from different planets will, thus, allow to decorrelate these two parameters. MEX and VEX tracking data have actually led to an important improvement of Mars and Venus orbits in INPOP08 (Fienga *et al.* 2009).

Indeed, we can evaluate the impact of the observations of a specific planet on the determination of J_2 and β by dividing the advance of the perihelion over the time span of observations by the angle uncertainty of INPOP (table 1). For the same observational accuracy, it appears that Venus data are seven times more efficient than Mars to test general relativity and to estimate the Sun J_2. Therefore, if the VEX mission is extended from 2 years to 4 years, and if VLBI observations of the spacecraft are done with an accuracy of about 1 mas, VEX data will be as important for the PPN and Sun J_2 estimations as the direct 800-meter accuracy radar ranging on Mercury. Besides, the Mars data are still very important because of the long time span of observations of very good quality obtained since the Viking mission in 1978.

Thanks to the information brought by the combination of very accurate tracking data of spacecraft orbiting different planets, the planetary ephemerides become thus an interesting tool for gravity testing. In the following, we give some examples of such tests.

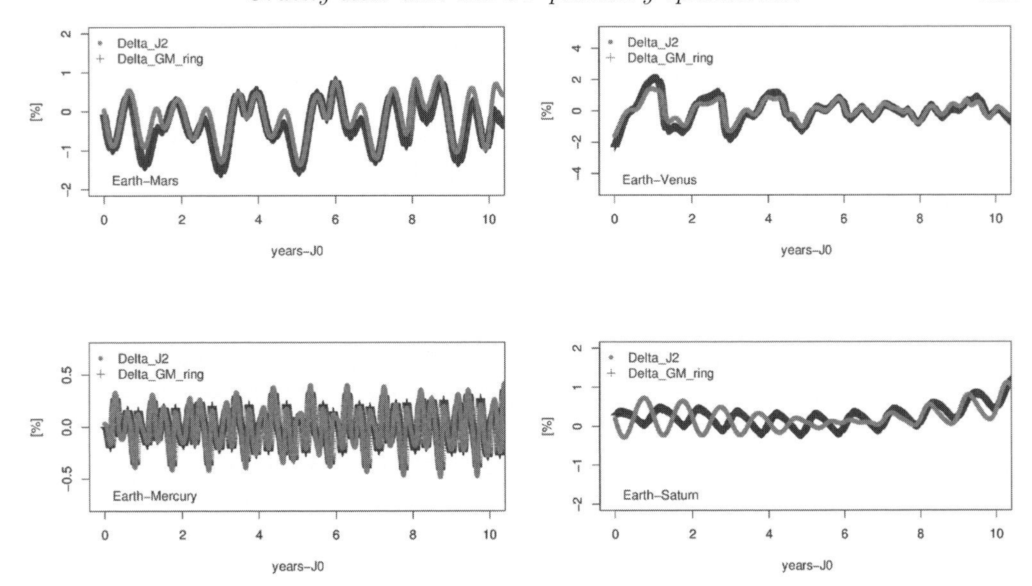

Figure 1. Residuals obtained by comparisons between Mercury direct range, MGS/MO, MEX, VEX and Cassini range tracking data and ephemerides perturbed by a small change in the Sun J_2 (12%) and by a small change in the mass of the asteroid ring (17%).

2.2. *Correlation between the Sun J_2 and asteroid modeling*

The advance of the perihelion induced by general relativity and the Sun J_2 has an impact very similar to the advance induced by the main-belt asteroids on the inner planet orbits. In INPOP08, a ring was added to average the perturbations induced by the main belt asteroids which cannot be fitted individually by tracking observations. This ring has its physical characteristics (mass and distance to the Sun) estimated independently from the fit by considering the albedos and physical properties of 24635 asteroids (for more details see Kuchynka *et al.* 2008).

As illustrated in figure 1, there is a correlation between the effect on the geocentric distance of the modeling of the ring as done in INPOP08, in one hand, and the effect of the Sun oblatness in the other hand. Indeed, on these plots, one may see how a small change in the value of the Sun J_2 (12%) induces, after the refit of the planet initial conditions a periodic effect very similar in amplitude and frequency as a change in the mass of the asteroid ring (17%). This effect is obvious on Mercury, Mars and Venus distances to the Earth, but not for Saturn. The Saturn-Earth distances are indeed not affected in the same way. We can also conclude that, when new accurate observations of outer planets will be obtained, they will be very useful to decorralate asteroid effects on planet orbits by combination with inner planet data. Finally, it stresses the crucial importance of having a model of the asteroid perturbations as a fixed ring, characterized independantly from the fit of planetary ephemerides.

By fixing the ring, we limit then an overestimation of the value of the Sun J_2 merging in this value some effects induced by the asteroids.

Two different but complementary analysis and determination of PPN β and the Sun J_2 are presented in the next sections with a fixed model of asteroid perturbations (same values of asteroid and ring masses and of densities as INPOP08).

2.3. *Estimations by least squares*

The first approach is based on the classic least square estimation of parameters during the fit of planetary equations of motion to observations. To check numerically the simplified assertion made in the introduction (section 1), we estimate here what is the impact of each data set in the determination of J_2 and β: several adjustments of the initial conditions of planets and the parameters J_2 and β are made using different sets of observations. This leads to 32 adjustments based on INPOP08. For each fit, changes were made in the selection of Mars and Venus data in order to estimate the impact of each important set of observations in the fit of the Sun J_2 and PPN β. We look at the variations in the estimation errors of the 2 parameters and we use the 1-σ given by the least squares as an indicator of this uncertainty. With this method, we are able to quantify the influence of each data set on the determination of the pair (β, J_2) as well as the stability of the determinations of the parameters. Indeed, these variations in the error's estimation of the pair (β, J_2) are a relevant indicator of the uncertainty of the fit of β and J_2.

To take into account the correlation between J_2 and β, we use two modes of adjustments: in *mode1*, β or J_2 are fitted alone with the initial conditions of planets; in *mode2*, both parameters are fitted simultaneously with the initial conditions of planets.

The results are summarized in table 2. As expected from the correlation of J_2 and β, the determinations of the Sun J_2 and β made separately (i.e. *mode1*) give better σ than fits including the simultaneous (β, J_2) determination (*mode2*). The best results for a correlated determination of J_2 and β (*mode2*) are obtained when only the most accurate observations of Mars (MGS/MO, MEX and Viking) and Venus (VEX) are used simultaneously.

Moreover, we note that the combined use of Venus ranging data and the complete data set for Mars does not really improve the separated determination (*mode1*) of β and J_2, mainly due to the low accuracy of these observations, but *a contrario*, it gives better correlated estimations (mode2). This is also consistent with the fact that fitting over observations from two different planets helps to decorrelate J_2 and β. Furthermore, the Viking data, by prolonging the fit interval with observations of rather good accuracies, allow a decreasing of the uncertainties of about 20 % for J_2 and about 10% for β. Finally, it appears that VEX data improve the determinations in a significant way: decreasing by 31 % the least squares σ of the J_2 estimation and 48 % for β. Less than 2 years of VEX data have a bigger influence than a large interval (more than 30 years) of accurate Mars observations. This is especially relevant for the PPN parameter β with an improvement of about 48% of the accuracy between a determination including only Mars data and another one with both Mars and VEX data. In the same time, the improvement induced by the addition of Viking data is about 20 % for J_2 and 10% for the PPN parameter β. These figures show the crucial role of the VEX data before the use of future data from the ongoing generation of Mercury orbiters.

2.4. *Incremental method and sensitivity estimation*

An alternate strategy to study the sensitivity of the planetary ephemerides to J_2 and PPN β is to estimate how does an ephemeris built using different values for J_2 and PPN β and fitted on the same set of observations as INPOP08 differ from INPOP08. Such differences give an indication on how observations are sensitive to these parameters and with which accuracy we can estimate a parameter such as β.

To test such sensitivity, we focus our attention on the postfit residuals of the most accurate data sets used in the INPOP08 adjustment: the Mercury direct range, because of its sensitivity to general relativity and to the Sun J_2; VEX, MEX and MGS/MO data, because of their high accuracy and simulated S/N presented on table 1; and Jupiter

Table 2. 1-σ least squares obtained for J_2 and β using several sets of observations.

	mode	J_2 $\times 10^7$	$(1-\beta)$ $\times 10^3$		mode	J_2 $\times 10^7$	$(1-\beta)$ $\times 10^3$
Modern Mars	1	0.181		Impact of VEX	1	0.144	
MEX + MGS/MO	1		0.042	Mars + VEX	1		0.025
	2	0.367	0.085		2	0.208	0.037
Impact of Vkg	1	0.161		Impact of old Venus	1	0.188	
MEX + MGS/MO + Vkg	1		0.040	Mars + old Venus	1		0.040
= Mars	2	0.302	0.076		2	0.283	0.060

Galileo data and Saturn Cassini normal points. These 2 latest data sets are selected because they induce a global improvement of the planetary ephemerides and especially of the Earth orbit.

To estimate the sensitivity of these 7 most accurate sets of data used in the INPOP08 adjustment to the variations of values of J_2 and PPN β, we have estimated and plotted the S/N ratio defined as:

$$S/N = \frac{\sigma_{i,j} - \sigma_{0,0}}{\sigma_{0,0}}$$

where $\sigma_{i,j}$ is the 1-sigma dispersion of the postfit residuals of an ephemeris based on INPOP08 but with values of J_2 and PPN β different from the ones used in INPOP08 (which are $\beta = 1.0$ and $J_2 = 1.82 \times 10^{-7}$) and fitted to all the INPOP08 data sets, and $\sigma_{0,0}$ is the 1-sigma dispersion of the postfit INPOP08 residuals. We have used 9 values of J_2 varying from 1.45×10^{-7} to 3.05×10^{-7} with a 0.2 step, and 24 values of PPN β, building then 192 different ephemerides. The 24 values of β are distributed over 2 windows: a global one based on 12 values of β varying from 0.997 to 1.003 with a 0.0005 step (window 1) and from 0.9996 to 1.0004 with a step of 0.0001 (window 2). Results presented as the S/N percentage, are plotted in figure 2. As one can see in figure 2, the impact of the PPN β is not symmetric with respect to $\beta = 1$. In figure 2, one notices also the direct correlation between the S/N obtained with MGS/MO and MEX data and the one obtained for VEX.

One may see in figure 2 that the S/N of the Jupiter and Saturn data sets are sensitive to changes in J_2 and PPN β. The sensitivity of these data sets are not crucial for the analysis, but they reflect the impact of the use of such observations in the improvement of the Earth orbit and then the sensitivity of the Earth orbit to the gravity testing. In table 3, we have gathered minimum and maximum values of PPN β defining the sensitivity interval of the different data sets. The sensitivity interval is the interval of PPN β for which the S/N remains below 5%. Values of PPN β greater than the maximum value given in table 3, or smaller than the minimum value, cannot be seen as realistic in comparison to modern observations. By considering figure 2 and table 3 it appears that MGS/MO and MEX data provide the most narrow interval of sensitivity with $0.99995 < \beta < 1.0002$. This interval is in agreement with the latest determinations done by Williams *et al.* (2009), Fienga *et al.* (2008) and Pitjeva (2006).

3. Secular advances of planetary perihelia

We are interested here in evaluating if the observations used to fit INPOP08 would be sensitive to supplementary precessions of the planet orbits. Such anomalous precessions that would be unexplained by general relativity have been recently investigated (Iorio,

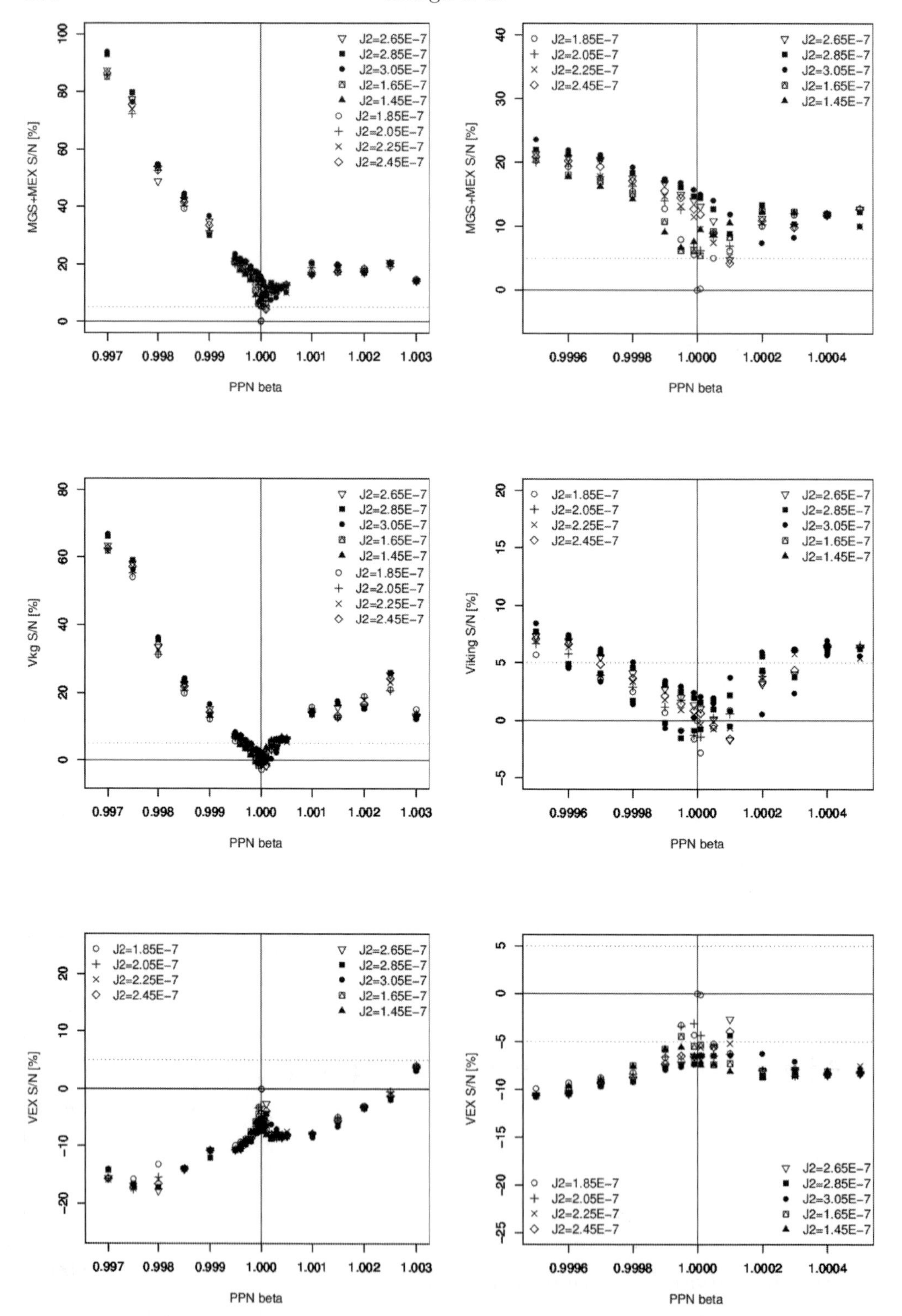

Figure 2. Residuals obtained by comparisons between observations and ephemerides estimated with different values of PPN β (values given on x-axis of each subframes) and different values of the Sun J_2.

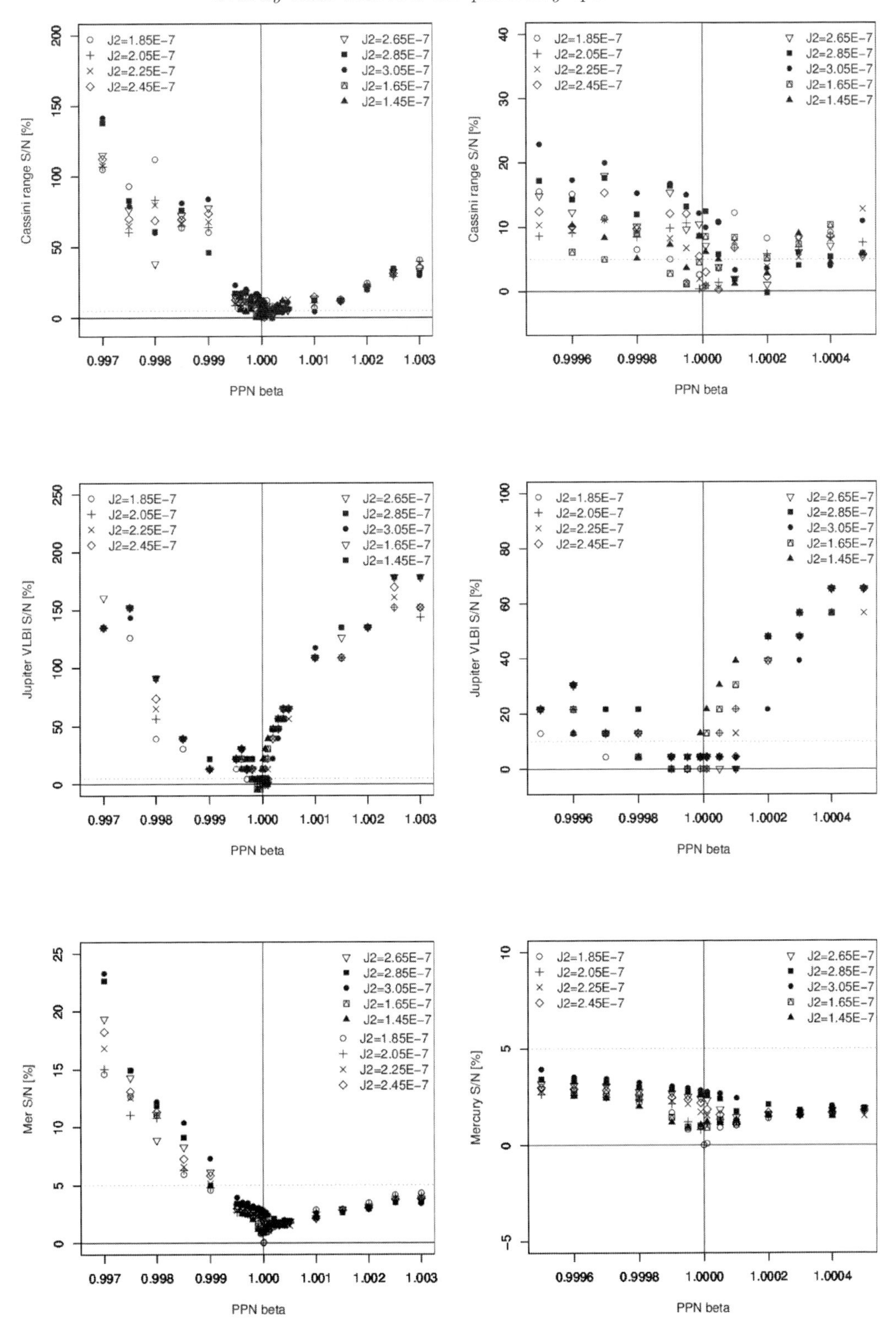

Figure 3. Residuals obtained by comparisons between Mercury, Jupiter VLBI and Saturn range observations and ephemerides estimated with different values of PPN β (values given on x-axis of each subframes) and different values of the Sun J_2.

Table 3. β intervals in which the residuals stay below the 5% limit. The values of β given here are estimated for $\gamma = 1$.

Data	β min	β max	Data	β min	β max	Data	β min	β max
MGS/MO+MEX	0.99995	1.0002	Jupiter VLBI	0.9996	1.0002	Viking	0.9995	1.0002
VEX	0.99990	1.0002	Saturn Cassini range	0.9998	1.0005	Mercury	0.9985	1.005

2009). To estimate the sensitivity of the modern tracking data, we first fix J_2=1.8×10^{-7}, $\beta = 1$ and $\gamma = 1$. By fixing the value of the Sun J_2, we then isolated the impact of the secular advance of the perihelion, $\dot{\varpi}_{\text{sup}}$, for one given value of J_2.

For each different value of $\dot{\varpi}_{\text{sup}}$, initial conditions of planets are fit to the INPOP08 observations and we compare the postfit residuals to the INPOP08 ones. We focused our study on the same sets of observations as for the J_2, β study. As one can see in figure 4, the behaviour of the obtained S/N (as defined in section 2.4) is symmetrical to a minimum value, this minimal value being centered around $\dot{\varpi}_{\text{sup}} = 0$ or not. This symmetry explains why in table 4 we give an interval of $\dot{\varpi}_{\text{sup}}$ for which the minimum of S/N is obtained. One can then compare these values to those published by (Pitjeva 2009). For all the planets, except Saturn, the values of $\dot{\varpi}_{\text{sup}}$ minimizing the residuals are not significantly different from zero. One can note that the best constraint on the Earth orbit is given by the Jupiter VLBI data set which gives the narrowest interval of $\dot{\varpi}_{\text{sup}}$. For Saturn, an offset in the minimum of the S/N is obtained for the Cassini tracking data set (-10 ± 8) and the VEX data set (200 ± 160). These estimations lead to determinations of a supplementary precession of the Saturn orbit that are only marginally statistically significant. By comparisons, (Pitjeva 2009) the value is very close to the one we obtain by considering only the S/N induced on the Cassini observations. This result shows how important the description of the method used for evaluating such quantities.

To test the stability of the estimations and as it is well-known that the asteroids induce a global precession of the inner planets perihelia, we operate the same computations with small changes in the mass of the ring (20%) and in the Sun J_2 (5%) values. The obtained variations of the S/N are plotted in figure 4 where the dash curves are the results obtained with the change in the mass of the ring and the longdash curves are the ones deduced from the J_2. Some changes are noticeable for Viking and Jupiter, however, for Cassini and VEX, the minimum are stable.

The investigation about a statistically significant advance in the Saturn perihelion has to be continued in using more Cassini and VEX data. Indeed, a prolongation of the interval of time covered by these two data sets will improve the accuracy of the estimations.

4. Does the Pioneer anomaly impact the ephemerides?

Since 2002 and the confirmation by several teams of the detection of acceleration anomalies in the tracking of several spacecrafts, three features of possible explanations can be given; first, the detected acceleration is not really an acceleration but is a manifestation of a mis-modeling in the Doppler and ranging signals taped by navigation teams. Second, the anomaly is a mis-modeling in the orbit of the probe itself induced by a technical problem or misunderstandings of the spacecraft techniques. The third cause invocated is a generalization of the second one by implying a mis-modeling in the dynamics of the probe but also of all objects in the solar system and beyond. Thus, if the equivalence

Table 4. $\dot{\varpi}_{\rm sup}$ intervals minimizing postfit residuals.

	$\dot{\varpi}_{\rm sup}$ [mas/cy]							
Data	Mer	Ven	EMB	Mars	Jup	Saturn	Ura $\times 10^{-4}$	Nep $\times 10^{-4}$
Mercury	-10 ± 30	30 ± 130	0 ± 40	> 2000	> 2000	0 ± 200	> 20	> 20
VEX	0 ± 200	18 ± 22	0 ± 4	0 ± 1.4	0 ± 200	200 ± 160	0 ± 2	> 20
MGS/MO+MEX	0 ± 200	-24 ± 34	-0.4 ± 0.8	0.4 ± 0.6	-20 ± 180	0 ± 60	0 ± 2	0 ± 10
Viking	0 ± 200	-24 ± 34	0 ± 0.2	0 ± 0.2	-200 ± 200	0 ± 10	(4 ± 4)	0 ± 10
Jupiter VLBI	0 ± 400	-4 ± 6	0 ± 0.016	0 ± 0.6	142 ± 156	0 ± 10	0 ± 2	0 ± 2
Saturn range Cassini	> 2000	0 ± 10	0.1 ± 0.1	0 ± 0.2	0 ± 400	-10 ± 8	0 ± 2	0 ± 2
Pitjeva 2009	-3.6 ± 5	-0.4 ± 0.5	-0.2 ± 0.4	0.1 ± 0.5		-6 ± 2		

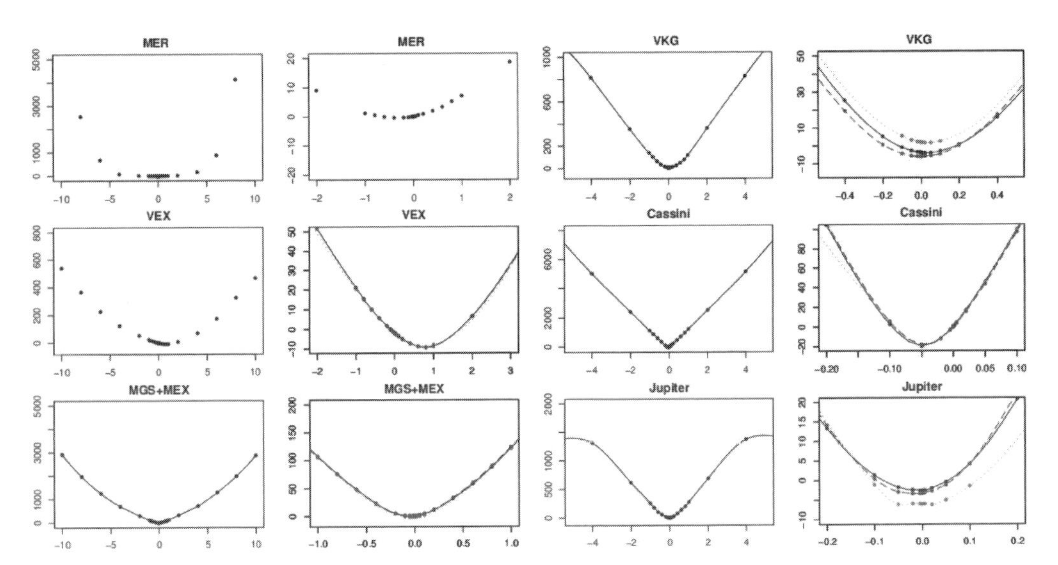

Figure 4. Residuals obtained by comparisons between observations and ephemerides estimated with different values of $\dot{\varpi}_{\rm sup}$. The dash curves represent residuals obtained with a 20% change in the mass of the asteroid ring whereas the longdash curses represent residuals obtained with a 5% change in the J2. The x-axis give the values of $\dot{\varpi}_{\rm sup} \times 10^8$ used in the simulations and the y-axis give the variations of the S/N in %.

principal is followed, the equations of motion of the major planets of our solar system have also to be modified in the same manner as the spacecraft dynamical equations are.

We investigate this question by using the INPOP08 planetary ephemerides as a test bed for some hypothesis describing the pioneer anomalies.

A classic description of the pioneer anomalies (PA) is the appearance of a constant acceleration of about $8.75 \times 10^{-10} \mathrm{m\,s^{-2}}$, Sun-oriented after 20 AU (Anderson *et al.* 2002). We, thus, add this constant acceleration in the equations of motions of Uranus, Neptune and Pluto.

We have then fit the modified ephemerides to observations usually used to built INPOP08. Residuals obtained after the fit are plotted in Figure 5. The value of the acceleration was changed in a way to obtain a minimum value for which the effect induced by such acceleration becomes detectable in the residuals. As it appears clearly in the residuals of Uranus right ascension, a constant acceleration of $8 \times 10^{-10} \mathrm{m\,s^{-2}}$ added to

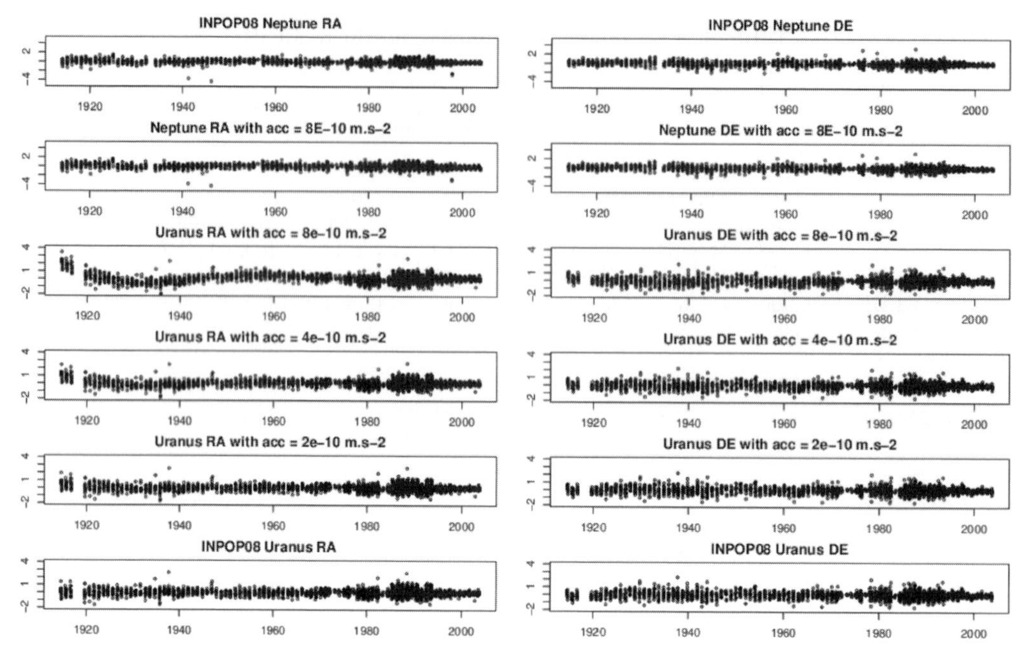

Figure 5. Residuals in right ascension and declination of Neptune and Uranus obtained with INPOP08 (solution of reference) and fitted ephemerides including PA of different magnitudes: from 8 to 2 $\times 10^{-10}\,\mathrm{m\,s^{-2}}$. The x-axis are years and y-axis is in arcseconds.

the classical Einstein-Hoffmann equations of motion can not be missed, even after the fit of the Uranus initial conditions. A systematic effect remains especially after 1930. This effect cannot be absorbed by the fit or by the noise of the old Uranus observations. By changing the value of the acceleration, one can see that the acceleration must be at least 4 times smaller than the one commonly adopted to be absorbed by the residuals. For Neptune and Pluto, the situation is different. For these planets, the effect of a constant acceleration is absorbed by the fit, as one can see on figure 5 with the postfit and prefit residuals of Neptune.

5. Conclusions

Concerning the determination of the PPN parameter β, an estimation of the sensitivity of planetary ephemerides to this parameter is done following two methods. Our results show that a global fit is needed in order to decorrelate parameters such as PPN β, the Sun J_2 and the asteroid pertubations.

We have tested possible detection of an anomalous advance of perihelia of planets. More investigations are needed for the analysis of the perihelion rate of Saturn and more observations of Cassini and VEX data are necessary.

Finally, the results obtained here for the Pioneer Anomaly conclude that no constant acceleration larger than 1/4 the PA can affect the planets of our solar system. If it was so, it would have been detected sooner. In the frame of the equivalence principle, this means that no constant acceleration larger than 1/4 the PA can be realistic.

References

Anderson, J. D., Laing, P. A., Lau, E. L., Liu, A. S., Nieto, M. M., & Turyshev, S. G., 2002, *Phys. Rev. D*, 65, 082004

Bertotti, B., Iess, L., & Tortora, P., 2003, *Nature*, 425, 374

Fienga, A., Manche, H., Laskar, J., & Gastineau, M., 2009, *JOURNEES-2008/ Astrometry, Geodynamics and Astronomical Reference Systems*

Fienga, A., Manche, H., Laskar, J., & Gastineau, M., 2008, *A&A*, 477, 315

Iorio, L., 2009, *AJ*, 137, 3615

Kopeikin, S. & Makarov, V., 2008, *IAU Symposium*, 248, 391

Lambert, S. & Le Poncin-Lafitte, C., 2009, *A&A*, 499, 331

Moyer, T. D., 1981, *Cel. Mech.*, 23, 33.

Kuchynka, P., Laskar, J., Fienga, A., Manche, H., & Somenzi, L., 2009, *JOURNEES-2008/ Astrometry, Geodynamics and Astronomical Reference Systems*

Pitjeva, E. V., 1986, *Byull. Inst. T. A., Ross. Akad. Nauk.*, 15, 538

Pitjeva, E. V., 2009, *JOURNEES-2008/ Astrometry, Geodynamics and Astronomical Reference Systems*

Shapiro, S. S., Davis, J. L., Lebach, D. E., & Gregory, J. S., 2004, *Phys. Rev. Lett.*, 92, 121101

Will, C. M., 2006, *Living Rev. Relativity*, 9, 3

Will, C. M., 1993, Cambridge University Press, New York, U.S.A.2nd edition

Will, C. M., 1971, *Astrophys. J.*, 163, 611

Williams, J. G., Turyshev, S. G., & Boggs, D. H., 2009, *Int. Jour. Mod. Phys. D*, arXiv, gr-qc0507083v2

Relativity in Fundamental Astronomy
Proceedings IAU Symposium No. 261, 2009
S. A. Klioner, P. K. Seidelman & M. H. Soffel, eds.

doi:10.1017/S1743921309990342

EPM ephemerides and relativity

E. V. Pitjeva

Institute of Applied astronomy RAS,
Kutuzov quay 10, 191187 St. Petersburg, Russia
email: evp@ipa.nw.ru

Abstract. In the seventies of the last century the EPM ephemerides (**E**phemerides of **P**lanets and the **M**oon) of IAA RAS originated and have been developed since that time. These ephemerides are based upon relativistic equations of motion of celestial bodies and light rays and upon relativistic time scales. The updated model of EPM2008 includes the new values of planet masses and other constants, the improved dynamical model with adding Trans–Neptunian Objects and the expanded database (1913–2008). More than 260 parameters have been determined while improving the planetary part of EPM2008 to 550000 observations. EPM2008 have been oriented to ICRF by including into the total solution the VLBI data of spacecraft near the planets. The real uncertainty of EPM ephemerides has been checked by comparison with the JPL's DE ephemerides. Some estimates of the post–model parameters have been obtained:
$|1-\beta| < 0.0002$, $|1-\gamma| < 0.0002$, $\dot{G}/G = (-5.9 \pm 4.4) \cdot 10^{-14}$ per year, the statistic zero corrections to the planet perihelion advances.

Keywords. Relativity, celestial mechanics, ephemerides, radar astronomy

1. Historical introduction: general relativity in EPM ephemerides

In the seventies of the last century to support space flights the EPM ephemerides (**E**phemerides of **P**lanets and the **M**oon) of IAA RAS originated at about the same time as DE ephemerides and have been developed since that time.

After the brilliant explanation by Einstein the strange ($43''$/cy) discrepancy between theoretical predictions and observations of the secular motion of Mercury perihelion, the planet ephemerides are to be constructed on the basis of General Relativity. The relativistic basis for constructing ephemerides was provided many years ago in the papers by Estabrook (1971), Will (1974), and it has been used for JPL (Standish, 1976), IAA RAS (Krasinsky *et al.*, 1978), and MIT (Ash *et al.*, 1967) ephemerides for more than 30 years. However, for the workers in Russia the main guide was the book of Brumberg (1972). Moreover, the relativistic equations of the planet motion may be given in different coordinate systems of the Schwarzschild metric (parameter α), namely, standard, harmonic, izotropic, etc. However, planet coordinates turned out to be essentially different for the standard and harmonic systems. Brumberg (1979) proved that ephemeris construction and processing of observations should be done in the same coordinate system, in which case the dependence on the coordinate system (parameter α) vanishes. Later on, the resolutions of IAU (1991, 2000) recommended to use harmonic coordinates for BCRS. Actually, harmonic coordinates have been used for all modern ephemerides since long ago.

Our first ephemerides of the inner planets which were analytical (Krasinsky *et al.*, 1978) in contrast to more perfect analytical ephemerides of the Moon and planets by Chapront and Bretagnon were compared with optical and radar observations. Simultaneously we also computed numerical planet ephemerides. Our comparison revealed that numerical ephemerides were able to present accurate observations much better than any analytical theories did.

In the eighties of the previous century (for example, Krasinsky *et al.*, 1986) we tested relativistic effects processing the observations available at that time. A purely Newtonian theory was developed and results were tested by both the relativistic and the Newtonian theories. It was proved that the relativistic ephemeris for any observed planets provides considerably better fit of the observations (by 10%) than the Newtonian theory even if latter incorporates the observed perihelion secular motions. Moreover, at that time attempts to estimate PPN parameters β, γ and the rate of changing of gravitation constant $\dot{G}/G$ were also made.

All the modern ephemerides: DE – JPL (Folkner *et al.*, 2008), EPM – IAA RAS (Pitjeva, 2009), INPOP – IMCCE (Fienga *et al.*, 2008) are based upon relativistic equations of motion for celestial bodies and light rays as well as relativistic time scales. The numerical integration of the equations of celestial bodies motion has been performed in the Parameterized Post–Newtonian metric for General Relativity in the TDB time scale; the relativistic effects of the signal delay (the Shapiro effect), and path-bending of the radio-signal propagation in the gravitation field of the Sun, Jupiter, Saturn and the reduction of observations from the proper time of the observer to the coordinate time of the ephemerides are taken into account while processing observations.

2. Present EPM2008 ephemerides

EPM ephemerides are computed by numerical integration of the equations of celestial bodies motion in the barycentric coordinate frame of J2000.0 by Everhart (1974) method over the 400 years interval (1800–2200) using the program package ERA-7 (ERA: **E**phemeris **R**esearch in **A**stronomy) developed to support scientific research in dynamical and ephemeris astronomy (Krasinsky & Vasilyev, 1997). This paper concerns a planet part of the EPM ephemerides; the group of George Krasinsky is now developing a lunar part of the EPM ephemerides and fitting it to the LLR data (Yagudina, 2009).

The mass values of the planets have been taken from the recent best determinations by different authors obtained from the data of spacecraft orbiting and passing near planets or from the observations of satellites of these planets (http://maia.usno.navy.mil/NSFA/CBE.html). All other constants have been obtained inside the EPM2008 ephemeris fitting process.

The updated model of EPM2008 includes Eris (which surpasses Pluto in the mass) and the other 20 largest Trans–Neptunian Objects (TNO) into the process of the simultaneous numerical integration in addition to nine planets, the Sun, 301 biggest asteroids, the Moon as well as the lunar physical libration, and takes into account perturbations due to the solar oblateness and perturbation from the massive ring of small asteroids.

Moreover, some tests have been made for estimating the effect of other TNO on the motion of planets. Their perturbations have been modeled by the perturbation from a circular ring having a radius of 43 AU and the five versions of different masses. The minimum mass of this ring is equal to the mass of 100000 bodies with 100 km in diameter and density is equal to 2 g/cm^3, it amounts to 110 masses of Ceres. The maximum mass of the ring is expected to be 100 times the minimum mass. Other test versions of the TNO ring surpass the minimum mass by 25, 50, and 75 times. The effect of the ring is only noticeable for more accurate observations – the spacecraft data, especially for ones from spacecraft near Jupiter and Saturn. The rms residuals and the weight unit errors for the data after fitting the standard and test EPM ephemerides have shown that all the test masses of the TNO ring except the minimum mass are too large and make the

Table 1. Mean values and rms residuals for radiometric observations.

Planet	Type of data	Time interval	N	$<O-C>$	σ
MERCURY	τ [m]	1964–1997	746	0	575
VENUS	τ [m]	1961–1995	1354	-2	584
	Magellan dr [mm/s]	1992–1994	195	0	0.007
	MGN,VEX VLBI [mas]	1990–2007	22	1.6	3.0
	Cassini τ [m]	1998–1999	2	4.0	2.4
	VEX τ [m]	2006–2007	547	0.0	2.6
MARS	τ [m]	1965–1995	403	0	719
	Viking τ [m]	1976–1982	1258	0	8.8
	Viking $d\tau$ [mm/s]	1976–1978	14978	-0.02	0.89
	Pathfinder τ [m]	1997	90	0	2.8
	Pathfinder $d\tau$ [mm/s]	1997	7569	0	0.09
	MGS τ [m]	1998–2006	7342	0	1.4
	Odyssey τ [m]	2002–2008	5257	0	1.2
	MRO τ [m]	2006–2007	380	0	2.5
	spacecraft VLBI [mas]	1984–2007	96	0.0	0.7
JUPITER	spacecraft τ [m]	1973–2000	7	0.0	11.8
	spacecraft VLBI [mas]	1996–1997	24	-1.8	9.5
SATURN	spacecraft τ [m]	1979–2006	33	1.0	20.2
URANUS	Voyager-2 τ [m]	1986	1	1.9	105
NEPTUNE	Voyager-2 τ [m]	1989	1	0.0	14

Notes: VEX, MGS, Odyssey, MRO data are normal points representing about 400000 original observations.

data residuals worse. Thus, the upper limit of the mass of the TNO ring ($5.26{\cdot}10^{-8}M_\odot$) has been obtained.

Database, to which EPM2008 have been adjusted includes (in addition to previous observations since 1913) the recent spacecraft measurements, namely, ranging to Venus Express (VEX), Odyssey, Mars Reconnaissance Orbiter (MRO) and VLBI data of Odyssey and MRO (2006–2008), three-dimensional normal point observations of Cassini (2004–2006), along with CCD Flagstaff and TMO data of the outer planets and their satellites (2006–2008). These measurements have resulted in a significant improvement of planet orbits, especially for Venus and Saturn and the orientation of the EPM2008 ephemerides to ICRF. The most part of observations has been taken from the database of the IAU Commission 4 created by Myles Standish and continuing by William Folkner.

About 260 parameters have been determined while improving the planetary part of EPM2008 to more than 550000 data:

- the orbital elements of all the planets and 18 satellites of the outer planets observations those have been used to improve the orbits of these planets;
- the value of the Astronomical Unit in m;
- three orientation angles of the ephemerides relative to the International Celestial Reference Frame (ICRF) and their velocities;
- 13 rotation parameters of Mars and the coordinates of the three landers on the martian surface;
- masses of the ten asteroids that perturb Mars most strongly, mean densities for three taxonomic classes of asteroids (C, S, M), the mass and the radius of the asteroid ring, the ratio masses of the Earth and the Moon;
- the solar quadrupole moment (J_2) and 21 parameters of the solar corona for different conjunctions with the Sun;
- eight coefficients of Mercury's topography and the corrections to the surface levels of Venus and Mars;
- five coefficients of the phase effect correction for the outer planets;
- constant bias for spacecraft and some radar planet observations, that were interpreted as calibration errors of the instruments or as systematic errors of unknown origin;

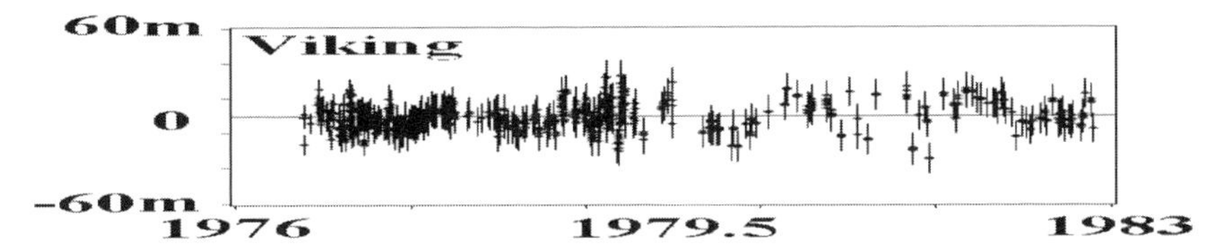

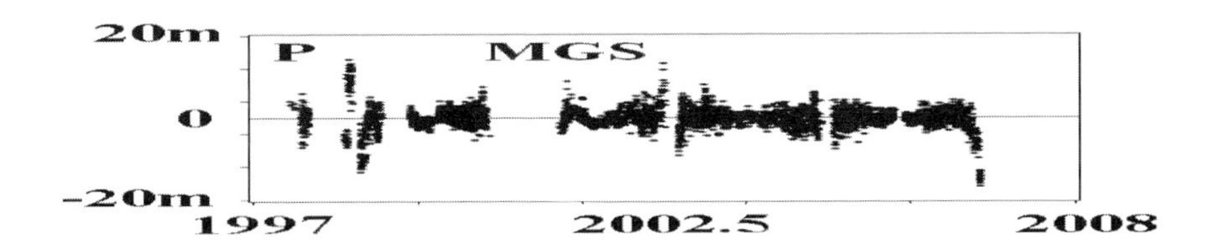

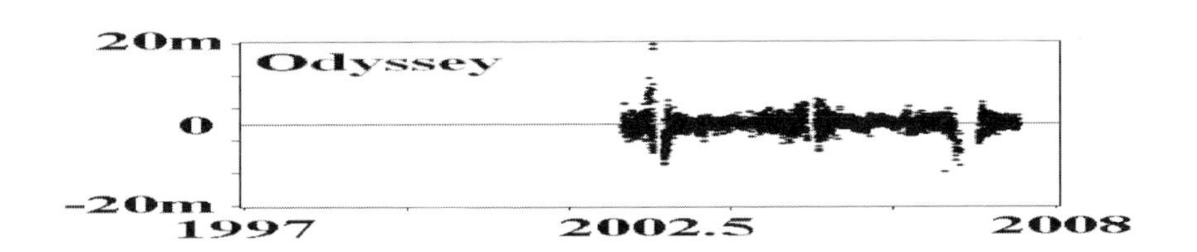

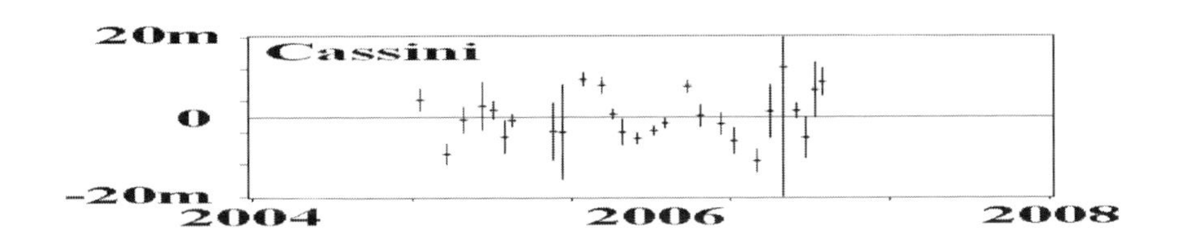

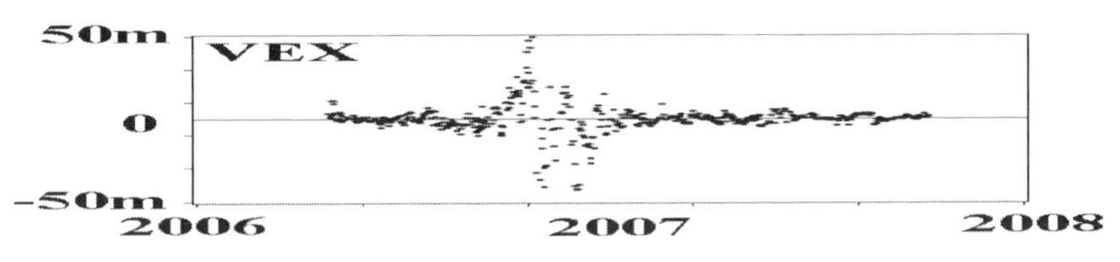

Figure 1. Viking, Pathfinder (P), MGS, Odyssey, Cassini, VEX range residuals

• the post-model parameters (β, γ, $\dot{G}/G$, secular trends of the planet perihelia and semi-major axes).

Mean values and rms residuals of observations are presented in Tables 1, 2 and on Fig. 1, 2. The data residuals don't exceed their *a priori* accuracies. The rms residuals of ranging for Viking are 8.8 m, for Pathfinder 2.8 m, for MGS and Odyssey 1.2–1.4 m, for Cassini (Saturn) 3.0 m, for VEX 2.6 m.

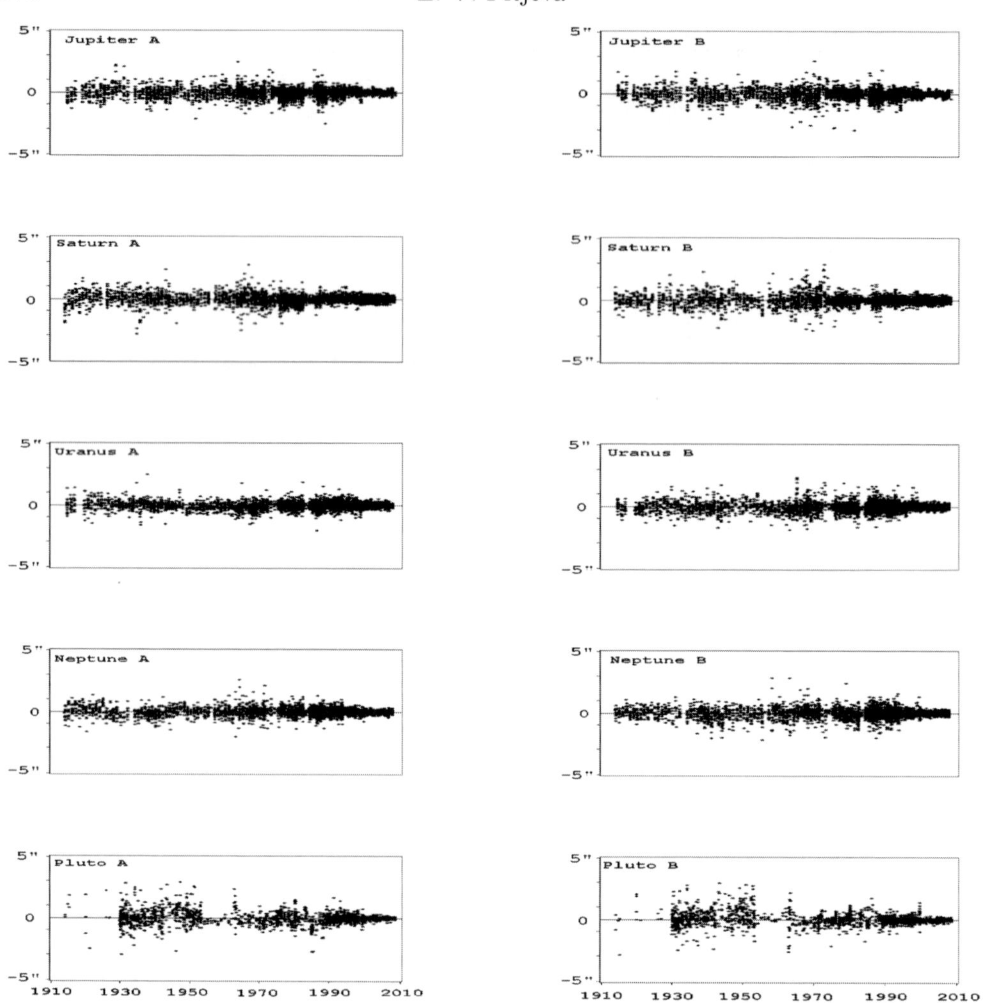

Figure 2. Residuals of the outer planets 1913–2008 in $\alpha \cos \delta$ (A) and in δ (B), the scale $\pm 5''$.

EPM2008 have been oriented to ICRF by including into the total solution the 118 ICRF-base VLBI measurements of spacecraft (Magellan, Phobos, MGS, Odyssey, VEX, and MRO) 1989–2007 near Venus and Mars. Several solutions for recent and previous data are given in Table 3.

The obtained values of the Astronomical Unit, the Moon-Earth mass ratio

$\mathrm{AU} = (149597870697 \pm 3)$ m, $M_{\mathrm{Earth}}/M_{\mathrm{Moon}} = 81.3005676 \pm 0.0000030$

and masses of several asteroids (Table 4) are presented with their real uncertainties estimated by comparing the values obtained in dozens of different test LS solutions that differed by the sets of observations, their weights, and the sets of parameters included in the solution, as well as by comparing parameter values produced by independent groups. The discussion of these values and their comparison with values obtained by other authors are given in the paper by Pitjeva & Standish (2009).

3. Comparison of DE and EPM ephemerides

The differences between various ephemerides are useful to know since they are indicative of the realistic accuracies of the ephemerides. The comparison of our recent EPM2008

Table 2. Mean values and rms residuals for optical observations and spacecraft encounters* α and δ in mas, 1913–2008.

Planet	N	$\|< O - C >_\alpha\|$	σ_α	$\|< O - C >_\delta\|$	σ_δ
VENUS*	4	1.5	2.0	1	6.5
JUPITER	12518	15	187	-30	199
JUPITER*	16	0.1	1.9	-4.1	6.1
SATURN	14296	-1	167	-3	160
SATURN*	68	2.2	2.9	4.2	5.9
URANUS	11446	6	178	2	208
URANUS*	2	-45	9	-25	12
NEPTUNE	10982	7	160	9	205
NEPTUNE*	2	-11	3.5	-14	4.0
PLUTO	5134	1	191	6	197

Table 3. The rotation angles for the orientation of EPM onto ICRF.

Time interval	Number of obs.	ε_x mas	ε_y mas	ε_z mas
1989–1994	20	4.5±0.8	−0.8±0.6	−0.6±0.4
1989–2003	62	1.9±0.1	−0.5±0.2	−1.5±0.1
1989–2007	118	-1.53±0.06	1.02±0.06	1.27±0.05

Table 4. Masses of Ceres, Pallas, Juno, Vesta, Iris, Bamberga in $(GM_i/GM_\odot)\cdot 10^{-10}$.

(1) Ceres	(2) Pallas	(3) Juno	(4) Vesta	(7) Iris	(324) Bamberga
4.71 ±0.03	1.06 ±0.03	0.129 ±0.008	1.32 ±0.03	0.040 ±0.008	0.046 ±0.008

ephemeris with the standard DE405 and the latest DE421 ephemerides has been made (Fig. 3, Table 5). The differences of heliocentric distances for the inner planets between EPM2008 and DE405 or DE421 are small. It is necessary to say about the real accuracy of DE405 ephemerides. Right now, 12 years after the DE405 construction and 27 years after observations of Viking-1 (with more accurate data included in this ephemeris) the residuals for modern data for Odyssey don't surpass 200 m (Konopliv *et al.*, 2006), and as Fig. 3, Table 5 demonstrate it. It is evident that modeling the Mars motion is more difficult than other planets because of a large number of asteroids perturbing its orbit. The availability of a number of spacecraft near Jupiter and Saturn (besides optical observations) allows their ephemerides to be known better than those of other outer planets. The Fig. 3, Table 5 show the significant progress in agreement (and in reduction of uncertainties) of the orbits of all the planets especially owing to the VEX data (they were kindly given to us by Dr. Fienga) and the Cassini data.

4. Relativistic tests and estimation of post–model parameters

At present, the relativistic terms from the Sun and all the planets are included into the motion equations for integration. However, it is interest to estimate how relativistic terms from different planets influence modern planet observations. In addition to the basic EPM2008 (case 1) with the total account of all the relativistic terms in motion equations, three test ephemerides have been constructed:

a) without Saturn relativistic terms (case 2);
b) without Jupiter relativistic terms (case 3);
c) taking into account only solar relativistic terms (case 4).

Then all these ephemerides were improved to all 550000 observations of planets of different types (1913–2008). All the angular observations (both classical and modern CCD

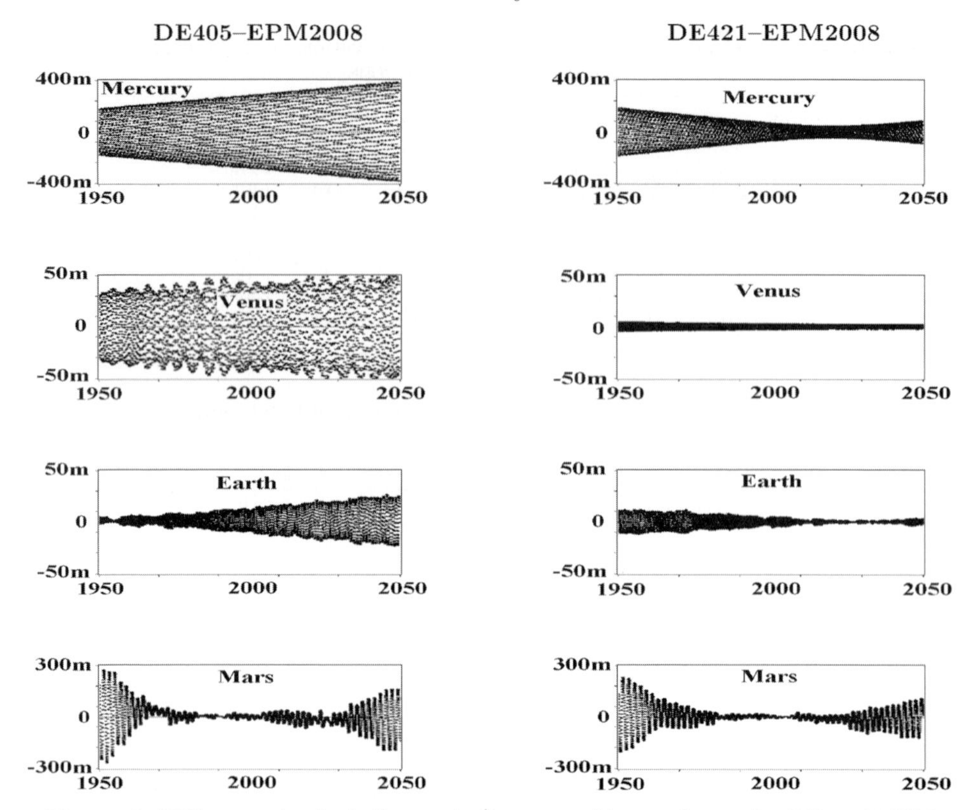

Figure 3. Differences in the heliocentric distances of inner planets for DE and EPM ephemerides, 1950–2050.

Table 5. Maximum differences in the heliocentric distances of planets for DE and EPM ephemerides, 1950–2050.

Planet	DE405–EPM2008	DE421–EPM2008
Mercury	384 m	185 m
Venus	53.7 m	4.6 m
Earth	26.8 m	11.9 m
Mars	272 m	233 m
Jupiter	19.7 km	4.8 km
Saturn	29.3 km	0.4 km
Uranus	864 km	310 km
Neptune	6100 km	848 km
Pluto	29000 km	1800 km

ones) and even VLBI data from spacecraft near planets don't show any differences in all cases. It is only the high-precision ranging that shows small differences. All the results are in given in Table 6. The results demonstrate that Saturn terms (from the comparison of the cases 1 and 2) don't affect the residuals of the observations. Comparison of the 3 and 4 cases show that all the other planet terms (except Jupiter) don't have effects also. Actually, for modern planet observations it is necessary to take into account only relativistic terms from the Sun and Jupiter; moreover, before the appearance of the recent spacecraft ranging data with 1–2 m accuracy it had been possible to take into account only the solar relativistic terms without introducting any errors.

The high-accuracy modern observations not only made it possible to improve the orbital elements of planets and values related to the ephemerides but enable to determine

Table 6. The rms residuals in m and the weight unit errors σ_0 for EPM ephemerides, accounting for different relativistic members.

Observations	planet ranging	Martian landers	Martian spacecraft	Venus Express	Cassini at Saturn	σ_0
Interval	1961–1997	1976–1997	1998–2008	2006–2007	2004–2006	1913–2008
Numbers n.p.	2504	1348	13903	547	31	97101
All relativity	612.20	11.75	2.03	2.59	3.29	0.874
Without Saturn	612.37	11.76	2.03	2.62	3.29	0.875
Without Jupiter	613.14	11.95	2.36	3.74	5.24	0.916
Only Sun	613.25	12.04	2.37	3.85	5.63	0.926

Table 7. Variations of $\dot{G}/G$ and $\dot{a}_i/a_i = \mathrm{R}_i$ per year with 3σ uncertainties.

$\dot{G}/G \cdot 10^{-14}$	$\mathrm{R}_{Ve} \cdot 10^{-14}$	$\mathrm{R}_{Ea} \cdot 10^{-16}$	$\mathrm{R}_{Ma} \cdot 10^{-16}$	$\mathrm{R}_{Ju} \cdot 10^{-12}$	$\mathrm{R}_{Sa} \cdot 10^{-13}$
-5.87	8.99	1.36	2.36	9.14	6.74
$\pm$4.44	$\pm$8.73	$\pm$0.99	$\pm$1.65	$\pm$ 69.48	$\pm$50.73

some small physical parameters characterizing the fundamental properties of our physical space. The EPM2008 ephemerides have been used to analyse these data. Unfortunately, the real accuracy of the parameters is reduced by order of magnitude or more because of systematic errors of observations of an unknown origin, impossibility to completely allow for the delay in the solar corona, and large correlations between parameters. However some estimations may be obtained, their real uncertainties were obtained from the comparison of many different versions of the solution. The PPN parameters and the quadrupole moment of the Sun (J_2) producing various secular and periodic effects in orbital elements of planets have been estimated from the simultaneous solution:

$$J_2 = (2.0 \pm 0.5) \cdot 10^{-7}, \ |\beta - 1| < 0.0002, \ |\gamma - 1| < 0.0002.$$

The variability of $G\dot{M}_\odot/GM_\odot$ should cause the corresponding variation of the semi-major axes of the planetary orbits. In this case the angular momentum integral holds: $GM_\odot(t) \cdot a(t) = \text{const}$, then $G\dot{M}_\odot/GM_\odot = -\dot{a}_\mathrm{i}/a_\mathrm{i}$.

As Dr. Nikolay Pitjev has proposed an attempt to estimate these values. The values of variation of the semi-major axes of the planetary orbits are found stable and have a quite good accuracy for planets covered by the high-accurate data from spasecraft. The results obtained simultaneously for $\dot{G}/G$ and $\dot{a}_\mathrm{i}/a_\mathrm{i}$ per year of the semi-major axes with their 3σ uncertainties are given in Table 7. It is to be noted that all the semi-major axes of the planets are increasing while $\mathrm{GM}_\odot$ decreases (the Sun is losing its mass), as it should be. The average weighted value obtained from $\dot{a}_\mathrm{i}/a_\mathrm{i}$ is

$$G\dot{M}_\odot/GM_\odot = (-1.63 \pm 1.50) \cdot 10^{-16}\text{per year}.$$

This result is preliminary, it demands further improvement and discussion. The obtained value is significantly less than the adjusted $\dot{G}/G$ value (Table 7) and the supposed mass reduction of the Sun owing to the solar radiation and wind (of the order $-8 \cdot 10^{-14}$/year). This discrepancy may to be part of the reason for comet falling on the Sun. The main result (Table 7) is

$$\dot{G}/G = (-5.9 \pm 4.4) \cdot 10^{-14}\text{per year}(3\sigma).$$

The corrections to the perihelion advances for the planets show to what extent the constructed model of the planet motion corresponds to the observations. In particular, the corrections for the inner planets demonstrate correspondence to General Relativity and to the value of the solar oblateness included into ephemerides. The corrections for the outer planets show agreement or non-agreement to the Newtonian theory of gravitation.

Table 8. Corrections to the perihelion advances of planets ($''$/ cy) and their real uncertainties.

Mercury	Venus	Earth	Mars	Author
42.98	8.62	3.84	1.35	Brumberg, 1972
0.11±0.22	−3.03±0.71	−0.12±0.16	−0.35±0.24	Pitjeva, 1986
−0.017±0.052	—	—	—	Pitjeva, 1993
−0.0040±0.0050	0.024±0.033	0.006±0.007	−0.007±0.007	Pitjeva, 2009
Jupiter	**Saturn**	**Uranus**	**Neptune**	**Pluto**
0.067±0.093	−0.010±0.015	− 3.89±3.90	-4.44±5.40	2.84±4.51

The obtained values (Table 8) are within the limits of their real uncertainties, in other words, the corrections to the planet perihelion advances are statistic zero.

5. Conclusion

Further improvement of the planet ephemerides and their parameters depends on the accuracy of modeling which results from the better knowledge of masses of celestial bodies including asteroids and TNO as well as decreasing errors of radiometrical data which originate from the ageing delay due to the solar corona and the spacecraft transponder.

References

Ash, M. E., Shapiro, I. I., & Smith, W. B. 1967, *AJ*, 72, 332

Brumberg, V. A. 1972, in: V. G. Demin (ed.), *Relativistic Celestial Mechanics* (Moscow)

Brumberg, V. A. 1979, *Celest. Mech.* 20, 329

Estabrook, F. B. 1971, in: *Derivation of Relativistic Lagrangian for n-Body Equations Containing Relativity Parameters β and γ*, JPL Internal Communication

Everhart, E. 1974, *Celest. Mech.*, 10, 35

Fienga, A., Manche, H., Laskar, J., & Gastineau, M. 2008, *A&A*, 477, 315

Folkner, W. M., Williams, J. G., & Boggs, D. H. 2008, *Interoffice Memorandum*, 343.R-08-003

Konopliv, A. S., Yoder, C. F., Standish, E. M., Yuan, D. N., & Sjogren, W. L., 2006, *Icarus*, 182, 23

Krasinsky, G. A., Pitjeva, E. V., Sveshnikov, M. L., & Sveshnikova, E. S. 1978, *Trudy Inst.Theoretical astronomy*, 17, 46, in Russian.

Krasinsky, G. A., Aleshkina, E. Yu., Pitjeva, E. V., & Sveshnikov, M. L. 1986, in: J. Kovalevsky, & V. A. Brumberg (eds.), *Relativity in Celestial Mechanics and Astrometry*, Proc. IAU Symposium No. 114 (Dordrecht: D. Reidel Publ.Com.), p. 315

Krasinsky, G. A. & Vasilyev, M. V. 1997, in I. M. Wytrzyszczak, J. H. Lieske & R. A. Feldman (eds.), *Dynamics and Astrometry of Natural and Artificial Celestial Bodies*, Proc. IAU Colloquim No. 165 (Dordrecht: Kluwer Academic Publishers), p. 239

Pitjeva, E. V. 1986, *Byull. Inst. T. A. Ross. Akad. Nauk*, 15, 538, in Russian

Pitjeva, E. V. 1993, *Celest. Mech. Dyn. Astr.*, 55, 313

Pitjeva, E. V. 2005, *Astron. Letters*, 31, 310

Pitjeva, E. V. 2009, in: M. Soffel & N. Capitane (eds.), *Astrometry, Geodynamics and Astronomical Reference Systems*, Proc. JOURNEES-2008 (Dresden), p. 57

Pitjeva, E. V. & Standish E. M. 2009, *Celest. Mech. Dyn. Astr.*, 103, 365

Standish, E. M. Jr., Keesey, M. W., & Newhall, XX 1976, *Technical Report, JPL*, 32-1603, 35 p.

Will, C. M. 1974, in: B. Bertotti (ed.), *The Theoretical Tools of Experimental Gravitation*, Experimental Gravitation (Academic Press)

Yagudina, E. I. 2009, in: M. Soffel & N. Capitane (eds.), *Astrometry, Geodynamics and Astronomical Reference Systems*, Proc. JOURNEES-2008 (Dresden), p. 61

Relativity in Fundamental Astronomy
Proceedings IAU Symposium No. 261, 2009
S. A. Klioner, P. K. Seidelman & M. H. Soffel, eds.

doi:10.1017/S1743921309990354

Testing alternate gravitational theories

E. M. Standish[1]

[1] Caltech/JPL, retired
519 Birchbark Court
Seneca, SC 29672 USA
ems @jpl.nasa.gov

Abstract. The planetary ephemerides are used to examine different suggested forms of the gravitational equations of motion which could possibly cause the observed Pioneer Anomaly. It is shown that most of the forms would be unacceptable, including that generally assumed – a constant acceleration directed toward the Sun. The tests show that three other forms could not exist within 10 au's of the Sun. Only one suggested form would be compatible with the Pioneer Anomaly affecting Saturn or any other more inward planet. Additional planetary observations in the future may possibly eliminate this form also.

Keywords. Pioneer Anomaly, Planetary Ephemerides

---

1. Introduction

In a previous paper (Standish, 2008), the planetary ephemerides were used to test different forms of the gravitational equations to see if any of them could possibly produce the observed effect known as the "Pioneer Anomaly", while still producing planetary ephemerides consistent with the planetary positional observations. Five different forms were considered, and for each, the addition to the equations of motion was used to generate a new ephemeris, subsequently adjusted to fit the complete set of planetary ephemeris observations currently used in the ephemeris production process. The successes or failures of the fitting processes were then judged by two criteria: 1) the reasonableness of the adjustment parameters (orbital parameters, masses of planets and asteroids, value of the au, etc.) and 2) the goodness of the fit as evidenced by plots and statistics of the residuals of the individual types of planetary observations. That paper showed that the generally-proposed form of the Pioneer Anomaly, a constant inward acceleration, could be easily ruled out; the planetary ephemerides could not be adjusted to properly fit that force. It was further shown that two other forms of the P.A. could exist, but not if the force applied to Saturn; only if the force were farther from the Sun than Saturn's orbit. The last two forms of the P.A. would be detectable if applied to Jupiter, but not if applied only outside of Jupiter.

In the present paper, the observational data set is augmented by a couple of years of available Cassini data and extensions to some Mars ranging and outer planet CCD data sets. The same process is followed as before: integrate an ephemeris with one of the five given forms of the P.A. force, applied to only planets beyond one of the three choices, 4, 6, or 11 a.u.; this gives 15 different examples in all, beside the base solution, the one without any P.A. force. The overall results are given in a Table.

2. Additions to the Data Sets

Available from the IAU Commission 4 (Ephemerides) website are the measurements fitted by planetary ephemeris improvement processes. Most importantly, for the purposes here, are

- Viking, MGS, and Odyssey ranges to Mars,
- MGS and Odyssey VLBI measurements of Mars,
- CCD observations of the outer planets, and
- Cassini orbit determination positions of Saturn.

Since the previous paper, the coverage of the MGS ranging data increased from 1999.1–2005.6 to 1999.1–2006.7, and the coverage of the Odyssey ranges, 2002.1–2005.6 to 2002.1–2008.5. The CCD observations, starting in 1996 have been extended from the end of 2006 to mid-2008. The available Cassini ODP points, not available for the previous paper, extend from late 2004 to the end of 2006. These latter data are far more accurate than any previous Saturn data and therefore provide an ability to test Saturn's orbit with a much greater sensitivity than previously possible.

3. The Forms of the Pioneer Anomaly

The generally assumed form of the Pioneer Anomaly is a constant acceleration upon a body directed toward the Sun, amounting to 8.74×10^{-10} m/sec^2 (Anderson *et al.*, 2002). The acceleration has been deduced from measurements of the Pioneer Spacecraft which have been obtained while the spacecraft were outside the orbit of Saturn. However, four other forms of the Anomaly were suggested to the author by (Laemmerzahl, 2007). Three versions of each of the five equation modifications are tried here: a version which applies to all planets farther than 4 au from the Sun (i.e., Jupiter through Pluto), a version outside of 6 au (Saturn through Pluto)., and a version outside 11 au (only Uranus, Neptune, and Pluto). The third version corresponds to the only region in the solar system in which the P.A. has actually been detected. Attempts to analyze data from spacecraft in closer regions to the Sun are ongoing (Turyshev, 2008).

The five tested modifications to the equations of motion are the following:

(*a*) 10% of the normally assumed form of the P.A., a constant acceleration directed toward the Sun (the full acceleration is 8.74×10^{-10} m/sec^2);

(*b*) $-\|\mathbf{V}_{\mathbf{rad}}\|\mathbf{C}_{\mathbf{P.A.}}$: an acceleration directed toward the Sun, proportional to the magnitude of the planet's heliocentric radial velocity;

(*c*) $-\mathbf{V}_{\mathbf{rad}}\mathbf{C}_{\mathbf{P.A.}}$: same as *(b)*, but with the same sign as the radial velocity itself (so that the acceleration alternates sign as the planet itself approaches and recedes from the Sun due to its orbital eccentricity);

(*d*) $-\mathbf{V}^2_{\mathbf{rad}}\mathbf{C}_{\mathbf{P.A.}}$: an acceleration directed toward the Sun, proportional to the magnitude of the heliocentric radial velocity *squared*; and

(*e*) $-\mathbf{V}_{\mathbf{rad}}\|\mathbf{V}_{\mathbf{rad}}\|\mathbf{C}_{\mathbf{P.A.}}$: same as *(d)*, but with the same sign as the radial velocity itself (as in *(c)* above).

In cases *(b)*–*(e)*, the velocities are normalized by dividing by 3.4 au/yr, the approximate radial velocities of the Pioneer Spacecraft. Thus, in all cases, the proposed force would produce the acceleration observed upon the Pioneer Spacecraft, whose radial velocities are nearly constant, aligned pretty much with the solar direction. The effects upon the planets from the five different forms, however, vary greatly from one form to another.

Table 1. Attempts to fit various data sources.
'X' ⇒ completely unsuccessful attempt
'o' ⇒ fair-to-poor attempt
'-' ⇒ normal fit

Form of P.A.	ρ_{min} [au]	parameter quality	Outer Pl. CCD	Mars VLBI	Mars Ranging	Cassini ODP at Saturn
No Added Force		-	-	-	-	-
10% of $\mathbf{C}_{\mathrm{P.A.}}$	4	X	X	X	X	X
	6	X	X	X	X	X
	11	-	X	-	-	X
$-\lVert\mathbf{V}_{\mathrm{rad}}\rVert\mathbf{C}_{\mathrm{P.A.}}$	4	X	o	X	X	X
	6	X	o	o	X	X
	11	-	o	-	-	o
$-\mathbf{V}_{\mathrm{rad}}\mathbf{C}_{\mathrm{P.A.}}$	4	X	X	X	X	X
	6	o	o	o	X	X
	11	-	o	-	-	o
$-\mathbf{V}^2_{\mathrm{rad}}\mathbf{C}_{\mathrm{P.A.}}$	4	o	-	-	-	X
	6	-	-	-	-	X
	11	-	-	-	-	-
$-\mathbf{V}_{\mathrm{rad}}\lVert\mathbf{V}_{\mathrm{rad}}\rVert\mathbf{C}_{\mathrm{P.A.}}$	4	o	-	o	o	o
	6	-	-	-	-	o
	11	-	-	-	-	-

4. Testing the Different Choices

All of the observational data sets were reduced against the base ephemeris, the one without any added force, which had been iterated to provide an optimum fit of the data. Then, each of the 15 different choices was used in 15 new iterations, providing best fits to the data sets. The final sets of observational residuals are analyzed and the adjustment parameters are examined in order to judge how well the data can be represented by each particular form and range of the force.

The results of the judging of each case are given in Table 1, where the form of the added force is shown in the first column and the minimum range over which the force applies is given in the second column. The third column shows the judgement of how reasonable the solution parameters seem to be (changes to the au, the densities of the asteroids, the masses of the planets, etc.), The final four columns show how well the data seems to be fit by that particular adjusted ephemeris. An upper case 'X' signifies a very bad fit, a lower case 'o' shows a marginal set of parameters or residuals, and a hyphen indicates that the parameters or residuals are virtually identical to those of the base solution.

• It is evident from Table 1 that the generally assumed form of the P.A., that of a constant acceleration directed toward the Sun, would completely distort the planetary ephemerides; in fact, as shown, even just 10 percent of such a force would be unacceptable. If applied only past Saturn, the data of Mars would be pretty much unaffected, but the CCD data of the outer planets and especially the Cassini data would not be fit well at all.

• The second and third forms of the P.A. could marginally exist past Saturn; they certainly could not apply to Saturn's orbit, for the Mars ranging and the Cassini data would show unacceptable signatures in their residuals.

• It is the Cassini data which rules out the fourth form of the P.A. affecting Jupiter's or Saturn's orbit. The other data sets are not much affected.

- The planetary ephemerides can only marginally rule out the fifth form of the P.A. if it is applied to Jupiter's orbit. The data residuals are slightly worse than those of the base solution. The Cassini residuals suffer slightly if this fifth form is applied to Saturn. If applied only past Saturn, the ephemerides are virtually unaffected.

5. Conclusions

Only certain forms of the Pioneer Anomaly can exist in the solar system and still be consistent with the planetary ephemerides. Further, all but one of the forms tested here can be eliminated from having the effect applied to any of the planets out to and including Saturn. These conclusions are similar to those presented previously with the exception that the addition of the Cassini Orbit Determination normal points has provided a more sensitive test for the fourth form considered.

References

Anderson, J. D., Laing, P. A., Lau, E. L., Liu, A. S., Nieto, M. M., & Turyshev, S. G. 2002, Phys. Rev. D, 65, 082004

Laemmerzahl, C. 2007, private communication

Standish, E. M. 2008, Recent Developments in Gravitation and Cosmology, 977, 254

Turyshev, S. G. 2008, APS Meeting Abstracts, 7001

Relativity in Fundamental Astronomy
Proceedings IAU Symposium No. 261, 2009
S. A. Klioner, P. K. Seidelman & M. H. Soffel, eds.

doi:10.1017/S1743921309990366

Probing general relativity with radar astrometry in the inner solar system

Jean-Luc Margot[1] and Jon D. Giorgini[2]

[1]University of California, Los Angeles,
595 Charles Young Drive East, Los Angeles, CA 90095, USA
email: `jlm@ess.ucla.edu`

[2]Jet Propulsion Laboratory,
4800 Oak Grove Drive, Pasadena, CA 91109, USA
email: `jdg@tycho.jpl.nasa.gov`

Abstract. We describe a long-term program designed to obtain and interpret high-precision radar range measurements of a number of near-Earth objects (NEOs) that have trajectories reaching deep inside the gravitational well of the Sun. Objects in our sample have perihelion shift rates 1.5 to 2.5 times that of (1566) Icarus (10"/cy) and span a wide range of inclinations and semi-major axes, allowing for an unambiguous separation of general relativistic and solar oblateness effects. Four objects have been observed at Arecibo on at least two apparitions since 2000, with typical uncertainties of a few hundred meters. Within the next three years, we anticipate securing a total of 15 observations of 5 different NEOs. This program is expected to provide a purely dynamical measurement of the oblateness of the Sun (J_2 at the 10^{-8} level) and to constrain the Eddington parameter β at the 10^{-4} level. Although our objects are selected to minimize Yarkovsky orbital drift, we also anticipate measuring Yarkovsky drift rates, which are orthogonal to the GR and J_2 signatures.

Keywords. general relativity, solar quadrupole moment, Yarkovsky drift, asteroids, radar

1. Motivation

Attempts to quantize gravity and to unify it with other forces indicate that general relativity (GR) cannot be the final theory on gravity (Will 2006). Testing metric theories of gravity to higher levels of precision is, therefore, critical and has resulted in new efforts in the solar system (e.g. Nordtvedt 2000; Margot 2003; Pireaux and Rozelot 2003; Iorio 2005; Folkner 2009). While the uncertainty on γ in the parametrized post-Newtonian (PPN) formalism is now of order 10^{-5} (Bertotti *et al.* 2003), there has been no comparable improvement in the knowledge of β. A *direct* constraint on β can be obtained by measurement of the perihelion shift. Anderson *et al.* (2002) have combined radar and spacecraft ranging data and find $|\beta - 1| < 1.2 \times 10^{-3}$, while Folkner (2009) recently reported $|\beta - 1| < 10^{-3}$. Our simulations show that Arecibo radar measurements obtained over a decade can discriminate changes in β at the 10^{-4} level.

A second motivation for our observations stems from the difficulties in reconciling helioseismological inferences with new solar abundance measurements. Confidence in the helioseismology inversions has been shaken as independent solar abundance measurements displaying a high degree of consistency (Asplund *et al.* 2004; Caffau *et al.* 2008) have ruined the previous agreement between helioseismological inferences and models of the solar interior. The new measurements place the oxygen abundance at ~60% of the Anders and Grevesse (1989) values, changing the opacity and depth of the base of the convective layer (Basu and Antia 2008). The quadrupole moment of the Sun is of fundamental importance to the internal structure of the Sun and warrants an independent determination that does not rely on inversion models of helioseismology data. Our simulations show that changes in the solar quadrupole moment J_2 at the 10^{-8} level are detectable and would put the preferred helioseismology value of $J_2 \sim 2 \times 10^{-7}$ (Pireaux and Rozelot 2003) to a very serious test with a direct dynamical measurement.

Finally we are motivated by the benefits of measuring the Yarkovsky orbital drift, which is due to the anisotropic reradiation of sunlight from asteroid surfaces (Bottke *et al.* 2006). This effect has been detected with Arecibo radar data (Chesley *et al.* 2003) and turns out to be the dominant source of uncertainty in near-Earth asteroid (NEA) trajectory predictions (Giorgini *et al.* 2002) for bodies smaller than 2 km. By far the largest influence on orbital parameters is a change in the semi-major axis of objects as a function of their spin, shape, orbit, and material properties. For asteroids of known sizes and spins, a measurement of the Yarkovsky drift rate can be interpreted in terms of bulk density and thermal properties (Chesley *et al.* 2003). Our goal is to obtain such measurements along with detailed physical characterizations, with a particular focus on the binaries in our sample for which independent density estimates can be established.

2. Theoretical background

The spacetime geometry around a spherical star is described by a metric that is static and spherically symmetric (Schwarzschild 1916). In isotropic coordinates,

$$ds^2 = -\left(1 - 2\frac{GM}{c^2 r} + 2\left(\frac{GM}{c^2 r}\right)^2\right)(cdt)^2 + \left(1 + 2\frac{GM}{c^2 r}\right)[dx^2 + dy^2 + dz^2], \quad (2.1)$$

where G is the gravitational constant, c is the speed of light, and M is the mass of the star. GR derives part of its elegance from the fact that it depends only on G and c, which are non-adjustable constants.

The PPN formalism is a framework to parametrize various theories of gravity in a systematic way. Of the ten parameters, β and γ are the most important. Their placement in the metric illuminates their physical significance:

$$ds^2 = -\left(1 - 2\frac{GM}{c^2 r} + 2\beta\left(\frac{GM}{c^2 r}\right)^2\right)(cdt)^2 + \left(1 + 2\gamma\frac{GM}{c^2 r}\right)[dx^2 + dy^2 + dz^2]. \quad (2.2)$$

Appearing in the spatial part of the metric, γ is related to the amount of curvature produced by a unit rest mass, and is tested by deflection of light and Shapiro delay experiments. The degree of non-linearity in the superposition law for gravity is captured by β. In GR, $\beta = \gamma = 1$.

Orbits of test particles in curved spacetime do not close and their perihelion precesses. A Keplerian orbit modified for perihelion precession can be written (Misner *et al.* 1973)

$$r = \frac{a(1-e^2)}{1 + e\cos[(1 - \delta\phi/2\pi)\phi]}, \quad (2.3)$$

where a is the semi-major axis, e is the eccentricity, and ϕ is the true anomaly. The perihelion shift per orbit is

$$\delta\phi = \frac{6\pi G M_\odot}{a(1-e^2)c^2}\frac{(2 - \beta + 2\gamma)}{3}. \quad (2.4)$$

Under the influence of an oblate Sun with quadrupole moment J_2, the perihelion shift contains an additional contribution:

$$\delta\phi = \frac{6\pi GM_\odot}{a(1-e^2)c^2}\left[\frac{(2-\beta+2\gamma)}{3}\right] + \frac{6\pi}{2}R_\odot^2\frac{(1-3/2\sin^2 i)}{a^2(1-e^2)^2}J_2, \tag{2.5}$$

with i the orbital inclination with respect to the solar equator and $R_\odot$ is the solar radius.

3. Previous studies

Shapiro *et al.* [1968, 1971, 1972] determined the perihelion precession of Mercury and that of asteroid (1566) Icarus (Gilvarry 1953) to test GR and to constrain values of the PPN parameters. Because Newtonian precession due to the oblate Sun affects the measurements, Shapiro emphasized the need to measure the precession of several bodies in order to separate, based on their heliocentric distance dependence, the general relativistic effects from those due to the gravitational quadrupole moment of the Sun (J_2). Another way of separating the two effects is to use several bodies with different orbital inclinations, since GR is a purely central effect whereas the precession due to the oblate Sun has a known dependence on inclination. The GR and J_2 influences cause no change in the semi-major axis, which is orthogonal to the effect of the Yarkovsky drift . We rely on these different signatures on orbital evolution to distinguish the Yarkovsky effect from the perihelion shift.

The perihelion shift of Mercury predicted from GR alone is 43 arcseconds per century ("/cy) (Nobili and Will 1986). The value measured with radar is known with 0.5% uncertainties (Shapiro *et al.* 1976; Anderson *et al.* 1991) and is consistent with GR predictions. The influence of the solar J_2 has not been detected and is ~0.1% of the GR influence for $J_2 \sim 2 \times 10^{-7}$.

We anticipate improvements over previous studies involving Mercury and Icarus because 1) Several newly-discovered asteroids have orbits offering a better sensitivity to the solar J_2; 2) Our sample incorporates a range of heliocentric distances and inclinations that can unambiguously separate GR and J_2 effects and provide more robust estimates in a joint solution; 3) The center of mass locations of the small bodies can be determined to 100–500 m, about an order of magnitude better than existing Mercury ranges. The Mercury determinations suffer from km-scale uncertainties due to unknown topography and possible center of mass/center of figure offset. Mercury topography is an important source of systematic errors in GR tests (Pitjeva 1993).

4. Observational strategy

In light of the large number of recent NEO discoveries, a search for asteroids that provide better opportunities than Mercury or Icarus to detect GR and J_2 effects was performed (Margot 2003). Roughly ten candidates with long astrometric arcs, repeated observability at Arecibo, and GR perihelion shifts larger than that of Icarus have been identified (Figure 1 and Table 1). We regularly update the target list to incorporate the NEOs most suited to the realization of our science objectives.

We rely primarily on optical astrometry (typically hundreds of measurements) to secure state estimates for each object, and we rely primarily on the radar measurements to expose the parameters of interest: one Yarkovsky rate per object, β, and J_2.

As shown in Table 1, we have now acquired observations on two apparitions for 4 objects (1999 KW4, 1999 MN, 2000 BD19, 2000 EE14), giving very roughly 8 independent data constraints for 6 solve-for parameters (In reality, six orbital parameters must also be determined - see previous paragraph). In the next three years a modest investment

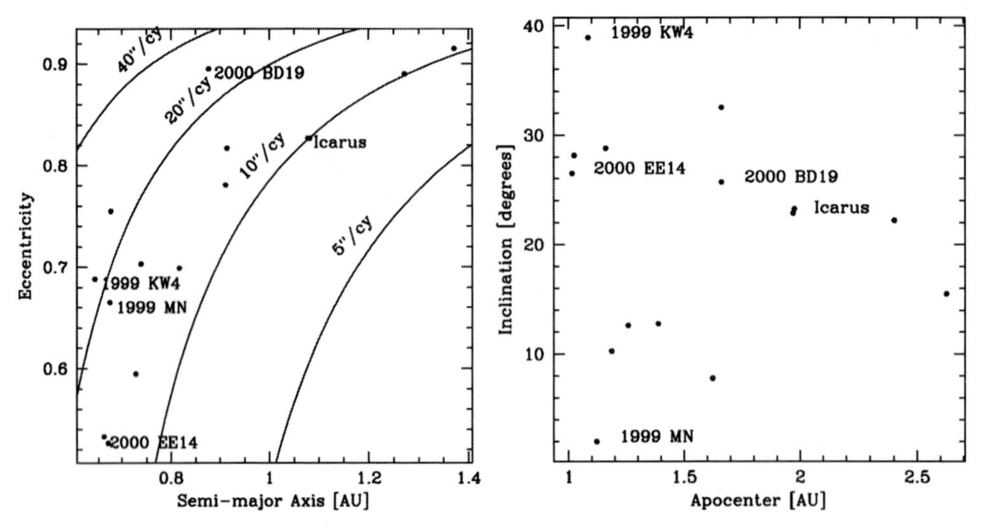

Figure 1. A. Predicted rates of perihelion shift due to GR alone for a number of newly discovered NEAs, compared to that of (1566) Icarus, shown in semi-major axis versus eccentricity space. Objects with two existing radar detections are labeled. B. Same objects shown in apocenter versus inclination space, illustrating the wide range of inclinations that can separate GR and J_2 effects.

Target	H	D [km]	P [h]	Arc [days]	N_o	N_r	a [AU]	e	i [deg]	$\dot{\tilde{\omega}}$ ("/cy)	App
1999 KW4	16.5	**1.32**	2.8	3735	1672	2	0.642	0.688	38.9	22.1	15 16 17
1999 MN	21.4	**0.12**	5.5	2204	75	2	0.674	0.665	2.0	18.5	9 **10**
2000 BD19	17.2	**0.90**	12.5	4332	359	2	0.876	0.895	25.7	26.8	**11** 15 20
2000 EE14	17.1	**0.60**	*5.0*	2952	242	2	0.662	0.533	26.5	15.0	14 15
2008 EA32	16.5	*2.02*	*5.0*	349	65		0.616	0.305	28.3	14.2	**10 11 12**
Icarus	16.9	1.30	2.3	20861	590	1	1.078	0.827	22.8	10.1	15 24
Phaethon	14.6	5.10	3.6	9231	1801	1	1.271	0.890	22.2	10.1	13 16 17
Talos	17.0	*1.60*	38.5	6234	415		1.081	0.827	23.2	10.0	**10 11** 19

Table 1. Subset of NEOs that are particularly well-suited for our program, based on the NEO population known as of 2009 May 30. H is the absolute magnitude, D is diameter, and P is spin period. Sizes and spin periods were obtained from the DLR NEA Data Base, unless superseded by our own radar estimates (in bold). For those objects that had no size/albedo information, the value was evaluated on the basis of H and the 11% average albedo of NEAs (italics). For those objects that had no spin period information, the period was fixed at 5 hours and italicized in the table. Arc and N_o refer to the interval between first and last optical observation and the total number of optical observations, respectively. All objects have arc lengths in excess of 300 days, guaranteeing recovery and small pointing uncertainties. N_r is the number of apparitions with existing ranging observations. Orbital elements a, e, i have their usual definition. The perihelion shift rate $\dot{\tilde{\omega}}$ is given in arcseconds per century, with a cutoff of 10"/cy. The last column indicates the years of future apparitions detectable at Arecibo.

of ~70 hours of telescope time can secure ranges for an additional 7 epochs (Table 1). In rough accounting terms, there will be a total of 15 independent data constraints (for 5 different objects) and 7 solve-for parameters. At least three objects will have data on at least three apparitions, giving good prospects for measuring Yarkovsky drift rates. Yarkovsky drift rates are roughly 15 m/y in semi-major axis for a 1 km object, and the rate scales roughly as size^{-1}. The range is affected quadratically with time and rapidly produces a signal of several km (Chesley *et al.* 2003).

We obtain accurate range astrometry to NEAs using procedures that have been refined to exquisite precision over the years. The basic idea is to send a waveform encoded with a pseudo-random code sequence and to cross-correlate the received echoes with a replica of the code (Evans and Hagfors 1968). Transmission occurs for the duration of the round-trip light time to the object, and reception occurs for an equivalent duration. Each transmit-receive cycle constitutes a *run*. The duration of an individual code element (baud) and the length of the code are chosen in combinations that provide unambiguous range measurements to distances of several astronomical units. So-called closed-loop tests are performed with identical system parameters to fully calibrate delays within the telescope and electronics. The internal consistency of the measurements and external verifications via orbit determination software are both excellent.

Figure 2 illustrates that for ranging uncertainties of $\sim$100 m, a 10^{-3} variation in β represents a 10-σ signal after a decade with a single NEO. Our goal is to constrain β at the 10^{-4} level (and J_2 at the 10^{-8} level) from a joint analysis of the entire data set.

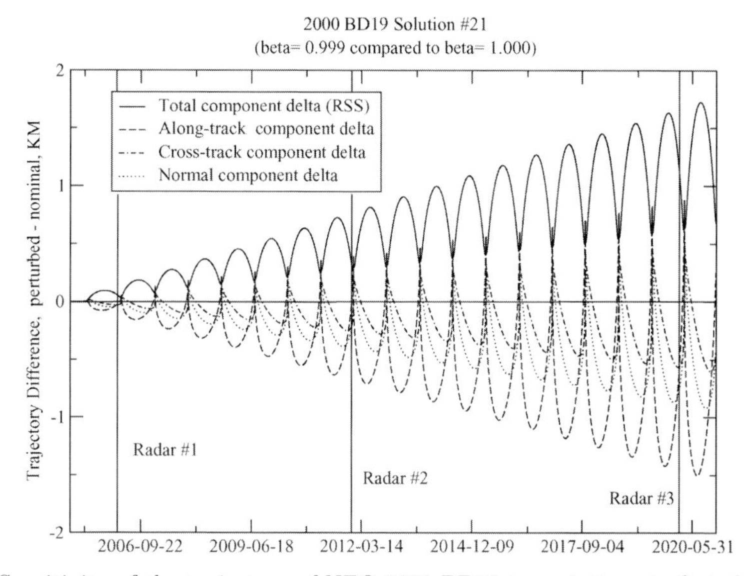

Figure 2. Sensitivity of the trajectory of NEO 2000 BD19 to variations in β at the 10^{-3} level.

5. Conclusions

The Yarkovsky and perihelion shift observations represent a long-term endeavor with little or no instant gratification. However, a modest investment in telescope time can improve our knowledge of asteroid densities and thermal properties, provide a dynamical measurement of the solar J_2, and test general relativity to new levels of precision.

References

E. Anders and N. Grevesse. Abundances of the elements - Meteoritic and solar. Geochim. Cosmochim. Acta, 53:197–214, January 1989.

J. D. Anderson, M. A. Slade, R. F. Jurgens, E. L. Lau, X. X. Newhall, and E. M. Standish. Radar and spacecraft ranging to Mercury between 1966 and 1988. *Proceedings of the Astronomical Society of Australia*, 9:324–+, 1991.

J. D. Anderson, E. L. Lau, S. Turyshev, J. G. Williams, and M. M. Nieto. Recent Results for Solar-System Tests of General Relativity. *Bulletin of the American Astronomical Society*, 34:660–+, May 2002.

M. Asplund, N. Grevesse, A. J. Sauval, C. Allende Prieto, and D. Kiselman. Line formation in solar granulation. IV. [O I], O I and OH lines and the photospheric O abundance. A&A, 417:751–768, April 2004.

S. Basu and H. M. Antia. Helioseismology and solar abundances. Phys. Rep., 457:217–283, March 2008.

B. Bertotti, L. Iess, and P. Tortora. A test of general relativity using radio links with the Cassini spacecraft. Nature, 425:374–376, September 2003.

W. F. Bottke, Jr., D. Vokrouhlický, D. P. Rubincam, and D. Nesvorný. The Yarkovsky and Yorp Effects: Implications for Asteroid Dynamics. *Annual Review of Earth and Planetary Sciences*, 34:157–191, May 2006.

E. Caffau, H.-G. Ludwig, M. Steffen, T. R. Ayres, P. Bonifacio, R. Cayrel, B. Freytag, and B. Plez. The photospheric solar oxygen project. I. Abundance analysis of atomic lines and influence of atmospheric models. A&A, 488:1031–1046, September 2008.

S. R. Chesley, S. J. Ostro, D. Vokrouhlický, D. Čapek, J. D. Giorgini, M. C. Nolan, J. Margot, A. A. Hine, L. A. M. Benner, and A. B. Chamberlin. Direct Detection of the Yarkovsky Effect by Radar Ranging to Asteroid 6489 Golevka. *Science*, 302:1739–1742, December 2003.

John V. Evans and Tor Hagfors, editors. *Radar Astronomy*. McGraw-Hill, New York, 1968.

W. M. Folkner. Relativistic Aspects of the JPL Planetary Ephemeris. *American Astronomical Society, IAU Symposium #261, 155. Relativity in Fundamental Astronomy: Dynamics, Reference Frames, and Data Analysis 27 April - 1 May 2009 Virginia Beach, VA, USA, #6.01*, 261, May 2009.

J. J. Gilvarry. Relativity Precession of the Asteroid Icarus. *Physical Review*, 89:1046–1046, March 1953.

J. D. Giorgini, S. J. Ostro, L. A. M. Benner, P. W. Chodas, S. R. Chesley, R. S. Hudson, M. C. Nolan, A. R. Klemola, E. M. Standish, R. F. Jurgens, R. Rose, A. B. Chamberlin, D. K. Yeomans, and J.-L. Margot. Asteroid 1950 DA's Encounter with Earth in 2880: Physical Limits of Collision Probability Prediction. *Science*, 296:132–136, April 2002.

L. Iorio. On the possibility of measuring the solar oblateness and some relativistic effects from planetary ranging. A&A, 433:385–393, April 2005.

J. L. Margot. Candidate Asteroids for Discerning General Relativity and Solar Oblateness. *AAS/Division of Dynamical Astronomy Meeting*, 34, August 2003.

Charles W. Misner, Kip S. Thorne, and John Archibald Wheeler. *Gravitation*. Freeman, 1973.

A. M. Nobili and C. M. Will. The real value of Mercury's perihelion advance. Nature, 320:39–41, March 1986.

K. Nordtvedt. Improving gravity theory tests with solar system "grand fits". Phys. Rev. D, 61 (12):122001–+, June 2000.

S. Pireaux and J.-P. Rozelot. Solar quadrupole moment and purely relativistic gravitation contributions to Mercury's perihelion advance. Ap&SS, 284:1159–1194, 2003.

E. V. Pitjeva. Experimental testing of relativistic effects, variability of the gravitational constant and topography of Mercury surface from radar observations 1964-1989. *Celestial Mechanics and Dynamical Astronomy*, 55:313–321, April 1993.

K. Schwarzschild. On the Gravitational Field of a Mass Point According to Einstein's Theory. *Abh. Konigl. Preuss. Akad. Wissenschaften Jahre 1906,92, Berlin,1907*, pages 189–196, 1916.

I. I. Shapiro, M. E. Ash, and W. B. Smith. Icarus - Further Confirmation of Relativistic Perihelion Precession. *Physical Review Letters*, 20(26):1517+, 1968.

I. I. Shapiro, W. B. Smith, M. E. Ash, and S. Herrick. General Relativity and the Orbit of Icarus. *Astron. J.*, 76:588–+, September 1971.

I. I. Shapiro, R. P. Ingalls, R. B. Dyce, M. E. Ash, D. B. Campbell, and G. H. Pettengill. Mercury's Perihelion Advance - Determination by Radar. *Physical Review Letters*, 28(24): 1594+, 1972.

I. I. Shapiro, C. C. Counselman, and R. W. King. Verification of the principle of equivalence for massive bodies. *Physical Review Letters*, 36:555–558, March 1976.

C. M. Will. The Confrontation between General Relativity and Experiment. *Living Reviews in Relativity*, 9:3–+, March 2006.

Relativity in Fundamental Astronomy
Proceedings IAU Symposium No. 261, 2009
S. A. Klioner, P. K. Seidelman & M. H. Soffel, eds.

doi:10.1017/S1743921309990378

Astrometric solar-system anomalies

John D. Anderson[1] **and Michael Martin Nieto**[2]

[1]Jet Propulsion Laboratory (Retired)
121 S. Wilson Ave., Pasadena, CA 91106-3017 U.S.A.
email: `jdandy@earthlink.net`

[2]Theoretical Division (MS-B285), Los Alamos National Laboratory
Los Alamos, New Mexico 87645 U.S.A.
email: `mmn@lanl.gov`

Abstract. There are at least four unexplained anomalies connected with astrometric data. Perhaps the most disturbing is the fact that when a spacecraft on a flyby trajectory approaches the Earth within 2000 km or less, it often experiences a change in total orbital energy per unit mass. Next, a secular change in the astronomical unit AU is definitely a concern. It is reportedly increasing by about 15 cm yr^{-1}. The other two anomalies are perhaps less disturbing because of known sources of nongravitational acceleration. The first is an apparent slowing of the two Pioneer spacecraft as they exit the solar system in opposite directions. Some astronomers and physicists, including us, are convinced this effect is of concern, but many others are convinced it is produced by a nearly identical thermal emission from both spacecraft, in a direction away from the Sun, thereby producing acceleration toward the Sun. The fourth anomaly is a measured increase in the eccentricity of the Moon's orbit. Here again, an increase is expected from tidal friction in both the Earth and Moon. However, there is a reported unexplained increase that is significant at the three-sigma level. It is prudent to suspect that all four anomalies have mundane explanations, or that one or more anomalies are a result of systematic error. Yet they might eventually be explained by new physics. For example, a slightly modified theory of gravitation is not ruled out, perhaps analogous to Einstein's 1916 explanation for the excess precession of Mercury's perihelion.

Keywords. gravitation, celestial mechanics, astrometry

1. Earth flyby anomaly

The first of the four anomalies considered here is a change in orbital energy for spacecraft that fly past the Earth on approximately hyperbolic trajectories (Anderson *et al.* 2008). By means of a close flyby of a planet, it is possible to increase or decrease a spacecraft's heliocentric orbital velocity far beyond the capability of any chemical propulsion system (see for example Flandro 1966 and Wiesel 1989). It has been known for over a century that when a small body encounters a planet in the solar system, the orbital parameters of the small body with respect to the Sun will change. This is related to Tisserand's criterion for the identification of comets (Danby 1988).

During a gravity assist, which is now routine for interplanetary missions, the orbital energy with respect to the planet is conserved. Therefore, if there is an observed energy increase or decrease with respect to the planet during the flyby, it is considered anomalous (Anderson *et al.* 2007).

Unfortunately, it is practically impossible to detect a small energy change with planetary flybys, both because an energy change is difficult to separate from errors in the planet's gravity field and because of the unfavorable Doppler tracking geometry of a distant planet. The more favorable geometry of an Earth flyby is needed. Also the Earth's gravity field is well known from the GRACE mission (Tapley *et al.* 2004). Earth's gravity is not a significant source of systematic error for the flyby orbit determination (Anderson *et al.* 2008).

The flyby anomaly was originally detected in radio Doppler data from the first of two Earth flybys by the Galileo spacecraft (for a description of the mission see Russell 1992). After launch on 1989-Oct-18, the spacecraft made one flyby of Venus on 1990-Feb-10, and subsequently two flybys of Earth on 1990-Dec-08 and two years later on 1992-Dec-08. The spacecraft arrived at Jupiter on 1995-Dec-07.

Without these planetary gravity assists, a propulsion maneuver of 9 km s^{-1} would have been needed to get from low Earth orbit to Jupiter. With them, the Galileo spacecraft left low Earth orbit with a maneuver of only 4 km s^{-1}. The first Earth flyby occurred at an altitude of 960 km. The second, which occurred at an altitude of 303 km, was affected by atmospheric drag, and therefore it was difficult to obtain an unambiguous measurement of an anomalous energy change on the order of a few mm s^{-1}.

The anomalistic nature of the flyby is demonstrated by Fig. 1. The pre-perigee fit produces residuals which are distributed about a zero mean with a standard error of 0.087 mm s^{-1}. However, when the pre-perigee fit is extrapolated to the post-perigee data, there is a clear asymptotic bias of 3.78 mm s^{-1} in the residuals. Further, the data immediately after perigee indicates that there is perhaps an anomalous acceleration acting on the spacecraft from perigee plus 2253 s, the first data point after perigee, to about 10 hr, the start of the asymptotic bias. (A discussion of these residuals and how they were obtained can be found in Antreasian & Guinn 1998.)

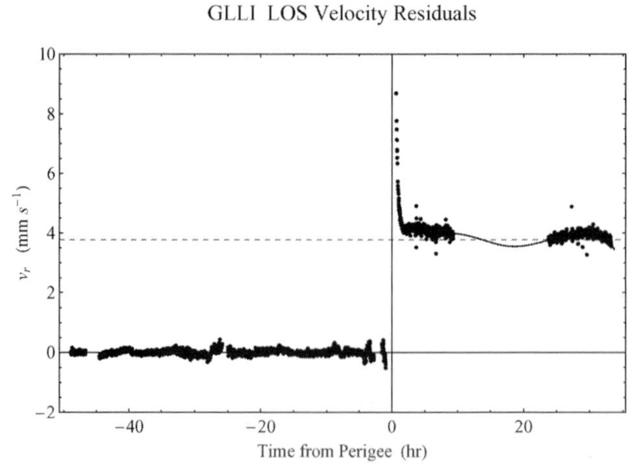

Figure 1. Doppler residuals (observed minus computed) converted to units of line of sight (LOS) velocity about a fit to the pre-perigee Doppler data, and the failure of this fit to predict the post perigee data. The mean offset in the post-perigee data approaches 3.78 mm s^{-1}, as shown by the dashed line. The solid line connecting the post-perigee data represents an eighth degree fitting polynomial to data after perigee plus 2.30 hours. The time of perigee is 1990-Dec-08 20:34:34.40 UTC.

A similar but larger effect was observed during an Earth flyby by the Near Earth Asteroid Rendezvous (NEAR) spacecraft. The spacecraft took four years after launch to reach the asteroid (433) Eros in February 2000 (Dunham *et al.* 2005). For the Earth gravity assist in January 1998, the pre-perigee Doppler data can be fit with a residual standard error of 0.028 mm s^{-1}. Note that the residuals are smaller for NEAR with its Doppler tracking in the X-Band at about 8.0 GHz, as opposed to Galileo in the S-Band at about 2.3 GHz. Scattering of the two-way radio signal by free ionospheric electrons is less at the higher frequency, although systematic and random effects from atmospheric refraction limit the X-Band tracking accuracy. Nevertheless, the post-perigee residuals (Antreasian & Guinn 1998) show a clear asymptotic bias of 13.51 mm s^{-1} (see Fig. 2). There is also some evidence from Fig. 2 that an anomalistic acceleration might be acting over perhaps plus and minus 10 hours of perigee.

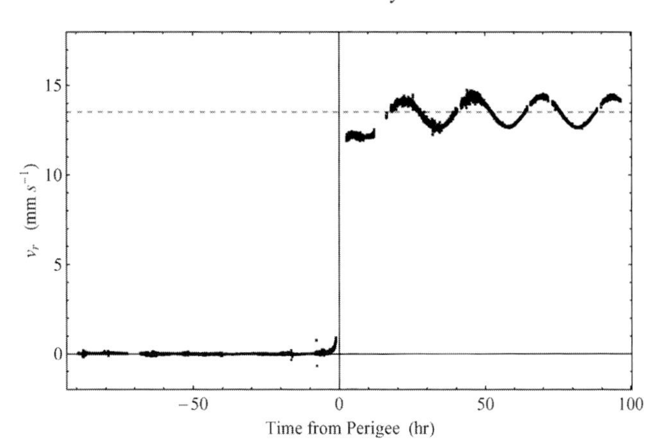

Figure 2. Similar to Fig. 1 but for the NEAR Doppler residuals. The mean offset in the post-perigee data approaches 13.51 mm s^{-1}, as shown by the dashed line. The post-perigee data start at perigee plus 2.51 hours. The time of perigee is 1998-Jan-23 07:22:55.60 UTC.

The anomalistic bias can also be demonstrated for both GLLI and NEAR by fitting the post-perigee data and using that fit to predict the pre-perigee residuals (Anderson *et al.* 2008). For both spacecraft, the two pre- and post-perigee fits are consistent with the same velocity increases shown in Fig. 1 and Fig. 2.

Earth flybys by the Cassini spacecraft on 1999-Aug-18 and the Stardust spacecraft in January 2001 yielded little or no information on the flyby anomaly. Both spacecraft were affected by thruster firings which masked any anomalous velocity change. However, on 2005-Mar-04 the Rosetta spacecraft swung by Earth on its first flyby and an anomalous energy gain was once again observed. Rosetta is an ESA mission with space navigation by the European Space Operations Center (ESOC). As such it provides an independent analysis at ESOC for both ESA and NASA tracking data for Rosetta (Morley & Budnik 2006). The Rosetta anomaly was confirmed independently at JPL with an asymptotic velocity increase of (1.80 ± 0.03) mm s^{-1} (Anderson *et al.* 2008). Similar data analysis by Anderson *et al.* (2008) yielded slightly different velocity changes than indicated by Fig. 1 and Fig. 2 but with error bars. The best estimates are (3.92 ± 0.03) mm s^{-1} for GLLI and (13.46 ± 0.01) mm s^{-1} for NEAR. Rosetta swung by the Earth again on 2007-Nov-13 (RosettaII), but this time no anomaly was reported.

There is most likely a distance dependence to the anomaly. The net velocity increase is 3.9 mm s^{-1} for the Galileo spacecraft at a closest approach of 960 km, 13.5 mm s^{-1} for the NEAR spacecraft at 539 km, and 1.8 mm s^{-1} for the Rosetta spacecraft at 1956 km. The altitude of RosettaII is 5322 km, perhaps too high for a detection of the anomaly. A third Rosetta Earth swing-by (RosettaIII) is scheduled for 2009-Nov-13 at a more favorable altitude of 2483 km. This third gravity assist, which possibly could reveal the anomaly, will place Rosetta on a trajectory to rendezvous with Comet 67P/Churyumov-Gerasimenko on 2014-May-22 and a lander will be placed on the comet on 2014-Nov-10.

The spacecraft bus will orbit the comet and escort it around the Sun until December 2015, when the comet will be at a heliocentric distance of about one AU.

Indeed there is a distance-independent phenomenological formula that models the anomaly quite accurately, at least for flybys at an altitude of 2000 km or less, be that fortuitous or not (Anderson *et al.* 2008). The percentage change in the excess velocity at infinity v_∞ is given by

$$\frac{\Delta v_\infty}{v_\infty} = K(\cos\delta_i - \cos\delta_f), \tag{1.1}$$

$$K = \frac{2\omega_\oplus R_\oplus}{c} = 3.099 \times 10^{-6}, \tag{1.2}$$

where $\delta_{\{i,f\}}$ are the initial (ingoing) and final (outgoing) declination angles given by

$$\sin\delta_{\{i,f\}} = \sin I \cos(\omega \mp \psi)\,. \tag{1.3}$$

The parameter $\omega_\oplus$ is the Earth's angular velocity of rotation, $R_\oplus$ is the Earth's mean radius, and c is the velocity of light.

The angle ψ is one half the total bending angle in the flyby trajectory, I is the osculating orbital inclination to the equator of date, and ω is the osculating argument of the perigee measured along the orbit from the equator of date. The angle ψ is related to the osculating eccentricity e by

$$\sin\psi = \frac{1}{e} \tag{1.4}$$

Alternatively, the total bending angle 2ψ can be obtained as the angle between the asymptotic ingoing and outgoing velocity vectors.

2. Increase in the Astronomical Unit

Radar ranging and spacecraft radio ranging to the inner planets result in a measurement of the AU to an accuracy of 3 m, or a percentage error of 2×10^{-11}, making it the most accurately determined constant in all of astronomy (Pitjeva 2007, Pitjeva & Standish 2009). In SI units the AU can be expressed by the constant A, or as the number of meters or seconds in one AU. The two SI units are interchangeable by means of the defining constant c, the speed of light in units m s^{-1}. In this form, and in combination with the IAU definition of the AU (Resolution No. 10 1976†), there is an equivalence between the AU and the mass of the Sun M_S given by

$$GM_S \equiv k^2 A^3, \tag{2.1}$$

where G is the gravitational constant and k is Gauss' constant.

According to IAU Resolution No. 10, k is exactly equal to 0.01720209895 $AU^{3/2}\ d^{-1}$, similar to c exactly equal to 299792458 m s^{-1}. The value of the AU is connected to the ranging observations by the time unit used for the time delay of a radar signal or a modulated spacecraft radio carrier wave, ideally the SI second, or equivalently the day d of 86400 s. The extraordinary accuracy in the AU is based on Earth-Mars spacecraft ranging data over an interval from the first Viking Lander on Mars in 1976 and continuing with Viking from 1976 to 1982, Pathfinder P (1997), MGS from 1998 to 2003, and Odyssey from 2002 to 2008 (Pitjeva 2009a, Pitjeva 2009b). In practice the AU is measured in units of Coordinated Universal Time (UTC), the time scale used by the Deep Space Network (DSN) in their frequency and timing system. Therefore the AU is given in

† http://www.iau.org/static/resolutions/IAU1976_French.pdf

SI seconds as determined by International Atomic Time TAI (Moyer 2003). The fitting models for the JPL ephemeris and for the IAA-RAS ephemeris (Pitjeva & Standish 2009) are relativistically consistent with ranging measurements in units of SI seconds. It seems that we really do know the AU to (149597870700 $\pm$ 3) m (Pitjeva & Standish 2009).

For purposes of deciding whether a measurement of a change in the AU is feasible, we simulate Earth-Mars ranging at a 40-day sample interval over a 27-year observing interval starting on 1976-July-01, for a total of 248 simulated normal points. We approximate the tracking geometry by means of a Newtonian integration of a four-body system consisting of the Sun, the Earth-Moon barycenter, the Mars barycenter, and the Jupiter barycenter, all treated as point masses. The initial conditions of the Earth and Mars are adjusted to give a best fit to the distance between the Earth-Moon barycenter and the Mars barycenter, as given by DE405. The rms error in this best fit is 2.6×10^{-5} AU, which is unacceptable as a fitting model, but sufficient for a covariance analysis. In the real analysis (Pitjeva 2009a, Pitjeva 2009b) the ranging data are represented by hundreds of parameters, only one of which is the AU.

The parameters for our covariance analysis consist of the 12 state variables for Earth and Mars, expressed as the Cartesian initial conditions at the July 1976 epoch, plus two parameters (k_1, k_2) for GM_S as given by $k^2\,[1 + k_1 + k_2(t - \bar{t})]$ in units of $\mathrm{AU}^3 d^{-2}$. This is the most direct way to express a bias in the AU and its secular time variation as a Newtonian perturbation. †The masses of the three planetary systems are constant at their DE405 values, and the initial conditions of the Jupiter system are not included in the covariance matrix, which makes it a 14×14 matrix. The rank of this matrix is actually 12. The mean Earth orbit defines the reference plane for the other orbits. Hence there are only four Earth elements that can be inferred from the data. A singular value decomposition (SVD) of the 14×14 matrix can be obtained and its pseudo inverse can be interpreted as the covariance matrix on the 14 parameters (Lawson & Hanson 1974). Actually all the information on k_1 and k_2 is obtained by the 8th singular value, so a rank 9 pseudo inverse is more than sufficient for a study of the AU and its time variation. The mean time $\bar{t}$ is introduced into the secular variation in GM_S such that k_1 and k_2 are uncorrelated. This mean time is 13.5 yr for the simulation, but in the real analysis it should be taken as the mean of all the observation times.

Taking account of the factor of three in Eq. 2.1, we normalize the result of the covariance analysis to a standard error in the AU of 3.0 m, represented by k_1 in the rank 12 matrix. The corresponding rank 9 standard error, where it is assumed that all the remaining five singular values are perfectly known, is 2.5 m. The corresponding error in the secular variation represented by k_2 is 2.9 cm yr^{-1} for full rank 12 and 2.7 cm yr^{-1} for rank 9.

We conclude that at least the uncertainty part of the reported increase in the AU (Krasinsky & Brumberg 2004) of (15 $\pm$ 4) cm yr^{-1} is reasonable. Any future work should be focused on checking the actual mean value of the secular increase and perhaps refining it. It is unlikely that its error bar can be decreased below 3.0 cm yr^{-1} with existing Earth-Mars ranging data. However, if the error in the AU can be reduced to $\pm$ 1.0 m with confidence, the error in its secular variation could perhaps be reduced to $\pm$ 1.0 cm yr^{-1}, with Earth-Mars ranging alone. Other than that, the Cassini spacecraft carries an X-Band ranging transponder (Kliore 2004). Range fixes on Saturn presumably can be obtained for each Cassini orbital period of roughly 14.3 days over an observing interval from July 2004 to July 2009, or for as long as the spacecraft is in orbit about Saturn

† The AU is not determined in ephemeris software by means of this physical approach (see Pitjeva (2007), Pitjeva (2009a) and Pitjeva (2009b) for details).

and ranging data are available. These data are not yet publicly available, but when they are released, we can expect a standard error in each ranging normal point of about 5 m. Spacecraft ranging to Mercury during the MESSENGER and BepiColombo missions could also add additional information for the AU and its secular variation. If the AU is really increasing with time, the planetary orbits by definition (Eq. 2.1) are shrinking and their periods are getting shorter, such that their mean orbital longitudes are increasing quadratically with t, the major effect that can be measured with Earth-planet ranging data.

However, rather than increasing, the AU should be decreasing, mainly as a result of loss of mass to solar radiation, and to a much lesser extent to the solar wind. The total solar luminosity is 3.845×10^{26} W (Livingston 1999). This luminosity divided by c^2 gives an estimated mass loss of 1.350×10^{17} kg yr^{-1}. The total mass of the Sun is 1.989×10^{30} kg (Livingston 1999), so the fractional mass loss is 6.79×10^{-14} yr^{-1}. Again with the factor of three from Eq. 2.1, the expected fractional decrease in the AU is 2.26×10^{-14} yr^{-1}, or a change in the AU of $-$ 0.338 cm yr^{-1}. A change this small is not currently detectable, and it introduces an insignificant bias into the reported measurement of an AU increase (Krasinsky & Brumberg 2004). If the reported increase is absorbed into a solar mass increase, and not into a changing gravitational constant G, the inferred solar mass increase is $(6.0 \pm 1.6) \times 10^{18}$ kg yr^{-1}. This is an unacceptable amount of mass accretion by the Sun each year. It amounts to a fair sized planetary satellite of diameter 140 km and with a density of 2000 kg m^{-3}, or to about 40,000 comets with a mean radius of 2000 m. If the reported increase holds up under further scrutiny and additional data analysis, it is indeed anomalous. Meanwhile it is prudent to remain skeptical of any real increase. In our opinion the anomalistic increase lies somewhere in the interval zero to 20 cm yr^{-1}, with a low probability that the reported increase is a statistical false alarm.

3. The Pioneer anomaly

The first missions to fly to deep space were the Pioneers. By using flybys, heliocentric velocities were obtained that were unfeasible at the time by using only chemical fuels. Pioneer 10 was launched on 1972-Mar-02 local time. It was the first craft launched into deep space and was the first to reach an outer giant planet, Jupiter, on 1973-Dec-04. With the Jupiter flyby, Pioneer 10 reached escape velocity from the solar system. Pioneer 10 has an asymptotic escape velocity from the Sun of 11.322 km s^{-1} (2.388 AU yr^{-1}).

Pioneer 11 followed soon after Pioneer 10, with a launch on 1973-Apr-06. It too cruised to Jupiter on an approximate heliocentric ellipse. This time a carefully executed flyby of Jupiter put the craft on a trajectory to encounter Saturn in 1979. So, on 1974-Dec-02, when Pioneer 11 reached Jupiter, it underwent a Jupiter gravity assist that sent it back inside the solar system to catch up with Saturn on the far side. It was then still on an ellipse, but a more energetic one. Pioneer 11 reached Saturn on 1979-Sept-01. Then Pioneer 11 embarked on an escape hyperbolic trajectory with an asymptotic escape velocity from the Sun of 10.450 km s^{-1} (2.204 AU yr^{-1}).

The Pioneer navigation was carried out at the Jet Propulsion Laboratory. It used NASA's DSN to transmit and obtain the raw radiometric data. An S-band signal ($\sim$2.11 Ghz) was sent up via a DSN antenna located either at Goldstone, California, outside Madrid, Spain, or outside Canberra, Australia. On reaching the craft the signal was transponded back with a (240/221) frequency ratio ($\sim$2.29 Ghz), and received back at the same station (or at another station if, during the radio round trip, the original station had rotated out of view). There the signal was compared with 240/221 times the recorded transmitted frequency and any Doppler frequency shift was measured directly by cycle count compared to an atomic clock. The processing of the raw cycle count produced a data record of Doppler frequency shift as a function of time, and from this a trajectory was calculated. This procedure was done iteratively for purposes of converging to a best fit by nonlinear weighted least squares (minimization of the chi squared statistic, see Lawson & Hanson 1974).

However, to obtain the spacecraft velocity as a function of time from this Doppler shift is not easy. The codes must include all gravitational and time effects of general relativity to order $(v/c)^2$ and some effects to order $(v/c)^4$. The ephemerides of the Sun, planets and their large moons as well as the lower mass multipole moments are included. The positions of the receiving stations and the effects of the tides on the exact positions, the ionosphere, troposphere, and the solar plasma are included.

Given the above tools, precise navigation was possible because, due to a serendipitous stroke of luck, the Pioneers were spin-stabilized. With spin-stabilization the craft are rotated at a rate of ~(4–7) rpm about the principal moment-of-inertia axis. Thus, the craft is a gyroscope and attitude maneuvers are needed only when the motions of the Earth and the craft move the Earth from the antenna's line-of-sight.

The Pioneers were chosen to be spin-stabilized because of other engineering decisions. As the craft would be so distant from the Sun solar power panels would not work. Therefore these were the first deep spacecraft to use nuclear heat from ^{238}Pu as a power source in Radioisotope Thermoelectric Generators (RTGs). Because of the then unknown effects of long-term radiation damage on spacecraft hardware, a choice was made to place the RTGs at the end of long booms. This placed them away from the craft and thereby avoided most of the radiation that might be transferred to the spacecraft.

Even so, there remained one relatively large effect on this scale that had to be modeled: the solar radiation pressure. This effect is approximately 1/30,000 that of the Sun's gravity on the Pioneers. It produced an acceleration of $\sim 20 \times 10^{-8}$ cm s^{-2} on the Pioneer craft at the distance of Saturn.

After 1976 small time-samples (approximately 6-month to 1-year averages) of the data were periodically analyzed. At first nothing significant was found, But when a similar analysis was done around Pioneer 11 's Saturn flyby, things dramatically changed. (See the first three data points in Fig. 3.) So people kept following Pioneer 11. They also started looking more closely at the incoming Pioneer 10 data.

By 1987 it was clear that an anomalous acceleration appeared to be acting on the craft with a magnitude $\sim 8 \times 10^{-8}$ cm s^{-2}, directed approximately towards the Sun. The effect was a concern, but the effect was small in the scheme of things and did not affect the necessary precision of the navigation. However, by 1992 it was clear that a more detailed look would be useful.

An announcement was made at a 1994 conference proceedings. The strongest immediate reaction was that the anomaly could well be an artifact of JPL's Orbit Determination Program (ODP), and could not be taken seriously until an independent code had tested it. So, a team was gathered that included colleagues from The Aerospace Corporation and their independent CHASMP navigation code. Their result was the same as that obtained by JPL's ODP.

The Pioneer anomaly collaboration's discovery paper appeared in 1998 (Anderson *et al.* 1998). and a detailed analysis appeared in 2002 (Anderson *et al.* 2002). The latter used Pioneer 10 data spanning 1987-Jan-03 to 1998-Jul-22 (when the craft was 40 AU to 70.5 AU from the Sun) and Pioneer 11 data spanning 1987-Jan-05 to 1990-Oct-01 (when Pioneer 11 was 22.4 to 31.7 AU from the Sun). The largest systematics were, indeed, from heat but the final result for the anomaly, is that there is an unmodeled

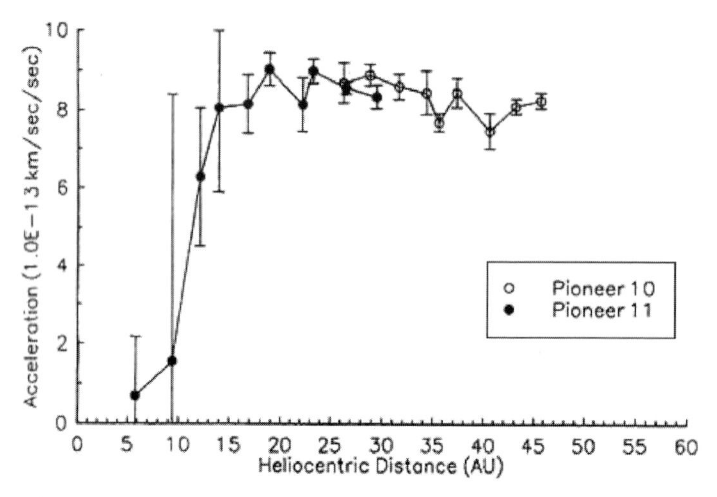

Figure 3. A JPL Orbit Determination Program (ODP) plot of the early unmodeled accelerations of Pioneer 10 and Pioneer 11, from about 1981 to 1989 and 1977 to 1989, respectively.

acceleration, directed approximately towards the Sun, of

$$a_P = (8.74 \pm 1.33) \times 10^{-8} \text{ cm s}^{-2}. \tag{3.1}$$

Two later and independent analyses of this data obtained similar results. The conclusion, then, is that this "Pioneer anomaly" is in the data. The question is (Nieto & Anderson 2007), "What is its origin?"

It is tempting to assume that radiant heat must be the cause of the acceleration, since only 63 W of directed power could cause the effect (and much more heat than that is available). The heat on the craft ultimately comes from the Radioisotope Thermoelectric Generators (RTGs), which yield heat from the radioactive decay of ^{238}Pu. Before launch, the four RTGs had a total thermal fuel inventory of 2580 W ($\approx$2070 W in 2002). Of this heat 165 W was converted at launch into electrical power, which decreased down to $\sim$ 70 W. So, heat as a mechanism yielding an approximately constant effect remains to be clearly resolved, but detailed studies are underway at JPL.

Indeed, from the beginning we observed that a most likely origin is directed heat radiation (Anderson *et al.* 1998, Anderson *et al.* 2002). However, suspecting this likelihood is different from proving it. Even so, investigation may well ultimately show that heat was a larger effect than originally demonstrated by Anderson *et al.* (2002). Their original estimate of the bias from reflected heat amounts to only 6.3% of the total anomaly. Nevertheless, a three-sigma error in the original estimate could amount to a 25% thermal effect. We would have difficulty accepting anything larger than this three-sigma limit.

On the other hand, if this is a modification of gravity, it is not universal; i.e., it is not a scale independent force that affects planetary bodies in bound orbits. The anomaly could, in principle be i) some modification of gravity, ii) drag from dark matter, or a modification of inertia, or iii) a light acceleration (Nieto & Anderson 2007);

Future study of the anomaly may determine which, if any, of these proposals are viable.

4. Increase in the eccentricity of the Moon's orbit

A detailed orbital analysis of Lunar Laser Ranging (LLR) data can be found in Williams & Boggs (2009). A total of 16,941 ranges were analyzed extending from 1970-Mar-16 to 2008-Nov-22. LLR can measure evolutionary changes in the geocentric lunar orbit over this interval of 38.7 years. Changes in both the mean orbital motion and eccentricity are observed. While the mean motion and semi-major axis rates of the lunar orbit are consistent with physical models for dissipation in Earth and Moon, LLR orbital solutions consistently reveal an anomalous secular eccentricity variation. After accounting for tides on the Earth that produce an eccentricity change of 1.3×10^{-11} yr^{-1} and tides on the Moon that produce a change of -0.6×10^{-11} yr^{-1}, there is an anomalous rate of $(0.9 \pm 0.3) \times 10^{-11}$ yr^{-1}, equivalent to an extra 3.5 mm yr^{-1} in perigee and apogee distance (Williams & Boggs 2009). This anomalous eccentricity rate is not understood and it presents a problem, both for a physical understanding of dissipative processes in the interiors of Earth and Moon, and for the modeling of dynamical evolution at the 10^{-11} yr^{-1} level.

References

Anderson, J. D., Laing, P. A., Lau, E. L., Liu, A. S., Nieto, M. M., & Turyshev, S. G. 1998, *Phys. Rev. Lett.*, 81, 2858

Anderson, J. D., Laing, P. A., Lau, E. L., Liu, A. S., Nieto, M. M., & Turyshev, S. G. 2002, *Phys. Rev. D*, 65, 082004

Anderson, J. D., Campbell, J. K., & Nieto, M. M. 2007, *New Astron.*, 12, 383

Anderson, J. D., Campbell, J. K., Ekelund, J. E., Ellis, J., & Jordan, J. F. 2008, *Phys. Rev. Lett.*, 100, 091102

Antreasian, P. G. & Guinn, J. R. 1998, *AIAA/AAS Paper No. 98–4287* (http://www2.aiaa.org/citations/mp-search.cfm)

Danby, J. M. A. 1988, *Fundamentals of Celestial Mechanics* (Richmond: Willmann-Bell), sec. 8.2

Dunham, D. W., Farquhar, R. W., & McAdams, J. V. 2005 *Ann. N.Y. Acad. Sci.*, 1065, 254

Flandro, G. A. 1966, *Astronaut. Acta*, 12, 329

Kliore, A. J., Anderson, J. D., Armstrong, J. W., & ten others 2004 *Space Science Reviews*, 115, 1

Krasinsky, G. A. & Brumberg, V. A. 2004 *Celest. Mech. Dynam. Astron.*, 90, 3

Lawson, C. J. & Hanson, R. J. 1974 *Solving Least Squares Problems* (Englewood Cliffs: Prentice-Hall)

Livingston, W. C. 1999, in *Allen's Astrophysical Quantities, Fourth Edition* ed. A. N. Cox, (New York, Berlin, Heidelberg: Springer-Verlag), Chap. 14

Morley, T. & Budnik, F. 2006, *19th Int. Symp. on Space Flight Dynamics*, Paper No. ISTS 2006-d-52

Moyer, T. D. 2003, *Formulation for Observed and Computed Values of Deep Space Network Data Types for Navigation* (Print ISBN: 9780471445357, Online ISBN: 9780471728474: John Wiley & Sons), chap. 2

Nieto, M. M. & Anderson, J. D. 2007 *Contemp. Phys.*, 48, 41

Pitjeva, E. V. 2007 in *Proceedings of the "Journées Systèmes de Référence Spatio-temporels 2007"* (Observatoire de Paris), p. 65.

Pitjeva, E. V. 2009 *this proceedings*, 170

Pitjeva, E. V. 2009 *JOURNEES-2008 Astrometry, Geodynamics and Astronomical Reference Systems* ed. M. Soffel & N. Capitaine, p. 57

Pitjeva, E. V. & Standish, E. M. 2009, *Celest. Mech. Dynam. Astron.*, 103, 365

Russell, C. T. 1992, *The Galileo Mission* (Dordrecht, Boston, London: Kluwer)

Tapley, B. D., Bettadpur, S., Watkins, M., & Reigber, C. 2004 *Geophys. Res. Lett.*, 31, L09607

Wiesel, W. E. 1989, *Spaceflight Dynamics* (New York: McGraw-Hill), sec. 11.5

Williams, J. G. & Boggs, D. H. 2009 in *Proceedings of 16th International Workshop on Laser Ranging* ed. S. Schillak, (Space Research Centre, Polish Academy of Sciences)

Relativity in Fundamental Astronomy
Proceedings IAU Symposium No. 261, 2009
S. A. Klioner, P. K. Seidelman & M. H. Soffel, eds.

doi:10.1017/S174392130999038X

The confrontation between general relativity and experiment

Clifford M. Will

Department of Physics, McDonnell Center for the Space Sciences,
Washington University, St. Louis MO 63130 USA
Gravitation et Cosmologie, Institut d'Astrophysique de Paris,
98 bis Bd. Arago, 75014 Paris France
email: cmw@wuphys.wustl.edu

Abstract. We review the experimental evidence for Einstein's general relativity. A variety of high precision null experiments confirm the Einstein Equivalence Principle, which underlies the concept that gravitation is synonymous with spacetime geometry, and must be described by a metric theory. Solar system experiments that test the weak-field, post-Newtonian limit of metric theories strongly favor general relativity. Binary pulsars test gravitational-wave damping and aspects of strong-field general relativity. During the coming decades, tests of general relativity in new regimes may be possible. Laser interferometric gravitational-wave observatories on Earth and in space may provide new tests via precise measurements of the properties of gravitational waves. Future efforts using X-ray, infrared, gamma-ray and gravitational-wave astronomy may one day test general relativity in the strong-field regime near black holes and neutron stars.

Keywords. gravitation, gravitational waves, relativity

Since the late 1960s, when it was frequently said that "the field of general relativity is a theorist's paradise and an experimentalist's purgatory" the field of gravitational physics has been completely transformed, and today experiment is a central, and in some ways dominant component. The breadth of current experiments, ranging from tests of classic general relativistic effects, to searches for short-range violations of the inverse-square law, to a space experiment to measure the relativistic precession of gyroscopes, attest to the ongoing vigor of experimental gravitation.

The great progress in testing general relativity during the latter part of the 20th century featured three main themes:

• The use of advanced technology. This included the high-precision technology associated with atomic clocks, laser and radar ranging, cryogenics, and delicate laboratory sensors, as well as access to space.

• The development of general theoretical frameworks. These frameworks allowed one to think beyond the narrow confines of general relativity itself, to analyse broad classes of theories, to propose new experimental tests and to interpret the tests in an unbiased manner.

• The synergy between theory and experiment. To illustrate this, one needs only to note that the LIGO-Virgo Scientific Collaboration, engaged in one of the most important general relativity investigations – the detection of gravitational radiation – consists of over 700 scientists. This is big science, reminiscent of high-energy physics, not general relativity!

Today, because of its elegance and simplicity, and because of its empirical success, general relativity has become the foundation for our understanding of the gravitational interaction. Yet modern developments in particle theory suggest that it is probably not the entire story, and that modifications of the basic theory may be required at some level.

However, any theoretical speculation along these lines *must* abide by the best current empirical bounds. Still, most of the current tests involve the weak-field, slow-motion limit of gravitational theory.

Putting general relativity to the test during the 21st century is likely to involve three main themes:

- Tests of strong-field gravity. These are tests of the nature of gravity near black holes and neutron stars, far from the weak-field regime of the solar system.
- Tests using gravitational waves. The detection of gravitational waves, hopefully during the next decade, will initiate a new form of astronomy but it will also provide new tests of general relativity in the highly dynamical regime.
- Tests of gravity at extreme scales. The detected acceleration of the universe, the observed large-scale effects of dark matter, and the possibility of extra dimensions with effects on small scales, have revealed how little precision information is known about gravity on the largest and smallest scales.

In this contribution to the Symposium, we reviewed selected highlights of testing general relativity during the 20th century and discussed the potential for new tests in the 21st century. We discussed the "Einstein equivalence principle", which underlies the idea that gravity and curved spacetime are synonymous, and describes its empirical support; solar system tests of gravity in terms of experimental bounds on a set of "parametrized post-Newtonian" (PPN) parameters; and tests of general relativity using binary pulsar systems. We also described tests of gravitational theory that could be carried out using future observations of gravitational radiation, and described the possibility of performing strong-field tests of general relativity. But because the content of this review has been published in various forms elsewhere, we did not submit a full article. For further discussion of topics in this paper, and for references to the primary literature, we encourage readers to consult *Theory and Experiment in Gravitational Physics* (Will 1993) and to the "living" review articles by Mattingly (2005), Psaltis (2008), Stairs (2003) and Will (2006).

Acknowledgements

This work was supported in part by the US National Science Foundation, Grant No. PHY 06-52448 and by the National Aeronautics and Space Administration, Grant No. NNG-06GI60G. We are grateful for the hospitality of the Institut d'Astrophysique de Paris, where this summary was prepared.

References

Mattingly, D. 2005, *Living Rev. Relativ.*, 8, 5 [On-line article] Cited on 1 June 2009, www.livingreviews.org/lrr-2005-5

Psaltis, D. 2008, *Living Rev. Relativ.*, 11, 9 [On-line article] Cited on 1 June 2009, www.livingreviews.org/lrr-2008-9

Stairs, I. H. 2003, *Living Rev. Relativ.* 6, 5 [On-line article] Cited on 1 June 2009, www.livingreviews.org/lrr-2003-5

Will, C. M. 1993, *Theory and Experiment in Gravitational Physics*, (Cambridge University Press, Cambridge)

Will, C. M. 2006, *Living Rev. Relativ.*, 9, 3 [Online article] Cited on 1 June 2009, www.livingreviews.org/lrr-2006-3

Relativity in Fundamental Astronomy
Proceedings IAU Symposium No. 261, 2009
S. A. Klioner, P. K. Seidelman & M. H. Soffel, eds.

doi:10.1017/S1743921309990391

APOLLO: A new push in solar-system tests of gravity

T. W. Murphy, Jr.[1], E. G. Adelberger[2], J. B. R. Battat[3], C. D. Hoyle[4], R. J. McMillan[5], E. L. Michelsen[1], C. W. Stubbs and H. E. Swanson[2]

[1]University of California, San Diego,
9500 Gilman Drive, La Jolla, CA, 92093-0424, USA
email: tmurphy@physics.ucsd.edu

[2]University of Washington, Seattle, WA 98195-1560, USA

[3]Massachusetts Institute of Technology, Cambridge, MA 02139, USA

[4]Humboldt State University, Arcata, CA 95521-8299, USA

[5]Apache Point Observatory, Sunspot, NM 88349-0059, USA

Abstract. Lunar laser ranging (LLR) has long provided many of our best measurements on the fundamental nature of gravity, including the strong equivalence principle, time -rate-of-change of the gravitational constant, the inverse square law, geodetic precession, and gravitomagnetism. This paper serves as a brief overview of APOLLO: a recently operational LLR experiment capable of millimeter-level range precision.

Keywords. gravitation

---

1. Introduction

Lunar laser ranging (LLR) has long provided the best constraints on a number of tests of gravity, including the strong equivalence principle, time variation of the gravitational constant, geodetic precession, the inverse square law, and gravitomagnetism (Williams *et al.* 1996; Murphy *et al.* 2007; Soffel *et al.* 2008).

APOLLO (the Apache Point Observatory Lunar Laser-ranging Operation) represents a new capability in the technique of lunar laser ranging (LLR), achieving millimeter-level range precision. Beginning in 2006, this order-of-magnitude improvement over previous capabilities will translate into similar gains in tests of general relativity, and provide a more comprehensive understanding of solar system dynamics as well as probe the interior structure of the moon.

What is required to reach a 1 mm goal? The random uncertainty associated with the lunar range measurement is dominated by the temporal spread induced by the finite-sized retroreflector trays tilted by angles up to 10-deg by lunar librations, amounting to 15–50 mm of one-way range error for an individual photon. Improvements in range precision must, therefore, come about by increasing the number of photons that constitute the range measurement. Improvements in laser pulse width, timing precision, etc. do not result in greater range precision. To reduce the 15–50 mm uncertainty introduced by the reflector, one must collect 225–2500 photons, which far exceeds typical performance of previous LLR stations. APOLLO, by virtue of its deployment on a large 3.5 m telescope at a site with good atmospheric seeing, is able to measure thousands of photon round trip times in a matter of minutes, exceeding previous records by a factor of $\sim$70. This boost in sensitivity is what allows APOLLO to push LLR into the millimeter regime.

A detailed description of the APOLLO science motivation, instrument design, and initial performance may be found in Murphy *et al.* (2008). Demonstration of one-millimeter

Table 1. APOLLO record performance for the different reflectors.

Reflector	photons per run	photons per minute	photons per shot (5 minute average)	photons per shot (15 second average)
Apollo 11	4497 (26×)[1]	1079 (65×)	0.90	1.4
Apollo 14	7606 (36×)	1825 (69×)	1.52	2.0
Apollo 15	15740 (26×)	3775 (67×)	3.15	4.5
Lunokhod 2	750 (11×)	180 (31×)	0.15	0.24

Notes:
[1] Ratios in parentheses denote the factor by which APOLLO exceeds previous records.

ranging performance is provided in Battat *et al.* (2009). In this paper, we present a few results not published elsewhere as a demonstration of APOLLO's promise to transform lunar ranging.

2. APOLLO Performance

APOLLO seeks to bring about an order-of-magnitude improvement in tests of gravity by way of a substantial enhancement in the rate of return photon detection. Not only are the random uncertainties tamed by the gain in $\sqrt{N}$, but systematic errors may be investigated in much greater detail when the signal is more robust. For example, strong runs may be split into shorter segments to look for scatter beyond that expected from random error, as was done in Battat *et al.* (2009). Model systematics may be checked by cycling among the various reflectors multiple times within a session—exposing lunar orientation offsets. Once confidence has been established in the short-term behavior of the system (within a one-hour session), any systematic drift in residuals may be used to inform earth orientation offsets.

Table 1 demonstrates APOLLO's photon collection capabilities. Though each run may vary in duration, the records represented in the first column all come from runs consisting of 5000 shots, lasting 250 seconds. The second column normalizes to photons collected in a 1 minute period. Here we see that APOLLO outperforms the previous records (in parentheses: all previous records set by the French LLR station at Grasse) by a factor of ~ 65 for all three Apollo reflectors. Lunokhod 2 is less responsive than it was in the past.

The higher return rate achieved by APOLLO means that we may obtain ranges at full moon for the first time since the McDonald 2.7 m telescope performed LLR measurements prior to 1985. This is especially important for the Equivalence Principle measurement, to which full moon measurements are most sensitive.

As an example of how the high return rate can impact model systematics, Fig. 1 shows the measured range offset from a prediction based on the DE421 ephemeris from the Jet propulsion Laboratory (JPL). The nine data points cover four reflectors over a 45 minute period. The four reflector residuals can be fit individually by a common slope, which is due mostly to inadequate treatment of earth orientation on the APOLLO prediction software, and may thus be ignored. The offset between the four reflectors indicates imperfect knowledge of the lunar orientation, as predicted in the DE421 ephemeris. An obvious coherence exists in the APOLLO data, so that the individual measurements are self-consistent within the error bars, allowing for arbitrary lunar orientation. Subtracting each reflector's individual fit-line puts all the residuals on the same footing, as seen in Fig. 2. The distribution of residuals is artificially small with relation to the error bars because the lunar orientation was "fixed" to minimize the residuals rather than appealing to a physical model of the moon subject to fitting requirements of data points on other nights.

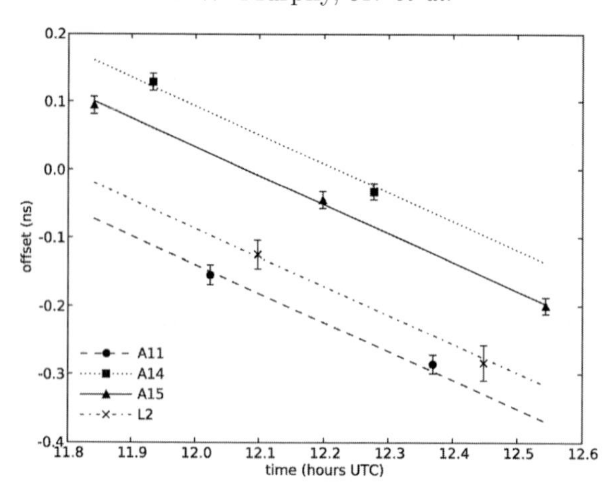

Figure 1. APOLLO residuals from the DE421 ephemeris on the night of 2008 September 24. 0.1 ns in the round-trip time is equivalent to 15 mm of one-way range. A common slope offset applies to all reflectors, indicating a likely earth orientation offset. The separation of the four reflectors is due to imperfect prediction of lunar orientation.

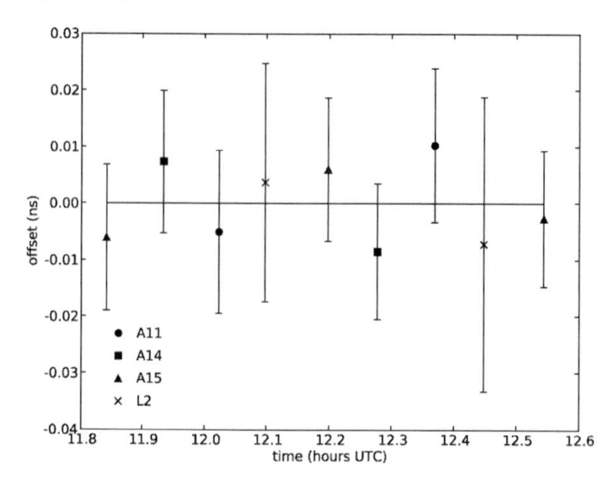

Figure 2. Residuals from Fig. 1 after subtracting the individual fit-lines for each reflector. The remaining offsets are artificially small with relation to the error bars because five degrees of freedom have been removed from 9 points (a common slope and four individually-adjusted offsets). Each vertical tick of 0.01 ns corresponds to 1.5 mm in one-way range.

Though Figures 1 and 2 represent a particular night, no deviation from this behavior has yet been seen in APOLLO data. Reflector-specific offsets are common, but the data points coherently follow a common trend-line among the reflectors. At the very least, then, APOLLO data points can be expected to have substantial power in establishing lunar orientation. As this is crucial for converting ranges to individual reflectors into effective ranges to the center of the moon, this capability should deliver also a direct improvement on the gravitational physics measured by LLR.

The overall residuals produced from APOLLO data points using the best models available do not yet indicate fidelity between model and observation. There could be many contributing reasons, and it is difficult to separate systematic measurement errors from model shortcomings. Thus far, the tests we can easily perform have not shown any obvious problems with data quality, but this is only possible over short timescales. Meanwhile, we know the models lack small-scale influences like ocean and atmospheric loading of the earth crust, earth center-of-mass motion, and the latest models on atmospheric refraction. Also, APOLLO data are not yet used at full weight to influence earth orientation solutions. It is, therefore, imperative that we advance the state-of-the-art in LLR modeling, so that we may use APOLLO's high data quality and advance the limits of our knowledge about gravitational physics.

References

Battat, J. B. R., Murphy, T. W., Adelberger, E. G., Gillespie, B., Hoyle, C. D., McMillan, R. J., Michelsen, E. M., Nordtvedt, K., Orin, A. E., Stubbs, C. W., & Swanson, H. E. 2009, *PASP* 121, 29

Murphy, T. W., Nordtvedt, K., & Turyshev, S. G. 2007, *Phys. Rev. Lett.*, 98, 071102

Murphy, T. W., Adelberger, E. G., Battat, J. B. R., Carey, L. N., Hoyle, C. D., LeBlanc, P., Michelsen, E. M., Nordtvedt, K., Orin, A. E., Strasburg, J. D., Stubbs, C. W., Swanson, H. E., & Williams, E. 2008, *PASP* 120, 20

Soffel, M., Klioner, S., Müller, J., & Biskupek, L. 2008, *Phys. Rev. D*, 78, 024033

Williams, J. G., Newhall, X. X., & Dickey, J. O. 1996, *Phys. Rev. D*, 53, 6730

Relativity in Fundamental Astronomy
Proceedings IAU Symposium No. 261, 2009
S. A. Klioner, P. K. Seidelman & M. H. Soffel, eds.

doi:10.1017/S1743921309990408

Tests of relativistic gravity from space

Slava G. Turyshev

Jet Propulsion Laboratory, California Institute of Technology,
4800 Oak Grove Drive, Pasadena, CA 91109-0899, USA
email: turyshev@jpl.nasa.gov

Abstract. Recent experiments have successfully tested Einstein's general theory of relativity to remarkable precision. We discuss recent progress in the tests of relativistic gravity in the solar system and present motivations for the new generation of high-accuracy gravitational experiments. We especially focus on the concepts aiming to probe parameterized-post-Newtonian parameter γ and evaluate the discovery potential of the recently proposed experiments.

Keywords. General relativity; modified gravity; scalar-tensor theories; space-based experiments

---

1. Status of Precision Tests of Gravity

Ever since its original publication in 1915, Einstein's general theory of relativity continues to be an active area of both theoretical and experimental research. Presently, the theory successfully accounts for all data gathered to date (Turyshev 2009a).

To describe the accuracy achieved in the solar system gravitational experiments, it is useful to refer to the parameterized post-Newtonian (PPN) formalism (Will 2006). A particular metric theory of gravity in the PPN formalism with a specific coordinate gauge is fully characterized by means of ten PPN parameters. The formalism uniquely prescribes the values of these parameters for the particular theory under study.† General relativity, when analyzed in the standard PPN gauge, gives $\gamma = \beta = 1$; other theories may yield different values of these parameters. Gravity experiments can be analyzed in terms of the PPN metric; an ensemble of experiments determine the unique value for these parameters and hence the metric field itself.

Over the years, a number of solar system experiments contributed to the phenomenological success of general relativity. Analysis of data obtained from microwave ranging to the Viking lander on Mars verified the prediction of this theory that the round-trip times of light signals traveling between the Earth and Mars are increased by the direct effect of solar gravity; the corresponding value of the PPN parameter γ was obtained at the level of 1.000 ± 0.002 (Reasenberg *et al.* 1979) (see Fig. 1). Analyses of very long baseline interferometry data have yielded result of $\gamma = 0.99983 \pm 0.00045$ (Shapiro *et al.* 2004). Lunar laser ranging constrained a combination of PPN parameters $4\beta - \gamma - 3 = (4.0 \pm 4.3) \times 10^{-4}$ via precision measurements of the lunar orbit (Williams *et al.* 2004). Finally, microwave tracking of the Cassini spacecraft on its approach to Saturn measured the parameter γ as $\gamma - 1 = (2.1 \pm 2.3) \times 10^{-5}$, thereby reaching the current best accuracy provided by tests of gravity in the solar system (Bertotti *et al.* 2003).

It is remarkable that even more than 90 years after general relativity was conceived, Einstein's theory has survived every test (Will 2006). Such longevity and success make general relativity the de-facto "standard" theory of gravitation for all practical purposes

† In a special case, when only two PPN parameters (γ, β) are considered, these parameters have a clear physical meaning. The parameter γ represents the measure of the curvature of space created by a unit rest mass; parameter β represents a measure of the non-linearity of the law of superposition of the gravitational fields in a theory of gravity.

involving spacecraft navigation and astrometry, astrophysics, cosmology and fundamental physics (Turyshev 2009a).

At the same time there are many important reasons to question the validity of general relativity and to determine the level of accuracy at which it is violated. On the theoretical front, problems arise from several directions, most concerning the strong-gravitational field regime. These challenges include the appearance of spacetime singularities and the inability of a classical description to describe the physics of very strong gravitational fields. A way out of this difficulty may be through gravity quantization. However, despite the success of modern gauge-field theories in describing the electromagnetic, weak, and strong interactions, we do not yet understand how gravity should be described at the quantum level. This continued inability to merge gravity with quantum mechanics indicates that the pure tensor gravity of general relativity needs modification or augmentation.

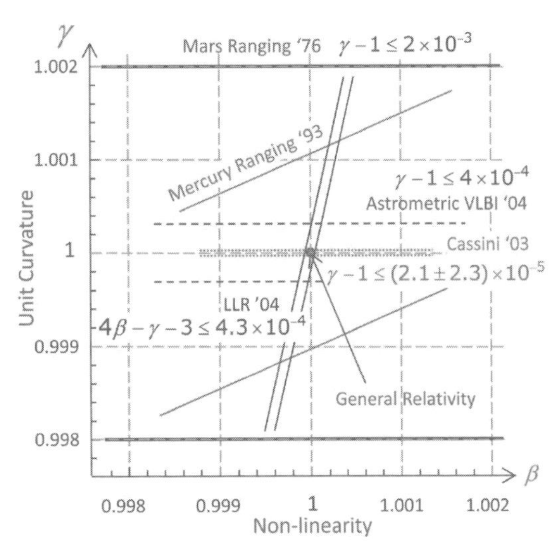

Figure 1. Progress in the measurement accuracy of the Eddington's parameters γ and β. So far, general theory of relativity survived every test (Turyshev 2009a), yielding $\gamma - 1 = (2.1 \pm 2.3) \times 10^{-5}$ (Bertotti *et al.* 2003) and $\beta - 1 = (1.2 \pm 1.1) \times 10^{-4}$ (Williams *et al.* 2004).

In addition, recent remarkable progress in observational cosmology has subjected the general theory of relativity to increased scrutiny by suggesting a non-Einsteinian scenario of the Universe's evolution. Researchers now believed that new physics is needed to resolve these issues. Theoretical models of the kinds of new physics that can solve the problems described above typically involve new interactions, some of which could manifest themselves as violations of the EP, variation of fundamental constants, modification of the inverse-square law of gravity at short distances, Lorenz symmetry breaking, or large-scale gravitational phenomena. Each of these manifestations offers an opportunity for space-based experimentation and, hopefully, a major discovery.

Given the immense challenge posed by the unexpected discovery of the accelerated expansion of the universe, it is important to explore every option to explain and probe the underlying physics. Theoretical efforts in this area offer a rich spectrum of new ideas, some discussed below, that can be tested by experiment.

Motivated by the dark-energy and dark-matter problems, long-distance gravity modification is one of the proposals that have recently gained attention (Deffayet *et al.* 2002). Theories that modify gravity at cosmological distances exhibit a strong coupling phenomenon of extra graviton polarizations (Dvali 2006). This phenomenon plays an important role in this class of theories in allowing them to agree with solar system constraints. In particular, the "brane-induced gravity" model (Dvali *et al.* 2000) provides an interesting way of modifying gravity at large distances to produce an accelerated expansion of the universe, without the need for a non-vanishing cosmological constant (Deffayet *et al.* 2002). One of the peculiarities of this model is means of recovering the usual gravitational interaction at small (i.e. non-cosmological) distances, motivating precision tests of gravity on solar system scales (Dvali *et al.* 2003; Magueijo & Bekenstein 2007).

2. Future Space-based Tests of Gravitational Theories

The Eddington parameter γ, whose value in general relativity is unity, is perhaps the most fundamental PPN parameter (Will 2006), in that $\frac{1}{2}(\gamma-1)$ is a measure, for example, of the fractional strength of the scalar-gravity interaction in scalar-tensor theories of gravity (Damour & Nordtvedt 1993). The current best accuracy of the Cassini† γ result of $\gamma - 1 = (2.1 \pm 2.3) \times 10^{-5}$ approaches the region where multiple tensor-scalar gravity models, consistent with the recent cosmological observations (Spergel *et al.* 2007), predict a lower bound for the present value of this parameter at the level of $\gamma - 1 \sim 10^{-6} - 10^{-7}$ (Damour & Nordtvedt 1993; Damour *et al.* 2002). Therefore, improving the measurement of this parameter would provide crucial information to separate modern scalar-tensor theories of gravity from general relativity, probe possible ways for gravity quantization, and test modern theories of cosmological evolution.

The current accuracy of modern optical astrometry, as represented by the Hipparcos Catalogue, is about 1 milliarcsecond, which gave a determination of γ at the level of 0.997 ± 0.003 (Froeschle *et al.* 1997). Future astrometric missions such as Gaia and the Space Interferometry Mission (SIM) will push the accuracy to the level of a few microarcseconds; the expected accuracy of determinations of the PPN parameter γ will be 10^{-6} to 5×10^{-7} (Turyshev 2009b).

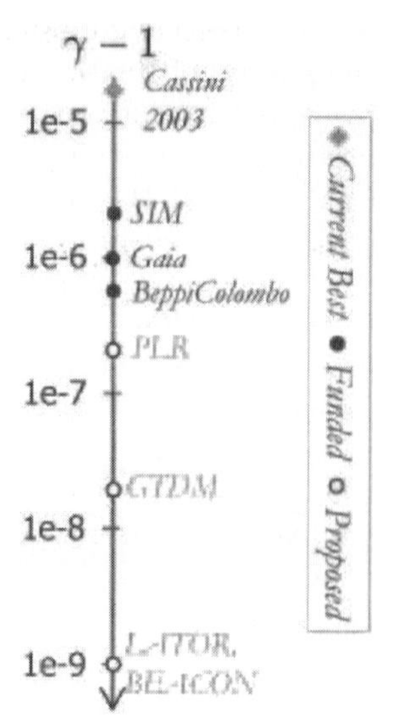

Figure 2. Anticipated progress in the tests of the PPN parameter γ (Turyshev 2009a).

Interplanetary laser ranging could lead to a significant improvement in the measurement accuracy of the PPN parameter γ. Thus, precision laser ranging between the Earth and a lander on Phobos (i.e., Phobos Laser Ranging – PLR) during solar conjunctions may offer a suitable opportunity. If the lander were equipped with a laser transponder capable of reaching a precision of 1 mm, a measurement of γ with accuracy of 2 parts in 10^7 would be possible.

The Gravitational Time Delay Mission (GTDM) (Ashby & Bender 2006) proposes to use laser ranging between two drag-free spacecraft (with spurious acceleration levels below 1.3×10^{-13} m/s$^2/\sqrt{\text{Hz}}$ at 0.4 μHz) to accurately measure the Shapiro time delay for laser beams passing near the Sun. One spacecraft will be kept at the L1 Lagrange point of the Earth-Sun system; the other one will be placed on a 3:2 Earth-resonant orbit. A high-stability frequency standard ($\delta f/f \lesssim 1 \times 10^{-13}$ Hz$^{-1/2}$ at 0.4 μHz) located on the L1 spacecraft will permit accurate measurement of the time delay. If requirements on the performance of the disturbance compensation system, the timing-transfer process, and the high-accuracy orbit determination are successfully addressed (Ashby & Bender 2006), then determination of the time delay of interplanetary signals to a 0.5 ps precision in terms of the instantaneous clock frequency could lead to an accuracy of 2 parts in 10^8 in measuring parameter γ.

The Laser Astrometric Test of Relativity (LATOR) (Turyshev *et al.* 2004) proposes to measure parameter γ with an accuracy of 1 part in 10^9, which is a factor of 30,000 beyond the best currently available, Cassini's 2003 result (Bertotti *et al.* 2003). The key element of LATOR is a geometric redundancy provided by the long-baseline optical interferometry and interplanetary laser ranging. By using a combination of independent time-series of gravitational deflection of light in immediate proximity to the Sun, along with

† A similar experiment is planned for the ESA's BepiColombo mission to Mercury (Iess & Asmar 2007) (see Fig. 2).

measurements of the Shapiro time delay on interplanetary scales (to a precision better than 0.01 picoradians and 3 mm, respectively), LATOR will significantly improve our knowledge of relativistic gravity and cosmology. LATOR's primary measurement, the precise observation of the non-Euclidean geometry of a light triangle that surrounds the Sun, pushes to unprecedented accuracy the search for cosmologically relevant scalar-tensor theories of gravity by looking for a remnant scalar field in today's solar system. LATOR could lead to very robust advances in the tests of fundamental physics. It could discover a violation or extension of general relativity or reveal the presence of an additional long range interaction.

Similar to LATOR, the Beyond Einstein Advanced Coherent Optical Network (BEACON) (Turyshev *et al.* 2009) is an experiment designed to reach a sensitivity of 1 part in 10^9 in measuring the PPN parameter γ. The superior sensitivity of BEACON is enabled by redundant optical-truss architecture which eliminates the need for expensive drag-free systems. The mission uses four identical commercially-available spacecraft that are placed on 80,000 km co-planar circular orbits around the Earth (Fig. 3). Each spacecraft is equipped with three sets of identical laser ranging transceivers to form a system of range measurements within the resulted flexible light-trapezoid formation. To enable its primary science objective, BEACON will precisely measure and monitor all six inter-spacecraft distances within the trapezoid using laser transceivers capable of achieving a nanometer resolution over distances of 160,000 km.

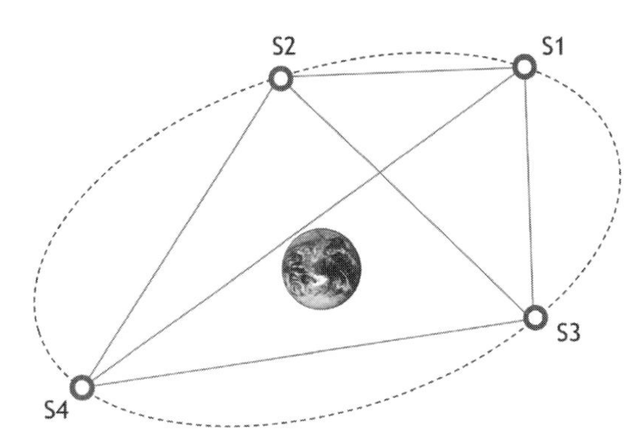

Figure 3. Schematic of the BEACON formation.

In a planar geometry this system is redundant; by measuring only five of the six distances one can compute the sixth one. The resulting geometric redundancy is the key element that enables BEACON's unique sensitivity in measuring a departure from Euclidean geometry. In the Earth's vicinity, this departure is primarily due to the curvature of the relativistic space-time around the Earth. It amounts to $\sim$10 cm for laser beams just grazing the surface of the Earth and then falls off inversely proportional to the impact parameter. The BEACON's laser measurements form a trapezoid with diagonal elements such that one of the legs in the trapezoid skims close to the Earth, picking up an additional gravitational delay; the magnitude of this signal is modulated by moving the position of one of the spacecraft relative to the others (thus changing the impact parameter of the trapezoid legs).

The BEACON architecture trades drag-free operation for redundancy in the optical truss. BEACON requires 4 Earth-orbiting satellites moving in the same orbital plane to participate in the metrology truss. This evolving light-trapezoid architecture is the fundamental requirement for BEACON; it enables geometric redundancy, thereby eliminating the need for drag-free spacecraft. Given range measurements of all 6 legs between the 4 fiducials, and assuming they are held in a planar configuration to within a few meters, it becomes possible to significantly improve the measurement of the PPN parameter γ. Simultaneous analysis of the resulting time-series of these distance measurements will allow BEACON to measure the curvature of the space-time around the Earth with an accuracy of better than 1 part in 10^9 (Turyshev *et al.* 2009).

3. Discussion and Outlook

Today physics stands at the threshold of major discoveries. Growing observational evidence points to the need for new physics, intensifying the efforts to resolve the challenges that fundamental physics faces today (Turyshev 2009a). We emphasize that modern-day optical technologies could lead to important progress in the tests of relativistic gravity.

The experiments discussed above necessitate important modeling work needed to enable the anticipated high accuracy results (Turyshev 2009b). The models would have to account for a number of relativistic-gravity effects on the light propagation in the solar system valid to the post-post-Newtonian order for a number of competing theories of gravity, including general relativity. Special attention must be paid to the static gravitational fields produced by the mass monopoles of the Sun, Earth and Jupiter. The model should also be able to account for quadrupole deflection of light for the observations conducted in close proximity to the Sun and some planets. Lastly, gravitomagnetic effects due translational motion of the Sun and planets are important and should be accounted for. The model would have to rely on a robust realization of the theory of relativistic reference frames, including relevant coordinate transformations and associated constants.

The work outlined here will require a coordinated community effort to develop a reliable and effective set of relativistic modeling tools and data analysis algorithms. The work has already begun in the context of the development of relativistic reference frames and their application for precision spacecraft navigation, astrometry and gravitational experiments (the recent IAU Symposium 261 is a good evidence of such efforts), but more efforts are needed; this short paper was intended to motivate such a work in the near future.

The work described here was carried out at the Jet Propulsion Laboratory, California Institute of Technology, under a contract with the National Aeronautics and Space Administration.

References

Ashby, N., & Bender, P. 2006, in Lasers, Clocks, and Drag-Free: Technologies for Future Exploration in Space and Tests of Gravity, ed. H. Dittus, C. Lämmerzahl, & S. G. Turyshev (Berlin: Springer Verlag), 219
Bertotti, B., Iess, L., & Tortora, P. 2003, Nature, 425, 374
Damour, T., & Nordtvedt, K. 1993, Phys. Rev. D, 48, 3436
Damour, T., Piazza, F., & Veneziano, G. 2002, Phys. Rev. D, 66, 046007
Deffayet, C., Dvali, G. R., & Gabadadze, G. 2002, Phys. Rev. D, 65, 044023
Dvali, G. 2006, New J. Phys., 8, 326
Dvali, G., Gabadadze, G., & Porrati, M. 2000, Phys. Lett. B, 485, 208
Dvali, G., Gruzinov, A., & Zaldarriaga, M. 2003, Phys. Rev. D, 68, 024012
Froeschle, M., Mignard, F., & Arenou, F. 1997, in ESA Special Publication, Vol. 402, ESA Special Publication, ed. R. M. Bonnet, E. Høg, & *et al.*, 49–52
Iess, L., & Asmar, S. 2007, Int. J. Mod. Phys. D, 16, 2117
Magueijo, J., & Bekenstein, J. 2007, Int. J. Mod. Phys. D, 16, 2035
Reasenberg, R. D., *et al.* 1979, Astrophys. J. Lett., 234, L219
Shapiro, S. S., Davis, J. L., Lebach, D. E., & Gregory, J. S. 2004, Phys. Rev. Lett., 92, 121101
Spergel, D. N., *et al.* 2007, Astrophys. J. Suppl., 170, 377
Turyshev, S. G. 2009a, Usp. Fiz. Nauk, 179, 3
—. 2009b, Astron. Letters, 35, 215
Turyshev, S. G., Lane, B., Shao, M., & Girerd, A. R. 2009, Int. J. Mod. Phys. D, 17, 1025
Turyshev, S. G., Shao, M., & Nordtvedt, K. L. 2004, Int. J. Mod. Phys. D, 13, 2035
Will, C. M. 2006, Living Rev. Relativity, 9
Williams, J. G., Turyshev, S. G., & Boggs, D. H. 2004, Phys. Rev. Lett., 93, 261101

Relativity in Fundamental Astronomy
Proceedings IAU Symposium No. 261, 2009
S. A. Klioner, P. K. Seidelman & M. H. Soffel, eds.

doi:10.1017/S174392130999041X

Open loop doppler tracking in Chinese forthcoming Mars mission

Kun Shang[1], **Jinsong Ping**[1], **Chunli Dai**[1], **and Nianchuan Jian**[1]

[1]Shanghai Astronomical Observatory,
80 Nandan Road, Shanghai 200030, China
email: shangkun@shao.ac.cn

Abstract. Using the radio telescopes in Chinese VLBI Network and the K5/VSSP32 VLBI system of NICT in Japan, we have developed algorithms that can extract open loop Doppler information from the Chang'E-1 and Mars Express radio tracking data. Our latest results indicate the Doppler accuracy of open loop three-way Doppler is about 1mm/s in 1 second integration time, relative to a 8.4GHz carrier. In the forthcoming joint Russian-Chinese Martian mission in 2009, the current software algorithms and hardware performance will be improved, and we will attempt to use the high precision Doppler shift and phase information to test gravitation theories.

Keywords. celestial mechanics, space vehicles: instruments, techniques: radial velocities, planets and satellites: general

1. Introduction

Lunar and the planetary exploration is to recovery the origin of the Moon and the planets, to explore their evolution, and to look for the possible life in other planetary bodies. Since the beginning of the new century, Mars exploration has attracted the huge attention from space communities. As a beginner in this area, China has launched her 1st lunar orbiter Chang'E-1 (CE-1) successfully, and has got some new scientific results from this exploration. Beyond this, a joint Russian-Chinese Martian mission, YingHuo-1 sub-satellite (YH-1) and Phobos-Grunt Spacecraft (FGSC), has been developed and promoted solidly. The two probes will be launched together in October 2009. After a successful launch, the joint spacecraft YH-1 and FGSC will be sent to a transfer orbit flying to Mars. After 10 - 11 months, the joint craft will arrive in the Martian system, and will be ejected into an equatorial orbit of 800 km periapsis altitude and 80,000 km apoapsis altitude, with a period of about 72 hours, inclination of 1°- 5°. The joint craft will fly in this orbit for about 3 circles, then they will be separated. FGSC will change its orbit in order to land on Phobos, and then take some soil (0.1 kg - 0.2 kg) back to the Earth. YH-1 will free-fly in this large elliptical orbit for 1 year.

YH-1 is a small sub-satellite focused on investigating the Martian space environment and the solar wind-Mars interaction. It will be combined with FGSC to form a two-point measurement configuration in the Martian space environment. The two spacecrafts will also carry out satellite-to-satellite radio link, so as to study the Martian ionosphere by using radio occultation links at UHF.

2. Open loop experiments and results

Considering that the Chinese deep space tracking system is still under construction, there will not be any uplink system in China which can meet the power requirement of uplink communication for a distance of AU. To simplify and minimize the design, an X-band receiver and X-band transmitter system have been adopted for onboard communication. There is not a common PLL transponder used for tracking. To solve the tracking and orbit determination problem, an USO-based one-way open loop concept will be used in YH-1, and the ground astronomical VLBI system (Chinese VLBI Network) will be used to receive the radio signal, so as to retrieve the Doppler information and then the differential Doppler information. The open loop Doppler observable will be applied in the positioning and orbit determination of YH-1.

Chinese VLBI Network (CVN), composed of four radio telescopes, acts as the deep space tracking system in China. Four radio telescopes located in four long-distance stations with different sizes, including: Shanghai station with a 25m telescope built in 1987, Urumqi station with a 25m telescope built in 1993, Beijing station with a 50m telescope built in 2006, and Kunming station with a 40m telescope built also in 2006. Recording devices have already been installed in two CVN stations - Shanghai and Urumqi. We use K5/VSSP32 sampling board as recording device, with a local H-maser atomic clock. K5/VSSP32 is a high-performance sampler dedicated to a geodetic VLBI system developed by the Kashima VLBI group of National Institute of Information and Communications Technology (NICT) in Japan.

During the nominal mission period of Chinese lunar orbiter CE-1, we used the three-way method to test the open loop tracking ability as a preview of future one-way method in YH-1 mission. Three-way method means that a tracking station sends uplink S-Band signal generated by Rb atomic clock to CE-1, and then the transponder in the orbiter locks the uplink carrier wave and sends it back to the Earth. The reason why a one-way Doppler experiment was not carried in CE-1 is that there is no stable oscillator. Similar experiments have been done together with ESOC deep space tracking stations for tracking the Mars Express. The radio telescopes used as the receiver station in three-way doppler experiment are located in Shanghai and Urumqi.

A software receiver using the post-processing method is developed to retrieve the Doppler frequency from the data recorded in K5/VSSP32 sampler. The kernel of this algorithm is phase counting (Ping *et al.* 2001), which can extract the precise main frequency from the raw data. Comparing the calculated received main frequency with the original transmitted frequency, we obtain the frequency shift in a certain integral time. The Doppler Effect presents the relationship between the frequency shift and the radial velocities.

$$f_R \approx \left(1 - \frac{v_1}{c}\right)\left(1 - \frac{v_2}{c}\right) M \cdot f_T \approx \left(1 - \frac{v_1 + v_2}{c}\right) M \cdot f_T \tag{2.1}$$

Where, f_R represents the calculated received frequency, f_T represents the original transmitted frequency, v_1 represents the range rate of the transmitting station and spacecraft, v_2 represents the range rate of spacecraft and receiving station, and M represents the transmit ratio. Define the average of v_1 and v_2 as the three-way velocity, which means $v_{3w} = (v_1 + v_2)/2$. Figure 1 and Figure 2 show the open loop three-way Doppler results of CE-1 mission and Mars Express mission, including the three-way velocity and the residuals after orbit determination.

3. Prospects

At present, the open loop Doppler accuracy of the Chinese tracking system is about 3mm/s in S-band and 1mm/s in X-band, both in 1 second integration time. Compared with the result of DSN in JPL (Thornton 2002) , the current accuracy of China is

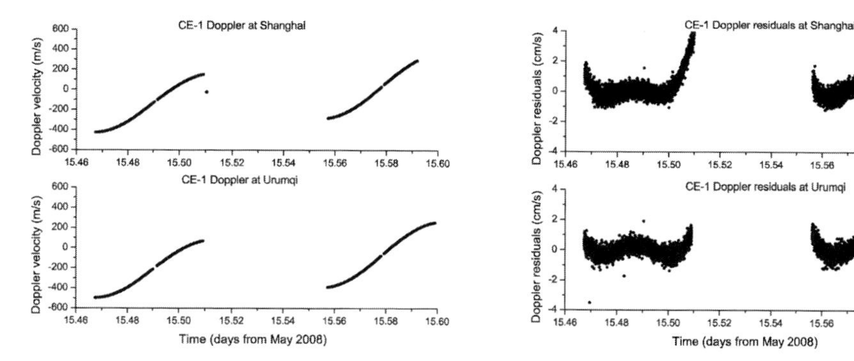

Figure 1. Open loop three-way Doppler result of CE-1 mission on 15 May 2008. The left panel shows the three-way Doppler velocity of Shanghai and Urumqi station. The right panel shows the Orbit Determination Doppler residual. The r.m.s of 1 second integration time is about 3mm/s, relative to a 2.3 GHz carrier.

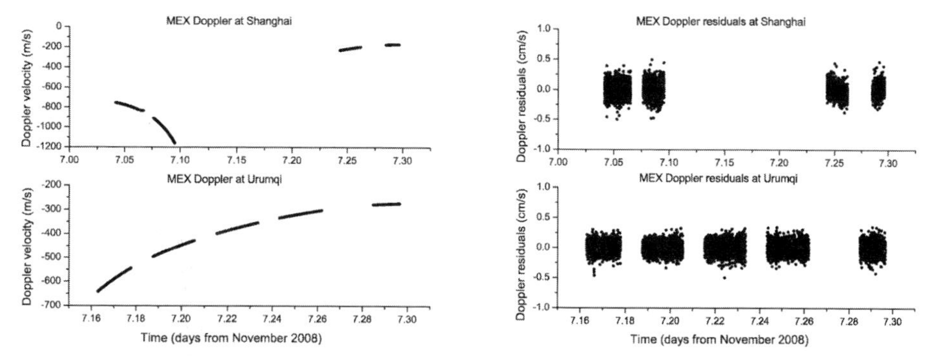

Figure 2. Open loop three-way Doppler result of MEX mission on 7 November 2008. The left panel shows the three-way Doppler velocity of Shanghai and Urumqi station. The right panel shows the Orbit Determination Doppler residual. The r.m.s of 1 second integration time is about 1mm/s, relative to an 8.4 GHz carrier.

better than that of JPL in 1980s but worst than that of JPL in 1990s. The limit of accuracy is restricted by the size of the radio telescopes in CVN. In 2014, a 65m radio telescope will be built by Shanghai Astronomical Observatory, which will improve the sensitivity and accuracy in the Chinese deep space missions, especially the open loop Doppler tracking accuracy. High accuracy Doppler tracking can be used for the test of gravitational theories, such as testing the PPN parameter γ (Will 1993). Considering the large elliptical orbit of the YH-1 satellite, other relativistic Doppler shift such as second order Doppler shift and gravitational shift can also be tested.

Acknowledgements

This work was supported by the VLBI system of Chinese Academic of Science and GEODYN software provided by NASA GFSC.

References

Ping, J. S., Frank, W., & Yusuke, K. 2001, *Journal of Planetary Geodesy*, 36, 15–22

Thornton, C. L. & Border, J. S. 2002, *Radiometric Tracking Techniques for Deep-Space Navigation* (Pasadena, California: JPL Publication), p. 9–46

Will, C. W. 1993, *Theory and Experiment in Gravitational Physics* (Cambridge: Cambridge University Press)

Relativity in Fundamental Astronomy
Proceedings IAU Symposium No. 261, 2009
S. A. Klioner, P. K. Seidelman & M. H. Soffel, eds.

doi:10.1017/S1743921309990421

The art of precision pulsar timing

Matthew Bailes[1]

[1]Centre for Astrophysics and Supercomputing,
Swinburne University of Technology, H39, PO Box 218,
Hawthorn Vic, Australia 3122
email: mbailes@swin.edu.au

Abstract. Pulsar timing has proven to be a wonderful tool with which to study neutron stars, providing insights into their ages, distances, proper motions, magnetic field strengths, internal structure, binary histories and evolution, and for tests of General Relativity. Here I describe how to optimise strategies for millisecond pulsar timing to enable the highest timing precision.

Keywords. Pulsars, Millisecond Pulsars, Precision Timing, General Relativity, Gravitational Waves.

---

1. Introduction

Almost all of the major advances in pulsar astrophysics have arisen as the result of the practice known as "pulsar timing". Pulsar timing is the practice of observing a known pulsar and "folding" the data at an assumed period to create an average pulse profile. The inherent assumption is that average pulsar profiles are stable (Lorimer & Kramer 2004), and that any given observed profile $O(x)$ is just a phase-shifted version of the intrinsic profile $P(x)$ multiplied by a constant plus noise $N(x)$, ie

$$O(x) = AP(x - \alpha) + N(x) \tag{1.1}$$

where A is a scale factor and α is the phase shift. One then cross correlates this observed profile with a template, and obtains an arrival time and associated error. By fitting a model to the arrival times of the form

$$\phi = \phi_0 + \nu(t - t_0) + (1/2)\dot{\nu}(t - t_0)^2 + (1/6)\ddot{\nu}(t - t_0)^3 \tag{1.2}$$

where ϕ is the phase of the pulsar, ν is the rotation frequency and the dots represent time derivatives, it is possible to parameterise the majority of pulsars' timing history by remarkably few parameters.

The vast majority of the almost 2000 known radio pulsars have been found in either large-scale surveys of our galaxy with the world's largest radio telescopes (Lyne *et al.* 1998, Manchester *et al.* 2001), or from targetted observations of globular clusters (Manchester *et al.* 1991, Ransom *et al.* 2005). Most of the known radio pulsars are young objects with characteristic (indicative) ages of a few million years and estimated magnetic field strengths of $\sim 10^{12}$G. These tend to be imperfect rotators, and deviate from their few-parameter spin-down models after a few years, exhibiting what we usually refer to as "timing noise".

A sub-class of pulsars however, the so-called "millisecond" pulsars, have remarkably small rotation periods ($\sim$1–$\sim$20) milliseconds, and fields that are much weaker ($\sim 10^8$G). For these pulsars special instrumentation is required to avoid the deleterious effects of pulse dispersion to obtain the highest precision, and their remarkably small torques and large angular momenta make them clocks of outstanding precision.

The "art" of pulsar timing concerns adopting techniques that maximise the yield from a given set of observations. Telescope time is a greatly sought-after resource, and almost all pulsars have their timing precision limited by it, due to their small radio fluxes that prohibit high signal-to-noise ratio (SNR) observations.

2. Pulsar Timing soon after Discovery

When a pulsar is first discovered, it is important to quickly establish "phase connection", a term which means that you can unambiguously establish the exact number of rotations between the observations. But very little is usually known about a pulsar when it is first discovered. The original detection provides an approximate period and dispersion measure, and a few arc minute position at best. It is only after a prolonged timing campaign, often extending for 1–2 years that these parameters become more refined, but with perserverence phase-coherent pulsar timing can yield parameters to astounding precision. Without phase connection, pulsar timing fails to capitalize on its greatest strength, and that is that for many of the most interesting parameters, the precision grows much faster than time $t^{-1/2}$. In most astrophysical experiments, continued observing reduces one's relative errors by the total integration time $t^{-1/2}$. In pulsar timing, provided you have maintained phase connection, the error in the pulsar period decreases as $T^{-3/2}$, where T is the entire time span between your first and last observations. This makes period derivatives $\dot{P}$ easy to establish, and these in turn provide both an *estimate* of the characteristic age (from $\tau = P/(2\dot{P})$) and magnetic field strength ($B^2 \propto P\dot{P}$).

In order to establish and keep phase connection, it is normal to observe the same pulsar ~3 times on a given day, then once a day for 2-3 days, then again, say weekly for a month. For the first year, it is important to maintain frequent ~ monthly observations, in order to decouple the covariant parameters of position (RA, Dec), P and $\dot{P}$.

Binary pulsars can prove more difficult to time, as they add another five Keplerian parameters (and sometimes more relativistic parameters) to the mix, and until a complete orbit is observed with phase connection the covariances make pulsar phase prediction often impossible to achieve. Obtaining phase connection early greatly shortens the time until you can report significant findings, hence, the "first law" of precision pulsar timing.

The first law of pulsar timing. *Establish phase connection as early as possible as for many of the most interesting parameters, your experiment's precision is often a strong function of the time you have maintained phase connection. A corollary is that profiles smeared by a poor folding model are often of diminished use in the highest precision timing and can lead to unfortunate systematic errors. It is wise to adopt your final timing system instrumentation as soon as practicable.*

3. Precision Timing Theory

The real power of pulsar timing relies on extending the time duration of your experiment to many years. Timing studies of relativistic binaries have provided extraordinarily convincing evidence that General Relativity is the correct theory of gravity, both for neutron star-neutron star binaries (Taylor and Weisberg 1982, Stairs *et al.* 2002, Kramer *et al.* 2006), as well as neutron stars with white dwarf companions (Bhat, Bailes & Verbiest 2008, van Straten *et al.* 2001).

Pulsars are intrinsically low-luminosity objects, and pulsar timing requires the world's largest telescopes in many cases just to detect them. From an astrometric point of view, we can think of pulsar timing as an interferometric experiment. The pulsar, due to its massive moment of inertia and small braking torque is a very precise clock. The pulses therefore represent a source of coherent radiation and the Earth-Sun system are immersed in radiation from this coherent source. The wavelength of the radiation can be thought of as having a wavelength $\lambda = cP$, where c is the speed of light. The baseline of our experiment is 2 astronomical units (AU). The position of the pulsar can therefore be determined to an accuracy of $\sim\lambda/(2AU)$ times the relative accuracy with which we can determine the arrival time of the pulse at any given epoch. For a gaussian profile this is approximately $w/(2 \times P \times SNR)$ where w is the half-width of the Gaussian and SNR is the signal-to-noise ratio of the pulsar.

The SNR is provided by a modified version of the radiometer equation:

$$SNR = \frac{SG\sqrt{BN_pt}}{T_{\rm rec} + T_{\rm sky}}\sqrt{\frac{P-w}{w}} \tag{3.1}$$

where S is the pulsar flux in Janskys, G the receiver gain in the units of K/Jy, B the receiver bandwith (Hz), N_p is the number of polarisations (usually 2), t the integration time in seconds, and $T_{\rm rec}$ and $T_{\rm sky}$ are the effective temperatures of the receiver and sky respectively.

This means the error (σ) in the arrival times or TOAs is approximately:

$$\sigma = \frac{w}{2}\sqrt{\frac{w}{P-w}\frac{T_{\rm rec} + T_{\rm sky}}{SG\sqrt{BN_pt}}} \tag{3.2}$$

and for the brightest millisecond pulsars can be below 100 nanoseconds. These types of errors permit not only positions to be determined to a precision of just tens of microarc-seconds, but also allow very accurate proper motions and even parallaxes, orbital period derivatives, and relativistic effects such as Shapiro delay to be observed.

For narrow pulsar profiles, $\sigma \propto w^{3/2}$, highlighting the importance of finding pulsars with narrow features or using instrumentation that minimizes any smearing.

The theory of pulsar timing relies on the pulse profile being an invariant. We know it isn't. Observations of slow pulsars with large fluxes demonstrate that pulsar profiles require many rotations to stabilize, but due to their small fluxes, this is difficult to establish for the millisecond pulsars.

Astronomers fit a template, often constructed of either the sum of the best observations or the addition of several Gaussian components that closely approximate the pulse shape to each observation. As we saw in the introduction the "theory" is that each observation is just a scaled version of the template plus *random* noise with a phase shift. Fitting to equation 1.1 for A and α (often in the Fourier domain) yield an arrival time (TOA) and error. These TOAs and errors are often placed into a least squares fitting program like tempo2 (Hobbs, Edwards and Manchester 2006) to yield pulsar parameter estimations from which physical interpretations are made.

Unfortunately, once this fit is performed, the reduced chi-squared is often far from unity, limiting the interpretation of the physical parameter errors, and ultimately the pulsar timing methodology. A conservative approach to this problem is to keep adding a systematic error term in quadrature to each TOA until the reduced chi-squared is unity. A more dangerous (but more often used approach) is to linearly increase the size of each error until the reduced chi-squared is unity. This is an optimistic assumption, but reduces the size of the errors on the physical parameters.

Investing in instrumentation that minimizes systematic errors can be greatly worthwhile. Pulsar fluxes often vary greatly (by factors of $\gg$10) and when such amplifications occur, capitalising on them is essential. The radiometer equation tells us that an observation at scintillation maximum with 10x the flux of the average theoretically has 100x the influence in the fit. Instrumentation that eliminates the need for artificially adjusting the error estimates can therefore be very powerful.

There are a number of factors which lead to poor residual chi-squares, and most are related to distortions of the profile. These fall into two categories, natural, and man-made.

Natural sources of error: For the younger pulsars the assumption that the pulsar is a consistent rotator is just flawed. Unfortunately, on moderate $\sim$ yearly timescales, the imperfect rotation can be masked by the fit for other parameters, such as position or proper motion. Fitting for these parameters often gives nonsensical values and should be avoided. Other sources of error are pulse shape variability, variable scattering and pulsar dispersion due to the changing interstellar medium. In principle variable delays due to changing dispersion measures can be removed by multi-frequency observing, thus observing with two or three different frequencies removes most of this source of error. Unfortunately, multi-path scattering means there is no "one true dispersion measure" for observations conducted at different frequencies, a point first made by Don Backer. Nevertheless when trying to get sub-microsecond timing one should observe the second law of pulsar timing:

Second law of pulsar timing: *Eliminate the effects of variable pulse dispersion measures by conducting observations at widely-spaced radio frequencies as closely as possible in time. This gives a history of the pulsar's dispersion measure, which can be removed by the fitting program.*

Man-made sources of error: As pulsar radiation traverses the ionized interstellar medium it gets delayed by a frequency-dependent amount. Integration of the pulsar's radiation over finite bandwidths smears the profile. Pulsar astronomers seek to eliminate this by either building multi-channel filterbanks with very narrow filters, or by sampling the voltages and performing the feat known as "coherent dedispersion" (Hankins and Rickett 1975), that removes the dispersive effects "perfectly". Until this century computers were too slow to perform coherent dedispersion on all but narrow bandwidths and filterbanks were too costly to mass produce. This greatly degraded the quality of pulsar timing data.

This has all changed rapidly with the advent of the field programmable gate array and the march of Moore's law producing cheap computer power. It is now possible to build 1024- or 2048-channel digital filterbanks at reasonable cost such as the Australia Telescope's DFB series. It is also possible to attempt real-time coherent dedispersion on large bandwidths using moderately priced clusters in tandem with an A/D convertor, an FPGA and 10 Gb ethernet. These systems are leading to greatly increased timing precision and a reduction in systematic errors (see Verbiest *et al.* 2009 for his analysis of 20 millisecond pulsars with the coherent dedispersion system CPSR2). CPSR2 is a 2x2x64 MHz coherent dedispersion system that achieves theoretical errors that provide reduced chi squared fits near unity for many pulsars. This enables the use of high SNR observations in the fit, and thus more reliable physical parameters. Early results from the ATNF's DFB series of instruments that exploit more bandwidth than CPSR2 are also yielding smaller reduced chi squares than pre-2000 instruments.

Third law of pulsar timing: *Eliminate as many distortions to the profile as you can by using digital (not analogue) electronics, and coherent dedispersion where practicable, or at worst multi-channel digital filterbanks at higher frequencies.*

It is very difficult to combine data from different telescopes and instrumental "backends", as each introduce their own unique distortions to the profile, leading to an arbitrary phase jump. This leads to the fourth law of pulsar timing that concerns the fact that we are conducting difference experiments.

Fourth law of pulsar timing: *Never change any equipment unless it leads to a very substantial improvement in timing, and where possible, overlap your timing to eliminate the need for arbitrary phase jumps. Your data reduction software is just as important to freeze as the hardware in this respect. Even subtle changes in binning algorithms can lead to undiagnosed DC offsets in timing which limit the ultimate timing precision.*

Unfortunately, even the best timing equipment in the world cannot undo smearing introduced by a bad pulsar ephemeris. Until one obtains phase connection, it is easy to artificially smear the pulse profile by observing with a bad ephemeris. Similarly, unnoticed changes to the observatory (a new cable can add many 10s of nanoseconds to an arrival time), clock errors etc, all conspire to reduce your knowledge of your true arrival time errors. The best timing experiments check the quality of their (reduced) data regularly, include at least one well-behaved pulsar for comparison purposes and inspect the data by eye as well as by automated reduction processes. Finally, be patient. Pulsar timing takes time to yield results.

4. The Future

We are still learning about how to get the most from our instruments. The effect of radio frequency interference (RFI), both periodic and impulsive are increasingly frustrating sources of error for pulsar astrophysicists. New instrumentation has higher dynamic range A/Ds than earlier models to reduce distortions of the profile. This however makes them subject to large distortions of the profile in literally microseconds during bursts of RFI. In the future, monitoring both the total power and the accuracy with which each subintegration obeys the fundamental assumptions about pulsar timing (equation 1.1) will help reduce these effects. Others are pioneering the use of reference antennae to subtract sources of known RFI (Kesteven *et al.* 2005, Manchester 2008). At the high fidelity end, van Straten (in preparation) has recently demonstrated that polarization can be used to increase timing precision by using a pulsar to calibrate the receiver, and hence remove the distortion of profiles induced by imperfect receivers. van Straten used the bright millisecond pulsar PSR J0437–4715 to calibrate the Parkes 20cm receivers over many years, determined appropriate coefficients of the Jones' matrices, and then improved the timing of the millisecond pulsar PSR J1022+1001 to below 1 microsecond RMS residuals.

Other second-order contributions to our systematic errors become increasingly important as our instrumentation and sensitivities improve. To maximize our signal-to-noise ratio, we integrate over increasingly large bandwidths. Since the pulse width is a function of frequency, time-dependent scintillation varies the width of the final profile leading to a systematic change in width after "scrunching" in frequency, which can lead to systematic errors in the shift and error estimate. Doppler shifts and the time delay between the top and bottom of the band make "labelling" an integration with a specific time and frequency something to be used with caution. Dispersion measures can appear to change if Doppler corrections are not properly taken into account.

We must not forget about the importance of an accurate solar system barycenter, nor the provision of accurate clocks (see various authors in these proceedings). All of these incremental improvements are leading astronomers to timing below 100 nanoseconds in precision. With the advent of the Square Kilometre Array (see Kramer *et al.* these proceedings), and lessons learned over the past 40 years, pulsar timing is set to search for even more subtle effects, such as the effect of a gravitational wave background on millisecond pulsar timing (see Hobbs these proceedings) and experiments on relativistic gravity (see Stairs these proceedings).

References

Bhat, N. D. R., Bailes, M., & Verbiest 2008, *Phys. Rev. D*, 77, 12, 124017

Hankins, T. H. & Rickett, B. J. 1975, *Methods in Computational Physics*,Vol 14, Radio Astronomy, 55.

Hobbs, Edwards & Manchester 2006, *MNRAS*, 369, 655

Kramer *et al.* 2006, *Sci.*, 314, 97

Kesteven *et al.* 2006, *Radio Science 40*, RS5S06, 1

Lorimer, D. R. & Kramer, M. 2004, *The Handbook of Pulsar Astronomy*, Cambridge University Press.

Lyne *et al.* 1998, *MNRAS*, 295, 743

Manchester *et al.* 1991, *Nature*, 352, 219

Manchester *et al.* 2001, *MNRAS*, 328, 17

Manchester 2006, *http://www.nanograv.org/presentations/IPTA/IPTAAug2008Presentations.aspx*

Ransom, S. *et al.* 2005, *Science*, 307, 892

Stairs *et al.* 2002, *ApJ*, 581, 501

Taylor, J. H. & Weisberg, J. M. 1982, *ApJ*, 253, 908

van Straten *et al.* 2001, *Nature*, 412, 158

Verbiest *et al.* 2009, *MNRAS*, in press

Relativity in Fundamental Astronomy
Proceedings IAU Symposium No. 261, 2009
S. A. Klioner, P. K. Seidelman & M. H. Soffel, eds.

doi:10.1017/S1743921309990433

Binary pulsars and tests of general relativity

I. H. Stairs

Dept. of Physics and Astronomy, University of British Columbia,
6224 Agricultural Road, Vancouver, B. C., V6T 1Z1, Canada, and

Centre for Astrophysics & Supercomputing, Swinburne University of Technology,
Mail 39, P. O. Box 218, Hawthorn, Vic 3122, Australia
email: stairs@astro.ubc.ca

Abstract. Binary pulsars are a valuable laboratory for gravitational experiments. Double-neutron-star systems such as the double pulsar provide the most stringent tests of strong-field gravity available to date, while pulsars with white-dwarf companions constrain departures from general relativity based on the difference in gravitational binding energies in the two stars. Future observations may open up entirely new tests of the predictions of general relativity.

Keywords. pulsars: general, pulsars:individual (J0737−3039A/B), gravitation

---

1. Introduction: Observing and Using Pulsars

The 1974 discovery of a pulsar in a binary system (Hulse & Taylor 1975) provided the experimental gravity community with a precision clock embedded in the distorted space near a compact object. The tremendous potential of such a system for testing gravitational theories was quickly recognized (Wagoner 1975; Eardley 1975; Damour & Ruffini 1974; Barker & O'Connell 1975). Since then, pulsars have been discovered in similar and in even more relativistic orbits, and many of these binary systems can be used to test aspects of general relativity (GR) and other theories of gravity. This article gives an overview of these tests; see also Stairs (2003) for a fuller though somewhat dated description of much of this material, and Stairs (2005, 2006) for slightly updated versions.

Pulsar observing and timing are discussed thoroughly in other sources (e.g., Lorimer & Kramer; Bailes 2009), but a few relevant points are worth mentioning as background to gravitational tests. Pulsar timing relies on the observed reproducibility of pulse profiles when averaged over timescales of minutes to hours. This allows the determination of pulse Times of Arrival (TOAs) by cross-correlation with a "standard profile" for a given pulsar. TOA precision improves for pulsars with high signal-to-noise ratio and sharp pulse-shape components. Pulsar timing is accomplished by enumerating the pulsar rotations between TOAs and fitting ephemerides that include spin, astrometric and dispersion parameters as well as full descriptions of any binary orbits. Many gravitational tests depend on thorough understanding of pulsar orbits.

By considering a pulsar's spin and spin derivative, estimates can be made of its surface dipolar magnetic field strength as well as its spin-down or "characteristic" age. Most of the pulsars used in relativity tests have been recycled by transfer of mass and angular momentum as their binary companions evolve (e.g., Tauris & van den Heuvel 2006), and have millisecond spin periods and surface magnetic fields of 10^8–10^9 G. A millisecond pulsar (MSP) in a nearly circular orbit typically has a white-dwarf companion, and there is an expectation that both stellar spin vectors will have been aligned with the orbital angular momentum during the mass transfer phase (e.g. Phinney 1992). A pulsar in an eccentric orbit is more likely to have a companion that is also a neutron star, such that the last event in the system was a supernova explosion that could have misaligned one or both spins with the post-explosion orbital angular momentum (e.g. Wex *et al.* 2000).

Pulsars in both types of orbits (and, indeed, some isolated pulsars as well) find uses in gravitational tests. It should be noted that, although the pulsar orbits known to date do not test strong-field gravity in the sense of approaching the event horizon, the use of pulsars and white dwarfs (and someday perhaps black holes) in these tests does mean that the tests must consider the theoretical predictions for objects that are strongly self-gravitating. For example, many of the tests rely on a extension of the definitions of the Parametrized Post-Newtonian (PPN) parameters (Will & Nordtvedt 1972; Will 2006) to generalized tensor-multiscalar gravitational theories (Damour & Esposito-Farèse 1992a; Damour & Esposito-Farèse 1996). In this description, for example, the PPN parameter α_1 becomes $\hat{\alpha_1} = \alpha_1 + \alpha_1'(c_1 + c_2) + \cdots$, where c_i describes the "compactness" of mass m_i. The compactness is $c_i = -2\partial \ln m_i / \partial \ln G \simeq -2(E^{\rm grav}/(mc^2))_i$, where G is Newton's constant and $E_i^{\rm grav}$ is the gravitational self-energy of mass m_i. Compactnesses are about -0.2 for a neutron star (NS) and -10^{-4} for a white dwarf (WD).

2. Equivalence Principle Violations

Equivalence principles are thoroughly described in, e.g.,Will (1993). General relativity incorporates even the Strong Equivalence Principle, predicting gravitational results independent of self-gravity, but this and other equivalence principles may be violated in alternate theories of gravity.

Pulsar timing sets limits on $\hat{\alpha_1}$, which would imply preferred-frame effects by violation of Lorentz invariance, and $\hat{\alpha}_3$, which would additionally imply non-conservation of momentum if non-zero. Pulsar can also constrain other SEP-violation effects produced by various combinations of the (modified) PPN parameters: the Nordtvedt effect, dipolar gravitational radiation, and changes in Newton's constant.

2.1. *Orbital Polarization Tests*

Nordtvedt (1968) proposed direct tests of the SEP through Lunar Laser Ranging (LLR) experiments. The principle behind these tests is that the different contributions of self-gravitation to the masses of the Earth and Moon would cause them to accelerate differently in the gravitational field of Sun, resulting in a "polarization" of the orbit in the direction of the Sun. LLR tests have set a limit of $\eta = (4.4 \pm 4.5) \times 10^{-4}$ (Williams *et al.* 2009), where η is a combination of PPN parameters.

Binary pulsar tests of the SEP and of $\hat{\alpha_1}$ and $\hat{\alpha_3}$ look for the same type of phenomenon: polarization of orbits of pulsar–white-dwarf systems in preferred directions given by the projection of "extra" acceleration vectors onto the planes of the orbits. The prototype of these tests is the Nordtvedt-equivalent SEP test, which considers a strong-field version of η labeled $\Delta_{\rm i}$, with pulsar–white dwarf systems constraining the difference $\Delta_{\rm net} = \Delta_{\rm pulsar} - \Delta_{\rm companion}$ (Damour & Schäfer 1991). The extra acceleration in this case would be in the direction of the Galactic acceleration of the system (Damour & Schäfer 1991), accessible through potential models of the Galaxy (e.g., Kuijken & Gilmore 1989). For the $\hat{\alpha_1}$ test, the direction is given by the velocity of the pulsar system relative to the Cosmic Microwave Background (Damour & Esposito-Farèse 1992b; Bell *et al.* 1996). For $\hat{\alpha_3}$ the direction comes from the cross-product of this absolute velocity and the pulsar spin direction (Nordtvedt & Will 1972; Bell & Damour 1996) and the test relies on the evolutionary assumption of alignment between the pulsar spin and the orbital angular momentum. All of the tests attempt to distinguish the strength of a forced eccentricity component against that of the "natural" eccentricity, which should fall within a certain range based on the orbital period (Phinney 1992) and whose direction evolves according to the general-relativistic advance of periastron. The time-dependent eccentricity vector may be written as (Damour & Schäfer 1991): $\mathbf{e}(t) = \mathbf{e}_F + \mathbf{e}_R(t)$, where $\mathbf{e}_R(t)$ is the rotating "natural" eccentricity vector, and $\mathbf{e}_F$ is the forced component.

Applying the tests requires making two important decisions. The first is the selection of pulsars. For the $\Delta_{\rm net}$ test, the figure of merit is $P_{\rm b}^2/e$ (Damour & Schäfer 1991), where $P_{\rm b}$ is the orbital period, while for $\hat{\alpha_1}$ it is $P_{\rm b}^{1/3}/e$ (Damour & Esposito-Farèse 1992b) and for $\hat{\alpha_3}$ it is $P_{\rm b}^2/(eP)$ (Bell & Damour 1996), where P is the pulsar's spin period. This makes MSPs in long-period orbits the logical choice for $\Delta_{\rm net}$ and $|\hat{\alpha}_3|$, and those in shorter-period, low-eccentricity orbits the choice for $\hat{\alpha_1}$. However, the pulsars must also satisfy other restrictions: they must be old enough that $\mathbf{e}_R(t)$ can be assumed to be randomly oriented, and have $\dot{\omega}$ (advance of periastron) larger than the rate of Galactic rotation (Damour & Schäfer 1991; Damour & Esposito-Farèse 1992b; Bell *et al.* 1996; Wex 1997). A related issue is ensuring that a set of pulsars is used representing the full related population, even those systems that might have a low figure of merit (Wex 2000). The set of pulsars to be used is reasonably clear for $\Delta_{\rm net}$ and $\hat{\alpha_3}$, for which the optimal pulsars have likely all followed the same evolutionary path (e.g. Rappaport *et al.* 1995), but the situation is less clear for the shorter-orbital period pulsars used for $\hat{\alpha_1}$.

The other decision is the method for obtaining the limit. Some sampling or averaging is typically needed for the orbital inclination angle i, the masses m_1 and m_2, the sky orientation of the binary and the pulsar distance, though these are reasonably well-constrained in some cases (Verbiest *et al.* 2008; Splaver *et al.* 2005). One place where progress has been made is the issue of how to treat the unknown angle between $\mathbf{e}_{\rm F}$ and the "natural" eccentricity $\mathbf{e}_{\rm R}$. Limits derived from "worst-case" cancellation scenarios have been obtained for the various tests using individual pulsars (Damour & Schäfer 1991; Damour & Esposito-Farèse 1992b; Bell *et al.* 1996) and an ensemble of pulsars (Bell & Damour 1996; Wex 1997, 2000); Wex (2000) attempted to account for population selection effects by using a larger set of pulsars as discussed above. An alternative analysis uses a Bayesian formalism and more of the known pulsar parameters such as the longitude of periastron, again operating on the full set of pulsars that likely have similar evolutionary histories, to obtain a 95% confidence upper limit on $|\Delta|$ of 5.6×10^{-3} and on $|\hat{\alpha_3}|$ of 4×10^{-20}. (Stairs *et al.* 2005). These limits will soon be updated with the inclusion of new pulsars, improved timing parameters for known pulsars and some small error corrections (Gonzalez *et al.*, in prep.). For $|\hat{\alpha_1}|$, the most recent limit is 1.4×10^{-4} at 95% confidence (Wex 2000). This test could be updated with the Bayesian analysis, but the selection of pulsars is less straightforward. It should be noted that isolated pulsars can also be used to set a limit on $\hat{\alpha_3}$ based on the average value of the observed period derivatives (Will 1993; Bell 1996; Bell & Damour 1996).

2.2. *Orbital Decay Tests*

The difference in self-gravitation contributions to the masses of pulsars and white dwarfs makes such binary systems targets of scrutiny for anomalous orbital decay other than that due to the quadrupolar gravitational radiation emission predicted by GR. Theories that violate the SEP may predict *dipolar* gravitational radiation, for example. The decrease of the period of a circular orbit due to dipolar emission can be written as (Will 1993; Damour & Esposito-Farese 1996):

$$\dot{P}_{\rm b\,Dipole} = -\frac{4\pi^2 G_*}{c^3 P_{\rm b}} \frac{m_1 m_2}{m_1 + m_2} (\alpha_{c_1} - \alpha_{c_2})^2, \tag{2.1}$$

where $G_* = G$ in GR, and α_{c_i} is the coupling strength of object "i" to a scalar gravitational field (Damour & Esposito-Farese 1996). In a complementary fashion, these observations may also be used to set limits on variations of Newton's constant, which would affect the binding energies of the stars as well as the angular momentum of the system. The expected orbital period derivative may be written as (Damour *et al.* 1988; Nordtvedt 1990):

$$\left(\frac{\dot{P}_{\rm b}}{P_{\rm b}}\right)_{\dot{G}} = -\left[2 - \left(\frac{m_1 c_1 + m_2 c_2}{m_1 + m_2}\right) - \frac{3}{2}\left(\frac{m_1 c_2 + m_2 c_1}{m_1 + m_2}\right)\right]\frac{\dot{G}}{G}. \tag{2.2}$$

The challenge is to quantify the various possible contributions to a measured or constrained orbital period derivative (e.g., Lazaridis *et al.* 2009). The quadrupolar prediction is expected but may be too small to be measurable. There may be terms corresponding to mass loss or other binary evolution effects in the system. The dipolar and $\dot{G}$ contributions can be estimated with the compactness or coupling strengths remaining as parameters. Another expected contribution to any observed value is the "Shklovskii effect" (Shklovskii 1970) due to the pulsar's transverse velocity: $\dot{P}_{\rm pm} = P\mu^2 d/c$, where μ is the proper motion and d is the distance to the pulsar. This term can be difficult to estimate well in the absence of a good distance measurement. A similar term comes from the differential acceleration of the solar system and the pulsar in the gravitational field of the Galaxy (Damour & Taylor 1991; Nice & Taylor 1995); its determination also requires good knowledge of the pulsar's spatial position.

Not surprisingly, then, the best tests come from pulsar–white-dwarf binaries with short-period orbits and reliable distance estimates, often through parallax. The best dipolar radiation limit in a recycled-pulsar system comes from PSR J1012+5307, which sets a 95% confidence limit of $(\alpha_{c_1} - \alpha_{c_2})^2 = (0.5 \pm 6.0) \times 10^{-5}$ (Lazaridis *et al.* 2009). For $\dot{G}/G$, there are several similar limits: PSR J1713+0747 yields a 95% confidence limit of $\dot{G}/G = (1.5 \pm 3.8) \times 10^{-12}\,{\rm yr}^{-1}$ (Splaver *et al.* 2005; Nice *et al.* 2005, D. Nice, private communication), while Deller *et al.* (2008) report a 95% upper limit of $(0.5 \pm 2.6) \times 10^{-12}\,{\rm yr}^{-1}$ using the Damour *et al.* (1988) expression for $\dot{G}/G$ and the pulsar J0437−4715 with slightly different error estimation. Recently, that result has been combined with timing of PSR J1012+5307 to yield a combined limit on $\dot{G}/G$ and dipolar radiation that does not need external input such as LLR limits (Lazaridis *et al.* 2009).

An especially interesting case is the young-pulsar–white-dwarf system PSR J1141−6545 (Kaspi *et al.* 2000), which is eccentric and should therefore be a strong emitter of gravitational radiation and and an excellent constrainer of dipolar gravational radiation (Gérard & Wiaux 2002; Esposito-Farese 2005). The orbital period derivative is well-measured (Bhat *et al.* 2008) and agrees with the GR predictions to within 6%, setting a limit on $(\alpha_{c_1} - \alpha_{c_2})^2 \simeq \alpha_0^2 < 3.4 \times 10^{-6}$ in the case of strongly non-linear coupling of the pulsar to the scalar field and $\alpha_0^2 < 2.1 \times 10^{-5}$ in the case of weakly non-linear coupling (Damour & Esposito-Farèse 1992a; Esposito-Farese 2005; Bhat *et al.* 2008).

While double-neutron-star (DNS) systems can also limit dipolar gravitational radiation (e.g., Will 1977; Will & Zaglauer 1989), this is of less interest because of their symmetry. However, their observed decrease in orbital periods can rule out those theories that predict an increase (e.g., Rosen 1973; Ni 1973; Lightman & Lee 1973; Weisberg & Taylor 1981). A different pulsar-derived limit to G variation uses pulsar spin-down as a limit on changes due to the gravitational binding of the neutron star (Counselman & Shapiro 1968; Goldman 1990), with MSPs being especially suitable for this task and yielding limits on $\dot{G}/G$ of a few $10^{-11}\,{\rm yr}^{-1}$ (e.g. Arzoumanian 1995). In the past, a limit also could be derived based on similarities in pulsar masses in old globular clusters and younger galactic field systems (Thorsett 1996), but the recent profusion of pulsars in globular clusters, some with possibly quite large mass estimates (e.g. Ransom *et al.* 2005; Freire *et al.* 2008), throws the validity of this test into question.

3. Highly Relativistic Systems

3.1. *Self-consistency Tests*

Measurement of orbital period derivatives is also a component of the best-known realization of binary pulsar tests of GR: verification of self-consistency between relativistic observables in DNS and similar systems. This is achieved by measuring the relativistic corrections to the Keplerian orbit (i.e., the "post-Keplerian" or "PK" parameters, which besides $\dot{P}_b$ and $\dot{\omega}$ include the time-dilation/gravitational-redshift term γ and the r and s Shapiro delay parameters) in a manner that does not assume that any particular theory of gravity is correct (Damour & Deruelle 1985). For any theory of gravity (e.g., GR, or a tensor-scalar theory with its parameters fixed), these PK parameters will be functions only of the two stellar masses, which are the two unknowns in any single-lined spectroscopic Keplerian orbit. Thus measurement of two PK parameters will yield the masses, and measurement of three or more will result in an over-constrained system that may be checked for internal consistency. In GR, the equations describing the PK parameters in terms of the stellar masses are (Damour & Deruelle 1986; Taylor & Weisberg 1989; Damour & Taylor 1992):

$$\dot{\omega} = 3\left(\frac{P_b}{2\pi}\right)^{-5/3} (T_\odot M)^{2/3}\,(1-e^2)^{-1}\,, \tag{3.1}$$

$$\gamma = e\left(\frac{P_b}{2\pi}\right)^{1/3} T_\odot^{2/3}\,M^{-4/3}\,m_2\,(m_1+2m_2)\,, \tag{3.2}$$

$$\dot{P}_b = -\frac{192\pi}{5}\left(\frac{P_b}{2\pi}\right)^{-5/3}\left(1+\frac{73}{24}e^2+\frac{37}{96}e^4\right)(1-e^2)^{-7/2}\,T_\odot^{5/3}\,m_1\,m_2\,M^{-1/3}\,, \tag{3.3}$$

$$r = T_\odot\,m_2\,, \tag{3.4}$$

$$s = x\left(\frac{P_b}{2\pi}\right)^{-2/3} T_\odot^{-1/3}\,M^{2/3}\,m_2^{-1}\,. \tag{3.5}$$

where x is the projected semi-major axis of the orbit of m_1 in light-seconds, $s \equiv \sin i$, $M = m_1 + m_2$ (all masses in units of solar mass), and $T_\odot \equiv GM_\odot/c^3 = 4.925490947\,\mu$s. Other theories of gravity will have somewhat different mass dependencies for these parameters.

The self-consistency test is often illustrated by a "mass-mass diagram" such as the one for the double pulsar J0737−3039A/B in GR in Figure 1. When the curves intersect in a common region, as they do in this case, the parameters are said to agree with the theory being tested. This type of test was pioneered for the first DNS PSR B1913+16 (Hulse & Taylor 1975; Taylor & Weisberg 1982, 1989) and has now also been accomplished for PSRs B1534+12 (Stairs *et al.* 2002), the double pulsar (Kramer *et al.* 2006), J1756−2251 (to low precision; Ferdman 2008) and the white-dwarf binary J1141−6545 (Bhat *et al.* 2008). Of these, B1534+12, J0737−3039A/B and (again to low precision) J1756−2251 allow tests that do not include the orbital period decay and hence provide an important "quasi-static" complement to the Hulse-Taylor case (Taylor *et al.* 1992). Combination of results from all the pulsars has potential to set significant constraints on deviations from GR (e.g., Taylor *et al.* 1992; Esposito-Farese 2005). Such systems can also be used to limit preferred-frame effects in a novel way (Wex & Kramer 2007).

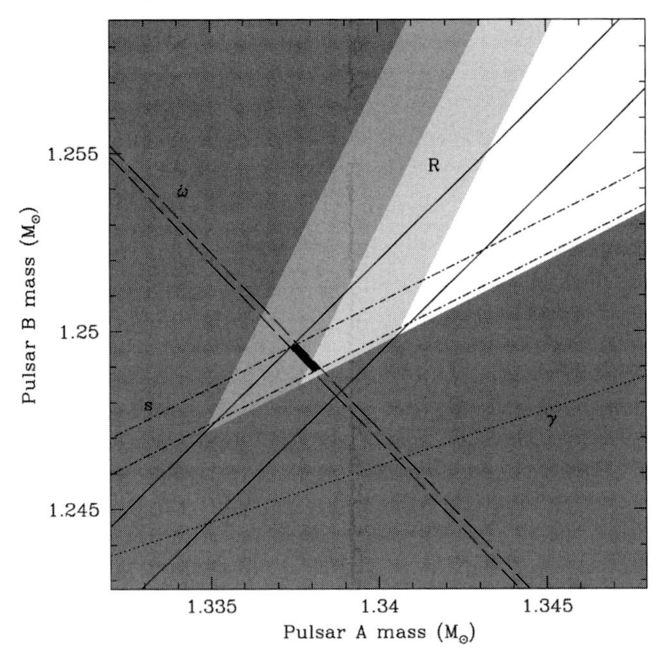

Figure 1. A section of the mass-mass diagram for the double pulsar J0737−3039A/B within GR, based on the measurement of 5 PK parameters in the recycled "A" pulsar and the mass ratio R, which is derived from the relative sizes of the"A" and "B" orbits (Kramer *et al.* 2006). Only the lower curve for γ falls within this plot. The grey-shaded regions are those forbidden by the mass functions of the two pulsars, with the deepening shades of grey illustrating the sizeable uncertainty in the mass function for B. The intersection of all the curves in one area (shaded black) indicates that GR passes the self-consistency test in this system. Note that 1) the similar plot in Stairs (2008) mistakenly used incorrect uncertainties for x_B and for s and 2) the s curves plotted here correspond directly to the asymmetric 68% range in s listed in Kramer *et al.* (2006), slightly different from the plotting approximation used in that paper.

The double pulsar has another crucial advantage in the measurement of the mass ratio R, which has the same dependence on the masses in all gravitational theories, at least to order $(v/c)^2$ (Damour & Taylor 1992). For this system, the combination of R and $\dot{\omega}$ allows an estimate of the masses and prediction of all the other PK parameters, with the Shapiro delay s parameter agreeing with the prediction to within 0.05%, the most stringent test to date of strong-field gravity (Kramer *et al.* 2006). In future, this system may allow measurement of high-order corrections to $\dot{\omega}$, such as that due to spin-orbit coupling of the recycled "A" pulsar; in turn this may provide a constraint on the NS equation of state (e.g,. Lattimer & Schutz 2005).

3.2. *Geodetic Precession*

As discussed above, the supernova explosion that creates a DNS (or similar) system is expected to leave the spin axes of the stars misaligned with the orbital angular momentum. This will result in geodetic precession of the NS spins about the total angular momentum (dominated by the orbital component) (Damour & Ruffini 1974; Barker & O'Connell 1975). Example precession periods are roughly 700 years for PSR B1534+12, and 75 and 71 years for PSR J0737−3039A and B, respectively.

One means of identifying this effect is through secular changes in the observed pulse profiles, as different parts of the emission region come into view. Such changes were soon noticed in the Hulse-Taylor binary (Weisberg *et al.* 1989): a difference in the relative

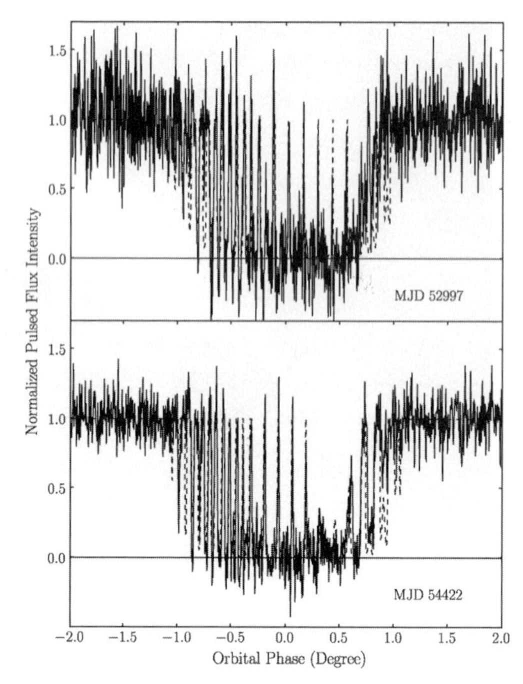

Figure 2. Changing eclipses in the double pulsar. The plots show the relative flux density of the A pulsar as a function of orbital phase at two different epochs (solid lines) as well as fits to the Lyutikov (2005) model (dashed lines). The secular changes in the short-term modulation indicate the changing precession phase of B's spin axis. Figure courtesy of René Breton.

heights of the two profile peaks over several years. In the 1990s, the peaks began to draw together, suggesting a spin-orbit misalignment angle of about 22° and allowing a prediction that the pulsar will disappear from view in about 2025 (Kramer 1998). Profile shape changes as well as polarization changes are seen in PSR B1534+12 (Arzoumanian 1995; Stairs *et al.* 2004); the profile changes are also reflected in orbital-timescale aberrational variations, allowing a beam-model-independent, though low-precision, test of the precession rate, which agrees with the GR prediction (Stairs *et al.* 2004). Geodetic precession has also been observed in the pulsar–white-dwarf system PSR J1141−6545 (Hotan *et al.* 2005, Manchester *et al.*, in prep.) and there is good evidence for the effect in the recently discovered PSR J1906+0746 (Lorimer *et al.* 2006, Kasian *et al.* 2008).

Once again, the double pulsar provides a unique and fascinating view on the problem. Profile shape changes are seen in the young "B" pulsar (Burgay *et al.* 2005), which also shows large orbitally-dependent profile variations (Lyne *et al.* 2004). The recycled "A" pulsar shows no changes, however, which may have interesting implications for the "B" supernova (Manchester *et al.* 2005; Ferdman 2008; Ferdman *et al.* 2008, and references therein). But the combination of the two pulsars allows an even more powerful probe. Because the orbit is nearly edge-on to the line of sight, the A pulsar is eclipsed by the magnetosphere of B for about 30 seconds per orbit (Lyne *et al.* 2004; Kaspi *et al.* 2004; Lyutikov & Thompson 2005), with the flux of A showing strong modulation at the spin period of B and its second harmonic (McLaughlin *et al.* 2004). The overall behaviour is nicely described by a simple dipolar field structure for the B pulsar with a well-defined orientation (Lyutikov 2005). Changes in the A flux modulation pattern over time (see Figure 2) can be matched up to changes in the B spin orientation, resulting in a 13% measurement of B's spin-orbit precession rate that agrees beautifully with the GR prediction (Breton *et al.* 2008). Moreover, because the sizes of both orbits are known, this test in fact applies to a large generic class of fully conservative, Lagrangian-based gravitational theories (Damour & Taylor 1992; Breton *et al.* 2008).

The planned Square Kilometre Array will vastly improve all of these tests, and furthermore has great potential for finding highly exotic systems such as pulsar–black-hole binaries, which would open new realms of gravitational exploration (Kramer *et al.* 2004).

Acknowledgements

Pulsar research at UBC is supported by an NSERC Discovery Grant. The author also acknowledges sabbatical support from the Swinburne University of Technology Visiting Distinguished Researcher Scheme. She thanks her many collaborators on some of the results discussed here, and especially Michael Kramer for a careful reading of the manuscript and René Breton for providing Fig. 2.

References

Arzoumanian, Z. 1995, PhD thesis, Princeton University
Bailes, M. in *this proceedings*, 212
Barker, B. M. & O'Connell, R. F. 1975, Phys. Rev. D, 12, 329
Bassa, C., Wang, Z., Cumming, A., & Kaspi, V., eds. 2008, AIP Conf. Proc. 983: 40 Years of Pulsars (Springer)
Bell, J. F. 1996, ApJ, 462, 287
Bell, J. F., Camilo, F., & Damour, T. 1996, ApJ, 464, 857
Bell, J. F. & Damour, T. 1996, Class. Quant Grav., 13, 3121
Bhat, N. D. R., Bailes, M., & Verbiest, J. P. W. 2008, Phys. Rev. D, 77, 124017
Breton, R. P. *et al.* 2008, Science, 321, 104
Burgay, M. *et al.* 2005, ApJ, 624, L113
Counselman, C. C. & Shapiro, I. I. 1968, Science, 162, 352
Damour, T. & Deruelle, N. 1985, Ann. Inst. H. Poincaré (Physique Théorique), 43, 107
—. 1986, Ann. Inst. H. Poincaré (Physique Théorique), 44, 263
Damour, T. & Esposito-Farèse, G. 1996, Phys. Rev. D, 53, 5541
Damour, T. & Esposito-Farèse, G. 1992a, Class. Quant Grav., 9, 2093
—. 1992b, Phys. Rev. D, 46, 4128
Damour, T. & Esposito-Farese, G. 1996, Phys. Rev. D, 54, 1474
Damour, T., Gibbons, G. W., & Taylor, J. H. 1988, Phys. Rev. Lett., 61, 1151
Damour, T. & Ruffini, R. 1974, Academie des Sciences Paris Comptes Rendus Ser. Scie. Math., 279, 971
Damour, T. & Schäfer, G. 1991, Phys. Rev. Lett., 66, 2549
Damour, T. & Taylor, J. H. 1991, ApJ, 366, 501
—. 1992, Phys. Rev. D, 45, 1840
Deller, A. T., Verbiest, J. P. W., Tingay, S. J., & Bailes, M. 2008, ApJ, 685, L67
Eardley, D. M. 1975, ApJ, 196, L59
Esposito-Farese, G. 2005, in The Tenth Marcel Grossmann Meeting, ed. M. Novello, S. Perez Bergliaffa, & R. Ruffini (World Scientific Publishing), 647
Ferdman, R. D. 2008, PhD thesis, University of British Columbia
Ferdman, R. D. *et al.* 2008, in Bassa *et al.* (2008), 474
Freire, P. C. C., *et al.* 2008, ApJ, 675, 670
Gérard, J.-M. & Wiaux, Y. 2002, Phys. Rev. D, 66, 1
Goldman, I. 1990, MNRAS, 244, 184
Hotan, A. W., Bailes, M., & Ord, S. M. 2005, ApJ, 624, 906
Hulse, R. A. & Taylor, J. H. 1975, ApJ, 195, L51
Kasian, L. & the PALFA Consortium. 2008, in in Bassa *et al.* (2008), 485
Kaspi, V. M. *et al.* 2000, ApJ, 543, 321

Kaspi, V. M. *et al.* 2004, ApJ, 613, L137
Kramer, M. 1998, ApJ, 509, 856
Kramer, M. *et al.* 2004, New Astr., 48, 993
Kramer, M. *et al.* 2006, Science, 314, 97
Kuijken, K. & Gilmore, G. 1989, MNRAS, 239, 571
Lattimer, J. M. & Schutz, B. F. 2005, ApJ, 629, 979
Lazaridis, K. *et al.* 2009, MNRAS, in press, arXiv:0908.0285
Lightman, A. & Lee, D. 1973, Phys. Rev. D, 8, 3293
Lorimer, D. R. & Kramer, M. 2005, Handbook of Pulsar Astronomy (Cambridge University Press)
Lorimer, D. R. *et al.* 2006, ApJ, 640, 428
Lyne, A. G. *et al.* 2004, Science, 303, 1153
Lyutikov, M. 2005, MNRAS, 362, 1078
Lyutikov, M. & Thompson, C. 2005, ApJ, 634, 1223
Manchester, R. N. *et al.* 2005, ApJ, 621, L49
McLaughlin, M. A. *et al.*, 2004, ApJ, 616, L131
Ni, W. 1973, Phys. Rev. D, 7, 2880
Nice, D. J. *et al.* 2005, ApJ, 634, 1242
Nice, D. J. & Taylor, J. H. 1995, ApJ, 441, 429
Nordtvedt, K. 1968, Phys. Rev., 170, 1186
—. 1990, Phys. Rev. Lett., 65, 953
Nordtvedt, K. & Will, C. M. 1972, ApJ, 177, 775
Phinney, E. S. 1992, Phil. Trans. Roy. Soc. A, 341, 39
Ransom, S. M. *et al.* 2005, Science, 307, 892
Rappaport, S. *et al.* 1995, MNRAS, 273, 731
Rosen, N. 1973, Gen. Relativ. Gravit., 4, 435
Shklovskii, I. S. 1970, Sov. Astron., 13, 562
Splaver, E. M., Nice, D. J., Stairs, I. H., Lommen, A. N., & Backer, D. C. 2005, ApJ, 620, 405
Stairs, I. H. 2003, Living Reviews in Relativity, 5, uRL (Cited on 2008/02/16): http://relativity.livingreviews.org/Articles/lrr-2003-5
Stairs, I. H. 2005, in Binary Radio Pulsars, ed. F. Rasio & I. H. Stairs (San Francisco: Astronomical Society of the Pacific), 3
Stairs, I. H. 2006, in Proceedings of 33rd SLAC Summer Institute on Particle Physics (SSI 2005): Gravity in the Quantum World and the Cosmos, 2005, Vol. eConf C0507252, L004
Stairs, I. H. 2008, in Bassa *et al.* (2008), 424
Stairs, I. H. *et al.* 2005, ApJ, 632, 1060
Stairs, I. H., Thorsett, S. E., & Arzoumanian, Z. 2004, Phys. Rev. Lett., 93, 141101
Stairs, I. H., Thorsett, S. E., Taylor, J. H., & Wolszczan, A. 2002, ApJ, 581, 501
Tauris, T. M. & van den Heuvel, E. P. J. 2006, Formation and Evolution of Compact Stellar X-ray Sources, ed. W. H. G. Lewin & M. van der Klis (Cambridge: Cambridge University Press), 623–665
Taylor, J. H. & Weisberg, J. M. 1982, ApJ, 253, 908
—. 1989, ApJ, 345, 434
Taylor, J. H., Wolszczan, A., Damour, T., & Weisberg, J. M. 1992, Nature, 355, 132
Thorsett, S. E. 1996, Phys. Rev. Lett., 77, 1432
Verbiest, J. P. W. *et al.* 2008, ApJ, 679, 675
Wagoner, R. V. 1975, ApJ, 196, L63
Weisberg, J. & Taylor, J. 1981, Gen. Relativ. Gravit., 13, 1
Weisberg, J. M., Romani, R. W., & Taylor, J. H. 1989, ApJ, 347, 1030
Wex, N. 1997, A&A, 317, 976
Wex, N. 2000, in Pulsar Astronomy - 2000 and Beyond, IAU Colloquium 177, ed. M. Kramer, N. Wex, & R. Wielebinski (San Francisco: Astronomical Society of the Pacific), 113
Wex, N., Kalogera, V., & Kramer, M. 2000, ApJ, 528, 401
Wex, N. & Kramer, M. 2007, MNRAS, 380, 455

Will, C. 2006, Living Reviews in Relativity, 9, 1, uRL (Cited on 2008/02/16): http://relativity.livingreviews.org/Articles/lrr-2006-3

Will, C. M. 1977, ApJ, 214, 826

—. 1993, Theory and Experiment in Gravitational Physics (Cambridge: Cambridge University Press)

Will, C. M. & Nordtvedt, K. J. 1972, ApJ, 177, 757

Will, C. M. & Zaglauer, H. W. 1989, ApJ, 346, 366

Williams, J. G., Turyshev, S. G., & Boggs, D. H. 2009, International Journal of Modern Physics D, 18, 1129

Relativity in Fundamental Astronomy
Proceedings IAU Symposium No. 261, 2009
S. A. Klioner, P. K. Seidelman & M. H. Soffel, eds.

doi:10.1017/S1743921309990445

Pulsar timing array projects

G. Hobbs[1]

[1]Australia Telescope National Facility, CSIRO, P.O. Box 76, Epping, NSW 1710, Australia
email: george.hobbs@csiro.au

Abstract. Pulsars are amongst the most stable rotators known in the Universe. Over many years some millisecond pulsars rival the stability of atomic clocks. Comparing observations of many such stable pulsars may allow the first direct detection of gravitational waves, improve the Solar System planetary ephemeris and provide a means to study irregularities in terrestrial time scales. Here we review the goals and status of current and future pulsar timing array projects.

Keywords. pulsars: general

---

1. Introduction

The pulsar timing technique (e.g. Lorimer & Kramer 2005 for an overview and Edwards, Hobbs & Manchester 2006 for a detailed description) has enabled some of the most exciting recent results in astrophysics. For instance, the first extra-Solar planets were discovered orbiting the pulsar PSR B1257+12 (Wolszczan & Frail 1992), the first evidence for gravitational waves was provided by the binary system PSR B1913+16 (Hulse & Taylor 1974) and the most stringent tests of General Relativity in the strong-field regime have been carried out using the double pulsar system (Kramer *et al.* 2006). Even though the PSR B1913+16 system has provided evidence for the existence of gravitational waves, most researchers regard this as indirect evidence and not a direct detection of such waves.

Sazhin (1978) and Detweiler (1979) showed that low-frequency gravitational waves will induce pulsar timing residuals† that could be detectable with high-precision observations of the pulse arrival times. However, with a single pulsar it is difficult, and perhaps impossible, to distinguish between the effects of gravitational waves and many other effects such as the irregular rotation of the neutron star, interstellar medium propagation effects or inaccuracies in the planetary ephemeris. Hellings & Downs (1983) showed that such effects could be distinguished by looking for correlated behaviour in the timing residuals of many pulsars. In brief, the timing residuals caused by irregular rotation or propagation effects should be uncorrelated between different pulsars. Irregularities in the terrestrial atomic time scale would produce completely correlated timing residuals. The angular correlation function expected for an isotropic, stochastic gravitational wave background (Hellings & Down 1983) is shown in Figure 1 (using simulated data described in Hobbs *et al.* 2009).

The first attempt to create a "pulsar timing array" in which enough pulsars are observed with sufficient timing precision to search for correlated signals was reported by Foster & Backer (1990). More recently Jenet *et al.* (2005) calculated the number of pulsars, the timing precision and data span required in order to make a significant detection

† Pulse times of arrival for a given pulsar are compared with a physical model of the pulsar. The differences between the measured arrival times and predicted arrival times using the model are known as the pulsar timing residuals and denote physical effects that are not included in the model.

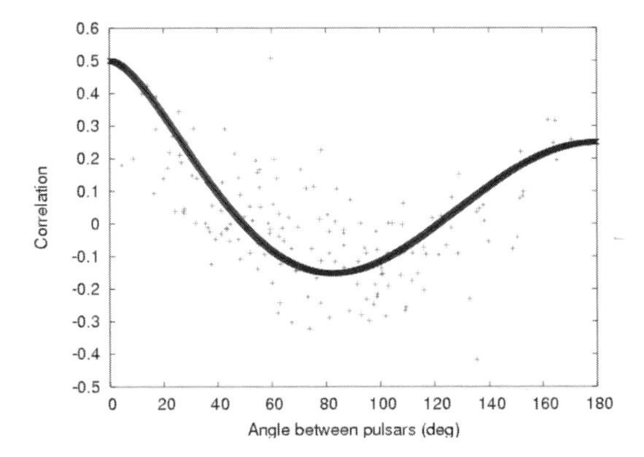

Figure 1. The expected correlation in the timing residuals of pairs of pulsars as a function of angular separation for an isotropic gravitational wave background. The solid line gives the theoretical curve. The dots are calculated by correlating simulated data sets for 20 pulsars in the presence of a gravitational wave background that produces white timing residuals.

of the expected stochastic background of gravitational waves. In summary, at least 20 millisecond pulsars are required, observed for five years with timing precision around 100 ns. In this review paper we discuss the current status of various existing pulsar timing array projects that aim to achieve this goal. In the future, telescopes such as the Square Kilometre Array may make gravitational wave detection using pulsars commonplace. The current status of this project is described in §3.

2. The pulsar timing array projects

The main aims of pulsar timing arrays are to search for correlated signals in the pulsar timing residuals in order to 1) improve the Solar System ephemeris, 2) look for irregularities in the terrestrial time standards and 3) make a direct detection of gravitational waves. Three projects have been started:

• *The Parkes Pulsar Timing Array* is the only timing array project (to date) in the Southern Hemisphere. This project makes use of the 64-m Parkes radio telescope. Every two–three weeks, each of 20 pulsars is observed for approximately one hour using a dual-band 10cm-50cm receiver and for another hour using a 20cm receiver. The data are processed using the PSRCHIVE suite of software (Hotan, van Straten & Manchester 2004) that allows polarimetric and flux density calibration. Pulsar timing models and the resulting timing residuals are obtained using TEMPO2 (Hobbs, Edwards & Manchester 2006).

• *The European Pulsar Timing Array* is making precise timing observations of approximately 15 pulsars. This project makes use of the existing four 100-m class telescopes in Europe (Jodrell Bank, Effelsberg, Westerbork and Nancay) and will also include observations from the 64-m Sardinian radio telescope after its completion. As well as using each of these telescopes as individual instruments, the Large European Array for Pulsars (LEAP) project will allow these telescopes to act as a phased array giving the equivalent sensitivity of a 200 m telescope.

• *The North American Pulsar Timing Array,* known as *Nanograv,* combines data collected with the 300-m Arecibo radio telescope and the 100 m GreenBank telescope. These telescopes are currently obtaining about monthly observations on 24 pulsars.

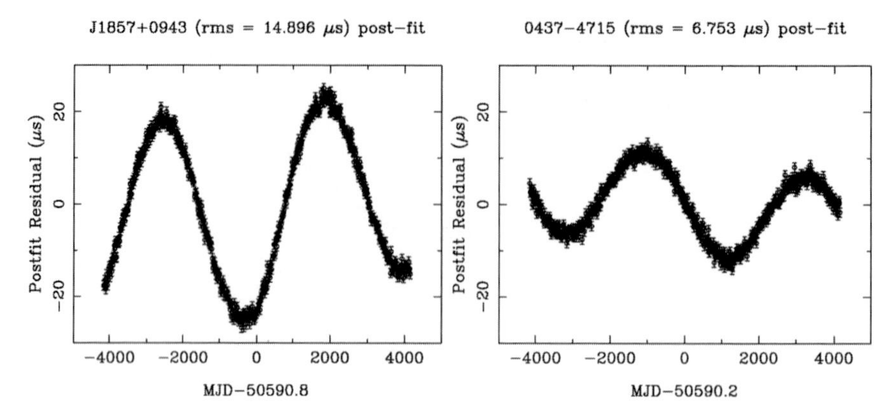

Figure 2. Simulated timing residuals for PSR J1857+0943 (left panel) and J0437−4715 induced by an error in the mass of Jupiter of 10^{-8} $M_{\odot}$.

Members of these three projects have agreed to share data to form the International Pulsar Timing Array (IPTA) project which hopefully will result in the detection of gravitational waves within 5–10 years.

2.1. *The planetary ephemeris*

The pulsar timing procedures requires that the observed pulse times-of-arrival are corrected to the Solar System barycentre. This correction is carried out using a planetary ephemeris that provides the position of the Earth with respect to the barycentre. Such ephemerides are complex and take into account the masses and orbital parameters of the planets and over 300 asteroids. If the mass of one planet was slightly incorrect, then the position of the barycentre will be incorrectly determined and sinusoidal timing residuals will be induced with the period of the planet's orbit (note that a change in the mass of Jupiter will affect the entire fit that is carried out to create the planetary ephemeris. The change in position of the barycentre is a first order effect and, as the change in Jupiter's mass considered here is small, is the only change in the ephemeris that is considered. A more detailed study will be published elsewhere.) In Figure 2 we simulate the timing residuals of PSR J1857+0943 (left panel) and PSR J0437−4715 (right panel) for a change in the mass of Jupiter of 10^{-8} $M_{\odot}$. The IPTA project is currently combining data from a selection of pulsars to place a limit on the error in Jupiter's mass which is expected to be more precise than measurements obtained using the *Voyager* space-craft, but not as precise as unpublished *Galileo* space-craft measurements. In the near future pulsar timing arrays should be able to rule out, or detect, the existence of an Earth-mass "planet-X" closer than 60 A.U., or a Jupiter-mass object closer than 300 A.U.

2.2. *Terrestrial time standards*

Millisecond pulsars are extremely stable. A measure of stability, σ_z (Matsakis *et al.* 1997), allows pulsars to be compared with terrestrial time scales. In Figure 3 we plot the stability of PSRs J0437−4715 (crosses) and J1909−3744 (squares). The arrival times for both these pulsars were referenced to the Bureau International des Poids et Mesures (BIPM) 2008 terrestrial time standard. The solid line indicates the stability of the difference between the Physikalisch-Technische Bundesanstalt (PTB) and National Institute of Standards and Technology (NIST) time standards. The dotted line corresponds to timing residuals with a flat spectrum. It is clear that on short time scales the atomic time scales are significantly more stable than pulsars, but on time scales of ~10 yr and longer the pulsars become as stable as the terrestrial clocks. In the near future it is expected that correlated

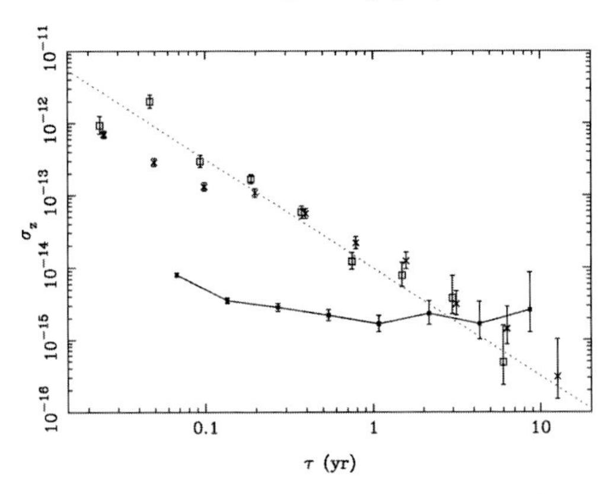

Figure 3. $\sigma_z(\tau)$ for PSR J0437−4715 and PSR J1909−3744 and for a data sequence of atomic clock differences: PTB-NIST.

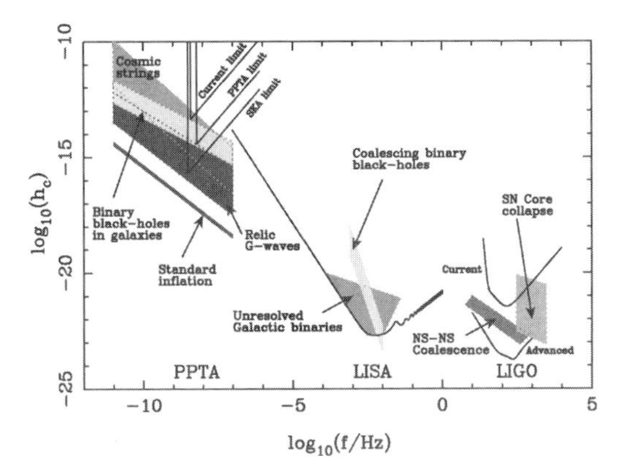

Figure 4. Characteristic strain sensitivity for existing and proposed GW detectors as a function of GW frequency. Predicted signal levels from various astrophysical sources are shown (Hobbs 2008).

signals in the timing residuals will indicate the presence of irregularities in the terrestrial time scales and will allow the formation of a pulsar-based time standard (e.g. Rodin 2008).

2.3. *Gravitational wave detection*

The most likely detectable gravitational wave signal is from an isotropic, unpolarized, gravitational-wave background caused by binary supermassive black holes at the center of merged galaxies (Sesana *et al.* 2008). The induced timing residuals induced by such a background can be simulated using the TEMPO2 software package (Hobbs *et al.* 2009) and lead to upper bounds on the amplitude of any such signal. Jenet *et al.* (2006) gave an upper limit on the energy density per unit logarithmic frequency interval of $\Omega_g[1/(8\text{yr})]h^2 \leqslant 1.9 \times 10^{-8}$ which can be used to constrain the merger rate of supermassive binary black hole systems at high redshift (Wen *et al.* 2009). A comparison between the sensitivity to gravitational waves for different detectors is shown in Figure 4 along with the predicted signal levels from various astrophysical sources.

3. Future timing array projects

With current telescopes it is a challenge to achieve a timing precision of ~100 ns for more than a few pulsars (the best published data set is for PSR J0437−4715 observed for more than 10 yr with rms timing residuals ~200 ns; Verbiest *et al.* 2008). Within ~10 yr it is expected that various new telescopes such as the Australian Square Kilometer Array Pathfinder (ASKAP) and the South African pathfinder (MEERKAT) telescopes will increase the number of known pulsars and provide a larger number of telescopes in the Southern Hemisphere for high-precision pulsar timing (Johnston *et al.* 2008, http://www.ska.ac.za). Also a new 500m-diameter telescope (FAST) should be completed in China (Nan *et al.* 2006). This telescope will be a highly efficient pulsar search and timing instrument and should significantly improve our timing precision for a large number of pulsars.

On a longer timescale, the Square Kilometer Array (SKA) telescope is planned. This telescope should be able to observe many hundreds of pulsars with the precision currently achieved for only a few pulsars. Such data sets should allow the gravitational-wave background and individual sources of such waves to be studied in detail (Cordes *et al.* 2004, Kramer *et al.* 2004). For instance a high S/N detection of a stochastic background would provide a test of general relativity (Lee *et al.* 2008) and allow a detailed understanding of the properties of the sources that form the background.

4. Conclusion

Many pulsars are currently being observed as part of pulsar timing array projects. Correlated timing residuals allow 1) irregularities in terrestrial time standards to be identified, 2) inaccurate planetary masses in the Solar System ephemeris to be corrected and 3) the detection of gravitational wave signals. Combining observations from Northern and Southern hemisphere telescopes should allow the detection of gravitational waves on a time scale of 5–10 years.

5. Acknowledgements

GH is supported by an Australian Research Council QEII Fellowship (project # DP0878388).

References

Cordes, J., Kramer, M., Lazio, T. J. W., Stappers, B. W., Backer, D. C., & Johnston, S. 2004, New Astronomy Reviews, 48, 1413
Detweiler, S. 1979, *ApJ*, 234, 1100
Edwards, R., Hobbs, G., & Manchester, R. 2006, *MNRAS*, 372, 1549
Foster, R. S. & Backer, D. C. 1990, *ApJ*, 361, 300
Hellings, R. W. & Downs, G. S. 1983, *ApJ*, 265, L39
Hobbs, G., Edwards, R., & Manchester, R. 2006, *MNRAS*, 369, 655
Hobbs, G. 2008, *Classical and Quantum Gravity*, 25, 11
Hobbs, G. *et al.* 2009, *MNRAS*, 394, 1945
Hotan, A. W., van Straten, W., & Manchester, R. N. 2004, *PASA*, 21, 302
Hulse, R. A. & Taylor, J. H. 1974, *ApJ*, 191, L59
Jenet, F. A., Hobbs, G., Lee, K. J., & Manchester, R. N. 2005, *ApJ*, 625, L123
Jenet, F. A. *et al.* 2006, *ApJ*, 653, 1571
Johnston, S.. *et al.* 2008, *ExA*, 22, 151

Kramer, M., Backer, B. W., Cordes, D. C., Lazio, J., Stappers, T. J. W., & Johnston, S. 2004, New Astronomy Reviews, 48, 993

Kramer, M. *et al.* 2006, *Sci*, 314, 97

Lee, K. J., Jenet, F. A. & Price, R. H. 2008, *ApJ*, 685, 1304

Lorimer, D. R. & Kramer, M. 2005, *Handbook of Pulsar Astronomy*, Cambridge University Press

Matsakis, D. N., Taylor, J. H, & Eubanks, T. M. 1997, *A&A*, 326, 924

Nan, R. D., Wang, Q. M., Zhu, L. C., Zhu, W. B., Jin, C. J., & Gan, H. Q. 2006, *Chin. J. Astron. Astrophys., Suppl.*, 6, 304

Rodin, A. 2008, *MNRAS*, 387, 1583

Sazhin, M. V. 1978, *SvA*, 22, 36

Sesana, A., Haardt, F., & Madau, P. 2008, *MNRAS*, 390, 192

Verbiest, J. *et al.* 2008, *MNRAS*, 679, 675

Wen, Z. *et al.* 2009, *"Constraining the coalescence rate of supermassive black-hole binaries using pulsar timing"*, submitted to MNRAS

Wolszczan, A. & Frail, D. A. 1992, *Nature*, 355, 145

Relativity in Fundamental Astronomy
Proceedings IAU Symposium No. 261, 2009
S. A. Klioner, P. K. Seidelman & M. H. Soffel, eds.

doi:10.1017/S1743921309990457

Astrometric and timing effects of gravitational waves

Bernard F. Schutz[1]
[1]Albert Einstein Institute, D-14424 Potsdam, Germany

Abstract. Gravitational wave detection can be done by precision timing of millisecond pulsars, and (with less likelihood) by precision astrometry on distant objects whose light or radio waves pass through gravitational waves on their way to our observatories. Underlying both of these is the relatively simple theory of light propagation in spacetimes with gravitational waves, which is also the basis of interferometric gravitational wave detectors. I review this theory and apply it to the timing and astrometric methods of detection. While pulsar timing might even be the first way that we directly detect gravitational waves, light deflection by gravitational waves seems out of reach.

Keywords. gravitational waves,gravitational lensing, astrometry,pulsars: general

---

1. Introduction

Gravitational wave detection is a challenge that is being pursued using many different technologies suited to finding waves in many different wavebands, with frequencies ranging from tens of kilohertz down to the Hubble frequency H_0. Although all the periods of these waves are long compared to the time-resolution needed for conventional astrometry, there are at least two ways in which astrometric considerations are relevant to gravitational wave detection. The first way is the detection of gravitational waves through the precise timing of millisecond pulsars using timing arrays; the second is the possibility that gravitational waves crossing the path of light or radio waves from a distant object might influence the observed frequencies or angular positions strongly enough to be detected. In order to understand both possible effects I will review the principles of "beam detectors", gravitational wave detectors based on the propagation of electromagnetic radiation, and then apply what we learn to the two problems just mentioned.

Ground-based gravitational wave detection is a very active field now. The LIGO Scientific collaboration (LSC 2009) operates the two 4-km interferometers of the LIGO project (Abbott *et al.* 2009) and the smaller GEO600 detector, a combined development platform and high-frequency observatory (Grote and the LIGO Scientific Collaboration 2008). The 3-km VIRGO interferometer (Acernese *et al.* 2008) shares data with the LSC detectors and does joint analysis of that data. All three large detectors were operating at the long-sought goal of a broadband strain sensitivity of 10^{-21} in 2007. The S6 observing run, due to start in the middle of 2009, has a non-negligible chance of making the first detection, if it can achieve a somewhat better sensitivity. Other detectors are in the planning stage, the most advanced of which is the 3-km LCGT cryogenically cooled detector in Japan (LCGT 2009). The existing long-baseline detectors will reach a sensitivity of around 10^{-22} when they upgrade to Advanced Detectors around 2016. In the medium future it is now possible to envision super-sensitive ground-based detectors, with ten times better sensitivity even than Advanced Detectors and with a range extending to high redshifts, as a result of the European Einstein Telescope (ET) design study (ETP 2009).

With a network of at least three major detectors observing an event, it is possible to fully reconstruct the event's polarization, source location, and intrinsic amplitude. As broad-band detectors, the ground-based systems are going to be the workhorses of gravitational wave astronomy. They will be able to reconstruct waveforms, measure distances to merging binary systems of neutron stars and black holes, perform strong tests of general relativity, and explore the world of compact stellar objects out to high redshifts. But they will not be able to observe below a few Hz, and this means that their sources will be stellar-mass objects, up to perhaps $10^3\ M_\odot$. More massive objects radiate only at lower frequencies.

For this reason the international community is also developing the LISA space-based gravitational wave detector (LISA 2009), which will have a sensitivity in the milliHertz range. With the geometry of an equilateral triangle, LISA will actually provide three interferometry signals, so it is in fact a complete network in space, able to reconstruct positions, polarizations, amplitudes, and distances. Currently scheduled for launch in 2010, LISA is a cooperation between NASA and ESA. The current decadal review of astronomy in the US may have a significant influence on LISA's launch date. Beyond LISA there is a strongly developing project in Japan called DECIGO (Kawamura *et al.* 2006), aimed at exploring the 0.1 Hz band between LISA and the ground-based projects. Beyond that, there is a NASA concept study of the Big Bang Observer (BBO 2009), which demonstrates that, with effort, it might be possible to detect directly the gravitational waves left over from the Big Bang.

LISA will have enormous sensitivity, being able to detect mergers of black holes with masses between 10^3 and $10^7\ M_\odot$ out to very high redshifts. By making observations of small black holes falling into larger ones (so-called Extreme Mass-Ratio Inspirals, or EMRIs), LISA will provide detailed maps of the geometry of the central black holes of many galaxies, testing general relativity and proving (or disproving!) that the central objects really are black holes. Although LISA's launch is far away, the mere existence of the project has already stimulated an unprecedented amount of scientific research, particularly into stellar populations in galaxies and in clusters around central black holes, as well into the dynamics of galaxy mergers and their subsequent black-hole mergers.

Coming along strongly in recent years is the search for gravitational waves using pulsar timing, e.g. (Jenet *et al.* 2006). Since millisecond pulsars are such good clocks, their arrival times at Earth record the effects of gravitational waves. By cross-correlating a suite of stable pulsars it is possible to beat down their noise further and look for a systematic variation in arrival times, which – if it has a quadrupolar pattern on the sky – could be due to a strong gravitational wave. The method requires averaging over several years, so it is only suited to frequencies below about 10^{-8} Hz. But in this range there are interesting sources, particularly binaries of supermassive black holes. When the SKA is constructed (Carilli and Rawlings 2004), the detection and study of gravitational waves in this frequency range will become routine. But even in the next five years it is possible that pulsar timing will be the first technique to directly detect gravitational waves.

Pulsar timing and interferometry on the ground and in space are all examples of beam detectors, where a beam of light or radio waves is used to measure the amplitudes of gravitational waves. They work on the basis of timing: even in interferometers, the primary effect of the waves is to change the round-trip light travel-time along each arm, so that when light returns to the interference point, the two beams have been shifted in time relative to one another. We will see how this works below. The remaining technique for detection is the search for random cosmological gravitational waves in the microwave background radiation. Here the effect being measured is in the polarization of the microwave radiation; small changes in polarization are imprinted on the waves as they decouple from the expanding but recombining hot plasma, so the effect measures the amplitude of the gravitational waves at the time of decoupling. The recently launched PLANCK spacecraft (PLANCK 2009) will measure polarization, but it is not clear whether it will be able to see the small effect.

2. Beam Detectors

All interferometers and timing experiments on gravitational waves can be understood by studying the effect of gravitational waves on a beam of light. For simplicity we start with a restricted geometry: the light beam is travelling along the x-axis and the gravitational wave along the z-axis. The wave is a plane wave moving perpendicular to the light-beam. The light is moving between two mirrors or detectors that are themselves freely-falling in the gravitational field.

The wave is described by the following metric in linearized theory and the TT-gauge (Schutz 2009)

$$\mathrm{d}s^2 = -\mathrm{d}t^2 + (1 + h_+)\mathrm{d}x^2 + (1 - h_+)\mathrm{d}y^2 + \mathrm{d}z^2, \quad h_+ = h_+(t - z), \tag{2.1}$$

where h_+ is the amplitude of the "+" polarization component of the wave; the orthogonal "×" component does not affect the light-beam in this geometry.

The properties of this gauge, which is essentially a choice of coordinates, are that the spatial coordinates are comoving with free particles, so that the end-points of the beam remain at fixed coordinate locations even as the wave passes. The time-coordinate of this gauge is proper time on clocks at rest, which is, therefore, the time measured by clocks on the end-points of the light beam.

The one-way travel-time of light along the x-direction is, to first order in h_+, just solved from the above equation by setting $\mathrm{d}s^2 = 0$ and $\mathrm{d}z^2 = 0$. This gives $\mathrm{d}t^2 = (1 + h_+)\mathrm{d}x^2$, or integrating:

$$t_{\mathrm{end}} = t_{\mathrm{start}} + \int [1 + \frac{1}{2}h_+(t(x))]\mathrm{d}x. \tag{2.2}$$

The argument of h_+ here is just t, since the integration takes place on $z = 0$. But the relevant time t is the time at which the light-beam is at the location x. So this is an implicit equation: we want to find t but need it in the integrand.

The way to solve this implicit equation is to recall that h_+ is small, so that we do not need corrections of order h inside the argument of h. We can set $t(x) = x$ in this argument, which is the relation between t and x when the wave is not present. This allows the integral to be done. We can see the result most easily if we differentiate the arrival time with respect to the initial time:

$$\frac{\mathrm{d}t_{\mathrm{end}}}{\mathrm{d}t_{\mathrm{start}}} = 1 + \frac{1}{2}h_+(t_{\mathrm{end}}) - \frac{1}{2}h_+(t_{\mathrm{start}}). \tag{2.3}$$

The travel time varies only because of the wave amplitude when the beam starts out and when it arrives. The wave amplitude in the intervening time is not relevant. This is an important conclusion to which we will come back when we discuss gravitational wave astrometry below. To measure gravitational waves with this one needs stable clocks at both ends of the light beam.

Of course, our geometry is restricted, but if the wave were travelling at an angle θ to the z-axis, the only thing that would change is that there would be a $\cos\theta$ factor in the wave amplitude: a wave travelling in the same direction as the beam has no effect on it.

This formula was used to look for a possible supermassive binary black hole in 3C66b, using timing effects on the pulsar B1855+09 (Jenet *et al.* 2004). The pulsar is the stable clock at the emitting location, and the observatory clock is the stable reference at the receiving location.

From this it is easy to see how pulsar timing works. The beam starts at the pulsar and arrives at the Earth. When the variations in arrival times of pulses are correlated between different pulsars, the amplitudes of the gravitational waves at the pulsars themselves will go away, because they are uncorrelated. But the amplitude at the Earth will correlate, and there will be the factor of $\cos\theta$ with respect to the direction of the wave at the Earth. This gives a characteristic sky pattern that can be used to dig further into the residual pulsar timing noise and clock noise at the detector.

In an interferometer the light is reflected back from the distant mirror, so that the relevant time is the two-way travel time along a path whose proper length is L when there is no gravitational wave. In our geometry this is just, to first order,

$$t_{\text{return}} = t_{\text{start}} + 2L + \frac{1}{2}\int [h_+(t_{\text{start}} + x) + h_+(t_{\text{start}} + L + x)]\mathrm{d}x. \tag{2.4}$$

To use this one needs only a stable clock in one location, namely at the emitter. The observable is the variation with time in this travel time. Interferometers essentially provide their second arm as the stable clock: the effect one looks for is the difference in light-travel-time variations between the two arms.

If we adopt a general geometry, with the wave arriving at an angle θ to the z-axis, then we get the following three-term formula for the variation in the return time:

$$\begin{aligned}\frac{\mathrm{d}t_{\text{return}}}{\mathrm{d}t_{\text{start}}} &= \frac{1}{2}\{(1-\sin\theta)h_+(t_{\text{start}}+2L) - (1+\sin\theta)h_+(t_{\text{start}}) \\ &\quad + 2\sin\theta h_+[t_{\text{start}} + (1-\sin\theta)L]\}\,. \end{aligned} \tag{2.5}$$

This formula was first derived by Estabrook and Wahlquist (Estabrook and Wahlquist 1975) to support their efforts to detect gravitational waves by timing the transponding time to solar-system spacecraft, which is another beam-detector method. The three-term signature helps to discriminate signals from noise.

By combining the three-term formula for the two arms of an interferometer one gets the general expression for the difference in the variation in the return times in the two arms, which translates into a phase shift in the light as it recombines. As soon as the second arm is introduced, one has to consider both polarizations, since the arm need not lie in the plane defined in our example by the beam and the wave directions. Ground-based detectors operate in the simple regime where the gravitational wavelength is much larger than the arm length L. This means that the time-arguments in Equation (2.5) can all be expanded around t_{start}. Timing accuracies must be very precise: Advanced Detectors will have to sense time-delays of only 10^{-27} s! LISA, on the other hand, has arms that become comparable to a wavelength at the high-frequency end of its observing band, so that its response pattern to waves is more complex. This has to be taken into account in LISA data analysis. Its timing resolution is more relaxed: only 10^{-22} s.

3. Light Deflection by Gravitational Waves

Gravitational waves produce more effects than just time-delays. They can also deflect the direction of travel of a light beam, which has led to suggestions that precision astrometry could perhaps detect waves, particularly if the waves pass through the light or radio beam when they are still strong, i.e. shortly after they have been emitted by a highly relativistic event like the merger of two black holes. Some initial studies of the effect, however, overestimated the size of the deflection. This was finally cleared up in a series of investigations by, among others, Damour, Schäfer, and Kopeikin (Damour and

Esposito-Farèse 1998; Kopeikin *et al.* 1999; Kopeikin and Schäfer 1999; Blanchet *et al.* 2001; Kopeikin and Korobkov 2005). I shall here give a simplified treatment based on linearized light propagation.

We need more than time-delays for this kind of astrometry, which involves the deflection of light beams. So we need the propagation equation for null geodesics, which is basically the transport equation for the momentum 4-vector p^μ of a photon moving through the wave. Consider the same geometry as before, where the photon moves along the x-axis and the wave along the z-direction. Then any change in the direction of the photon will be seen as a change in its momentum in the z-direction. The equation of propagation is

$$\frac{\mathrm{d}p^\mu}{\mathrm{d}\lambda} = \Gamma^\mu{}_{\alpha\beta} p^\alpha p^\beta, \tag{3.1}$$

which leads at lowest order to

$$\frac{\mathrm{d}p^z}{\mathrm{d}t} = \frac{1}{2} p^0 \frac{\mathrm{d}h_+}{\mathrm{d}t}. \tag{3.2}$$

This can be integrated from the start to the end, giving, again to lowest order, a deflection angle

$$\Delta\Theta = \frac{\Delta p^z}{p^z} = \frac{1}{2}[h_+(t_{\mathrm{end}}) - h_+(t_{\mathrm{start}})]. \tag{3.3}$$

The important conclusion from this is that – just as in our earlier discussion of time-delays – the effect depends only on the amplitude of the wave at the end points, i.e. where the wave starts and where it is observed. It does not matter if the light beam travels through a strong-wave region in between. The effects of this region cancel out, and the residual depends only on the (presumed weak) wave field at the end-points of the beam.

This is, of course, just a discussion at the lowest order. We can expect that the amplitude of strong waves might come in at the next order. In particular, if the spacetime through which the light beam travels is curved, say by a Newtonian gravitational field, then there will not be complete cancellation. So we could try to estimate a reasonable size for effects of order $(GM/r)h_+$ for a beam passing a distance r from a radiating source of mass M giving off (possibly strong) waves of amplitude h_+.

Now, for wave detection, the distance r must exceed one wavelength λ of the gravitational wave, for otherwise we are not in the wave zone. This can have a serious effect. The wavelength is determined from the natural frequency of the radiating mass, $f \sim (GM/R^3)^{1/2}/2\pi \sim \lambda^{-1}$, where R is the size of the source. Now, there is an upper bound on the amplitude, $h_+ < (GM/r)(GM/R)$ (Schutz 2009). Combining these results gives

$$h_+ < \left(\frac{GM}{R}\right)^{5/2} \sim v^5, \tag{3.4}$$

where v is a typical velocity inside the source. When multiplied by GM/λ to get the size of the nonlinear effect, one gets in the end

$$\Delta\Theta < h_+ \frac{GM}{\lambda} < \left(\frac{GM}{R}\right)^4 \sim v^8. \tag{3.5}$$

Clearly these effects are smaller than the linear deflection in Equation (3.3). For a gravitational wave source like the Binary Pulsar, for example, this means that the maximum nonlinear effect that could be detected by this method is 10^{-24}. There are of course intrinsically stronger sources, with shorter timescales and wavelengths, but they will be transient, so the chance of observing them are reduced.

If the waves that do the deflection are not from any localized source, but from a cosmological or astrophysical random background, the conclusion is not much different.

The characteristic amplitude of a wave of frequency f in a random background whose energy density is a fraction $\Omega_{\rm gw}$ of the cosmological closure density ρ_c is

$$h_c = (G\rho_c)^{1/2} f^{-1} \Omega_{\rm gw}^{1/2}/2\pi. \tag{3.6}$$

If the background comes from inflation, and we normalize the observation to a period of 3 years or a frequency of 10^{-8} Hz, then we get

$$h_c = 10^{-18} \left(\frac{\Omega_{\rm gw}}{10^{-14}}\right)^{1/2} \left(\frac{f}{10^{-8}\,{\rm Hz}}\right)^{-1}. \tag{3.7}$$

This still needs to be multiplied by GM/R appropriate to some deflecting mass, so this will be small as well. Of course, it is likely at some frequencies that there are astrophysical backgrounds with higher energy densities, but even if we go up to $\Omega_{\rm gw} = 10^{-6}$, the nucleosynthesis upper bound, we gain only four orders of magnitude in amplitude.

It follows that the prospects for astrometric observation of gravitational waves are very slim.

References

LSC (2009), *Ligo scientific collaboration home page*, URL `http://ligo.org/`.

B. P. Abbott, R. Abbott, R. Adhikari, *et al.*, Reports on Progress in Physics **72**, 076901 (2009).

H. Grote and the LIGO Scientific Collaboration, Classical and Quantum Gravity **25**, 114043 (2008).

F. Acernese, M. Alshourbagy, P. Amico, F. Antonucci, S. Aoudia, P. Astone, S. Avino, L. Baggio, F. Barone, L. Barsotti, *et al.*, Journal of Physics Conference Series **120**, 032007 (2008).

LCGT (2009), *Large-scale cryogenic gravitational-wave telescope project*, URL `http://www.icrr.u-tokyo.ac.jp/gr/LCGT.html`.

ETP (2009), *Einstein telescope*, URL `http://www.et-gw.eu/`.

LISA (2009), *Laser interferometer space antenna*, URL `http://www.esa.int/esaSC/120376_index_0_m.html`.

S. Kawamura, T. Nakamura, M. Ando, N. Seto, K. Tsubono, *et al.*, Class. Quantum Grav. **23**, S125 (2006), URL `http://stacks.iop.org/0264-9381/23/S125`.

BBO (2009), *Nasa vision missions*, URL `http://universe.nasa.gov/program/vision.html#big%20bang%20observer`.

F. Jenet, G. Hobbs, W. van Straten, R. Manchester, M. Bailes, J. Verbiest, R. Edwards, A. Hotan, J. Sarkissian, and S. Ord, Astrophys. J. (2006), astro-ph/0609013.

C. Carilli and S. Rawlings, New Astronomy Reviews **48** (2004).

PLANCK (2009), *Planck home page*, URL `http://www.rssd.esa.int/index.php?project=PLANCK`.

B. F. Schutz, *A First Course in General Relativity, 2nd edition* (Cambridge University Press, 2009).

F. Jenet, A. Lommen, S. Larson, and L. Wen, Astrophys. J. **606**, 799 (2004), arXiv:astro-ph/0310276.

F. B. Estabrook and H. D. Wahlquist, Gen. Rel. & Grav. **6** (1975).

T. Damour and G. Esposito-Farèse, Phys.Rev.D **58**, 044003 (1998), arXiv:gr-qc/9802019.

S. M. Kopeikin, G. Schäfer, C. R. Gwinn, and T. M. Eubanks, Phys.Rev.D **59**, 084023 (1999), arXiv:gr-qc/9811003.

S. M. Kopeikin and G. Schäfer, Phys.Rev.D **60**, 124002 (1999), arXiv:gr-qc/9902030.

L. Blanchet, S. Kopeikin, and G. Schäfer, in *Gyros, Clocks, Interferometers ...: Testing Relativistic Gravity in Space*, edited by C. Lämmerzahl, C. W. F. Everitt, and F. W. Hehl (2001), vol. 562 of *Lecture Notes in Physics, Berlin Springer Verlag*, pp. 141–+.

S. Kopeikin and P. Korobkov, ArXiv General Relativity and Quantum Cosmology e-prints (2005), arXiv:gr-qc/0510084.

Relativity in Fundamental Astronomy
Proceedings IAU Symposium No. 261, 2009
S. A. Klioner, P. K. Seidelman & M. H. Soffel, eds.

doi:10.1017/S1743921309990469

Gravitational wave astronomy, relativity tests, and massive black holes

Peter L. Bender
(For the LISA International Science Team)
JILA, Univ. of Colorado and NIST

Abstract. The gravitational wave detectors that are operating now are looking for several kinds of gravitational wave signals at frequencies of tens of Hertz to kilohertz. One of these is mergers of roughly 10 $M_\odot$ BH binaries. Sometime between now and about 8 years from now, it is likely that signals of this kind will be observed. The result will be strong tests of the dynamical predictions of general relativity in the high field regime. However, observations at frequencies below 1 Hz will have to wait until the launch of the Laser Interferometer Space Antenna (LISA), hopefully only a few years later. LISA will have 3 main objectives, all involving massive BHs. The first is observations of mergers of pairs of intermediate mass (100 to $10^5 M_\odot$) and higher mass BHs at redshifts out to roughly z=10. This will provide new information on the initial formation and growth of BHs such as those found in most galaxies, and the relation between BH growth and the evolution of galactic structure. The second objective is observations of roughly 10 $M_\odot$ BHs, neutron stars, and white dwarfs spiraling into much more massive BHs in galactic nuclei. Such events will provide detailed information on the populations of such compact objects in the regions around galactic centers. And the third objective is the use of the first two types of observations for testing general relativity even more strongly than ground based detectors will. As an example, an extreme mass ratio event such as a 10 $M_\odot$ BH spiraling into a galactic center BH can give roughly 10^5 observable cycles during about the last year before merger, with a mean relative velocity of 1/3 to 1/2 the speed of light, and the frequencies of periapsis precession and Lense-Thirring precession will be high. The LISA Pathfinder mission to prepare for LISA is scheduled for launch in 2011.

Keywords. Gravitational waves, massive black holes, structure formation, astrophysics, relativity tests

---

1. Introduction

A number of ground-based gravitational wave detectors based on laser interferometers are now in operation. This includes the following observatories: the LIGO observatories in Hanford, WA and Livingston, LA; the joint Italian-French VIRGO observatory near Pisa, Italy; the German-British GEO-600 observatory near Hannover, Germany; and the TAMA observatory in Mitaka, Japan. Major upgrades of the LIGO observatories over the next six years or so have been approved, and comparable improvements in sensitivity are expected at other sites. (See also preceeding article: [Schutz (2009b)].)

With the improvements in sensitivity that are planned, the predicted event rate for gravitational wave detections range from a few per month to tens of events per day. Most of the events are expected to be from mergers of roughly 10 $M_\odot$ black holes or of neutron stars. However, some may also be from supernova explosions. Despite the large uncertainty in the event rate, some black hole-black hole mergers are likely to have high enough signal-to-noise ratios (SNRs) so that their gravitational wave signals will provide strong tests of the predictions of general relativity in strong fields.

In the future, gravitational wave observations also are expected from the Laser Interferometer Space Antenna (LISA) [Stebbins (2006); Sallusti *et al.* (2009)]. The design, launch and operation of LISA are being planned as a joint mission of the European Space Agency (ESA) and NASA. The scientific objectives of LISA include both astrophysical studies of several types and detailed testing of gravitational theory. The general design of the LISA mission and its scientific objectives will be described briefly in this article. In addition, the preparations for LISA through the LISA Pathfinder mission will be discussed.

2. General Design of the LISA Mission

In two-armed laser interferometer gravitational wave detectors, the quantity that is measured is essentially the difference of the light travel times between two arms. When gravitational waves are present, they change the light travel times differently for the two arms. LISA will have three arms, so observations can be made on two different polarizations of the gravitational waves simultaneously.

The LISA antenna consists of three spacecraft that form a nearly equilateral triangle with side lengths of about 5 million km. The center of the triangle follows a circular path around the Sun, roughly 50 million km behind the Earth. Each spacecraft orbit has an eccentricity e of about 0.01 and an inclination i equal to $\sqrt{3}$ times the eccentricity. The initial conditions for the orbits are arranged so that the triangle maintains an inclination of 60 degrees to the ecliptic as it goes around the Sun.

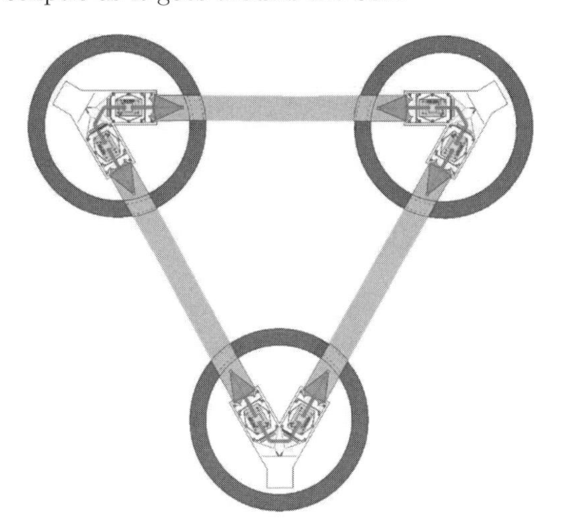

Figure 1. The LISA antenna.

Laser beams are sent both ways along each side of the triangle between the different spacecraft, as shown in Fig. 1. There are two optical assemblies in each spacecraft that transmit the laser beams through 40 cm diameter telescopes, and also receive the beams from the two distant spacecraft. The phase of the beat between the received beam and a portion of the transmitted beam is recorded as a function of time in each optical assembly. The resulting six records of the beat phase as a function of time are the main science output from the antenna, and are sent down to the ground every two days for analysis. The noise in the phase measurement over one second is equivalent to about 20 picometers in displacement, or 4×10^{-21} in terms of the fractional change in arm length.

The other noise source that becomes important at frequencies below about 3 millihertz is spurious accelerations of the test masses that are used to correct for non-gravitational forces on the spacecraft [Carbone *et al.* (2007); Armando *et al.* (2009)]. Each optical assembly has a subsystem called a Gravitational Reference Sensor (GRS) mounted on the back of its telescope and optical bench, as shown in Fig. 2. At the center of each GRS is a test mass that is kept as free as possible of locally generated forces. Each test mass is a 4.6 cm cube of a gold-platinum alloy that has high density but low magnetic susceptibility. A cubical housing surrounds the test mass, with capacitive electrodes on its inside to sense displacements of the test mass with respect to the center of the cavity via ac bridge measurements. These measurements are used to servo-control the spacecraft to follow each test mass in its sensitive direction along the laser beam. Fairly weak electrical forces can also be applied via the capacitor plates to keep the test masses centered in their housings, and to avoid rotation of the test masses in their housings.

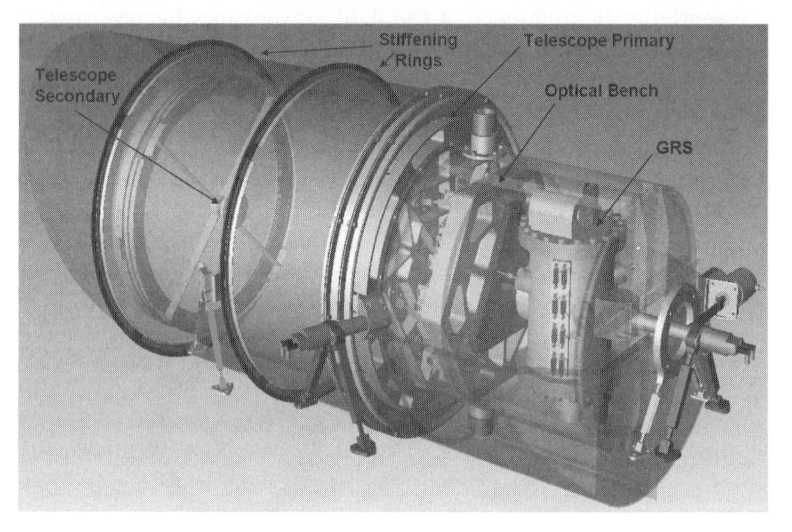

Figure 2. Optical assembly for LISA.

In contrast to previous missions, the gaps between the test masses and the housings around them are made quite large, about 4 mm, in contrast to roughly 0.04 mm gaps in some previous missions. This greatly reduces possible spurious electrical forces due to work function differences between different surfaces. In addition, in order to minimize variations in forces acting on the test masses, the noise in the displacement of the housing with respect to each test mass is kept down to a level of 2 nanometers/($\mathrm{Hz}^{0.5}$) at frequencies down to 0.1 millihertz (mHz). (This notation is to be interpreted as meaning that the mean square displacement noise power has to be less than $[(2 \times 10^{-9}\ \mathrm{m})^2]/\mathrm{Hz}$ down to 0.1 mHz.)

Such position-dependent perturbing forces could come from gravity gradients due to the mass distribution in the spacecraft, from work function differences between different surfaces, from magnetic field gradients, etc. An extensive error allocation budget for the various spurious acceleration sources is maintained, with a total level of 3×10^{-15} $\mathrm{m}/(\mathrm{s}^2)/(\mathrm{Hz}^{0.5})$ down to at least 0.1 mHz, including a 35% margin.

In ground-based gravitational wave detectors, the arms of the interferometer are kept equal to suppress noise due to laser frequency fluctuations. For LISA, since each spacecraft is in a nearly geodesic orbit around the Sun, it turns out that roughly 1% variations in the differences in the arm lengths cannot be avoided. However, these variations are essentially at annual, semi-annual and tri-annual periods, and thus are very smooth.

The resulting Doppler shifts in the laser beams sent between the spacecraft can be as high as 15 Megahertz (MHz), but highly accurate phase measurements as a function of time can still be made on the laser beat notes. This is done by specially designed phase meters that have been designed and thoroughly tested both in the US and in Europe [Shaddock *et al.* (2006); Wand *et al.* (2006)]. These phase measurements are made with respect to stable on-board oscillators, and the results are then sent down to the ground, where the data from the different spacecraft are combined and analyzed.

The threshold sensitivity curve for LISA is shown in Fig. 3 for the case of a monochromatic signal from a random direction that is observed for one year and that gives an amplitude signal-to-noise ratio (SNR) of 5. The vertical axis is gravitational wave strain h. At frequencies above about 3 mHz the curve is dominated by noise in measuring variations in the distance between the test masses at the ends of the arms, with comparable contributions from photon detection shot noise in the photo-detectors for the laser beat notes and from other possible sources of distance measurement noise. At lower frequencies the dominant noise source is the residual small spurious accelerations of the test masses.

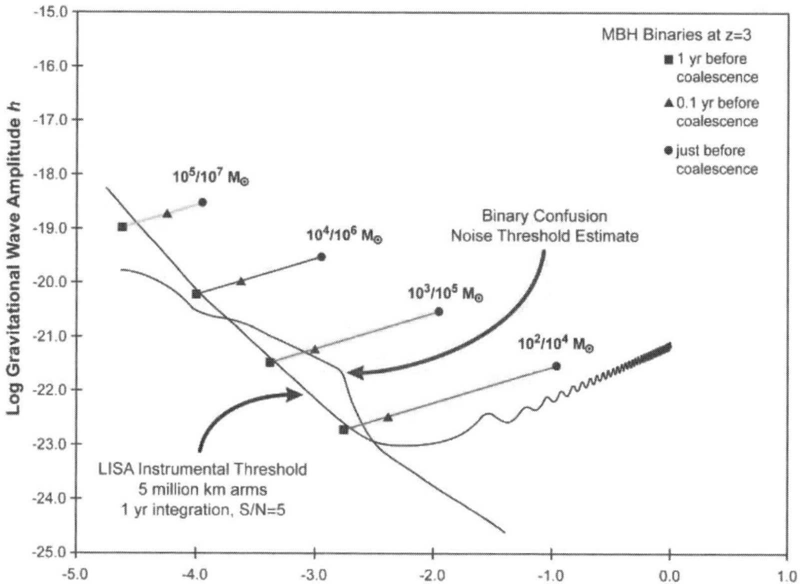

Figure 3. Threshold sensitivity curve and MBH binary signal strengths for LISA.

Also shown in Fig. 3 are the signal strengths that would be produced for 100:1 ratios of masses in massive black hole (MBH) binaries merging at $z = 3$. The signal strength is shown at times of 1 year, 0.1 year, and 1 hour before reaching the last stable circular orbit. Since the threshold sensitivity curve is for 1 year of observations at a given frequency and the signal frequencies are changing rapidly, a separate calculation is needed to determine the integrated S/N for the whole merger event. However, even the lowest mass merger shown would have an integrated $S/N > 5$.

It should be noted that the threshold sensitivity curve shown at frequencies of 0.03 to 0.1 mHz is based on the goal for the LISA mission, subject to verification by analysis. The additional curve shown is a rough estimate of the confusion noise threshold level likely to be present in the data because of unresolved signals from many close binary stars.

3. Primary Scientific Objectives of LISA

The first primary objective of the LISA mission, as discussed earlier, is the detection and detailed observation of gravitational wave signals from the mergers of intermediate mass and larger mass black holes throughout the universe. Current information indicates that most fairly large galaxies contain MBHs with masses of roughly 10^6 $M_\odot$ or higher at their centers, and that the MBH mass depends quite directly on the mass of the central bulge and the dispersion of stellar velocities near the center. This indicates coevolution of the MBHs and the galaxies [Volonteri (2006); Di Matteo *et al.* (2008)]. When smaller pre-galactic structures merged, according to present theories of structure formation in the universe, the MBHs at their centers are likely to have spiraled down to the center of the resulting galaxy, formed a binary, and in most cases, merged. Thus observations of the signals from such mergers can provide valuable constraints on the history of structure formation and the growth of galaxies.

One constraint on our knowledge of this process is that we don't know how or when the intermediate mass black holes that later grew much larger in galactic nuclei were formed. One idea is that the stars that formed first in the universe had higher masses, and some of them experienced supernova explosions that left roughly 50 $M_\odot$ or larger black holes as remnants that later absorbed gas rapidly and grew [see e.g. Volonteri, Haardt, & Madau (2003)]. Another is that mass segregation in dense star clusters may have led to such rapid accretion of a star at the center that normal stellar evolution was delayed [Freitag, Gürkan, & Rasio (2006)]. A supernova explosion after an unusually high mass was reached could then result in an intermediate mass black hole, that later managed to get to the galactic center and grow. A third of the many suggested scenarios is that the very rapid accretion happened in a gas cloud of much higher mass and with very low metalicity, where stars hadn't yet formed. This would lead to the growth of an intermediate mass black hole within what is called a quasi-star. This is an object where the mass inflow rate is high enough to prevent normal stellar evolution for a quite long period. In this case, a black hole of perhaps 10^3 or 10^4 $M_\odot$ could be formed before the rest of the surrounding mass was blown away by radiation [Begelman, Rossi, & Armitage (2008)].

Various scenarios for when and how the intermediate mass black holes formed have been considered in what are called merger tree simulations of galaxy formation, including the mergers of pregalactic structures and the growth and mergers of the galactic center black holes in them, as a function of cosmic time (see e.g., [Sesana, Volonteri & Haardt (2009)]). The resulting rates of events that LISA could detect are expected to be roughly tens to about 100 black hole merger events per year, depending on the intermediate mass black hole formation scenario that is assumed. The mass ratios tend to be roughly 10:1 or larger, and most events are expected at redshifts in the range of about 2 to 10.

The second primary scientific objective for LISA is to observe the mergers of compact objects such as roughly 10 $M_\odot$ black holes, which are believed to be present in high numbers in galactic nuclei, with galactic center MBHs (see e.g., [Gair (2009)]). The rate of such events from which LISA could observe signals is again quite uncertain, but is estimated to be roughly tens to about 100 events per year. Such events would be observable out to redshifts of about $z = 1$, and are usually called Extreme Mass Ratio Inspirals, or EMRIs. Observing such events involving neutron stars and white dwarfs as well as black holes would provide new information on the conditions at the centers of galaxies containing MBHs, as well as on the scattering processes that lead to the compact objects being fed down to low angular momentum orbits around the black holes.

The third primary objective is closely related to the first two. It is to carry out detailed studies of whether the observed merger signals are in complete agreement with the predictions of general relativity (see e.g., [Schutz (2009a)]). For MBH binary mergers with comparable masses, advances in numerical relativity in the last few years indicate that accurate general relativistic templates for the signals will be available by the time they are needed, even for the merger and ringdown phases of the process. In some cases, the S/N will be high, and the dynamics of the process will be tested, since both masses will be moving very rapidly.

For EMRIs, the tests will be even more stringent. For 10 $M_{\odot}$ and 10^5 or 10^6 $M_{\odot}$ as the black hole masses, there will be about 100,000 cycles of the signal that can be observed in the last year before merger. And even small deviations from the predictions of general relativity can be observed, as long as they don't have the same time signatures as the main parameters describing the masses and their spins. Because of the way that the 10 $M_{\odot}$ black holes get scattered into orbits that can evolve mainly by gravitational wave energy loss, the initial orbits will be highly elliptical, and eccentricities of roughly 0.3 to 0.7 are expected right up to reaching the last stable orbit.

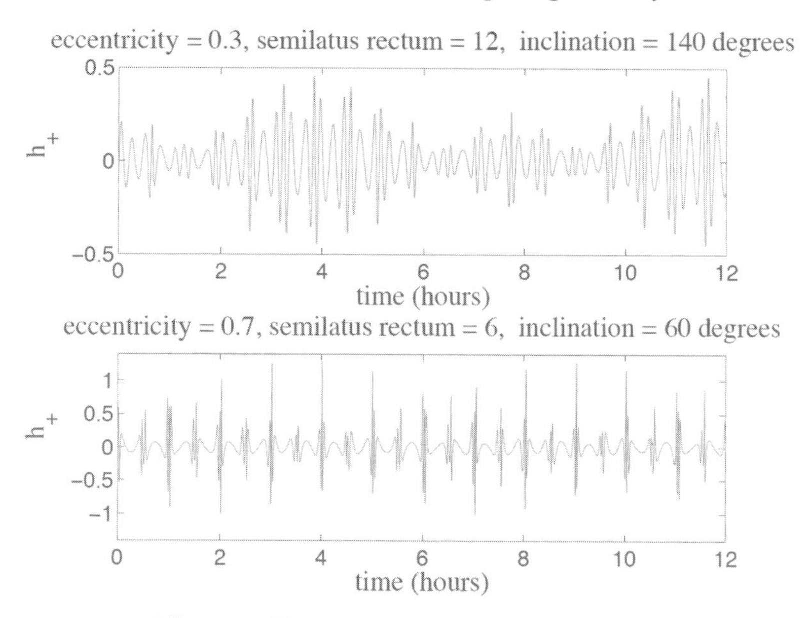

Figure 4. Examples of expected EMRI signals.

The speed at periapsis for 10 $M_{\odot}$ black hole EMRIs during the last year of inspiral is roughly 1/3 or 1/2 the speed of light, as mentioned earlier. The rate of precession of periapsis will be nearly as large as the radial motion frequency, and for rapidly spinning MBHs the Lense-Thirring precession can be comparable. Thus the observed signals will be complicated, involving perhaps 10^4 cycles or more each of periapsis precession and Lense-Thirring precession during the last year. If there are even small errors in the predictions of general relativity, such as some small interaction between the above two types of precession or with the radial motion, it seems likely that this would show up in the EMRI results, but would have been missed in weak field or slow motion tests of the theory.

To illustrate the complexity of the relativistic effects discussed, two examples of possible signals as a function of time are shown in Fig. 4. This figure is from a White Paper [Schutz *et al.* (2009)] that was submitted in connection with the ASTRO2010 Decadal Survey to describe the scientific objectives for testing general relativity that can be carried out in the near future by a mission such as LISA. The possible tests are described in more detail there.

4. LISA Pathfinder Mission

The spurious acceleration level that is being planned on for the LISA mission is roughly four orders of magnitude better than has been aimed for or achieved on previous missions. Thus ESA decided in 2002 to provide a technology demonstration mission called LISA Pathfinder [McNamara (2006); McNamara *et al.* (2008); Racca & McNamara (2009)] to validate the performance in space of the Gravitational Reference Sensors that were being developed to be flown on LISA [Armano *et al.* (2009)]. Initial design studies for the LISA GRSs were done at ONERA in Paris, but most of the later development of the GRSs for LISA Pathfinder and for LISA has been carried out at the University of Trento (see e.g., [Carbone *et al.* (2007)]).

Figure 5. Electrode housing from Gravitational Reference Sensor designed for the LISA Pathfinder and LISA missions.

The electrode housing from one of the GRS units for LISA Pathfinder is shown in Fig. 5. The housing is made from molybdenum, which has good thermal conductivity. The electrodes mounted on its inner surface are thin sapphire plates with gold coatings on them. Holes in the housing permit a clamping mechanism to hold the test mass firmly in place against stops during launch, and then release it gently at the center of the housing when the spacecraft is in its final orbit. The test mass, housing, and clamping mechanism are contained in a vacuum enclosure, with sufficient getters to maintain a low pressure for more than the nominal mission lifetime of somewhat under a year.

The basic LISA Pathfinder payload is shown in Fig. 6. It consists of two GRS units mounted on opposite ends of an optical bench made of low-expansion glass. Optical elements mounted on the optical bench form several interferometers that are used to monitor changes in the separations of the test masses in the GRS units with high precision. The GRS test mass spurious acceleration level is expected to be less than 3×10^{-14} $\mathrm{m/s^2/(Hz^{0.5})}$ at frequencies down to 1 mHz. A series of experimental investigations will be performed on LISA Pathfinder to validate the noise model that will permit extrapolation to the factor 10 better performance expected down to 0.1 mHz for LISA because of the lower environmental disturbances that will be present on the LISA spacecraft.

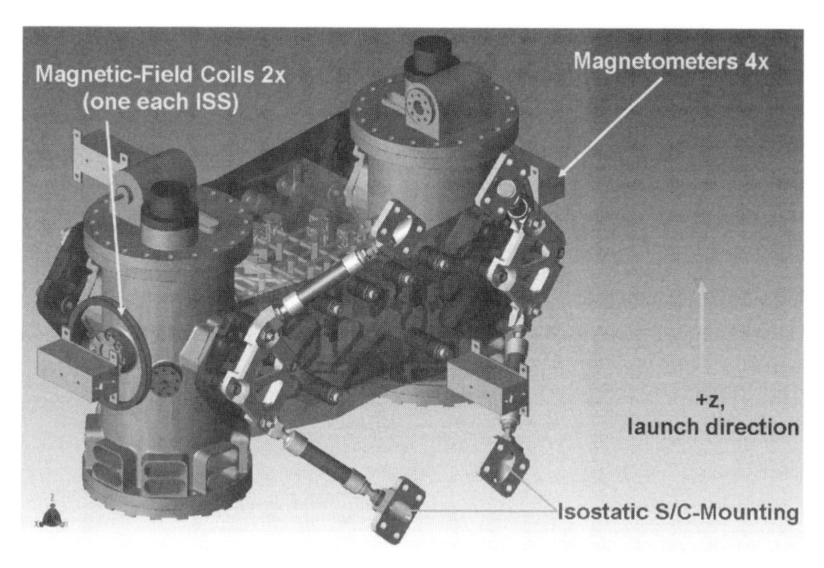

Figure 6. Basic science payload for the LISA Pathfinder mission.

In parallel with the development of the GRS for LISA Pathfinder, extensive laboratory tests have been done on a wide variety of possible spurious surface forces that could act on the test mass. These measurements have been carried out both at the University of Trento (see e.g., [Carbone *et al.* (2007)]) and at the University of Washington [Schlamminger *et al.* (2006); Pollack, Schlamminger, & Gundlach (2006)]. Torsion pendulums supporting replicas of the test mass were used, and perturbing surfaces could be brought up close to the test mass to enhance the interactions. By operating near the thermal noise level for the torsion pendulums, limits on spurious accelerations from surface forces were brought down close to the goals set for LISA Pathfinder. However, when LISA Pathfinder is launched in 2011, the combination of space and ground tests of the GRS performance will provide even stronger evidence that no significant sources of spurious accelerations have been overlooked.

LISA Pathfinder also will provide tests of two types of micronewton thrusters that have been developed for use in LISA and in other missions. The main force acting on the LISA spacecraft will be roughly 30 micronewtons, due almost completely to solar radiation pressure. However, torques also will be present, so that rotations of the spacecraft as well as displacements with respect to one of the test masses inside will have to be controlled. Thus at least six small thrusters are needed, as well as an equal or larger number to provide sufficient redundancy. A similar number of micronewton thrusters will be flown on LISA Pathfinder. They will include both cesium ion thrusters developed under ESA support and thrusters that emit small but highly charged liquid droplets developed under NASA support.

5. Conclusions

The LISA mission is expected to provide unique new information on the development of structure in the universe, the origin and growth of intermediate mass and larger mass black holes, the co-evolution of galaxies and of the massive black holes at their centers, and the population of stellar mass compact objects around the galactic center black holes. In addition, it will provide nearly ideal tests of the predictions of general relativity from the gravitational wave signals observed during the mergers of different mass combinations of massive black holes. The LISA Pathfinder mission is scheduled for launch in 2011, to validate the performance of the new technology to be used on the LISA mission.

References

Armano, M. *et al.*, 2009, *Class & Quantum Grav.* 26(9), 094001

Begelman, M. C., Rossi, E. M., & Armitage, P. J., 2008, Mon. Not. R. Astron. Soc. 387, 1649

Carbone, L., *et al.*, 2007, *Phys. Rev. D* 75(4), 042001

Di Matteo, T., *et al.*, 2008, *ApJ* 676, 33

Freitag, M., Gürkan, M. A., & Rasio, F. A., 2006, Mon. Not. R. Astron. Soc. 368, 141

Gair, J. R., 2009, *Class & Quantum Grav.* 26(9), 094034

Hughes, S. A., 2006, in *Laser Interferometer Space Antenna*, AIP Conf. Proc. 873 (Melville, N. Y., Eds. S. M. Merkowitz and J. C. Livas), 13

McNamara, P. W., 2006, in *Laser Interferometer Space Antenna*, loc cit, 49

McNamara, P., Vitale, S., & Danzmann, K., 2008, *Class. & Quantum Grav.* 25(11), 114034

Pollack, S. E., Schlamminger, S., & Gundlach, J. H., 2006, in *Laser Interferometer Space Antenna*, loc cit, 158

Racca, G. & McNamara, P., 2009, *Space Sci. Rev.*, (submitted)

Sallusti, M., *et al.*, 2009, *Class. & Quantum Grav.* 26(9), 094015

Schlamminger, S., *et al.*, 2006, in *Laser Interferometer Space Antenna*, loc cit, 151

Schutz, B. F., 2009a, *Class. & Quantum Grav.* 26(9), 094020

Schutz, B. F., 2009b, *this proceedings*, 234

Schutz, B. F., *et al.*, 2009, "Will Einstein Have the Last Word on Gravity?" White Paper submitted to the ASTRO2010 Decadal Survey on Astronomy and Astrophysics

Sesana, A., Volonteri, M., & Haardt, F., 2009, *Class. & Quantum Grav.* 26(9), 094033

Shaddock, D., *et al.*, 2006, in *Laser Interferometer Space Antenna*, loc cit, 654

Stebbins, R. T., 2006, in *Laser Interferometer Space Antenna*, loc cit, 3

Volonteri, M., 2006, in *Laser Interferometer Space Antenna*, loc cit, 61

Volonteri, M., Haardt, F., & Madau, P., 2003, Astrophys. J. 582, 599

Wand, V., *et al.*, 2006, in *Laser Interferometer Space Antenna*, loc cit, 689

Relativity in Fundamental Astronomy
Proceedings IAU Symposium No. 261, 2009
S. A. Klioner, P. K. Seidelman & M. H. Soffel, eds.

doi:10.1017/S1743921309990470

Strong gravitational lensing: relativity in action

Joachim Wambsganss[1,2]

[1]Astronomisches Rechen-Institut, Zentrum für Astronomie der Universität Heidelberg, Mönchhofstr. 12-14, 69120 Heidelberg, Germany
email: jkw@uni-hd.de

[2]Bohdan Paczynski Visitor, Dept. of Astrophysical Sciences, Princeton University, Princeton, NJ 08540, USA

Abstract. Deflection of light by gravity was predicted by Einstein's Theory of General Relativity and observationally confirmed in 1919. In the following decades, various aspects of the gravitational lens effect were explored theoretically, among them measuring the Hubble constant from multiple images of a background source, making use of the magnifying effect as a gravitational telescope, or the possibility of a "relativistic eclipse" as a perfect test of GR. Only in 1979, gravitational lensing became an observational science when the first doubly imaged quasar was discovered. Today lensing is a booming part of astrophysics and cosmology. A whole suite of strong lensing phenomena have been investigated: multiple quasars, giant luminous arcs, Einstein rings, quasar microlensing, and galactic microlensing. The most recent lensing application is the detection of extrasolar planets. Lensing has contributed significant new results in areas as different as the cosmological distance scale, mass determination of galaxy clusters, physics of quasars, searches for dark matter in galaxy halos, structure of the Milky Way, stellar atmospheres and exoplanets. A guided tour through some of these applications will illustrate how gravitational lensing has established itself as a very useful universal astrophysical tool.

Keywords. light deflection, gravitational lensing, cosmology, extrasolar planets

---

1. The History of Strong Lensing

Gravitational lensing is considered a relatively new field in astrophysics. However, the history of light deflection is more than 200 years old (see in more detail in Wambsganss [1998] or in Schneider, Kochanek & Wambsganss [2006]). As early as 1784, Michell considered the deflection of light by the gravity of other bodies. In 1801, Soldner published a paper on light deflection, in which he determined – based on Newtonian mechanics – the deflection of a light ray just passing the solar limb to

$$\alpha_{\odot,\mathrm{Soldner}} = \frac{2GM_\odot}{c^2 R_\odot} = 0.87 \text{ arcsec}$$

(with G - gravitational constant, c - velocity of light, $M_\odot$ - mass of the sun, $R_\odot$ - radius of the sun). More than 100 years later, Einstein worked on the same problem and derived the same value (Einstein 1911). In fact, an expedition headed by Erwin Freundlich from Potsdam was set up to test this prediction during the solar eclipse in September 1914 on the Crimean Peninsula. However, another eclipse had started to darken Europe, World War I had begun and the scientists plus their equipment were captured by Russian soldiers. No harm was done to either scientists or equipment, they were later released and sent home, but the solar eclipse had passed and the measurement could not be made. Only after the General Theory of Relativity was finished, Einstein published the value of

$$\alpha_{\odot,\mathrm{Einstein1915}} = \frac{4GM_\odot}{c^2 R_\odot} = 1.74 \text{ arcsec},$$

for the light deflection at the solar limb, which was measured and proven to be correct in the famous solar eclipse expeditions led by Eddington in 1919 (Dyson *et al.* 1920).

In the 1920s/1930s, there were a few papers dealing with lensing, e.g., Chwolson investigated the situation of double imaging. In particular he figured out that for perfect alignment between lens and source the result would be a ring-like image (Chwolson 1924). Einstein looked again into this issue and derived the magnifications for the double images of a background star lensed by an intervening foreground star, but he was very sceptical about the possibility of observing this gravitational lensing effect (Einstein 1936). Henry Norris Russell had obtained Einstein's 1936 paper as a preprint. It impressed him quite a bit, so he immediately wrote a column for *Scientific American* entitled "A Relativistic Eclipse" (Russell 1937), in which he emphasized that the "Einstein effect" provides a perfect (but unavailable) test for General Relativity, and that the effects would be "conspicuous to the immediate gaze". He described the consequences of an eclipse in the Sirius A and B system, imagining an observer sitting on a planet around the white dwarf Sirius B, at a distance such that the angular size of Sirius B just matches the angular size of Sirius A. He illustrates the image shapes and describes what we call "arcs" today as "bright crescents" and images have "developed pointed horns". Fritz Zwicky immediately applied Einstein's idea on galaxies, he was convinced that "nebulae" should and would act as gravitational lenses, for him this appeared to be an unavoidable consequence (Zwicky 1937a,b) of the light deflection theory. Zwicky recognized the potential of applying gravitational lensing to extragalactic nebulae: he proposed that this would offer an additional test for general relativity, that the gravitational lens effect could be used to study fainter objects (effectively increasing the aperture of our telescopes), he emphasized that lensing would provide a powerful method to measure the masses of the "nebulae" (i.e. dark matter!), and he proposed that splittings up to 30 arcsecond should be expected.

In the 1960s there was another wave of theoretical investigations of the lensing effect. In particular, Refsdal showed that one can determine the Hubble constant from the time delay between the images of a multiply lensed quasar (Refsdal 1964). And finally in 1979, Dennis Walsh and colleagues discovered the first doubly imaged quasar Q0957+561 (Walsh *et al.* 1979). Although the deflection of light at the solar limb – hailed as the first experiment to confirm a prediction of Einstein's theory of General Relativity – happened already in 1919, it took more than half a century to establish this phenomenon observationally in some other environment.

2. The Basics of Strong Lensing

The path, the size and the cross section of a light bundle propagating through spacetime, in principle, are affected by all the matter between the light source and the observer. For most practical purposes one can assume that the lensing action is dominated by a single matter inhomogeneity at some location between source and observer. This is usually called the "thin lens approximation": all the action of deflection is thought to take place at a single distance. Here the basics of lensing will be briefly derived and explained in the thin lens approxmation: lens equation, Einstein radius, image positions and magnifications, time delay. More detailed reviews/introductions on lensing can be found in, e.g., Schneider *et al.* (2006).

2.1. *Lens Equation*

The basic setup for such a simplified gravitational lens scenario involving a point source and a point lens is displayed in Figure 1. The three ingredients in such a lensing situation are the source S, the lens L, and the observer O. Light rays emitted from the source are deflected by the lens. For a point-like lens, there will always be (at least) two images S_1 and S_2 of the source. With external shear – due to the tidal field of objects outside but near the light bundles – there can be more images. The observer sees the images in directions corresponding to the tangents to the real incoming light paths.

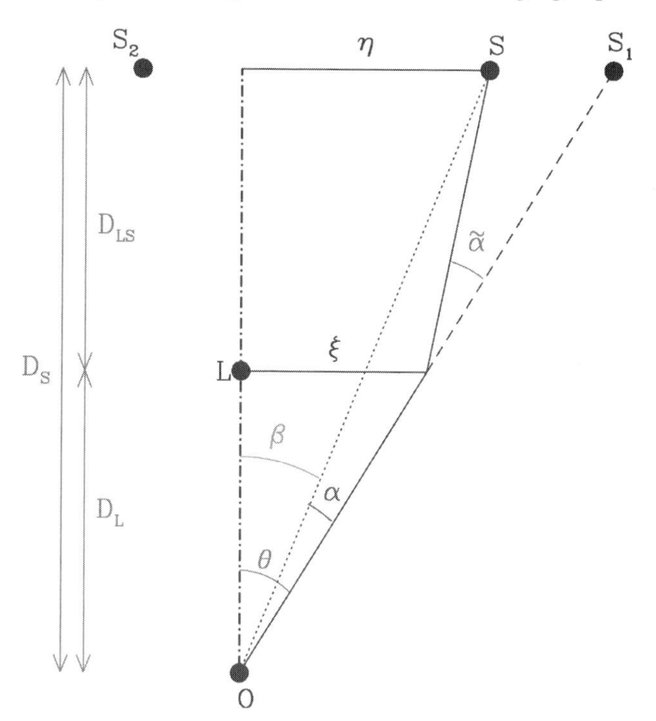

Figure 1. The relations between the various angles and distances involved in the lensing setup can be derived for $\tilde{\alpha} \ll 1$ and formulated in eq.(2.3), the lens equation. The symbols 'O', 'L', and 'S' mean 'observer', 'lens', and 'source', respectively. 'S_1' and 'S_2' are the two apparent positions of the doubly imaged source. The angular diameter distances D_L, D_S, and D_{LS} are between observer-lens, observer-source, and source-lens. All angles involved are small compared to one.

In Figure 1, the corresponding angles and angular diameter distances D_L, D_S, D_{LS} are indicated. In the thin-lens approximation the hyperbolic paths are approximated by their asymptotes. In the circular-symmetric case the deflection angle is given as

$$\tilde{\alpha}(\xi) = \frac{4GM(\xi)}{c^2}\frac{1}{\xi}, \tag{2.1}$$

where $M(\xi)$ is the mass of the lens inside a radius ξ. In this depiction the origin is chosen at the observer. From the diagram it can be seen that the following relation holds:

$$\theta D_S = \beta D_S + \tilde{\alpha} D_{LS} \tag{2.2}$$

(for θ, β, $\tilde{\alpha} \ll 1$; this condition is fulfilled in practically all astrophysically relevant situations). With the definition of the reduced deflection angle as $\alpha(\theta) = (D_{LS}/D_S)\tilde{\alpha}(\theta)$, this *lens equation* can be expressed as:

$$\beta = \theta - \alpha(\theta). \tag{2.3}$$

For real situations, all the astrophysics is "hidden" in the deflection angle $\alpha(\theta)$, which can become a quite complicated function for many lensing objects or an extended mass distributions.

2.2. *Einstein Radius*

For a point lens of mass M the deflection angle is given by equation (2.1). Plugging it into equation (2.3) and using the relation $\xi = D_L\theta$ (cf. Figure 1) one obtains:

$$\beta(\theta) = \theta - \frac{D_{LS}}{D_L D_S}\frac{4GM}{c^2\theta}. \tag{2.4}$$

For the special case in which the source lies exactly behind the lens ($\beta = 0$), due to the symmetry a ring-like image occurs whose angular radius is called *Einstein radius* θ_E:

$$\theta_E = \sqrt{\frac{4GM}{c^2}\frac{D_{LS}}{D_L D_S}}. \tag{2.5}$$

The Einstein radius defines the angular scale for a lens situation. For a massive galaxy with a mass of $M = 10^{12} M_\odot$ at a redshift of $z_L = 0.5$ and a source at redshift $z_S = 2.0$ (here $H = 50$km sec^{-1} Mpc^{-1} is used as the value of the Hubble constant and an Einstein-deSitter universe) the Einstein radius is

$$\theta_E \approx 1.8\sqrt{\frac{M}{10^{12} M_\odot}} \text{ arcsec} \tag{2.6}$$

(note that for cosmological distances in general $D_{LS} \neq D_S - D_L$!). For a galactic microlensing scenario in which stars in the disk of the Milky Way act as lenses for bulge stars close to the center of the Milky Way, the scale defined by the Einstein radius is

$$\theta_E \approx 0.5\sqrt{\frac{M}{M_\odot}} \text{ milliarcsec.} \tag{2.7}$$

It is obvious from these values that lensing by galaxies can be resolved by normal optical telescopes, whereas the angular scale for lensing by stars is much smaller than the resolution of even the best optical telescopes.

2.3. *Image Positions and Magnifications*

The lens equation (2.3) can be re-formulated in the case of a single point lens:

$$\beta = \theta - \frac{\theta_E^2}{\theta}. \tag{2.8}$$

Solving this for the image positions θ, one finds that an isolated point source always produces two images of a background source. The positions of the images are given by the two solutions:

$$\theta_{1,2} = \frac{1}{2}\left(\beta \pm \sqrt{\beta^2 + 4\theta_E^2}\right). \tag{2.9}$$

The magnification of an image is defined by the ratio between the solid angles of the image and the source, since the surface brightness is conserved. Hence the magnification μ is given as

$$\mu = \frac{\theta}{\beta}\frac{d\theta}{d\beta}. \tag{2.10}$$

In the symmetric case above the image magnification can be written as (by using the lens equation):

$$\mu_{1,2} = \left(1 - \left[\frac{\theta_E}{\theta_{1,2}}\right]^4\right)^{-1} = \frac{u^2+2}{2u\sqrt{u^2+4}} \pm \frac{1}{2}. \tag{2.11}$$

Here u is defined as the "impact parameter", the angular separation between lens and source in units of the Einstein radius: $u = \beta/\theta_E$. The magnification of one image (the one inside the Einstein radius) is formally negative. This means it has negative parity: it is mirror-inverted. For $\beta \to 0$ the magnification diverges: in the limit of geometrical optics the Einstein ring of a point source has infinite magnification†! The sum of the absolute values of the two image magnifications is the total magnification μ:

$$\mu = |\mu_1| + |\mu_2| = \frac{u^2+2}{u\sqrt{u^2+4}}. \tag{2.12}$$

Note that this value is (always) larger than one‡!

2.4. *Time delay and Fermat's Theorem*

The deflection angle is the gradient of an effective lensing potential ψ (see Schneider 1985). Hence the lens equation can be rewritten as

$$(\boldsymbol{\theta} - \boldsymbol{\beta}) - \boldsymbol{\nabla}_\theta \psi = 0 \tag{2.13}$$

or

$$\boldsymbol{\nabla}_\theta \left(\frac{1}{2}(\boldsymbol{\theta} - \boldsymbol{\beta})^2 - \psi\right) = 0. \tag{2.14}$$

The term in brackets appears as well in the physical time delay function for gravitationally lensed images:

$$\tau(\boldsymbol{\theta}, \boldsymbol{\beta}) = \tau_{\text{geom}} + \tau_{\text{grav}} = \frac{1+z_L}{c}\frac{D_L D_S}{D_{LS}}\left(\frac{1}{2}(\boldsymbol{\theta} - \boldsymbol{\beta})^2 - \psi(\theta)\right). \tag{2.15}$$

This time delay surface is a function of the image geometry ($\boldsymbol{\theta}$, $\boldsymbol{\beta}$), the gravitational potential ψ, and the distances D_L, D_S, and D_{LS}. The first part – the geometrical time delay τ_{geom} – reflects the extra path length compared to the direct line between observer and source. The second part – the gravitational time delay τ_{grav} – is the retardation due to the gravitational potential of the lensing mass (known and confirmed as Shapiro delay in the solar system). From equations (2.14) and (2.15) it follows that the gravitationally lensed images appear at locations that correspond to extrema in the light travel time, which reflects Fermat's principle in gravitational-lensing optics.

The (angular-diameter) distances that appear in equation (2.15) depend on the value of the Hubble constant (Weinberg 1972); therefore, it is possible to determine the latter

† Due to the fact that physical objects have a finite size, and also because at some limit wave optics has to be applied, in reality the magnification stays finite.

‡ This does not violate energy conservation, since this is the magnification relative to an "empty" universe and not relative to a "smoothed out" universe.

by measuring the time delay between different images and using a good model for the effective gravitational potential ψ of the lens.

2.5. *The Effects of strong lensing*

The effects of strong lensing can be summarized in four aspects:

- **change of position:** This effect was used as the first confirmation of General Relativity in 1919 with the offset of stellar positions close to the solar limb (Dyson *et al.* 1920). However, in general this is normally not observable, because one needs a "before-and-after"-scenario in order to compare two angular positions. There is some hope that in the future astrometric microlensing can help measure this effect (Treyer & Wambsganss 2004).
- **distortion:** Extended sources that are affected by lensing are distorted; to first order, circular sources are deformed into arclets, arcs or even Einstein rings.
- **(de)magnification:** Typically only the magnification effect of lensing is considered. In stellar microlensing a few events have been measured with magnifications as large as a factor 1000. However, since lensing does not create any photons, in order to compensate for this, some sources will appear fainter than without lensing. The distribution, however, is very skewed: almost all images appear slightly demagnified, whereas a small fraction of objects are highly magnified. This means, that in fact there are no standard candles in the universe (Wambsganss *et al.* 1997).
- **multiple images:** Multiple images of course are the most dramatic effect of lensing. So far, we know of more than 100 multiple quasar systems, and even more galaxy clusters that produce giant luminous arcs.

In Figure 2 all these effects are visualized assuming a particular source profile with some internal structure.

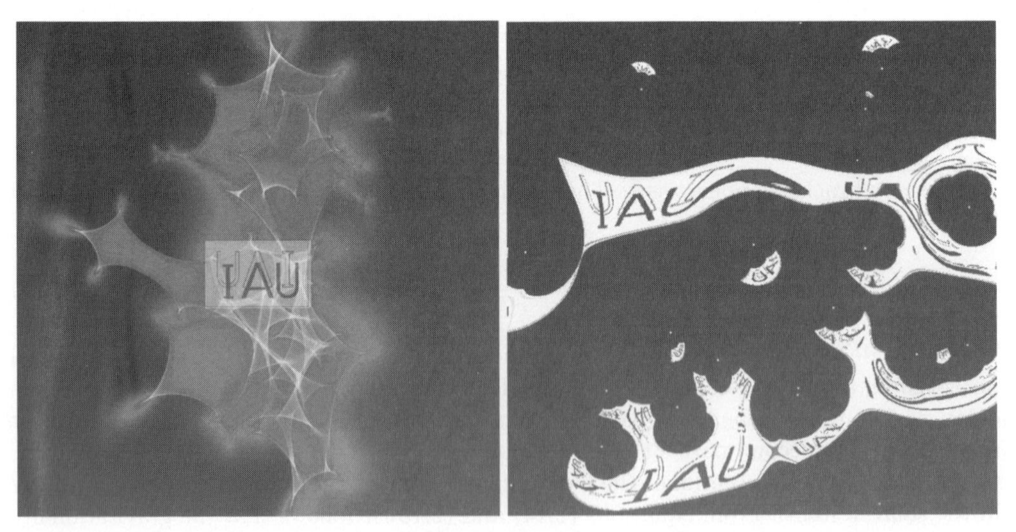

Figure 2. Magnification pattern produced by the stars in a lensing galaxy, with a particular source profile superposed (left); resulting configuration with distorted, offset, magnified multiple images (right).

2.6. *How can we observe strong lensing phenomena?*

Depending on the mass scale of the lens, there are two very different ways to observe strong lensing:

- "Statically" (images): For lensing objects with masses $m \geqslant 10^8 M_\odot$, the angular Einstein radius gets comparable to the resolution of the telescope. This means the lensing effect can be detected by resolving the multiple images and/or the shapes of the distorted images. This regime is occasionally called *macro*-lensing.
- "Dynamically" (brightness, positions): For lensing objects with masses roughly comparable to a solar mass, the angular Einstein radius is much smaller than the resolution of the telescope. However, due to the relative change between the (angular) positions of lens and source, the magnification (and also the center of light) can change with time. The time scale of these changes (Einstein radius divided by transverse velocity) turns out to be of order years/months/weeks and, hence, the phenomenon is detectable by measuring the apparent brightness frequently over time (in the future, we hope to be able also to measure the positional change of the center-of-light). This regime is called *micro*-lensing.

3. The Usefulness of Strong Lensing

In this section, three examples are presented in which strong lensing is being applied: Galaxy clusters as lenses providing a record time delay and arc statistics as a cosmological tool, galaxies as lenses to be used as microscopes for quasar accretion disks, and finally stars and planets as lenses in the Milky Way.

3.1. *Galaxy Clusters as Lenses: Time Delays and Arc Statistics*

Gravitational lensing directly measures mass density fluctuations along the lines of sight to very distant objects. No assumptions need to be made concerning bias, the ratio of fluctuations in galaxy density to mass density. Hence, strong lensing is a very useful tool to study the statistics of giant luminous arcs. This was done, e.g., regarding the frequency of giant luminous arcs predicted by various cosmological models. In Bartelmann *et al.* (1998) it was stated that a Lambda-dominated flat cosmological model ($\Omega_{\text{matter}} = 0.3$, $\Omega_\Lambda = 0.7$ known as "concordance cosmology") would underpredict luminous arcs by about an order of magnitude. Later, Wambsganss *et al.* (2004) showed that this result was based on the assumption that the sources are all at redshift unity. The probability for arcs, however, is a steep function of source redshift. If one allows for sources at redshift two or three (some of the observed arcs are even at higher redshifts), then the discrepancy disappears. These results were based on cosmological parameters derived from the WMAP-1 data. More recently, WMAP-3 results published lower values of the normalization, σ_8, which reduces the predicted number of arcs by a factor of 8 or 10 (Wambsganss *et al.*, in preparation). The most recent values for the cosmological parameters based on WMAP-5 results (in particular the slight increase in the normalization) lead to an increase in the frequency of arcs again. Since all three values of σ_8 published by the WMAP mission agree within each other at a level of about 2.5 σ, there is no significant discrepancy between the predicted and measured arc frequencies. The models mainly show the strong dependence on the normalization of the cosmology. If better statistics can be obtained from a larger number of observed arcs, this could be used in turn to predict an interval of σ_8 which is consistent with these observations.

A second example for the usefulness of galaxy clusters as lenses was obtained in the strongly lensing galaxy cluster SDSS J1004+4112, in which a quintuple quasar system was discovered. By monitoring this system over many years, finally two of the three relative time delays could be measured, and a lower limit on the third one could be obtained (see Fohlmeister *et al.* 2007, 2008): the time delay between images A and B was measured to $\Delta t_{BA} = 40.6 \pm 1.8$ days, the wide image pair C and A has a relative time delay of $\Delta t_{CA} = 821.6 \pm 2.1$ days, which is the longest gravitational lens time delay on record so far. For the image pair A and D only a lower limit could be obtained: $\Delta t_{AD} > 1250$ days (cf. Figure 3 and for more details the papers by Fohlmeister *et al.* 2007, 2008).

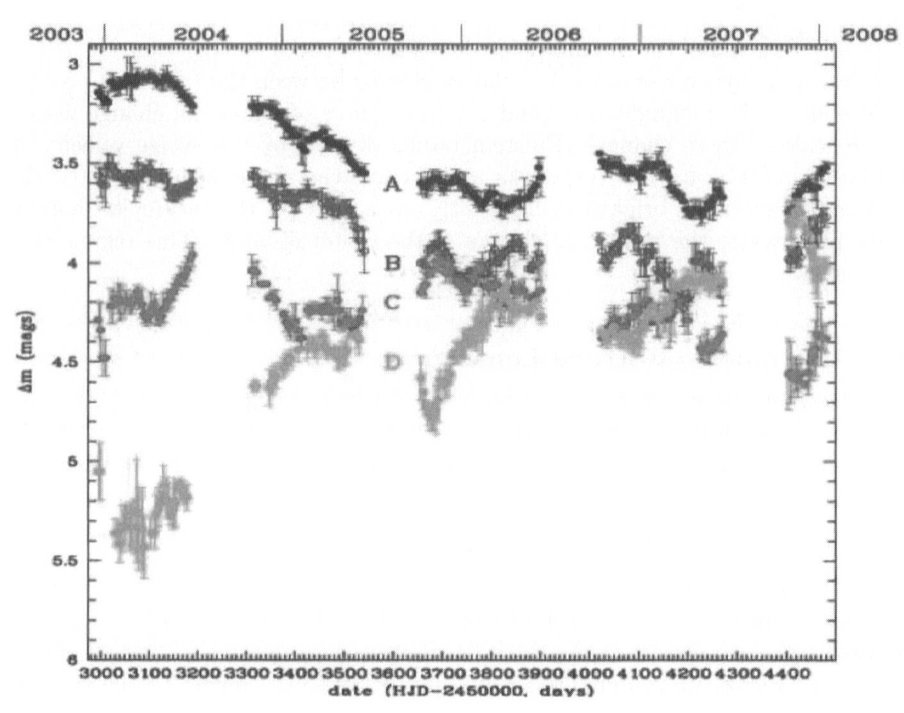

Figure 3. Lightcurves of the four brightest images of the multiple quasar system SDSS J1004+4112 (after Fohlmeister *et al.* 2008)

3.2. *Galaxies as lenses: Dark Matter and Quasar Accretions Disks*

A number of quasar systems have been explored in detail for microlensing, the effects of stellar mass objects along the line of sight to the quasar images. The most dramatic effects have been seen in the quadruple system Q2237+0305, where four quasar images are centered around the core of the lensing galaxy. Since its discovery, this system has shown uncorrelated fluctuations between the images, which were interpreted as microlensing, see Irwin *et al.* (1989) and Wambsganss *et al.* (1990). The problem was, however, the poor coverage in time. Only with the dedicated telescope and the dedicated scientists of the OGLE team, a good time coverage could be reached, with more than 100 data points per year. This observing strategy resulted in a dramatic increase in data quality, see Wozniak *et al.* (2000). The individual images fluctuate by as much as a factor of two in a few months, and these fluctuations are very well resolved. In this multiple quasar system there is no need to invoke potential dark matter objects as lenses: the four images are seen through the central part of the lensing galaxy, which is full of ordinary main sequence stars. The interpretation of the data is consistent with low-mass stars in the core of the galaxy being the lenses, and the size of the quasar continuum emission region to be of order 10^{15} cm or smaller (see, e.g., Wambsganss *et al.* (1990), Wyithe *et al.* (2000))

Recently, this quasar could be monitored spectro-photometrically over an extended period of time. Eigenbrod *et al.* (2008) analyzed the data with particular emphasis on

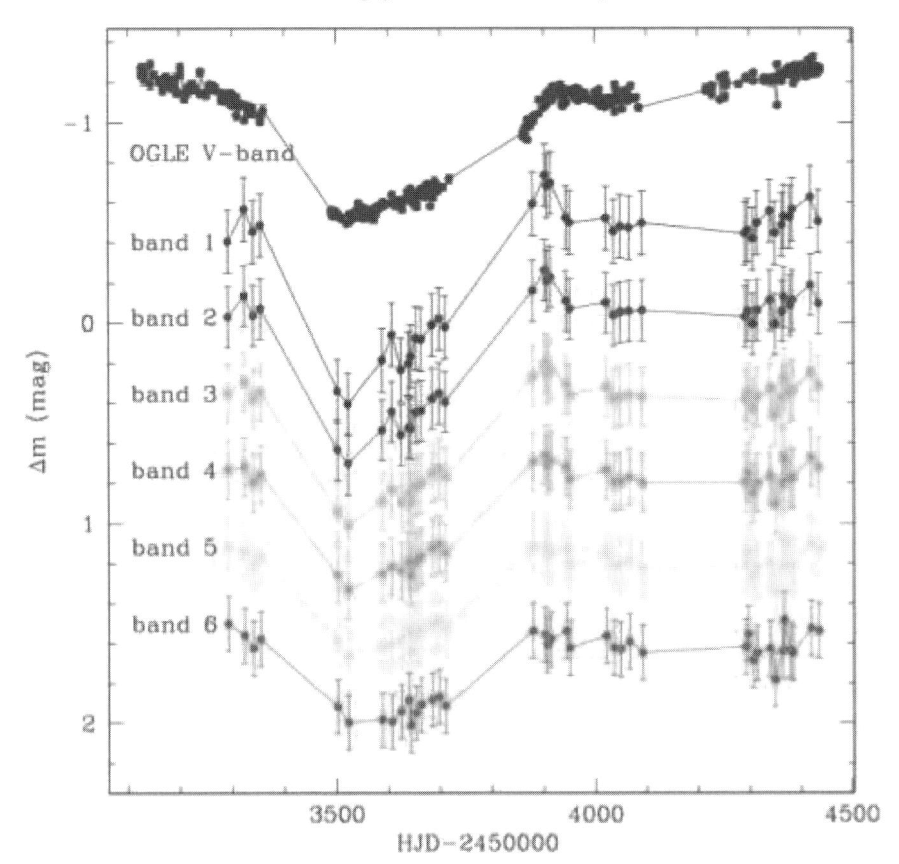

Figure 4. Microlensing lightcurves of the quadruple quasar Q2237+0305 in six bands, clearly indicating the dependence of the microlensing amplitude on the wavelength (from Eigenbrod *et al.* 2008).

the microlensing amplitudes as a function of wavelength/filter. As can be seen in Figure 4, the amplitudes increase with decreasing wavelength: in the blue wavelength range the microlensing effect is stronger than in the red part. This was originally predicted by Wambsganss & Paczynski (1991). Here this effect could be analyzed quantitively for the first time. The results show that the accretion disk in the quasar Q2237+0305 is consistent with a standard Shakura-Sunyaev accretion disk, details can be found in Eigenbrod *et al.* (2008).

3.3. *Stars as Lenses: In Search of extrasolar Planets*

In 1991, it was suggested that a fair fraction of stellar microlensing events towards the Galactic bulge should display signatures of binarity, and that even planetary companions should be detectable (Mao & Paczynski 1991). Starting in 1993, a number of teams (MACHO, EROS, OGLE, MOA) monitored of order 10^7 stars in the bulge in order to detect microlensing effects of intermediate stars or dark compact objects (Alcock *et al.* 2000; Laserre *et al.* 2000; Udalski *et al.* 1997). By now, more than 3000 microlensing events towards the galactic bulge have been found, currently over 800 events are detected per season. About 10% of them show the signature of binary lenses. This data set allows, among other things, to study the mass distribution of the Galactic disk with unprecedented accuracy. But one of the main goals of these monitoring experiments is still the detection of planets around the lensing stars.

In addition to the groups mentioned above, there are two other teams (PLANET, MicroFUN) who specialized in following up of current stellar lensing events with good photometric accuracy and very high temporal coverage, in order to find possible small deviations from the smooth single-lens lightcurve, which would be the signature of a planet. The signatures of planets are of short duration (of order hours) and typically have small amplitudes (a few percent), as was shown, e.g., in Wambsganss (1997). But the main aspect are: such planetary deviations at stellar microlensing lightcurves are rare: even if all stars had planets, only a small fraction of the microlensing lightcurves would show their signatures, due to the geometric path of the background star with respect to the planetary caustic.

Finally, in April 2004 the first microlensing planet was announced: The MOA/OGLE/MicroFUN teams announced their detection of a microlensing event which can be explained only with a very low mass companion to the primary star: OGLE 2003-BLG-235 or MOA 2003-BLG-53. The result is published in Bond *et al.* (2004). In the original words of the authors:

> "A short-duration ($\sim$7 days) low-amplitude deviation in the light curve due to a single-lens profile was observed in both the MOA and OGLE survey observations. We find that the observed features of the light curve can only be reproduced using a binary microlensing model with an extreme (planetary) mass ratio of 0.0039^{+11}_{-07} for the lensing system. If the lens system comprises a main-sequence primary, we infer that the secondary is a planet of about 1.5 Jupiter masses with an orbital radius of $\sim$3 AU."

By now, seven microlensing planets have been published (Bond *et al.* 2004, Udalski *et al.* 2005, Beaulieu *et al.* 2006, Gould *et al.* 2006, Gaudi *et al.* 2008, Bennett *et al.* 2008, Dong *et al.* 2009). These first unambiguous microlensing planet detections prove: Microlensing as a planet search technique has stepped out of its infancy. It is a viable method which is complementary to other techniques. Microlensing remains the most promising method for the detection of low-mass planets with ground-based techniques. It has been known all along that microlensing is sensitive to the low planet mass regime, that even Earth-mass objects are within the reach. However, even for microlensing the sensitivity is stronger for higher mass planets. So the fact that about half of the planets detected with microlensing are in the low planet mass regime, Neptune-mass to few Earth-mass range, allows to draw a robust conclusion: low mass planets must be abundant (more details see in references above).

4. The Future of Strong Lensing

In the 30 years of its existence, strong gravitational lensing has changed from an exotic subfield in astronomy into a very useful astrophysical tool which is applied in a large range of mass scales (20 orders of magnitude: from galaxy clusters to Earth-like planets), distance scales (few Gigaparsec to few kiloparsec, or even microparsec, if we include the Sun ...) and angular scales (arcminutes to microarcseconds). In the coming years, strong lensing will definitely help measure dark matter, resolve quasar and stellar luminosity profiles, and find many exoplanets, possibly Earth-like planets, and potentially even moons around extrasolar planets. So my conclusion is that the future of strong lensing is bright and promising.

References

Alcock, C., *et al.* (The MACHO collaboration), ApJ 542, 281 (2000)

Bartelmann, M. *et al.* A&A 330, 1 (1998)

Beaulieu, J. P., Bennett, D. P., Fouque, P., *et al.*, Nature, 439, 437 (2006)

Bennett, D. P. *et al.*, ApJ 684, 663 (2008)

Bond, I. A., Udalski, A., Jaroszynski, M., Rattenbury, N. J., Paczynski, B., *et al.*, ApJ 606, L155 (2004)

Chwolson, O. Astron. Nachr. 221, 329 (1924)

Dong, S. *et al.*, ApJ 698, 1826 (2009)

Dyson, F., Eddington, A., Davidson, C., Mem. Roy. Astr. Soc. 62, 291 (1920)

Eigenbrod, A. *et al.* Astron. Astrophys. 490, 933 (2008)

Einstein, A., Annalen der Physik 35, 898 (1911)

Einstein, A., Science 84, 506 (1936)

Fohlmeister, J., *et al.* , ApJ, 662, 62 (2007)

Fohlmeister, J., Kochanek, C. S., Falco, E. E., Morgan, C. W., & Wambsganss, J., ApJ, 676, 761 (2008)

Gaudi, B. S. *et al.*, Science , 319, 927 (2008)

Gould, A., Udalski A., An J., *et al.*, ApJ, 644, L 37 (2006)

Irwin, M. J., Webster, R. L., Hewett, P. C., Corrigan, R. T., & Jedrzejewski, R. I., ApJ 98, 1989 (1989)

Lasserre, T., *et al.* (The EROS collaboration), A&A 355, L39 (2000)

Mao, S., Paczyński, B. ApJ 374, L37 (1991)

Refsdal, S. MNRAS 128, 307 (1964)

Russel, H. N. Scientific American (February 1937)

Schneider, P., A&A 143, 413 (1985)

Schneider, P., Kochanek, C. S., & Wambsganss, J. "Gravitational Lensing: Strong, Weak, Micro" (Saas-Fee Advanced Course 33, Editors G. Meylan, P. Jetzer, P. North (Springer-Verlag, Berlin, 2006)

Treyer, M. & Wambsganss, J. A&A, 416, 19 (2004)

Udalski, A., Kubiak, M., & Szymanski, M. Acta Astron. 47, 319 (1997)

Udalski, A., Jaroszy, Å., ski, M., Paczy, Å., ski, B., *et al.*, ApJ, 628, L109 (2005)

Walsh, D., Carswell, R. F., & Weymann, R. J., Nature 279, 381 (1979)

Wambsganss, J., Paczyński, B., & Schneider, P. ApJ 358, L33 (1990)

Wambsganss, J. & Paczyński, B. AJ 102, 864 (1991)

Wambsganss, J., Bode, P. & Ostriker, J. P. ApJ 606, L93 (2004)

Wambsganss, J., MNRAS 284, 172 (1997)

Wambsganss, J., *Living Reviews in Relativity* 1998-12, `http://relativity.livingreviews.org/Articles/lrr-1998-12` (1998)

Weinberg, S.: *Gravitation and Cosmology* (Wiley, New York, 1972)

Wozniak, P. R., Udalski, A., Szymanski, M., *et al.* ApJ 540, L65 (2000)

Wyithe, J. S. B., Webster, R. L., & Turner, E. L. MNRAS 318, 762 (2000)

Zwicky, F., Phys. Rev. 51, 290 (1937a)

Zwicky, F., Phys. Rev. 51, 679 (1937b)

Relativity in Fundamental Astronomy
Proceedings IAU Symposium No. 261, 2009
S. A. Klioner, P. K. Seidelman & M. H. Soffel, eds.

doi:10.1017/S1743921309990482

Black holes in active galactic nuclei

M. J.Valtonen[1], S. Mikkola[1], D. Merritt[2], A. Gopakumar[3,8], H. J. Lehto[1], T. Hyvönen[1], H. Rampadarath[4,5], R. Saunders[6], M. Basta[7] and R. Hudec[7,9]

[1]Tuorla Observatory, Department of Physics and Astronomy, University of Turku, 21500 Piikkiö, Finland
[2]Centre for Computational Relativity and Gravitation, Rochester Institute of Technology, 78 Lomb Memorial Drive, Rochester, NY 14623, USA
[3]Tata Institute of Fundamental Research, Mumbai 400005, India
[4]Joint Institute for VLBI in Europe (JIVE), Postbus 2, 7990 AA Dwingeloo, The Netherlands
[5]Leiden Observatory, Leiden University, P.O. Box 9513, NL-2300 RA Leiden, The Netherlands
[6]Department of Physics, University of the West Indies, St. Augustine, Trinidad & Tobago
[7]Astronomical Institute, Academy of Sciences, Fricova 298, 25165 Ondrejov, Czech Republic
[8]Theoretisch-Physikalisches Institut, Friedrich-Schiller-Universität Jena, Max-Wien-Platz 1, 07743 Jena, Germany
[9]Czech Technical University in Prague, Faculty of Electrical Engineering, Technick 2, 166 27 Praha 6, Czech Republic

Abstract. Supermassive black holes are common in centers of galaxies. Among the active galaxies, quasars are the most extreme, and their black hole masses range as high as to $6 \cdot 10^{10} M_{\odot}$. Binary black holes are of special interest but so far OJ287 is the only confirmed case with known orbital elements. In OJ287, the binary nature is confirmed by periodic radiation pulses. The period is twelve years with two pulses per period. The last four pulses have been correctly predicted with the accuracy of few weeks, the latest in 2007 with the accuracy of one day. This accuracy is high enough that one may test the higher order terms in the Post Newtonian approximation to General Relativity. The precession rate per period is $39^{\circ}.1 \pm 0^{\circ}.1$, by far the largest rate in any known binary, and the $(1.83 \pm 0.01) \cdot 10^{10} M_{\odot}$ primary is among the dozen biggest black holes known. We will discuss the various Post Newtonian terms and their effect on the orbit solution. The over 100 year data base of optical variations in OJ287 puts limits on these terms and thus tests the ability of Einstein's General Relativity to describe, for the first time, dynamic binary black hole spacetime in the strong field regime. The quadrupole-moment contributions to the equations of motion allows us to constrain the 'no-hair' parameter to be 1.0 ± 0.3 which supports the black hole no-hair theorem within the achievable precision.

Keywords. gravitation — relativity — quasars: general — quasars: individual (OJ287) — black hole physics — BL Lacertae objects: individual (OJ287)

---

1. Introduction

Centers of galaxies typically host dark massive bodies which are thought to be black holes. The black hole concept plays a particularly strong role in theories of quasars and other active galactic nuclei. Some models use the assumed black hole properties of the dark bodies, such as the frame dragging around a spinning black hole (e.g. Camenzind 1989), but on the whole the activity is largely explained by accretion onto a central heavy body, as was first pointed out by Salpeter (1964) and Zeldovich (1964). Theories of accretion disks have been developed, in particular the α disk theory of Shakura and Sunyaev (1973) and its extension to magnetic disks by Sakimoto and Corotini (1981) which describe well the origin of activity around massive central bodies, whatever the exact nature of these bodies may be.

In order to prove that the central body is actually a black hole we have to probe the gravitational field around it. One of the most important characteristics of a black hole is that it must satisfy the so called no-hair theorem or theorems (Israel 1967, 1968, Carter 1970, Hawking 1971, 1972; see Misner, Thorne and Wheeler 1973). A practical test was suggested by Thorne (1980) and Thorne, Price and Macdonald (1986). In this test the quadrupole moment Q of the spinning body is measured. If the spin of the body is S and its mass is M, we determine the value of q in

$$Q = -q\frac{S^2}{Mc^2}. \tag{1.1}$$

For black holes $q = 1$, for neutron stars and other possible bosonic structures $q > 2$. This is an important test for stellar mass bodies where stable configurations of various kinds may exist (Wex and Kopeikin 1999), but it is also a prime test for the supermassive black hole concept in Active Galactic Nuclei and even in our Galactic Center (Will 2008).

2. OJ287

BL Lacertae object OJ287 is known to have a quasiperiodic pattern of outbursts at 12 year intervals (Sillanpää *et al.* 1988,Valtonen *et al.* 2008c). Further, the light curve has a double peak structure, with the two peaks separated by 1–2 years. The available information is that the radiation at the peaks is thermal bremsstrahlung radiation, in contrast to the non-thermal synchrotron emission at 'normal' times. The origin of the thermal bremsstrahlung is thought to be an impact of a secondary black hole on the accretion disc of the primary (Lehto & Valtonen 1996). The model has been successful in predicting future outbursts: the predictions for the beginning of 1994, 1995 and 2005 outbursts were correct within a few weeks. In addition to timing the impacts on the accretion disk, the model includes delays of outbursts relative to the disk crossing. The outburst begins when a bubble of gas torn off the accretion disk becomes transparent at optical wavelengths. For the 2005 outburst, it also was necessary to consider the bending of the accretion disk caused by the secondary (Sundelius *et al.* 1996,Sundelius *et al.* 1997); when combined with the earlier model for the radiation burst delay (Lehto & Valtonen 1996) the beginning of the 2005 outburst was expected at 2005.74. This is very close to the actual starting time of the outburst (Valtonen *et al.* 2006a, Valtonen *et al.* 2008a). It confirmed the need for relativistic precession since without the precession the outburst would have been a year later. Finally, the gravitational radiation energy loss was included in the prediction for the next outburst in September 2007 (Valtonen 2007, Valtonen *et al.* 2008a). Observations confirmed the correctness of the prediction within the accuracy of one day (Valtonen *et al.* 2008c). The model that does not incorporate the gravitational radiation reaction effect is clearly not tenable since it predicted the outburst three weeks too late. The probability of so many major outbursts happening at the predicted times by chance is negligible.

Recently, models have been developed which include the spin of the primary black hole (Valtonen *et al.* 2009). Here we will review this work and discuss its general implications.

3. Observations

The model requires accurate timing of the outbursts. The outburst is thought to arise from a hot bubble of gas which has cooled to a point where it becomes suddenly optically transparent. When such a bubble is viewed from a distance, the emission is seen to grow in a specific way as the observational front advances into the bubble. The size of the

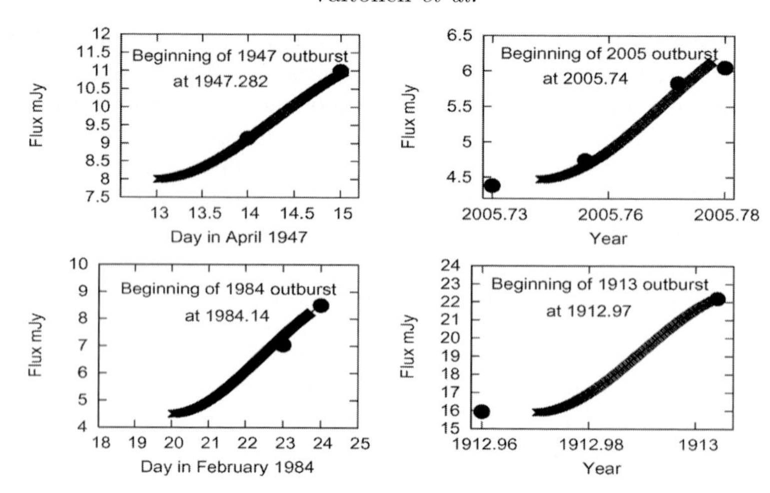

Figure 1. The observation of the brightness of OJ287 in four different seasons: 1912/3, 1947, 1984 and 2005. The heavy curve is a fit through the observed points (dots). Each panel labels the deduced start of the outburst.

bubble, and thus also the rate of development of the outburst light curve, is a known function of distance from the center of the accretion disk. In Figures 1 and 2 we display theoretical light curves for different outbursts and overlay them with observations. The labels inside the panels tell the timing which is related to the image.

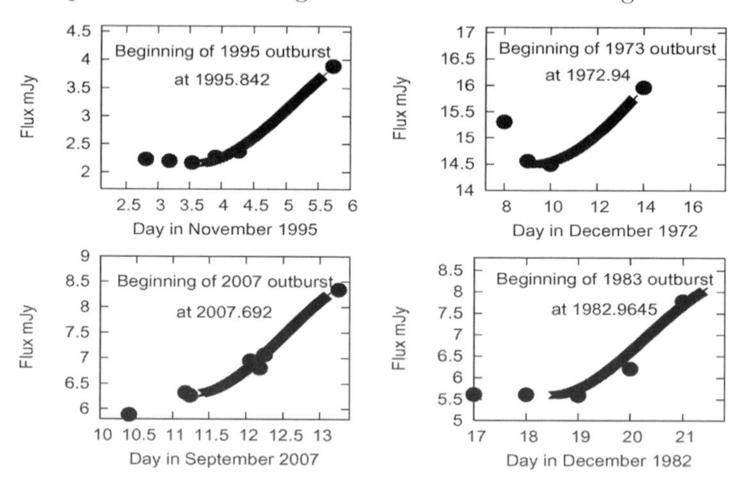

Figure 2. The observation of the brightness of OJ287 in four different seasons: 1972/3, 1982/3, 1995 and 2007. See caption of Fig. 1.

4. PN-accurate orbital description

We invoke the 2.5PN-accurate orbital dynamics that includes the leading order general relativistic, classical spin-orbit and radiation reaction effects for describing the temporal evolution of a binary black hole (Kidder 1995).

The 2.5PN-accurate equations of motion can be written schematically as

$$\ddot{\boldsymbol{x}} \equiv \frac{d^2\boldsymbol{x}}{dt^2} = \ddot{\boldsymbol{x}}_0 + \ddot{\boldsymbol{x}}_{1PN} + \ddot{\boldsymbol{x}}_{SO} + \ddot{\boldsymbol{x}}_Q + \ddot{\boldsymbol{x}}_{2PN} + \ddot{\boldsymbol{x}}_{2.5PN}, \tag{4.1}$$

where $\boldsymbol{x} = \boldsymbol{x}_1 - \boldsymbol{x}_2$ stands for the center-of-mass relative separation vector between the black holes with masses m_1 and m_2 and $\ddot{\boldsymbol{x}}_0$ represents the Newtonian acceleration given by $\ddot{\boldsymbol{x}}_0 = -\frac{G\,m}{r^3}\,\boldsymbol{x}$; $m = m_1 + m_2$ and $r = |\boldsymbol{x}|$. The PN contributions occurring at the conservative 1PN, 2PN and the reactive 2.5PN orders, denoted by $\ddot{\boldsymbol{x}}_{1PN}$, $\ddot{\boldsymbol{x}}_{2PN}$ and $\ddot{\boldsymbol{x}}_{2.5PN}$ respectively, are non-spin by nature. The explicit expressions for these contributions, suitable for describing the binary black hole dynamics, were derived for the first time in the harmonic gauge (Damour 1982). These are well known and they are not repeated here.

The leading order spin-orbit contributions to $\ddot{\boldsymbol{x}}$, appearing at 1.5PN order (Barker & O'Connell 1975), reads

$$\ddot{\boldsymbol{x}}_{SO} = \frac{G^2\,m^2}{c^3\,r^3}\left(\frac{1+\sqrt{1-4\,\eta}}{4}\right)\chi \cdot \tag{4.2}$$
$$\left\{\left[12\,[\boldsymbol{s}_1\cdot(\boldsymbol{n}\times\boldsymbol{v})]\right]\boldsymbol{n} + \left[(9+3\,\sqrt{1-4\,\eta})\,\dot{r}\right](\boldsymbol{n}\times\boldsymbol{s}_1) - \left[7+\sqrt{1-4\,\eta}\right](\boldsymbol{v}\times\boldsymbol{s}_1)\right\},$$

where the vectors $\boldsymbol{n}$ and $\boldsymbol{v}$ are defined to be $\boldsymbol{n} \equiv \boldsymbol{x}/r$ and $\boldsymbol{v} \equiv d\boldsymbol{x}/dt$, respectively, while $\dot{r} \equiv dr/dt = \boldsymbol{n}\cdot\boldsymbol{v}$, $v \equiv |\boldsymbol{v}|$ and the symmetric mass ratio $\eta = m_1\,m_2/m^2$. The Kerr parameter χ and the unit vector $\boldsymbol{s}_1$ define the spin of the primary black hole by the relation $\boldsymbol{S}_1 = G\,m_1^2\,\chi\,\boldsymbol{s}_1/c$ and χ is allowed to take values between 0 and 1 in general relativity. Further, the above expression for $\ddot{\boldsymbol{x}}_{SO}$ implies that the covariant spin supplementary condition is employed to define the center-of-mass world line of the spinning compact object in the underlying PN computation (Kidder 1995). Finally, the quadrupole-monopole interaction term $\ddot{\boldsymbol{x}}_Q$, entering at the 2PN order (Barker & O'Connell 1975), reads

$$\ddot{\boldsymbol{x}}_Q = -q\,\chi^2\,\frac{3\,G^3\,m_1^2\,m}{2\,c^4\,r^4}\left\{\left[5(\boldsymbol{n}\cdot\boldsymbol{s}_1)^2 - 1\right]\boldsymbol{n} - 2(\boldsymbol{n}\cdot\boldsymbol{s}_1)\boldsymbol{s}_1\right\},$$

where the parameter q, whose value is 1 in general relativity, is introduced to test the black hole 'no-hair' theorem. The precessional motion for the primary black hole spin is dominated by the leading order general relativistic spin-orbit coupling and the relevant equation reads

$$\frac{d\boldsymbol{s}_1}{dt} = \boldsymbol{\Omega}\times\boldsymbol{s}_1,$$
$$\boldsymbol{\Omega} = \left(\frac{G\,m\,\eta}{2c^2\,r^2}\right)\left(\frac{7+\sqrt{1-4\,\eta}}{1+\sqrt{1-4\,\eta}}\right)(\boldsymbol{n}\times\boldsymbol{v})\,. \tag{4.3}$$

It should be noted that the precessional equation for the unit spin vector $\boldsymbol{s}_1$ enters the binary dynamics at 1PN order, while the spin contribution enters $\ddot{\boldsymbol{x}}$ at the 1.5PN order.

The main consequence of including the leading order spin-orbit interactions to the dynamics of a binary black hole is that it forces both the binary orbit and the orbital plane to precess. Moreover, the orbital angular momentum vector, characterising the orbital plane, precesses around the spin of the primary in such a way that the angle between the orbital plane and the spin vector $\boldsymbol{s}_1$ remains almost constant (roughly within $\pm 0.^\circ 5$ in our model). The spin-vector itself precesses drawing a cone with an opening angle of about 12 degrees.

The precessional period for both the orbital plane and the spin of the binary is about 2400 years. The orbital inclination relative to the plane of symmetry of the spinning black hole (and presumably relative to the accretion disk of the primary as well) is taken to be 90 degrees.

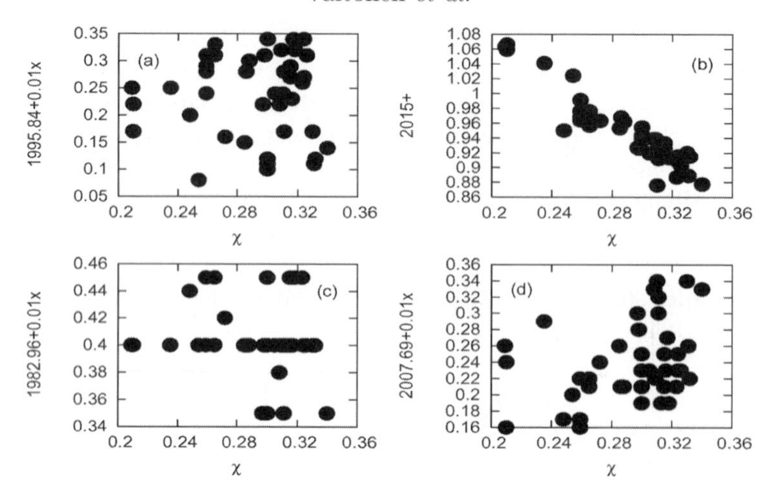

Figure 3. The dots represent solutions with different values of χ (horizontal axis) and outburst time (vertical axis). On the y-axis of panel (a) the points scatter around 1995.842, in panel (c) around 1982.964 and in panel (d) around 2007.692. In contrast, panel (b) shows a strong correlation between the outburst time in 2015 and χ.

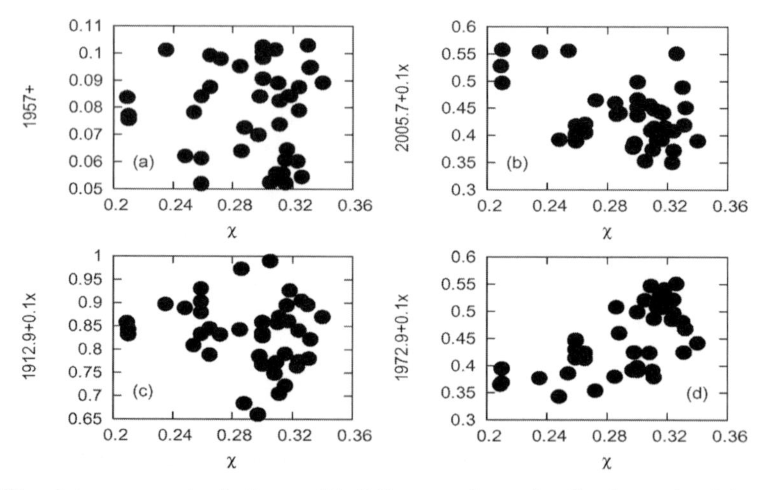

Figure 4. The dots represent solutions with different values of χ (horizontal axis) and outburst time (vertical axis). On the y-axis of panel (a) the points scatter around 1957.08, in panel (b) around 2005.745, in panel (c) around 1912.98 and in panel (d) around 1972.945.

5. Timing experiments

In previous work (e.g. Valtonen 2007) we do the timing experiments using the 1913, 1947, 1973, 1983, 1984 and 2005 outbursts as fixed points. We cannot determine the Kerr parameter χ and the no-hair parameter q without considering more fixed points. Here we also make use of the 1995 and 2007 outbursts which have been observed with high time resolution (see Figure 2) and they are therefore suitable for further refinement of the model. In addition, the newly discovered 1957 outburst is taken as another fixed point (Valtonen *et al.* 2006b, Rampadarath *et al.* 2007). The observed outburst times are listed in Table 1.

With this full set of outbursts no solutions were found unless the parameter χ is in the range 0.2–0.36. The majority of solutions cluster around $\chi = 0.29$, with a one standard deviation of about 0.04. The timing experiments give a unique solution for the system

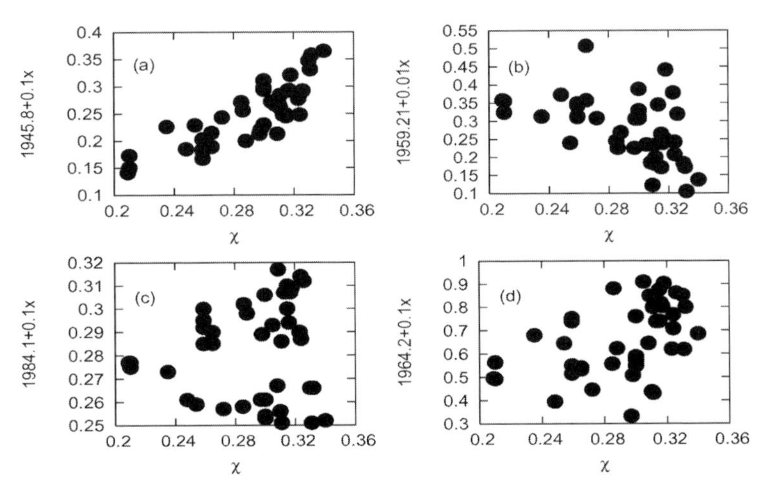

Figure 5. The dots represent solutions with different values of χ (horizontal axis) and outburst time (vertical axis). On the y-axis of panel (b) the points scatter around 1959.213, in panel (c) around 1984.129, and in panel (d) around 1964.27. Panel (a) displays a strong correlation between the timing of the 1945 outburst and χ. There are not enough data at present to verify the timings of 1959, 1964 and 1945 outbursts, while the 1984 outburst is used as a fixed point in our solutions. If new data points are found from years 1945–1964, the model can be further refined.

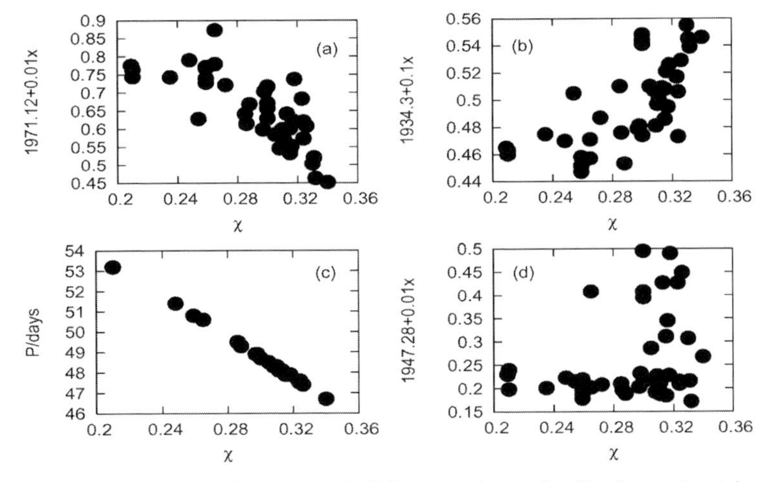

Figure 6. The dots represent solutions with different values of χ (horizontal axis) and outburst time (vertical axis). On the y-axis of panel (a) the points scatter around 1971.126, in panel (b) around 1934.35 and in panel (d) around 1947.283. In contrast, panel (c) shows a dependence of the half-period of the innermost stable orbit on χ. The observed half-period is between 43 and 49 days, implying that χ is likely to be at least 0.29. The 1947 outburst is one of our fixed points, while there are not enough data to verify 1934 and 1971 outbursts at present.

parameters: precession in the orbit plane per period $\Delta\phi$, masses m_1 and m_2 of the two black holes, spin parameter χ, initial phase ϕ_0, initial apocenter eccentricity e_0, 'no-hair' parameter q and time-delay parameter t_d. The results are shown in Table 2.

Here the precession rate is defined as the average change of the apocenter phase angle over the last 150 yrs. The time delay parameter depends on the structure of the accretion disk, and obtains the value of unity in the model of Lehto & Valtonen 1996. Its value is related to the thickness of the accretion disk, as explained in Valtonen *et al.* 2006b. Figures 3–6 show the distribution points, each representing a solution, in four panels per

Table 1. Outburst times with estimated uncertainties. These are starting times of the outbursts, normalised to the event of 1982.964.

1912.980	$\pm$ 0.020
1947.283	$\pm$ 0.002
1957.080	$\pm$ 0.030
1972.945	$\pm$ 0.012
1984.130	$\pm$ 0.005
1995.842	$\pm$ 0.0015
2005.745	$\pm$ 0.012
2007.692	$\pm$ 0.0015

Table 2. Solution parameters.

$\Delta\phi^\circ$	39.1 ± 0.1
m_1	$(1.83 \pm 0.01) \cdot 10^{10} M_\odot$
m_2	$(1.4 \pm 0.1) \cdot 10^{8} M_\odot$
χ	0.28 ± 0.08
ϕ_0	$56^\circ.5 \pm 1^\circ.2$
e_0	0.6584 ± 0.001
q	1.0 ± 0.3
t_d	0.75 ± 0.04

figure. Each panel displays the outburst time as a function of χ. In Figure 3 the panels (a), (c) and (d) are essentially scatter diagrams, demonstrating that the solutions cover the allowed range rather well. On the contrary, the panel (b) shows a strong correlation between the time of the 2015 outburst and χ. The range of possible outburst times extends from early November if $\chi = 0.36$ to late January 2016 if $\chi = 0.2$.

6. More OJ287's?

How unique is OJ287? Can we hope to observe more similar systems? OJ287 is highly variable mostly due to strong beaming. At its faintest it goes to $m_B = 18$ which may be taken as its intrinsic (unbeamed) brightness. There are about $2 \cdot 10^4$ quasars in the sky brighter than this magnitude limit (Arp 1981). Many of these quasars host binary black holes, perhaps as many as 50% (Comerford *et al.* 2009). Thus potentially there are about 10^4 bright binary quasars in the sky to be discovered. But are they all similar to OJ287? OJ287 is in the last stages of inspiral, with only 10^4 yr left out of its potential 10^7 yr lifetime (Volonteri *et al.* 2009). Thus the number of bright short period binaries which could be discovered by techniques similar to the discovery of OJ287 is only of the order of ten over the whole sky. Another property of OJ287 which may be relevant to the discovery rate, is its mass at the upper end of black hole mass range. Even though a high mass generally means high luminosity and sampling of large volumes of space (observed numbers increasing roughly as $M^{1.5}$ for luminosity being a constant fraction of Eddington luminosity), the frequency function of black hole mass falls with increasing mass. The power law slope at the high mass end is about -2.6, i.e. the probability density for mass being greater than M has a power law index of about -1.6 (Vestergaard and Osmer 2009). Thus the effect of frequency function is to compensate for the volume effect (within the uncertainties). For a binary system with a constant period, there is also another factor. The remaining lifetime of the binary goes down as $M^{-2/3}$. The most likely value for a mass from this distribution is one quarter of the upper limit (which is of the order of $6 \cdot 10^{10} M_\odot$, Vestergaard *et al.* 2008). It is not too far from our measured mass value. The main reason for the discovery of the OJ287 binary in a sample of about 50 quasars studied by the Tuorla monitoring group is the fact that it is a blazar. Because of this, it was a prime target for variability studies. The double peak optical outbursts would have easily gone unnoticed if it did not attract attention by its variable jet brightness which is seen all the time. Therefore one should extend the periodicity searches to quasars which are not highly variable, in hopes of detecting occasional double peaked outbursts.

From the measured frequency of over $10^{10} M_{\odot}$ black holes in quasars one easily estimates that there must be a dozen or more such quasars in the local volume encompassing OJ287. Since their jets would not typically point toward us, they would only appear bright during the short intervals of disk impacts ; it would require a major program of historical plate analysis and observational monotoring to catch these potentially OJ287-type quasars in action.

7. Parameter values

We have already pointed out that the primary mass $m_1 = (1.83 \pm 0.01) \cdot 10^{10} M_{\odot}$ is quite consistent with what might be expected on general grounds. What about the secondary mass? Its value was calculated by Lehto and Valtonen (1996) as $m_2 \approx 1.23 \cdot 10^8 M_{\odot}$, updated to today's Hubble constant, while Sundelius *et al.* (1997) preferred a 30% greater value. These estimates are based on the astrophysics of the disk impacts and on the amount of radiation produced in these events, and thus they are totally independent of the orbit model. These estimates agree well with our more exact value $m_2 = (1.4 \pm 0.1) \cdot 10^8 M_{\odot}$.

The eccentricity of the orbit, if defined by using pericenter and apocenter distances as for a Keplerian orbit, is $e = 0.7$. The initial eccentricity at the beginning of the final inspiral, say, 10^5 yr prior to merger, must have been about $e = 0.9$, a reasonable value at the initial stage of merging black holes (Aarseth 2008). Then by today the eccentricity would have evolved to its current value.

The parameter t_d gives the ratio of α_g to the accretion rate in Eddington units. The value $t_d = 0.75 \pm 0.04$ implies 14 ± 0.5 for this ratio. The mass accretion rate may be about 0.005 of the Eddington rate (Bassani *et al.* 1983) which gives $\alpha_g = 0.07 \pm 0.03$, a reasonable value.

We infer that the primary black hole should spin approximately at one quarter of the maximum spin rate allowed in general relativity. There is an additional observation which supports this spin value. OJ287 has a basic 46±3 day periodicity (Wu *et al.* 2006), which may be related to the innermost stable orbit in the accretion disk. However, since we presumably observe the accretion disk almost face on, and there is an $m = 2$ mode wave disturbance in the disk, this is likely to refer to one half of the period. Considering also the redshift of the system, and the primary mass value given in Table 2 this corresponds to the spin of $\chi = 0.35 \pm 0.06$ (McClintock *et al.* 2006). The uncertainty of 0.06 units is related to the width of a trough in the structure function of the flux variations. For comparison, it has been estimated that $\chi \sim 0.5$ for the black hole in the Galactic center (Genzel *et al.* 2003).

The values of q cluster around $q = 1.0$ with a standard deviation of 0.3. This is the first indication that the 'no-hair' theorem is valid. Even though there are no other known stable configurations than a black hole in the supermassive scale, it is still interesting that our result converges at the proper value for general relativity.

References

Aarseth, S. J. 2008, in Dynamical Evolution of Dense Stellar Systems, Proceedings of IAU Symp. 246, Eds. E. Vesperini, M. Giersz & A. Sills, Cambridge Univ. Press, Cambridge, p. 437

Arp, H. 1981, Astrophysical Journal 250, 31

Barker, B. M. & O'Connell, R. F. 1975, PhRvD, 12, 329
Bassani, L., Dean, A. J., & Sembay, S., Astronomy & Astrophysics, 125, 52
Camenzind, M. 1989, in Accretion Disks and Magnetic Fields in Astrophysics, Ed. G. Belvedere, Kluwer, Dordrecht, p. 129
Carter, B. 1970, Phys. Rev. Lett. 26, 331
Comerford, J. M., Gerke, B. F., Newman, J. A., Davis, M., Yan, R., Cooper, M. C., Faber, S. M., Koo, D. C., Coil, A. L., Rosario, D. J., & Dutton, A. A. 2009, Astrophysical Journal, 698, 956
Damour, T. 1982, C. R. Acad. Sci. Paris 294, (II), 1355.
Genzel, R., Schodel, R., Ott, T., Eckart, A., Alexander, T., Lacombe, F., Rouan, D., & Aschenbach, B. 2003, Nature 425, 934
Hawking, S. W. 1971, Phys. Rev. Lett. 26, 1344
Hawking, S. W. 1972, Commun. Math. Phys. 25, 152
Israel, W. 1967, Phys. Rev. 164, 1776
Israel, W. 1968, Commun. Math. Phys. 8, 245
Kidder, L. E. 1995, PhRvD, 52, 821
Lehto, H. J. & Valtonen, M. J. 1996, Astrophysical Journal, 460, 207
Misner, C. W., Thorne, K. S., & Wheeler, J. A. 1973, Gravitation, W. H. Freeman & Co, New York, p. 876
Rampadarath, H., Valtonen, M. J., & Saunders, R. The Central Engine of Active Galactic Nuclei, Eds. Luis C. Ho & Jian-Min Wang, ASP Conf. Ser., 373, 243
Sakimoto, P. J. & Corotini, F. V. 1981, Astrophysical Journal, 247, 19
Salpeter, E. E. 1964, Astrophysical Journal, 140, 796
Shakura, N. I. & Sunyaev, R. A. 1973, Astronomy & Astrophysics, 24, 337
Sillanpää, A., Haarala, S., Valtonen, M. J., Sundelius, B., & Byrd, G. G. 1988, Astrophysical Journal, 325, 628
Sillanpää *et al.* 1996a, Astronomy & Astrophysics, 305, L17
Sillanpää *et al.* 1996b, Astronomy & Astrophysics, 315, L13
Sundelius, B., Wahde, M., Lehto, H. J., & Valtonen, M. J. 1996, Blazar Continuum Variability, ASP Conf. Ser., 110, 99
Sundelius, B., Wahde, M., Lehto, H. J., & Valtonen, M. J. 1997, Astrophysical Journal, 484, 180
Thorne, K. S. 1980, Rev. Mod. Phys. D, 31, 1815
Thorne, K. S., Price, R. M., & Macdonald, D. A. 1986, in Black Holes: The Membrane Paradigm, Yale Univ. Press, New Haven
Valtonen, M. J. & Lehto, H. J. 1997, Astrophysical Journal, 481, L5
Valtonen, M. J. *et al.* 2006a, Astrophysical Journal, 643, L9
Valtonen, M. J. *et al.* 2006b, Astrophysical Journal, 646, 36
Valtonen, M. J. 2007, Astrophysical Journal, 659, 1074
Valtonen, M. J., Kidger, M., Lehto, H., & Poyner, G. 2008b, Astronomy & Astrophysics, 477, 407
Valtonen, M. J. 2008, RevMexA&Ap, 32, 22
Valtonen, M. J. *et al.*, 2008c, Nature, 452, 851
Valtonen, M. J., Mikkola, S., Merritt, D., Gopakumar, A., Lehto, H. J., Hyvönen, T., Rampadarath, H., Saunders, R., Basta, M., & Hudec, R. 2009, Astrophysical Journal, in press
Vestergaard, M., Fan, X., Tremonti, C. A., Osmer, P. O., & Richards, G. T. 2008, Astrophysical Journal, 674, L1
Vestergaard, M. & Osmer, P. S. 2009, Astrophysical Journal, 699, 8
Volonteri, M., Miller, J. M., & Dotti, M. 2009, Astrophysical Journal, 703, L86
Wex, N. & Kopeikin, S. M. 1999, Astrophysical Journal, 514, 388
Will, C. M. 2008, Astrophysical Journal 674, L25-L28.
Wu, J., Zhou, X., Wu, X.-B., Liu, F.-K., Peng, B., Ma, J., Wu, Z., Jiang, Z., & Chen, J., 2006, Astronomical Journal, 132, 1256.
Zeldovich, Ya. B. 1964, Sov. Phys. Doklady, 9, 195

Relativity in Fundamental Astronomy
Proceedings IAU Symposium No. 261, 2009
S. A. Klioner, P. K. Seidelman & M. H. Soffel, eds.

doi:10.1017/S1743921309990494

The galactic center: The ideal laboratory for studying supermassive black holes

Frank Eisenhauer

Max-Planck-Institut für extraterrestrische Physik,
Giessenbachstrasse, 85748 Garching, Germany,
email: eisenhau@mpe.mpg.de

Abstract. The Galactic Center constitutes the best astrophysical evidence for the existence of black holes, and it is the ideal laboratory for studying physics in the vicinity of such objects. The combination of infrared observations of three dimensional orbits of stars within the central light days and the extreme compactness and motionlessness of the radio-counterpart of the gravitational center have shown beyond any reasonable doubt that the Galactic Center harbors a supermassive black hole. The flaring activity from the black hole gives first insights to the physical processes close to the last stable orbit. Here I review the current state of observations and theory of the Galactic Center black hole and give an update on the latest results. I also outline the next steps towards even higher angular resolution observations, which give promise to directly probe the physics and space-time curvature just outside the event horizon.

Keywords. Galaxy: center, black hole physics, relativity, techniques: high angular resolution, interferometric

---

References

Backer, D. C. & Sramek, R. A., 1999, ApJ, 524, 805
Balick, B. & Brown, R. L., 1974, ApJ, 194, 265
Bower, G. C., Falcke, H., Herrnstein, R. M., Zhao, J.-H., Goss, W. M., & Backer, D. C., 2004, Sci, 304, 704
Bower, G. C., Goss, W. M., Falcke, H., Backer, D. C., & Lithwick, Y., 2006, ApJ, 648, L127
Broderick, A. E., Fish, V. L., Doeleman, S. S., & Loeb A., 2009, ApJ, 697, 45
Broderick, A. E. & Loeb, A., 2005, MNRAS, 363, 353
Broderick, A. E., Loeb, A., & Narayan, R., 2009, arXiv:0903.1105
Broderick, A. E. & Narayan, R., 2006, ApJ, 638, L21
Chapline, G., 2005, arXiv:astro-ph/0503200
Do, T., Ghez, A. M., Morris, M. R., Yelda, S., Meyer, L., Lu, J. R., Hornstein, S. D., & Matthews, K., 2009, ApJ, 691, 1021
Doeleman, S. S., *et al.*, 2008, Natur, 455, 78
Eckart, A. & Genzel, R., 1996, Natur, 383, 415
Eckart, A., Schödel, R., Meyer, L., Trippe, S., Ott, T., & Genzel, R., 2006, A&A, 455, 1
Eisenhauer, F., *et al.*, 2005, ApJ, 628, 246
Eisenhauer, F., *et al.*, 2008, SPIE, 7013, 69
Eisenhauer, F., Perrin, G., Rabien, S., Eckart, A., Lena, P., Genzel, R., Abuter, R., & Paumard, T., 2005, AN, 326, 561
Falcke, H., Melia, F., & Agol, E., 2000, ApJ, 528, L13
Fish, V. L., Broderick, A. E., Doeleman, S. S., & Loeb, A., 2009, ApJ, 692, L14
Fragile, P. C. & Mathews, G. J., 2000, ApJ, 542, 328
Fujii, Y., & Maeda, K.-I., 2003, The Scalar-Tensor Theory of Gravitation, Cambridge University Press, ISBN 0521811597
Genzel, R., Hollenbach, D., & Townes, C. H., 1994, RPPh, 57, 417

Genzel, R., Schödel R., Ott, T., Eckart, A., Alexander, T., Lacombe, F., Rouan, D., & Aschenbach, B., 2003, Natur, 425, 934

Ghez, A. M., Klein, B. L., Morris, M., & Becklin, E. E., 1998, ApJ, 509, 678

Ghez, A. M., *et al.*, 2008, ApJ, 689, 1044

Gillessen, S., Eisenhauer, F., Trippe, S., Alexander, T., Genzel, R., Martins, F., & Ott, T., 2009, ApJ, 692, 1075

Hamaus, N., Paumard, T., Müller, T., Gillessen, S., Eisenhauer, F., Trippe, S., & Genzel, R., 2009, ApJ, 692, 902

Huang, L., Cai, M., Shen, Z.-Q., & Yuan, F., 2007, MNRAS, 379, 833

Jaroszynski, M., 1998, AcA, 48, 653

Lacy, J. H., Townes, C. H., Geballe, T. R., & Hollenbach, D. J., 1980, ApJ, 241, 132

Levin, Y. & Beloborodov, A. M., 2003, ApJ, 590, L33

Mazur, P. O. & Mottola, E., 2001, arXiv:gr-qc/0109035

Meyer, L., Do, T., Ghez, A., Morris, M. R., Yelda, S., Schödel, R., & Eckart, A., 2009, ApJ, 694, L87

Narayan, R., Garcia, M. R., & McClintock, J. E., 1997, ApJ, 478, L79

Paumard, T., *et al.*, 2005, AN, 326, 568

Psaltis, D., 2004, AIPC, 714, 29

Reid, M. J. & Brunthaler, A., 2004, ApJ, 616, 872

Reid, M. J., Readhead, A. C. S., Vermeulen, R. C., & Treuhaft, R. N., 1999, ApJ, 524, 816

Rubilar, G. F. & Eckart, A., 2001, A&A, 374, 95

Shen, Z.-Q., Lo, K. Y., Liang, M.-C., Ho P. T. P., & Zhao, J.-H., 2005, Natur, 438, 62

Townes, C. H., Lacy, J. H., Geballe, T. R., & Hollenbach, D. J., 1983, Natur, 301, 661

Trippe, S., Paumard, T., Ott, T., Gillessen, S., Eisenhauer, F., Martins, F., & Genzel R., 2007, MNRAS, 375, 764

Weinberg, N. N., Milosavljević M., & Ghez, A. M., 2005, ApJ, 622, 878

Will, C. M., 2008, ApJ, 674, L25

Wollman, E. R., Geballe, T. R., Lacy, J. H., Townes, C. H., & Rank, D. M., 1977, ApJ, 218, L103

Yuan, F., Quataert, E., & Narayan, R., 2004, ApJ, 606, 894

Zucker, S., Alexander, T., Gillessen, S., Eisenhauer, F., & Genzel, R., 2006, ApJ, 639, L21

Relativity in Fundamental Astronomy
Proceedings IAU Symposium No. 261, 2009
S. A. Klioner, P. K. Seidelman & M. H. Soffel, eds.

doi:10.1017/S1743921309990500

Observing a black hole event horizon: (sub)millimeter VLBI of Sgr A*

Vincent L. Fish and Sheperd S. Doeleman

MIT Haystack Observatory
Off Route 40
Westford, MA 01886, USA
email: vfish@haystack.mit.edu, sdoeleman@haystack.mit.edu

Abstract. Very strong evidence suggests that Sagittarius A*, a compact radio source at the center of the Milky Way, marks the position of a super massive black hole. The proximity of Sgr A* in combination with its mass makes its apparent event horizon the largest of any black hole candidate in the universe and presents us with a unique opportunity to observe strong-field GR effects. Recent millimeter very long baseline interferometric observations of Sgr A* have demonstrated the existence of structures on scales comparable to the Schwarzschild radius. These observations already provide strong evidence in support of the existence of an event horizon. (Sub)Millimeter VLBI observations in the near future will combine the angular resolution necessary to identify the overall morphology of quiescent emission, such as an accretion disk or outflow, with a fine enough time resolution to detect possible periodicity in the variable component of emission. In the next few years, it may be possible to identify the spin of the black hole in Sgr A*, either by detecting the periodic signature of hot spots at the innermost stable circular orbit or parameter estimation in models of the quiescent emission. Longer term, a (sub)millimeter VLBI "Event Horizon Telescope" will be able to produce images of the Galactic center emission to the see the silhouette predicted by general relativistic lensing. These techniques are also applicable to the black hole in M87, where black hole spin may be key to understanding the jet-launching region.

Keywords. black hole physics, accretion, accretion disks, Galaxy: center, submillimeter, techniques: interferometric, techniques: high angular resolution

---

1. Introduction

The Galactic center radio source Sagittarius A* is believed to host a massive ($\sim 4 \times 10^6$ M_{sun}) black hole. Due to its proximity at ∼8 kpc (e.g., Reid 2008), Sgr A* has the largest apparent event horizon of any known black hole candidate: $r_{Sch} = 2r_G \approx 10$ μas.

There is active debate in the scientific community on whether the emission from Sgr A* is predominantly due to an accretion disk or a jet. Further data are necessary in order to disentangle the accretion/outflow physics from general relativistic effects.

There are several reasons to observe Sgr A* at millimeter wavelengths. The spectrum of Sgr A* peaks in the millimeter. Interstellar scattering, which varies as λ^2, dominates over intrinsic source structure at longer wavelengths, and the emission from Sgr A* transitions from optically thick to optically thin near $\lambda = 1$ mm (Doeleman *et al.* 2001). Finally, millimeter wavelengths permit observations by the technique of very long baseline interferometry (VLBI). The spatial scales accessible to millimeter VLBI range down to a few r_G, providing angular resolution currently unachievable by any other means.

Recent VLBI observations by Doeleman *et al.* (2008) were successful in detecting Sgr A* on the long baseline between the James Clerk Maxwell Telescope (JCMT) in Hawai‘i and the Arizona Radio Observatory's Submillimeter Telescope (SMT). Combining this result with a shorter-baseline detection from the SMT to the Combined Array for Research in Millimeter-wave Astronomy (CARMA) in California, Doeleman *et al.* were able to place a size limit of 37 μas on the emission if its distribution is a circular Gaussian. They conclude that the emission must be partially optically thin and offset from the center of Sgr A*, since an optically thick sphere centered on the black hole would be lensed to a larger size.

With only detections on two baselines, the data are insufficient to distinguish between other models such as a "doughnut" of emission as might be expected from a face-on accretion disk. In actuality, the structure of the emission from Sgr A* is likely to be much more complicated due to the inclination of the system and general relativistic effects, such as Doppler boosting of approaching emission and weakening of emission from in front of the black hole due to gravitational redshift. It is this complexity of source structure that drives the need for more data. The eventual goal is to produce an image of the quiescent emission in Sgr A*, from which many physical parameters (such as the spin of the black hole) can be derived. In the nearer term, nonimaging methods of data analysis can be used to produce scientific output from millimeter VLBI observations.

2. Parameter Estimation

The Doeleman *et al.* (2008) detections measure correlated flux densities on certain spatial scales and orientations, which can be used to constrain parameters in ensembles of models. For instance, Broderick *et al.* (2009) generate an ensemble of radiatively inefficient accretion flow (RIAF) models of disk emission from Sgr A* meeting certain prior constraints (such as the multiwavelength spectrum of emission from Sgr A*) and then use the Doeleman *et al.* (2008) detections to estimate probable values of the black hole spin, inclination of the spin axis, and orientation of the accretion disk on the plane of the sky. Within the RIAF ensemble, it is found that low-inclination (i.e., nearly face-on) models are strongly disfavored. Prospects are excellent for observations in the near future, which are likely to include a telescope in Chile, to be able to place strong constraints on the orientation of the disk within the RIAF context (Fish *et al.* 2009). As additional telescopes are added to the millimeter VLBI observing array, the constraints provided by the detections may be able to obtain a value for the black hole spin even before a high-quality image can be created.

It will be necessary to distinguish amongst different models of the mechanism of emission from Sgr A* (e.g., jets, fully general relativistic magnetohydrodynamic simulations, etc.), both in order to understand the accretion/outflow physics in general and because each model will have different particular predictions for parameter values such as the black hole spin. Thus, there is a need for similar analyses to be carried out in order to interpret present and future millimeter VLBI results.

3. Imaging with an Event Horizon Telescope

As additional telescopes are added to the millimeter VLBI array, it will become possible to produce an image of the emission from Sgr A*. Current observations have used the SMT, CARMA dishes, and a phased array of the JCMT, Caltech Submillimeter Observatory, and Submillimeter Array dishes on Mauna Kea. In the near future, it may be possible to extend this array by including the Atacama Submillimeter Telescope Experiment (ASTE) or Atacama Pathfinder Experiment (APEX) in Chile, the Large Millimeter Telescope (LMT) under construction in Mexico, the Institut de Radioastronomie Millimétrique (IRAM) 30 m telescope at Pico Veleta in Spain, and the IRAM Plateau de Bure interferometer in France.

With upgrades to some of these telescopes (§5), it will be possible to produce an image that shows the general morphology of the emission from Sgr A* (Figure 1), possibly including the "shadow" of the black hole (e.g., Falcke *et al.* 2000). Unfortunately, image fidelity will be limited by the placement of these telescopes. Effectively, VLBI baselines are sensitive to the Fourier transform of the sky emission, where the Fourier components are parameterized by u and v, the projected baseline lengths (in wavelengths) as viewed from Sgr A*. Longer baselines provide higher angular resolution. As the Earth rotates, the projected baselines sweep out arcs in the (u, v) plane, allowing an image to be produced by an inverse Fourier transform and deconvolution. However, the lack of intermediate-spacings in some directions means that large parts of the (u, v) plane are unsampled. Reconstructing a high-fidelity, model-independent image of Sgr A* will require the addition of a few telescopes to fill in these holes in the (u, v) plane. Several existing telescopes, such as the South Pole Telescope, the Haystack 37 m in Massachusetts, and the Swedish-ESO Submillimeter Telescope in Chile, could be upgraded for use in a millimeter VLBI array. It may also be desirable to place additional telescopes in geographically favorable locations such as South Africa, Kenya, and New Zealand (Figure 2). Expanding the millimeter VLBI array by upgrading existing telescopes and adding a few new ones would allow very high quality imaging of Sgr A*.

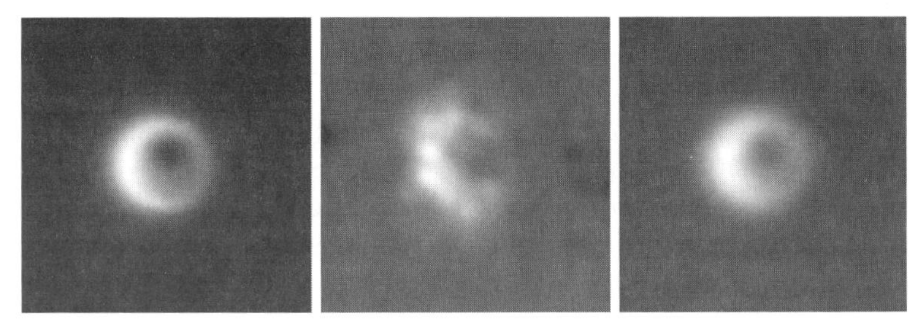

Figure 1. *Left:* Model of RIAF emission (345 GHz, $a = 0$, $i = 30°$), courtesy A. Broderick. *Center:* Simulated image from 7-telescope array of the near future. The black hole shadow is easily detected. *Right:* Simulated image from the 13-telescope array described in the text.

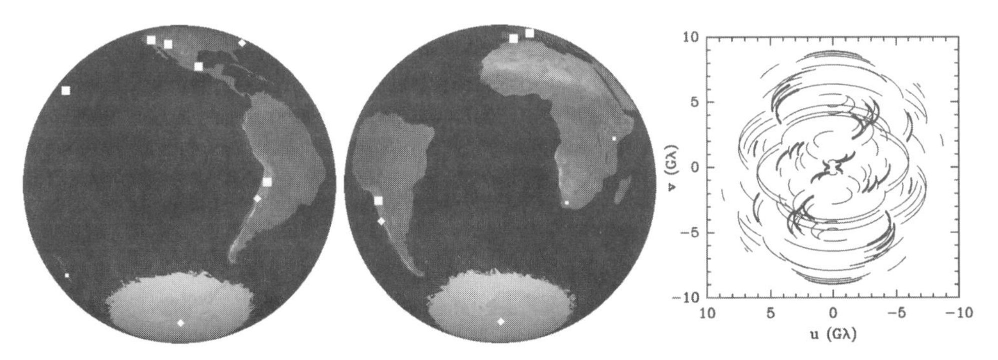

Figure 2. *Left two panels:* Current and future millimeter VLBI arrays. Large squares show telescopes that exist or are under construction. Medium diamonds show existing telescopes that could be refitted for VLBI. Small squares show other potential sites. *Right:* The (u, v) coverage produced by these arrays at 230 GHz. The bold curves show the coverage obtainable from the 7-telescope array described in the text. The curves in normal weight show the additional coverage that would be provided by the 13-station Event Horizon Telescope. High-fidelity imaging requires minimizing the gaps in (u, v) coverage.

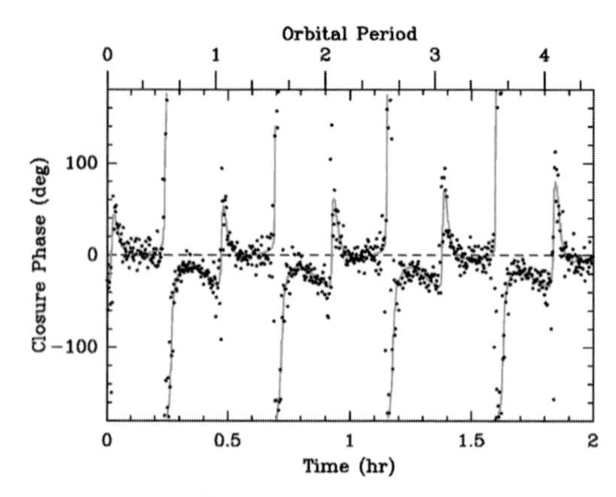

Figure 3. Simulated 230 GHz closure phases of a hot spot orbiting at the ISCO embedded in a quiescent RIAF disk ($a = 0, i = 30°$). The grey curve shows the model values for an array consisting of phased Mauna Kea (JCMT + CSO + 6 SMA dishes), phased CARMA (8 dishes), and the SMT. Black points represent simulated 10 second integrations at 8 Gbit s^{-1}.

4. Tracking Flaring Structures in Real Time with VLBI

While very high angular resolution millimeter imaging awaits the availability of additional VLBI telescopes, millimeter VLBI visibilities are already producing useful science. Indeed, the scientific implications of the Doeleman *et al.* (2008) detection stem from measuring the visibility amplitude on the JCMT-SMT baseline, which is a measure of the correlated flux density on small spatial scales. As Sgr A* is detected with larger arrays, it will also be possible to produce closure quantities, good observables that are very sensitive to the source structure of Sgr A* but independent of most calibration errors. The closure phase $\phi_{123} = \phi_{12} + \phi_{23} + \phi_{31}$, the sum of visibility phases around a closed triangle of three antennas, is robust against most phase errors, especially those caused by a variable troposphere. A nonzero closure phase implies source structure asymmetry. Similarly, a closure amplitude $A_{1234} = |A_{12}||A_{34}|/(|A_{13}||A_{24}|)$, the ratio of visibility amplitudes at 4 telescopes, is robust against systematic gain calibration errors. The number of independent closure quantities grows quickly with the number of antennas in an array.

The frequent flaring of Sgr A* implies that its source structure changes rapidly. Simulations by Doeleman *et al.* (2009) show that the sensitivity of VLBI combined with a time resolution much shorter than the natural orbit period make it very likely that time-variable structures during flares can be detected over the course of a single night of observing. Within the next few years, millimeter VLBI will have the sensitivity to probe $r_{\rm Sch}$-scale structural changes on time scales of 10 seconds (Figure 3). This time resolution will be critical for detecting source structure periodicity, as might be expected in orbiting hot spot models (Broderick & Loeb 2005). Rapid periodicity may provide strong evidence that the black hole in Sgr A* has spin, since the period of the innermost stable circular orbit (ISCO) decreases rapidly with increasing spin. Depending on the spin of the black hole, the ISCO period may be as long as nearly half an hour or as short as approximately 4 minutes.

5. Future Enhancements

Several technological advancements are currently in progress to increase the sensitivity of the millimeter VLBI array. A phased array processor to sum the collecting area on Mauna Kea (Weintroub 2008) has been tested. Similar hardware could be used to increase sensitivity at CARMA, Plateau de Bure, and the Atacama Large Millimeter/submillimeter Array (ALMA) in Chile. Digital backends (DBEs) have been developed to process 1 GHz of data (4 Gbit s^{-1} with 2-bit Nyquist sampling), and next-generation DBEs will improve upon this by a factor of four. Mark 5B+ recorders can already record 2 Gbit s^{-1} data streams (presently requiring two at each site per DBE), and the Mark 5C recorders currently being developed will be able to handle even faster data rates. Cryogenic sapphire oscillators are being examined as a possible frequency standard to supplement or replace hydrogen masers to provide greater phase stability, which may improve coherence at higher frequencies.

Future observations will initially focus on improving sensitivity by observing a wider bandwidth and using phased array processors. Dual polarization observations will become a priority not only for the $\sqrt{2}$ improvement in sensitivity for total-power observations but also to allow full polarimetric VLBI of Sgr A*. Higher frequency observations, such as in the 345 GHz atmospheric window, will provide even greater spatial resolution in a frequency regime where interstellar scattering and optical depth effects are minimized.

The timing is right to move forward on building an Event Horizon Telescope to produce high-fidelity images of Sgr A* as well as other scientifically compelling sources, such as M87. Receivers currently being produced en masse for ALMA could be procured for other millimeter VLBI stations, in many cases providing substantial improvements in sensitivity. Studies of climate and weather will be necessary to provide information on the astronomical suitability of prospective sites for future telescopes, such as those at the present ALMA Test Facility or additional telescopes constructed specifically for millimeter VLBI (which would mesh well with present ALMA construction). Some existing telescopes will require improvements to their systems, such as increasing the bandwidth of the intermediate frequency signal after mixing. It will also be highly desirable to install permanent VLBI hardware at all sites to allow turnkey VLBI observing in order to maximize the efficiency of VLBI observations in terms of personnel time and transportation costs.

6. Conclusions

Millimeter VLBI offers an unparalleled ability to probe the emission from Sgr A* at angular scales of a few $r_{\rm G}$ and on timescales of a few seconds. Current 1.3 mm VLBI observations have established that the millimeter emission emanates from a compact region offset from the center of the black hole. These data are already being used to constrain key physical parameters (e.g., spin, inclination, orientation) in models of the emission (e.g., RIAF models). Future additions to the VLBI array would allow the millimeter emission to be imaged directly. In the meantime, closure quantity analysis may allow the spin of the black hole to be inferred from source structure periodicity. The technical advancements necessary to realize these goals are already in progress.

Acknowledgements

VLBI work at MIT Haystack Observatory is supported through grants from the National Science Foundation. VLBI at (sub)millimeter wavelengths is made possible through broad international collaborative efforts and the support of staff and scientists at all participating facilities.

References

Broderick, A. E. & Loeb, A. 2005, *MNRAS*, 363, 353
Broderick, A. E., Fish, V. L., Doeleman, S. S., & Loeb, A. 2009, *ApJ*, 697, 45
Doeleman, S. S., Fish, V. L., Broderick, A. E., Loeb, A., & Rogers, A. E. E. 2009, *ApJ*, 695, 59
Doeleman, S. S., *et al.* 2001, *AJ*, 121, 2610
Doeleman, S. S., *et al.* 2008, *Nature*, 455, 78
Falcke, H., Melia, F., & Agol, E. 2000, *ApJ* (Letters), 528, L13
Fish, V. L., Broderick, A. E., Doeleman, S. S., & Loeb, A. 2009, *ApJ* (Letters), 692, L14
Reid, M. J. 2008, *Internat. J. Modern Phys. D*, in press, arXiv:0808.2624
Weintroub, J. 2008, *J. Phys. Conf. Ser.*, 131, 012047

Relativity in Fundamental Astronomy
Proceedings IAU Symposium No. 261, 2009
S. A. Klioner, P. K. Seidelman & M. H. Soffel, eds.

doi:10.1017/S1743921309990512

Optical interferometry from the Earth

Andreas Quirrenbach

Landessternwarte
Zentrum für Astronomie der Universität Heidelberg
Königstuhl 12
D-69117 Heidelberg, Germany
email: A.Quirrenbach@lsw.uni-heidelberg.de

Abstract. Ground-based optical interferometers can perform astrometric measurements with a precision approaching $10\,\mu$as between pairs of stars separated by $\sim 10''$ on the sky. These narrow-angle measurements can be used to search for extrasolar planets and to determine their orbital parameters, to characterize microlensing events, and to measure the orbits of stars around the black hole at the center of our Galaxy.

Keywords. instrumentation: interferometers, techniques: high angular resolution, techniques: interferometric, astrometry, gravitational lensing, Galaxy: center

1. Introduction

Astronomical interferometry at visible and infrared wavelengths has become an important tool in a number of fields, ranging from the measurement of fundamental stellar parameters and the determination of the distribution of circumstellar material to studies of the central regions of active galactic nuclei. Most of these applications rely on measurements of the visibility amplitudes (and sometimes closure phases) with a small number of baselines, and the parametric fitting of models to these data. This paper focuses mostly on a different interferometric technique, namely precise astrometry. Several instruments are currently under construction that will use this method to determine the orbits of extra-solar planets, to observe stars orbiting the black hole at the center of our Galaxy, and to characterize microlensing events. The first of these instruments to enter operation is PRIMA (Quirrenbach *et al.* 1998, Delplancke *et al.* 2000) at the Very Large Telescope Interferometer (VLTI), operated by the European Southern Observatory (ESO) on Cerro Paranal in Chile.

2. Interferometric Astrometry

2.1. *The Basic Principle of Interferometric Astrometry*

Astrometric observations by interferometry are based on measurements of the delay $D = D_{\rm int} + (\lambda/2\pi)\phi$, where $D_{\rm int} = D_2 - D_1$ is the internal delay measured by a metrology system (see Fig. 1), and ϕ the observed fringe phase (see e.g. Quirrenbach 2001 and references therein). D is related to the baseline $\vec{B}$ by

$$D = \vec{B} \cdot \hat{s} = B \cos\theta, \tag{2.1}$$

where $\hat{s}$ is a unit vector in the direction towards the star, and θ the angle between $\vec{B}$ and $\hat{s}$. Each data point is thus a one-dimensional measurement of the position of the star θ, provided that the length and direction of the baseline are accurately known. The second coordinate can be measured with a separate baseline at a roughly orthogonal orientation.

2.2. *Atmospheric Limitations of Ground-Based Astrometry*

The Earth's atmosphere imposes serious limitations on the precision that can be achieved with astrometric measurements from the ground. The first-order terms of the atmospheric wavefront distortions (frequently referred to as *tip* and *tilt*) are global wavefront gradients, which correspond to a motion of the centroid of the stellar light in the two coordinates. Because most of the power of atmospheric turbulence is in these low-order modes, the amplitude of this image motion is similar to the width of the stellar images, i.e., $\approx \lambda/r_0 \approx 0''.5...1''$. One can obviously reduce this error by taking many exposures and thus averaging over many independent realizations of the atmospheric turbulence, but achieving a precision of a small fraction of a milliarcsecond in this way is clearly not possible.

It helps, however, to make differential measurements over small angles on the sky, i.e., to measure the position of the target star with respect to that of a nearby reference. It can be shown that the variance σ_θ^2 of measurements of the angle θ is given by (Shao & Colavita 1992)

$$\sigma_\theta^2 \approx \frac{16\pi^2}{B^2 t} \int_0^\infty dh\, v^{-1}(h) \int_0^\infty d\kappa\, \Phi(\kappa, h) \cdot [1 - \cos(B\kappa)] \cdot [1 - \cos(\theta h \kappa)], \qquad (2.2)$$

if the integration time $t \gg \max(B, \theta h)/v$. Here $v(h)$ is the wind speed at altitude h, and $\Phi(\kappa, h)$ denotes the three-dimensional spatial power spectrum of the refractive index. It may at first seem surprising that stronger winds should give a smaller measurement error, but within the frozen-turbulence picture a higher wind speed means that one averages faster over independent realizations of the stochastic refractive index fluctuations. Inserting a Kolmogorov power spectrum in Eqn. 2.2 one obtains the two limiting cases

$$\sigma_\theta^2 \approx \begin{cases} 5.25\, B^{-4/3}\theta^2 t^{-1} \int_0^\infty dh\, C_N^2(h) h^2 v^{-1}(h) & \text{for } \theta \ll B/h, \quad t \gg B/v \\ 5.25\, \theta^{2/3} t^{-1} \int_0^\infty dh\, C_N^2(h) h^{2/3} v^{-1}(h) & \text{for } \theta \gg B/h, \quad t \gg \theta h/v \end{cases} \qquad (2.3)$$

for long and short baselines, respectively. In particular one can see that for sufficiently small angles θ the important scaling relations $\sigma_\theta \propto \theta$ and $\sigma_\theta \propto B^{-2/3}$ hold for the astrometric error σ_θ. For a good site such as Mauna Kea or Cerro Paranal astrometric measurements with a precision of $\sim 10\,\mu$as are possible over angles of $\sim 10''$. It is also apparent from the factor h^2 under the integral in this equation that the astrometric error is dominated by the turbulence at high altitudes. The low level of high-altitude turbulence at the South Pole would therefore make an astrometric interferometer at a site on the high Antarctic plateau an attractive possibility (Lawrence *et al.* 2004).

2.3. *Dual-Star Interferometry*

Because of the short coherence time of the atmosphere, precise astrometry from the ground requires simultaneous observations of the target and astrometric reference; it is not possible to alternate between the two as in the case of radio interferometry. In a dual-star interferometer, each telescope accepts two small fields and sends two separate beams through the delay lines. The delay difference between the two fields is taken out with an additional short-stroke differential delay line; an internal laser metrology system is used to monitor the delay difference. For astrometric observations, this delay difference ΔD is the observable of interest, because it is directly related to the coordinate difference between the target and reference stars; from Eqn. 2.1 it follows immediately that

$$\Delta D \equiv D_t - D_r = \vec{B} \cdot (\hat{s}_t - \hat{s}_r) = B(\cos\theta_t - \cos\theta_r), \qquad (2.4)$$

where the subscript t is used for the target, and r for the reference.

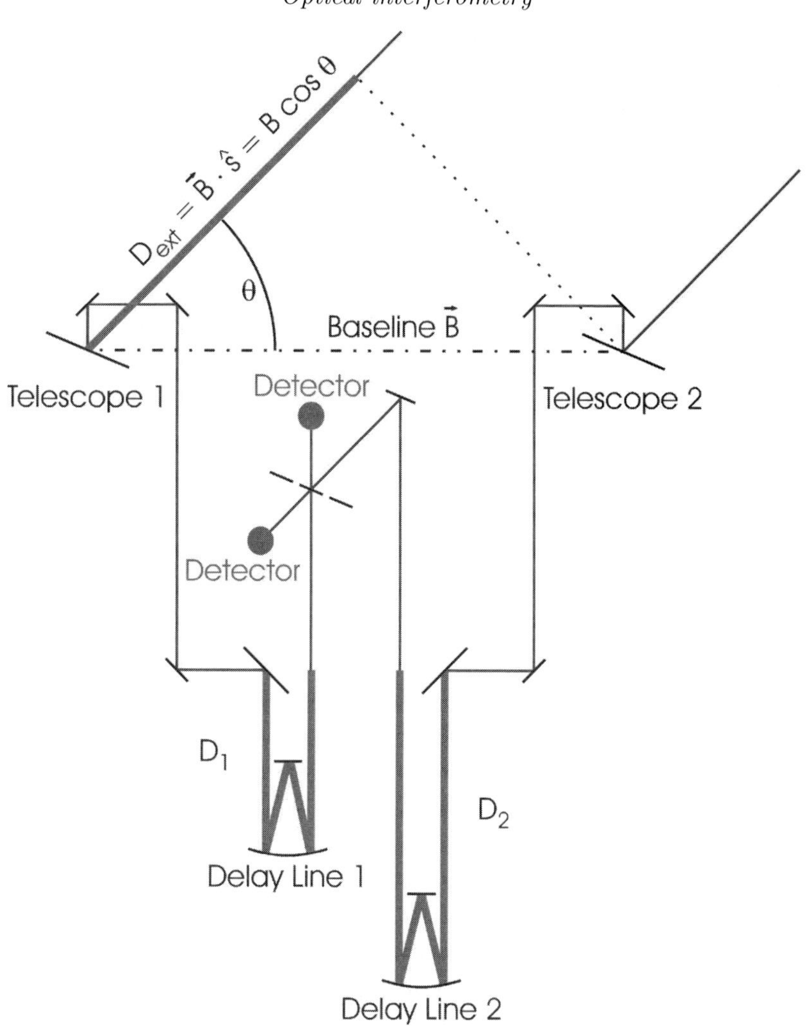

Figure 1. Schematic drawing of the light path through a two-element interferometer. The external delay $D = \vec{B} \cdot \hat{s}$ is compensated by the two delay lines. The pathlengths D_1, D_2 through the delay lines are monitored with laser interferometers. The zero-order interference maximum occurs when the delay line positions are such that the internal delay $D_{\rm int} = D_2 - D_1$ is equal to D.

Measurements of the delay difference between two stars give *relative* astrometric information; this means that the position information is not obtained in a global reference frame, but only with respect to the nearby comparison stars, which define a local reference frame on a small patch of sky. This approach greatly reduces the atmospheric errors, and some instrumental requirements are also relaxed (see below). The downside is that the information that can be obtained in this way is more restricted, because the local frame may have a motion and rotation of its own. This obviously makes it impossible to measure proper motions. Moreover, all parallax ellipses have nearly the same orientation and axial ratio, which allows only "relative parallaxes" to be measured.

2.4. *Astrometric Precision*

The photon noise limit for the precision σ of an astrometric measurement is given by the expression

$$\sigma = \frac{1}{\mathrm{SNR}} \cdot \frac{\lambda}{2\pi B}. \tag{2.5}$$

Since high signal-to-noise ratios can be obtained for bright stars, σ can be orders of magnitude smaller than the resolution λ/B of the interferometer. With an SNR ~ 50, it is thus possible to attain an astrometric error of $\sim 10\,\mu$as on the longest baselines of the VLTI, comparable to the atmospheric contribution expected for an angular separation of $10'' \ldots 20''$ and half-hour integrations (Shao & Colavita 1992, von der Lühe *et al.* 1995).

The fundamental instrumental requirements can be derived directly from the basic expression of the geometric delay, which can be written as

$$\Delta D \equiv D_t - D_r = \vec{B} \cdot (\hat{s}_t - \hat{s}_r) \equiv \vec{B} \cdot \Delta\vec{s}. \tag{2.6}$$

Here D_t and D_r denote the delay of the target and reference, respectively, $\vec{B}$ is the baseline vector, and $\hat{s}_t$ and $\hat{s}_r$ are unit vectors in the directions towards the two stars. The propagation of systematic errors in measurements of the differential delay $\delta\Delta D$ and of the baseline vector δB to errors in the derived position difference $\delta\Delta s$ can be estimated from the total differential

$$\delta\Delta s \approx \frac{\delta\Delta D}{B} + \frac{\Delta D}{B^2}\delta B = \frac{\delta\Delta D}{B} + \Delta s\frac{\delta B}{B}. \tag{2.7}$$

This formula allows one to draw two important conclusions. First, the systematic astrometric error is inversely proportional to the baseline length. Together with the $B^{-2/3}$ scaling of the atmospheric differential delay r.m.s. this clearly favors longer baselines, up to the limit where the target star gets resolved by the interferometer. The second important conclusion from Eqn. 2.7 is that the relative error of the baseline measurement gets multiplied with Δs; this means that the requirement on the knowledge of the baseline vector is sufficiently relaxed to make calibration schemes possible that rely primarily on the stability of the telescope mount. For a $10\,\mu$as (50 prad) contribution to the error budget for a measurement over a $20''$ angle, with an interferometer with a 100 m baseline, the metrology system must measure $\delta\Delta D$ with a 5 nm precision; the baseline vector has to be known to $\delta B \approx 50\,\mu$m (Quirrenbach *et al.* 1998). For PRIMA it is foreseen that the baseline vector will be determined from repeated observations of stars in the same way that is also customary in radio interferometry.

2.5. *PRIMA Observing and Data Reduction Strategy*

As explained above, astrometric observations with interferometers are equivalent to measurements of delays, i.e., to measurements of the difference in optical pathlength of light from a star at infinity to the two telescopes forming the interferometer†. The accuracy goal of $10\,\mu$as $=$ 50 prad corresponding to a total allowable error of 5 nm for a 100 m baseline can only be achieved through a quadruple-differential technique (Quirrenbach *et al.* 2004, Elias *et al.* 2008):

(*a*) Two stars with small angular separation are observed simultaneously to reduce the effects of atmospheric turbulence.

(*b*) The optical pathlength within the interferometer is monitored with a laser interferometer. The terms entering the error budgets are thus the differential effects between

† There are additional complications if the delay lines are not evacuated, see Daigne & Lestrade (1999).

the starlight and metrology beams, due e.g. to misalignments or dispersion between the effective observing wavelength and the wavelength of the metrology system.

(*c*) The paths of the two stars through the instrument are exchanged periodically by rotating the field by 180°. In this way many systematic errors caused by asymmetries are canceled.

(*d*) The orbits of extra-solar planets are determined from variations of the positions of their parent stars with time; only differences with respect to the position at some reference epoch matter.

It is important to realize that the raw delays have eleven (!) significant digits; astrometric planet detection implies taking differences of large and nearly equal numbers. The implementation of this quadruple-differential technique therefore requires unusual attention to detail in the understanding and calibration of varied astrophysical, atmospheric, and instrumental effects, in the construction of error budgets, in planning the operations, and in specifying and coding the data reduction software.

In particular, the desired accuracy can only be achieved if all systematic sources that can possibly affect the data are understood properly, and removed in a systematic way. While the magnitude of some astrometric errors can be predicted quite reliably (e.g., those related to atmospheric turbulence), others defy simple analysis and may have to be described with parameterized models (e.g., dynamic temperature gradients in the interferometer light ducts). Experience with other forefront astrometric facilities (e.g., the HIPPARCOS spacecraft, the Mark III Interferometer, the automated Carlsberg Meridian Circle) also shows that completely unanticipated systematic effects almost inevitably show up in the actual data. The ability to detect, diagnose, and remove such unanticipated effects is of paramount importance for the success of astrometric programs.

It is therefore necessary to perform a careful a priori analysis of the errors, and to design and implement systems for a posteriori analysis of remaining trends in the data. One further needs an operation and calibration strategy that takes full advantage of, and optimizes the use of the quadruple-differential technique described above. Finally one needs software to perform the initial steps of the data reduction, including carrying out said differences with appropriate corrections, and conversion of delays to angles on the sky. This data reduction software has to allow inspection of the residuals and to enable searches for remaining systematic trends over several years. The latter capability is required because the integrity of the data can only be checked after the quadruple-differencing process, and because the residuals are dominated by stellar parallax (which has a period of one year) and proper motion.

3. Astronomical Goals of Ground-Based Interferometric Astrometry

3.1. *Astrometric Planet Detection*

The first discovery of a planet orbiting a star similar to our Sun (Mayor & Queloz 1995) has opened a completely new field of astronomy: the study of extra-solar planetary systems. More than 350 planets outside our own Solar System are known to date, and new discoveries are announced at an increasing pace. These developments have started to revolutionize our view of our own place in the Universe. We know now that other planetary systems can have a structure that is completely different from that of the Solar System. Moreover, the existential question whether other habitable worlds exist can for the first time in human history be addressed in a scientific way.

Nearly all known extra-solar planets have been found with an indirect technique, the radial-velocity method. What is actually detected is not the planet itself, but the motion

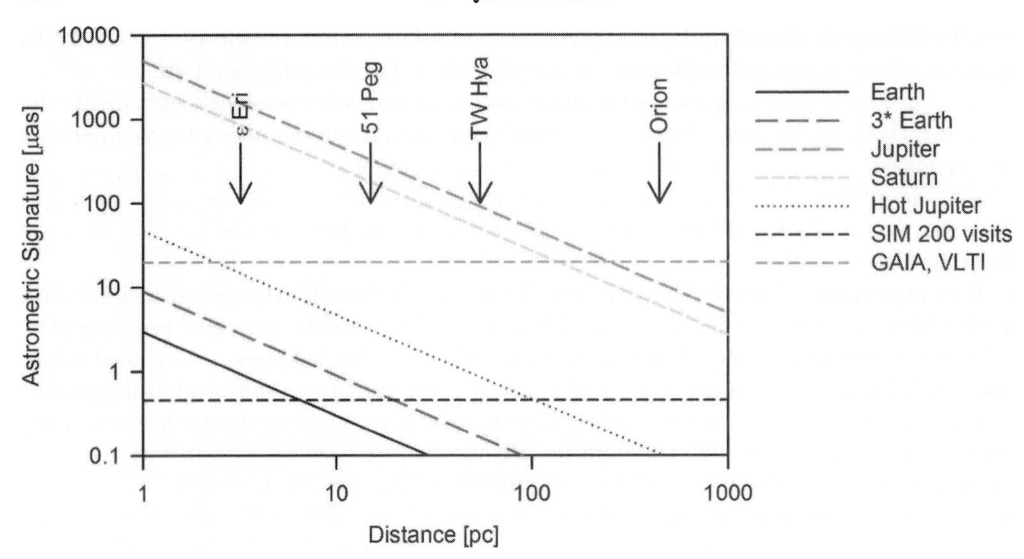

Figure 2. Astrometric signature (semi-amplitude) for five sample planets orbiting a Solar-mass star, as a function of distance. Anticipated detection limits for ground-based (VLTI PRIMA) and space-based (Space Interferometry Mission) instruments are also shown. Adopted from Quirrenbach (2003).

of its parent star around the common center of gravity. The Doppler shift due to the line-of-sight component of this motion can be detected with spectroscopic methods. While radial-velocity surveys have had tremendous successes, it must not be forgotten that they have technical and astrophysical limitations, which necessarily lead to a biased view of exo-planetary astrophysics. It is therefore important to develop complementary techniques, which can give additional information on the systems already detected, and find planets in situations where the radial-velocity technique cannot be used.

The principle of planet detection with astrometry is similar to that behind the Doppler technique: one infers the presence of a planet from the motion of its parent star around the common center of gravity. In the case of astrometry one observes the two components of this motion in the plane of the sky; this gives sufficient information to solve for the orbital elements without $\sin i$ ambiguity. Astrometry also has advantages for a number of specific questions, because this method is applicable to all types of stars, and more sensitive to planets with larger orbital semi-major axes.

From simple geometry and Kepler's Laws it follows immediately that the astrometric signal θ of a planet with mass m_p orbiting a star with mass m_* at a distance d in a circular orbit of radius a is given by

$$\begin{aligned} \theta &= \frac{m_p}{m_*}\frac{a}{d} = \left(\frac{G}{4\pi^2}\right)^{1/3} \frac{m_p}{m_*^{2/3}} \frac{P^{2/3}}{d} \\ &= 3\,\mu\text{as} \cdot \frac{m_p}{M_\oplus} \cdot \left(\frac{m_*}{M_\odot}\right)^{-2/3} \left(\frac{P}{\text{yr}}\right)^{2/3} \left(\frac{d}{\text{pc}}\right)^{-1} . \end{aligned} \tag{3.1}$$

This signature is shown in Fig. 2 for five sample planets (analogs to Earth, a "Super-Earth", Jupiter, Saturn, and a "Hot Jupiter" with $m_p = 1\,M_{\text{jup}}$ and $P = 4\,\text{days}$) orbiting a $1\,M_\odot$ star.

The specific strengths of the astrometric method enable it to answer a number of questions that cannot be addressed by any other planet detection method. Among the most prominent goals of astrometric planet surveys are the following projects:

• Mass determination for planets detected in radial velocity surveys (without the $\sin i$ factor). The RV method gives only a lower limit to the mass, because the inclination of the orbit with respect to the line-of-sight remains unknown. Astrometry can resolve this ambiguity, because it measures two components of the orbital motion, from which the inclination can be derived.

• Confirmation of hints for long-period planets in RV surveys. Many of the stars with detected short-period planets also show long-term trends in the velocity residuals. These are indicative of additional long-period planets, whose presence can be confirmed astrometrically.

• Inventory of planets around stars of all masses. The RV technique works well only for stars with a sufficient number of narrow spectral lines, i.e., fairly old main-sequence stars with $m_* \lesssim 1.2\,M_\odot$, and around G and K giants. Astrometry can detect planets around intermediate-mass main sequence stars and complete a census of gas and ice giants around stars of all types.

• Detection of gas giants around pre-main-sequence stars, signatures of planet formation. Astrometry can detect giant planets around young stars, and thus probe the time of planet formation and migration. Observations of pre-main-sequence stars of different ages can provide a critical test of the formation mechanism of gas giants. Whereas gas accretion on $\sim 10\,M_\oplus$ cores requires $\sim 10\,$Myr, formation by disk instabilities would proceed rapidly and thus produce an astrometric signature even at very young stellar ages (Boss 1998).

• Detection of multiple systems with masses decreasing from the inside out. Whereas the astrometric signal increases linearly with the semi-major axis a of the planetary orbit, the RV signal scales with $1/\sqrt{a}$. This leads to opposite detection biases for the two methods. Systems in which the masses increase with a are easily detected by the RV technique because the planets' signatures are of similar amplitudes. Conversely, systems with masses decreasing with a are more easily detected astrometrically.

• Determine whether multiple systems are coplanar or not. Many of the known extrasolar planets have highly eccentric orbits. A plausible origin of these eccentricities is strong gravitational interaction between two or several massive planets (Lin & Ida 1997, Papaloizou & Terquem 2001). This could also lead to orbits that are not aligned with the equatorial plane of the star, and to non-coplanar orbits in multiple systems.

• Search for massive terrestrial planets orbiting low-mass stars in the Solar neighborhood. With a $10\,\mu$as precision goal and operating in the K band, PRIMA at the VLTI will be able to look for rocky planets down to a limit of a few Earth masses around nearby M stars.

In summary, astrometry is a unique tool for dynamical studies of extrasolar planetary systems; its capabilities to determine masses and orbits are not matched by any other technique. Astrometric surveys of young and old planetary systems will therefore give unparalleled insight into the mechanisms of planet formation, orbital migration and evolution, orbital resonances, and interaction between planets. The first such program will be carried out with PRIMA at the VLTI, and pave the way towards future more precise astrometric surveys from space.

3.2. *Astrometry of Microlensing Events*

Photometric observations of microlensing events in rich stellar fields — such as the bulge of our Galaxy — have become a widely used tool in several fields of astrophysics.

Microlensing has been used to constrain the mass contained in massive compact halo objects (MACHOs), as suggested by Paczyński (1986), and to search for extra-solar planets (e.g., Beaulieu *et al.* 2006). Parameters of the lensing object can be derived from an analysis of the light curve, but unfortunately there are degeneracies between lens mass, relative proper motion, and relative parallax. These degeneracies can be broken by astrometric observations, determining the position of the center of light as a function of time during the encounter (Miyamoto & Yoshii 1995).

The astrometric signature of microlensing events is of the order of the Einstein radius, i.e., typically $\approx 100\,\mu$as, well within the reach of ground-based interferometry. The differential nature of ground-based astrometry prevents one from measuring absolute parallaxes. This introduces some complications, but such measurements are nevertheless sufficient for the present purpose (Boden *et al.* 1998). For relatively high lens masses (a few $M_{\odot}$), when the Einstein radius is large, the two images may be resolved by the interferometer, thus producing a binary signature in the observed visibilities (Deplancke *et al.* 2001). The main limitation of ground-based interferometry for observations of microlensing events is the fringe tracking sensitivity, which is sufficient only for events reaching exceptionally high brightness.

3.3. *Astrometry of the Galactic Center Cluster*

Beautiful images and spectral data cubes of the Galactic Center region obtained with speckle techniques and adaptive optics on large telescopes have provided a surprising wealth of information on the central stellar cluster (e.g., Ghez *et al.* 2005, Eisenhauer *et al.* 2005). In addition to providing probes of the gravitational field of the black hole, the stars in the central parsec of the Milky Way pose many interesting questions about their properties, formation, and dynamics.

Interferometry will enable measurements of more subtle effects such as the general-relativistic precession of stellar orbits, as well as the hypothetical precession due to an extended mass distribution (Rubilar & Eckart 2001, Eckart *et al.* 2002). In addition, it has been suggested that flaring infrared emission from the position of the black hole can be used to trace the potential well at a few Schwarzschild radii (Trippe *et al.* 2007). Interferometric astrometry of this source can then used to test general relativity in the strong-field regime. These scientific goals will be pursued with PRIMA (see Bartko *et al.* 2008); the ASTRA project at the Keck Interferometer (Pott *et al.* 2008) and GRAVITY at the VLTI are designed primarily to address these questions. More details can be found in the paper by Eisenhauer (2009) in this volume.

References

Bartko, H., Pfuhl, O., Eisenhauer, F., Genzel, R., Gillessen, S., Rabien, S., Abuter, R., v. Belle, G., Delplancke, F., Menardi, S., & Sahlmann, J. (2008). *Study of the science capabilities of PRIMA in the Galactic Center.* In *Optical and Infrared Interferometry.* Eds. Schöller, M., Danchi, W. C. & Delplancke, F., SPIE Vol. 7013, pp. 70134K-70134K-7

Beaulieu, J. P., Bennett, D. P., Fouqué, P., *et al.* (2006). *Discovery of a cool planet of 5.5 Earth masses through gravitational microlensing.* Nature **439**, 437–440

Boden, A. F., Shao, M., & van Buren, D. (1998). *Astrometric observation of MACHO gravitational microlensing.* ApJ **502**, 538–549

Boss, A. P. (1998). *Astrometric signatures of giant-planet formation.* Nature 393, 141–143

Daigne, G. & Lestrade, J. F. (1999). *Astrometric optical interferometry with non-evacuated delay lines.* A&AS **138**, 355–363

Delplancke, F., Górski, K. M., & Richichi, A. (2001). *Resolving gravitational microlensing events with long-baseline optical interferometry. Prospects for the ESO Very Large Telescope Interferometer.* A&A **375**, 701–710

Delplancke, F., Leveque, S. A., Kervella, P., Glindemann, A., & D'Arcio, L. (2000). *Phase-referenced imaging and micro-arcsecond astrometry with the VLTI.* In *Interferometry in Optical Astronomy.* Ed. Quirrenbach, A. & Léna, P., SPIE Vol. 4006, pp. 365–376

Eckart, A., Mouawad, N., Krips, M., Straubmeier, C., & Bertram, T. (2002). *Scientific potential for interferometric observations of the Galactic Center.* In *Future research direction and visions for astronomy.* Ed. Dressler, A. M., SPIE Vol. 4835, p. 12–21

Eisenhauer, F. (2009). This proceeding, 269

Eisenhauer, F., Genzel, R., Alexander, T., *et al.* (2005). *SINFONI in the Galactic Center: young stars and infrared flares in the central light-month.* ApJ **628**, 246–259

Eisenhauer, F., Perrin, G., Brandner, W., *et al.* (2009). *GRAVITY: Microarcsecond astrometry and deep interferometric imaging with the VLT.* In *Science with the VLT in the ELT Era.* Ed. Moorwood, A., pp. 361–365

Elias, N. M., Köhler, R., Stilz, I., Reffert, S., Geisler, R., Quirrenbach, A., de Jong, J., Delplancke, F., Tubbs, R. N., Launhardt, R., Henning, T., Mégevand, D., & Queloz, D. (2008). *The astrometric data-reduction software for exoplanet detection with PRIMA.* In *Optical and infrared interferometry.* Eds. Schöller, M., Danchi, W. C., & Delplancke, F., SPIE Vol. 7013, pp. 70133V-70133V-9

Ghez, A. M., Hornstein, S. D., Lu, J. R., *et al.* (2005). *The first laser guide star adaptive optics observations of the Galactic Center: Sgr A*'s infrared color and the extended red emission in its vicinity.* ApJ **635**, 1087–1094

Lawrence, J. S., Ashley, M. C. B., Tokovinin, A., & Travouillon, T. (2004). *Exceptional astronomical seeing conditions above Dome C in Antarctica.* Nature 431, 278–281

Lin, D. N. C. & Ida, S. (1997). *On the origin of massive eccentric planets.* ApJ 477, 781–791

Mayor, M. & Queloz, D. (1995). *A Jupiter-mass companion to a Solar-type star.* Nature **378**, 355-359

Miyamoto, M. & Yoshii, Y. (1995). *Astrometry for determining the MACHO mass and trajectory.* AJ **110**, 1427–1432

Paczyński, B. (1986). *Gravitational microlensing by the galactic halo.* ApJ **304**, 1–5.

Papaloizou, J. C. B. & Terquem, C. (2001). *Dynamical relaxation and massive extrasolar planets.* MNRAS 325, 221–230

Pott, J. U., Woillez, J., Akeson, R. L. *et al.* (2008). *Astrometry with the Keck-Interferometer: the ASTRA project and its science.* astro-ph 0811.2264

Quirrenbach, A. (2001). *Optical interferometry.* ARAA **39**, 353–401

Quirrenbach, A. (2003). *Astrometry as a precursor to DARWIN/TPF.* In *Towards other Earths – DARWIN/TPF and the search for extrasolar terrestrial planets.* Eds. Fridlund, M. & Henning, T., ESA SP-539, pp. 19–30

Quirrenbach, A., Coudé du Foresto, V., Daigne, G., Hofmann, K. H., Hofmann, R., *et al.* (1998). *PRIMA — study for a dual-beam instrument for the VLT Interferometer.* In *Astronomical interferometry.* Ed. Reasenberg, R. D., SPIE Vol. 3350, pp. 807–817

Quirrenbach, A., Henning, T., Queloz, D., Albrecht, S., Bakker, E., *et al.* (2004). *The PRIMA astrometric planet search project.* In *New frontiers in stellar interferometry.* Ed. Traub, W. A., SPIE Vol. 5491, pp. 424–432

Rubilar, G. F. & Eckart, A. (2001). *Periastron shifts of stellar orbits near the Galactic Center.* A&A **374**, 95–104

Shao, M. & Colavita, M. M. (1992). *Potential of long-baseline interferometry for narrow-angle astrometry.* A&A 262, 353–358

Trippe, S., Paumard, T., Ott, T., Gillessen, S., Eisenhauer, F., Martins, F., & Genzel, R. (2007). *A polarized infrared flare from Sagittarius A* and the signatures of orbiting plasma hotspots.* MNRAS **375**, 764–772

von der Lühe, O., Quirrenbach, A., & Koehler, B. (1995). *Narrow-angle astrometry with the VLT Interferometer.* In *Science with the VLT.* Ed. Walsh, J. R. & Danziger, I. J., pp. 445–450

Relativity in Fundamental Astronomy
Proceedings IAU Symposium No. 261, 2009
S. A. Klioner, P. K. Seidelman & M. H. Soffel, eds.

doi:10.1017/S1743921309990524

Very long baseline interferometry: accuracy limits and relativistic tests

Robert Heinkelmann[1] **and Harald Schuh**[2]

[1]Deutsches Geodätisches Forschungsinstitut DGFI,
Alfons-Goppel-Str. 11, 80539 München, Germany
email: heinkelmann@dgfi.badw.de

[2]Institute of Geodesy and Geophysics, Vienna University of Technology,
Gusshausstr. 27-29, 1040 Wien, Austria
email: harald.schuh@tuwien.ac.at

Abstract. We present a review on relativistic effects and best estimates of the relativistic PPN parameter γ obtained by analysis of data from the International VLBI Service for Geodesy and Astrometry (IVS). Relativistic implications are also considered in view of the upcoming new generation VLBI System: VLBI2010.

Keywords. VLBI, PPN parameter γ, VLBI2010

---

1. Introduction

Since the foundation of relativity by Einstein (1908) many groups have sought to verify this remarkable theory. Einstein (1911) himself was the first proposing an experimental test of the deflection of light by the Sun by observing star light very close to the Sun during a total solar eclipse. Such tests were actually carried out during almost all solar eclipses from 1919 on, e.g. by the Texas Mauritanian Eclipse Team (1976) and led to first experimental proofs of Einstein's theory within about ±20% uncertainty. Some years earlier, Very Long Baseline Interferometry (VLBI) came up and it was Shapiro (1967), who had the idea to utilize this new technique for the detection of light deflection. Seielstad *et al.* (1970) were among the first to realize radio interferometry for gravitational deflection in practice.

2. Determination of the PPN parameter γ by VLBI

The relativity in this context is usually described by the parameterized post-Newtonian (PPN) formalism (cf. Soffel, 1989), which defines a number of PPN parameters numerically expressing certain interactions between time, space, and e.g. mass. One of those variables, the parameter γ, describes how much space is curved by unit mass and equals unity in Einstein's theory. Expressed in the PPN formalism, the angle θ through which a radiowave is deflected is given approximately by (e.g. Shapiro *et al.*, 2004)

$$\theta \approx \frac{(1+\gamma)\,GM}{c^2 b}\,(1+\cos\phi) \tag{2.1}$$

where GM denotes the product of the Newtonian gravitational constant and the mass of the considered gravitating body, c the vacuum speed of light, b the distance of closest approach of the radiowave's path to the center of the gravitating body, and ϕ is the angle between the gravitating body and the radio source as viewed from Earth.

Besides the bending effect, interferometry is capable of measuring the temporal delay of radiowaves, also known as Shapiro delay (Fig. 1). In the case of VLBI the gravitational time delay is considered as

$$\tau_{\text{grav}} = (1+\gamma) \cdot \frac{GM}{c^3} \cdot \ln \left[\frac{|\mathbf{x_1}| + \mathbf{x_1} \cdot \mathbf{k}}{|\mathbf{x_2}| + \mathbf{x_2} \cdot \mathbf{k}} \right] \tag{2.2}$$

where $\mathbf{x_i}$ denotes the position vector of the i-th antenna with respect to the center of the gravitating body and $\mathbf{k}$ is the unit vector towards the radiosource as viewed from the Earth-bound baseline.

The partial derivative of the delay with respect to γ is easily comprehensible:

$$\frac{\partial \tau}{\partial \gamma} = \frac{GM}{c^3} \cdot \ln \left[\frac{|\mathbf{x_1}| + \mathbf{x_1} \cdot \mathbf{k}}{|\mathbf{x_2}| + \mathbf{x_2} \cdot \mathbf{k}} \right] \tag{2.3}$$

Up to now several groups have determined the parameter γ using the geodetic VLBI observations, which are provided by the International VLBI Service for Geodesy and Astrometry (IVS) (Tab. 1). All of those tests focus on the effects imposed by the Sun. However, there were also several efforts to observe Jupiter's deflection. Treuhaft & Lowe (1991) tried to find the deflection by Jupiter experimentally using a single long baseline DSN experiment during a near-occultation event, which was proposed by Schuh *et al.* (1988). A comparable near-occultation happened in 2002 and was investigated by several groups, e.g. by Fomalont & Kopeikin (2003).

In about mid 2002 a more stringent cut-off Sun elongation angle of about 14 deg was used for the IVS scheduling, significantly lowering the sensitivity of the VLBI observables during high solar activity, but unfortunately also for the gravitational time delay determination. Even with the new cut-off elongation angle geodetic VLBI remains competitive to γ-determinations by spacecraft ranging (Shapiro *et al.*, 1971; Reasenberg *et al.*, 1979; Bertotti *et al.*, 2003), in particular due to the large amount of involved data observing at many epochs and in most directions of the universe. Without commenting on the reported formal errors of the spacecraft ranging analyses, it can be stated, that geodetic VLBI tends to handle commited and ommited errors rather conservatively. Thus, relativity remains a point of interest to the IVS and its Observing Program Committee (OPC) is open for suggestions to carry out special Research and Development (R&D) observing sessions during relativistic scenarios.

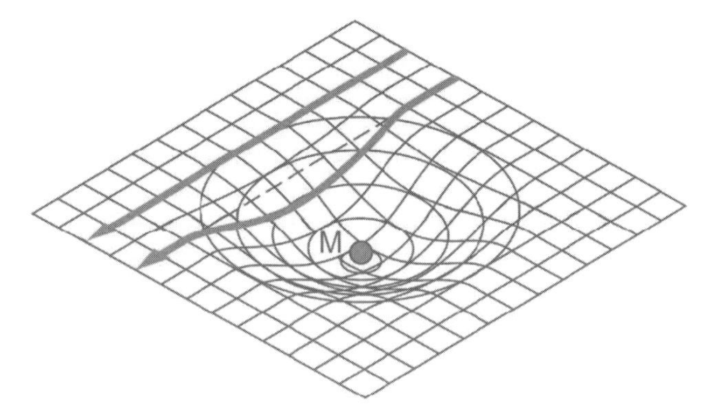

Figure 1. Gravitational time delay of radiowaves; 'Shapiro delay'.

Table 1. Review of γ-determination using geodetic VLBI data.

Author(s)	Year	γ	σ	Data
Counselman *et al.*	1974	0.98	±0.06	1972 occultation of 3C279 by the Sun
Fomalont & Sramek	1975	1.0075	±0.022	1974 occultation of 3C279 by the Sun
Fomalont & Sramek	1976	1.0035	±0.018	1974 and 1975 occultation of 3C279 by the Sun
Robertson & Carter	1984	1.008	±0.005	MERIT, POLARIS, IRIS
Carter, Robertson & MacKay	1985	1.000	±0.003	POLARIS, IRIS since 1980
Robertson, Carter & Dillinger	1991	1.000	±0.002	POLARIS, IRIS, CDP since 1980
Lebach *et al.*	1995	0.9996	±0.0017	1987 occultation of 3C279 by the Sun
Eubanks *et al.*	1997	0.99994	±0.00031	Geodetic VLBI sessions from 1979 to 1997
Shapiro *et al.*	2004	0.9998	±0.0002[1]	Geodetic VLBI sessions from 1979 to 1999
Lambert & Le Poncin-Lafitte	2009	0.99984	±0.00015[2]	Geodetic VLBI sessions from 1979 to 2008

Notes:
[1]Shapiro *et al.* (2004) estimate the standard error to 0.0004, however their reported formal error is 0.0002, which is used here to achieve a consistent comparison.
[2]Lambert & Le Poncin-Lafitte (2009) claim an estimated limit of 0.0002 to the standard error.

3. VLBI2010 and the relativistic delay model

To fulfill the requirements of space geodesy, in particular specified through GGOS, the Global Geodetic Observing System of the IAG, and to overcome the current limitations of the existing VLBI infrastructure, the IVS, in 2003, established Working Group 3 (WG3) named VLBI2010. The goal of VLBI2010 is to significantly modernize the existing VLBI system towards

- a positional acuracy of 1 mm,
- a velocity accuracy of 0.1 mm $\cdot$ yr^{-1},
- continuous monitoring of the Earth orientation parameters (EOP), and
- near real-time availibility of the results.

Working Group 3 closed its efforts presenting a Final Report to the IVS in 2005 (Niell *et al.*, 2005). The progress towards a new VLBI system, however, was taken over by the VLBI2010 Committee (V2C), which was established to continue and realize the studies recommended by WG3. To move to these targets, V2C proposed a number of strategies including VLBI antenna system considerations (Tab. 2) and extensive simulation studies were carried out (Petrachenko *et al.*, 2009).

Table 2. VLBI2010: System characteristics.

	Current VLBI	VLBI2010
Antenna size	5 to 100 m dish	about 12 m dish
Slew speed	about 20 to 200 deg $\cdot$ min^{-1}	$\geqslant$ 360 deg $\cdot$ min^{-1}
Sensitivity	200 to 15,000 SEFD	$\leqslant$ 2,500 SEFD
Frequency range	S/X band	about 2 to 15 GHz
Recording rate	128 or 256 Mbps	8 to 16 Gbps
Data transfer	usually ship disks, some e-transfer	e-transfer, e-VLBI ship disks when required

The current conventional relativistic VLBI model (IERS, 2004, chapter 11) is based on the 'consensus model' (Eubanks, 1991), which united various VLBI delay models prior to 1991. Klioner (1991) presented an additional model, which considers sources at finite distances, e.g. artificial satellites or spacecrafts and baselines including space-borne or orbiting telescopes as well. At that time a precision of 1 ps was aimed for and remained sufficient until now. For the target accuracy of VLBI2010, however, it will be necessary to revisit the consensus model and to assess, whether it includes all terms down to the order of 0.3 ps.

4. Conclusion and outlook

Astrometric and geodetic VLBI is able to measure effects of general relativity, the relativistic deflection and the relativistic time delay (Shapiro delay). With its huge amount of observations in all directions of the universe, currently about 5 million delays over about 30 years, the IVS maintains universal, reliable, and robust data for applied relativity. Simulations have shown, that a network of VLBI2010 stations provides the same amount of data (5 million observations) in a few weeks. Future developments in view of VLBI2010 foresee significant improvements and demand the relativistic VLBI delay model to hold for a slightly better (0.3 ps) accuracy. Besides the Sun and the Earth, the gravitational time delays of Jupiter, Saturn, Venus, and the Moon will have to be considered in standard VLBI2010 analyses.

The main factors of uncertainty of VLBI are
- the intrinsic source structure,
- the wet neutral atmosphere contribution,
- the uneven North-South distribution of the network, and
- the solar coronal plasma for smaller Sun elongations.

With structure corrections and special VLBI sessions dedicated to general relativity the accuracy of the γ-determination could be significantly better than 10^{-4}.

References

Bertotti, B., Iess, L., & Tortora, P. 2003, A test of general relativity using radio links with the Cassini spacecraft. *Nature*, 425, 374–376

Carter, W. E., Robertson, D. S., & MacKay, J. R. 1985, Geodetic Radio Interferometric Surveying: Applications and Results. *Journal of Geophysical Research*, 90, B6, 4577–4587

Counselman, C. C., Kent, S. M., Knight, C. A., Shapiro, I. I., Clark, T. A., Hinteregger, H. F., Rogers, A. E. E., & Whitney, A. R. 1974, Solar Gravitational Deflection of Radio Waves Measured by Very-Long-Baseline Interferometry. *Physical Review Letters*, 33, 27, 1621–1623

Einstein, A. 1908, Über das Relativitätsprinzip und die aus demselben gezogenen Folgerungen. *Jahrbuch der Radioaktivität und Elektronik*, IV., 4., 411–462

Einstein, A. 1911, Über den Einfluß der Schwerkraft auf die Ausbreitung des Lichtes. *Annalen der Physik*, 35, 898–908

Eubanks, T. M. 1991, A Consensus Model for Relativistic Effects in Geodetic VLBI. In: Eubanks, T. M. (ed.), *Proceedings of the USNO Workshop on Relativistic Models for use in Space Geodesy*, 60–82

Eubanks, T. M., Martin, J. O., Archinal, B. A., Josties, F. J., Klioner, S. A., Shapiro, S., & Shapiro, I. I. 1997, Advances in Solar System Tests of Gravity. American Physical Society, APS/AAPT Joint Meeting, April 18–21, 1997, abstract #K11.05

Fomalont, E. B. & Sramek, R. A. 1975, A confirmation of Einstein's general theory of relativity by measuring the bending of microwave radiation in the gravitational field of the Sun. *The Astronomical Journal*, 199, 3, 749–755

Fomalont, E. B. & Sramek, R. A. 1976, Measurement of the Solar Gravitational Deflection of Radio Waves in Agreement with General Relativity. *Physical Review Letters*, 36, 25, 1475–1478

Fomalont, E. B. & Kopeikin, S. M. 2003, The Measurement of the Light Deflection from Jupiter: Experimental Results. *arXiv:astro-ph/0302294v2*, 1–10

IERS: IERS Conventions (2003). 2004, in: McCarthy D. D. & Petit, G. (eds.) *IERS Technical Note*, 32

Klioner, S. A. 1991, General relativistic model of VLBI observables. In: U. S. Department of Commerce. National Oceanic and Atmospheric Administration. National Ocean Service (eds.), *Proceedings of the AGU Chapman Conference on Geodetic VLBI: Monitoring Global Change*, NOAA Technical Report NOS 137 NGS 49, 188–202

Lambert, S. B. & Le Poncin-Lafitte, C. 2009, Determination of the relativistic parameter γ using very long baseline interferometry. *Astronomy & Astrophysics*, arXiv:0903.1615v1, 1–6

Lebach, D. E., Corey, B. E., Shapiro, I. I., Ratner, M. I., Webber, J. C., Rogers, A. E. E., Davis, J. L., & Herring, T. A. 1995, Measurement of the Solar Gravitational Deflection of Radio Waves Using Very-Long-Baseline Interferometry. *Physical Review Letters*, 75, 8, 1439–1442

Niell, A. *et al.* 2005, VLBI2010: Current and future requirements for Geodetic VLBI Systems. `http://ivscc.gsfc.nasa.gov/about/wg/wg3/IVS_WG3_report_050916.pdf`, 32

Petrachenko, B. *et al.* 2009, Design Aspects of the VLBI2010 System. Progress Report of the IVS VLBI2010 Committee. In: Behrend D. & K. Baver (eds.), *2008 IVS Annual Report*, `ftp://ivscc.gsfc.nasa.gov/pub/misc/V2C/PR-V2C_090417.pdf`, 56

Reasenberg, R. D. *et al.* 1979, *VIKING* relativity experiment: verification of signal retardation by solar gravity. *The Astronomical Journal*, 234, L219–L221

Robertson, D. S. & Carter, W. E. 1984, Relativistic deflection of radio signals in the solar gravitational field measured with VLBI. *Nature* 310, 572–574

Robertson, D. S., Carter, W. E., & Dillinger, W. H. 1991, New measurement of the solar gravitational deflection of radio signals using VLBI. *Nature*, 349, 768–770

Schuh, H., Fellbaum, M., Campbell, J., Soffel, M., Ruder, H., & Schneider, M. 1988, On the deflection of radio signals in the gravitational field of Jupiter. *Physics Letters A*, 129, 5–6, 299–300

Seielstad, G. A., Sramek, R. A., & Weiler, K. W. 1970, Measurement of the deflection of 9.602-GHz radiation from 3C279 in the solar gravitational field. *Physical Review Letters*, 24, 24, 1373–1376

Shapiro, I. I. 1967, New Method for the Detection of Light Deflection by Solar Gravity. *Science*, 157, 806–808

Shapiro, I. I. *et al.* 1971, Fourth Test of General Relativity: New Radar Result. *Physical Review Letters*, 26, 18, 1132–1135

Shapiro, S. S., Davis, J. L., Lebach, D. E., & Gregory, J. S. 2004, Measurement of the Solar Gravitational Deflection of Radio Waves using Geodetic Very-Long-Baseline Interferometry Data, 1979–1999. *Physical Review Letters*, 92, 12, 121101 1–4

Soffel, M. H. 1989, Relativity in Astrometry, Celestial Mechanics and Geodesy. *Springer*, Berlin, Heidelberg, New York, A&A library, ISBN 3-540-18906-8, 208

Texas Mauritanian Eclipse Team (Brune, R. A. Jr., Cobb, C. L., DeWitt, B. S. *et al.*) 1976, Gravitational deflection of light: solar eclipse of 30 June 1973 I. Description of procedures and final results. *The Astronomical Journal*, 81, 6, 452–454

Treuhaft, R. N. & Lowe, S. T. 1991, A measurement of planetary relativistic deflection. *The Astronomical Journal*, 102, 5, 1879–1888

Relativity in Fundamental Astronomy
Proceedings IAU Symposium No. 261, 2009
S. A. Klioner, P. K. Seidelman & M. H. Soffel, eds.

doi:10.1017/S1743921309990536

Recent VLBA/VERA/IVS tests of general relativity

Ed Fomalont[1], Sergei Kopeikin[2], Dayton Jones[3], Mareki Honma[4] & Oleg Titov[5]

[1]National Radio Astronomy Observatory,
520 Edgemont Rd, Charlottesville, VA 22903, USA
email: efomalon@nrao.edu

[2]Dept. of Physics & Astronomy, University of Missouri-Columbia,
223 Physics Bldg., Columbia, MO 65211, USA
email: kopeikins@missouri.edu

[3]Jet Propulsion Laboratory,
4800 Oak Grove Ave, Pasadena, CA 91109, USA
email: dayton.jones@jpl.nasa.gov

[4]VERA Project Office, Mizusawa VLBI Observatory,
NAOJ, 181-8588, Tokyo, Japan
email: mareki.honma@nao.ac.jp

[5]Geoscience, Australia, GPO Box 378, Canberra, ACT 2601, Australia
email: Oleg.Titov@ga.gov.au

Abstract. We report on recent VLBA/VERA/IVS observational tests of General Relativity. First, we will summarize the results from the 2005 VLBA experiment that determined gamma with an accuracy of 0.0003 by measuring the deflection of four compact radio sources by the solar gravitational field. We discuss the limits of precision that can be obtained with VLBA experiments in the future. We describe recent experiments using the three global arrays to measure the aberration of gravity when Jupiter and Saturn passed within a few arcmin of bright radio sources. These reductions are still in progress, but the anticipated positional accuracy of the VLBA experiment may be about 0.01 mas.

Keywords. gravitation—quasars: individual (3C279)—relativity techniques: interferometric

---

1. The VLBA 2005 Solar-Bending Experiment

The history of the experiments to measured the deflection of light by the solar gravitational field is well-known. The 1919 optical experiment during a solar eclipse in Brazil established the viability of GR, and Shapiro (1964) suggested that radio techniques would provide much better accuracy. The precision of radio experiments over the last 40 years $\gamma < 0.001$ was achieved (e.g. Lebach *et al.* 1995, Shapiro *et al.* 2004).

Since the inception of operation in 1990, the positional accuracy that the Very Long Baseline Array (VLBA) can achieve has increased markedly, and it is now capable of measuring the angular separation between compact radio sources that are a few degrees apart in the sky with an accuracy of about 0.01 mas. These advances were made from increased stability and sensitivity of the array, more accurate determination of astrometric and geodetic parameters from the International VLBI Service (IVS), and the use of phase delay referencing even with antenna baselines greater than 5000 km. We, thus, observed on eight days in October 2005 in order to determined the change in relative

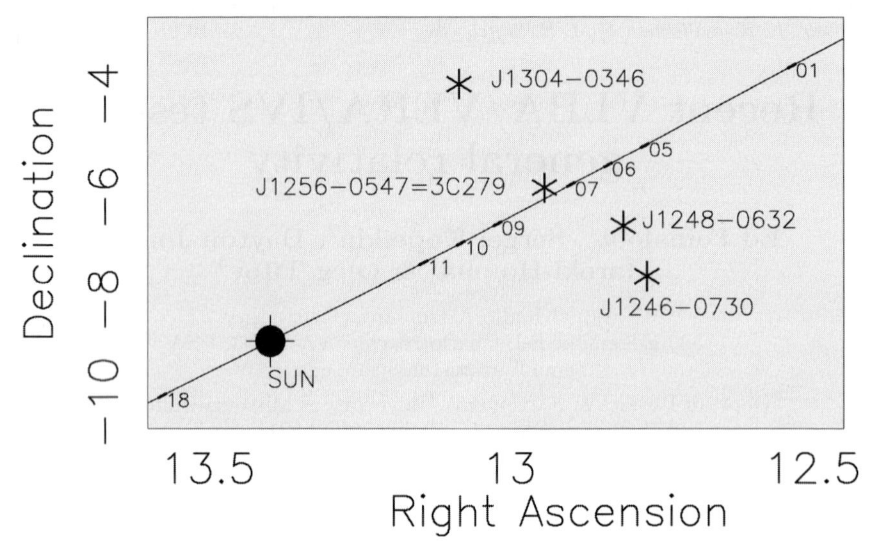

Figure 1. The Source Configuration for the Solar Deflection Experiment of 2005: The solar trajectory between October 1 and 18 is shown by the diagonal line, with the eight observing days superimposed. The location of the four radio sources are indicated.

position among four sources as the sun moved through the area in order to measure γ with high precision. The observing days and sources are shown in Fig. 1.

In order to minimize the effects of the solar corona, we used the highest routinely used VLBA frequency of 43 GHz. In order to lessen the effects of the changing tropospheric refraction above each telescope, we alternated source observations every 40 seconds. The choice of sources was a compromise among three criteria: sufficiently strong and compact to be detectable by the VLBA in about 20 sec; sufficiently close in the sky to avoid large tropospheric phase changes; but not closer than about two degrees in order to have a significant differential solar gravitational bending. We also observed at 23 GHz and 15 GHz (switching frequencies every 40 min) in order to estimate and remove the long-term coronal refraction.

The description of the experimental parameters, the reduction methods, and the analysis technique to estimate γ and its uncertainty are given in Fomalont *et al.* (2009). The results are $\gamma = 0.9998 \pm 0.0003$ (standard error). This is the most accurate radio interferometric result to date, although it is less accurate than the 2002 Doppler-tracking Cassini experiment (Bertotti *et al.* 2003) if one postulates that the translational gravito-magnetic field affects propagation of radio waves exactly as predicted in general relativity. This postulate was not parameterized in NASA ODP code and could not be directly tested in the Cassini experiment (Kopeikin *et al.* 2004, Bertotti *et al.* 2008, Kopeikin *et al.* 2009).

The accuracy of future VLBA experiments can be increased by a factor of four with several improvements: observing on all days when the sources were between 4° to 7° from the sun; choosing available sources in April to August when the sun is at its most northern declinations; increasing the relative observing time at 43 GHz since the coronal refraction correction was small; and scheduling an experiment as often as possible since about ten groups of sources near the ecliptic are available for a high precision experiment.

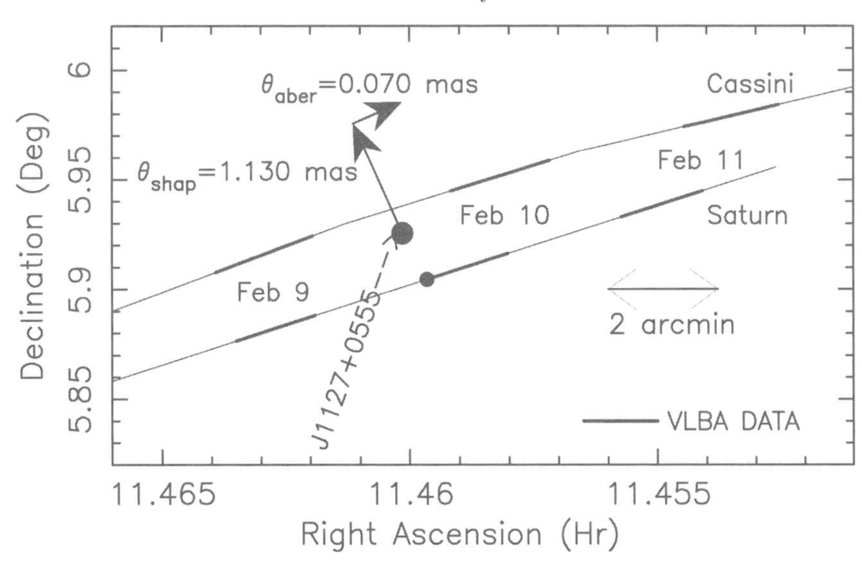

Figure 2. The Source Configuration for the February 2009 Aberration Experiment: The sky motion of Saturn and the Cassini spacecraft are shown by the lines from the lower left to upper right. The VLBA observing periods are shown by the bold part of the lines. At the closest approach of Saturn to J1127 + 0555, the predicted source deflection is 1.13 mas and the aberrational deflection is 0.070 mas.

2. The Planetary Gravitational Aberration Experiments

On September 8, 2002, Jupiter passed within 4′ of the quasar J0842+1835. Because of the motion of Jupiter, the gravitational bending of the quasar position was not precisely radial from the planet. For this experiment, the GR prediction of the non-radial deflection is 0.05 mas in the direction of Jupiter's motion in the sky at closest Jupiter/quasar encounter. (The radial deflection component at this time was 1.1 mas.) This aberrational-type deflection varies as $(v_J/c)\, d^{-2}$, where d is the angular separation in the sky between Jupiter and the quasar, and v_J is the *heliocentric* velocity of Jupiter (Kopeikin 2001). This aberrational deflection was measured with the VLBA with an accuracy of 20% (Fomalont & Kopeikin 2003). It demonstrated the gravito-magnetic effect caused by translational mass-current of a moving body, and also initiated many discussions about the experimental interpretation in the framework of the general theory of relativity.

Two planetary near encounters with bright radio sources have recently occurred. On November 19, 2008, Jupiter passed within 1.4′ of J1925 − 2219, and on February 10, 2009, Saturn pass within 1.3′ of J1127 + 0555. The arrays that observed these encounters were: International VLBI Service (IVS) array for both dates, the Japanese VLBI Exploration of Radio Astronomy (VERA) array for the November encounter (Honma *et al.* 2003), and the VLBA for the February encounter. The IVS observing programs are coordinated with the international community and use a variety of telescopes for semi-weekly to bi-monthly observations to monitor the earth orientation (see http://www.iers.org). For the November 19 experiment, the observing array included Parkes, Hobart26, Kokee, and Tsukub32. The nominal IVS 24-hour schedule (session OHIG60) was modified so that additional 8-hour observations of J1925 − 2219 and a nearby source J2000 − 1748 were included in order to determine the change of their relative position during the experiment. For the February 10 experiment (session RD0902), the observing array consisted of Parkes, Hobart26, Medicina, Matera, Badary, LA-VLBA, and KP-VLBA. The calibrator

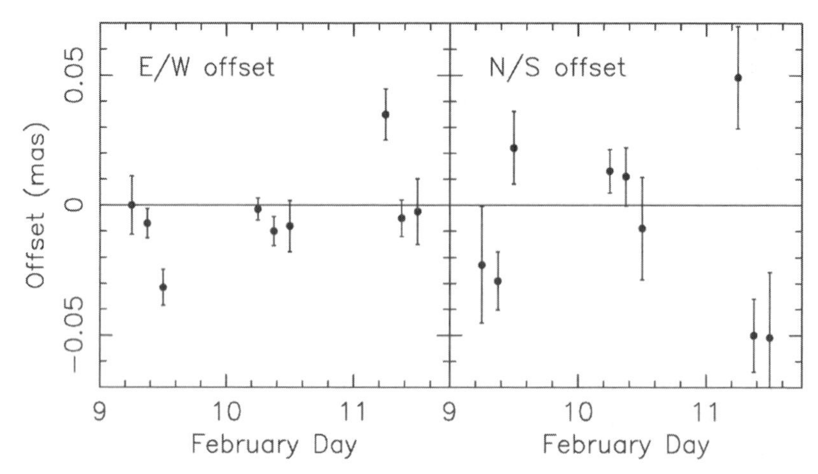

Figure 3. The Residual Differential Separation Between Cassini and J1127+0555: The separation between the two sources, after removal of the best-fit offset, velocity and acceleration from the measured offsets. The predicted gravitational bending of both Cassini and J1127 + 0555 has been removed, so the difference should be zero if the GR prediction is correct. The left plot is the east-west difference, the right plot is the north-south difference. The position difference estimates were obtained for the three 2.5-hour intervals on each day, and the error are the one-sigma errors expected from signal-to-noise considerations alone.

source for J1127 + 0555 was J1112 + 0724. The analysis of these two experiments are progressing.

The VERA observations were made on November 17, 19 (day of close encounter) and 22, each day for seven hours. This array has a dual-beam system so that two sources within 2.2° can be observed simultaneously—J1923−2104 and J1925−2219 are separated by 1.4°—and the expectation is that their relative positions can be determined to an accuracy of about 0.05 mas every two hours. The analysis is also in progress.

The reductions for the February VLBA experiment are nearly completed. The source configuration during the experiment is shown in Fig. 2. An interesting aspect of this experiment is that J1127+0555 was measured with respect to Cassini. (The emission from Saturn is too extended to be detected by the VLBA.) which was within 5′ of the source between February 9 to 11. With such a close encounter, the effects of the tropospheric refraction are small so the relative position of Cassini and the source could be measured with high accuracy, with the limit imposed by the signal-to-noise of the experiment of 0.004 mas. Both objects were sufficiently close to be observed simultaneously, rather than by switching between sources, further decreasing the tropospheric effects. However, the measurement of the deflection of the source requires that the orbit of Cassini is precisely known.

Because the VLBA experiment was correlated in early March using an approximate orbit of Cassini, the assumed position of Cassini used in the reductions produced a position difference with that of J1127+0555 that slowly drifted during the 3-day experiment. Cassini's orbital parameters cannot be predicted more than one month in advance because of interactions with the Saturnian moons and also by the occasional Cassini thrusts to optimize the orbit. We expect to have a much more accurate orbit for Cassini from the JPL Cassini navigation group by mid-2009.

Nevertheless, it is possible to surmise the positional sensitivity of the VLBA experiment. A good assumption is that the orbital model error of Cassini used in the VLBA reductions can be approximated by constant offset, velocity and acceleration over the observing period. If we further remove the radial and aberration deflection prediction by GR, then the resultant relative position of Cassini with respect to J1127 + 0555 should be zero. This position difference with the above adjustments is shown in Fig. 3. The departure of the residuals from zero has an error of about 0.01 mas E/W, and 0.02 mas N/S (this resolution is twice as poor). Since the aberrational deflection is 0.07 mas at closest encounter on February 10, this experiment may be a more accurate measure of the aberration deflection than that of the 2002 Jupiter experiment (Fomalont & Kopeikin 2003).

For the actual position comparison, we will use the observations on February 9 and 11 to determine the residual offset and velocity of Cassini with respect to J1127 + 0555. The acceleration residial of the Cassini orbit, however, must be known to less than 2.6×10^{-6} m^{-2}, corresponding to angular change of 0.02 mas over 48 hours, in order to interpolate the spacecraft quasar offset accurately on February 10 when the gravitational deflection is large.

3. Conclusion

Using the VLBA at 43 GHz with phase referencing observation, we have measured γ with an rms precision of 0.0003. With improved experimental strategies and observations of many source clusters near the ecliptic, the precision can be increased by at least a factor of four.

Recent experiments with the VLBA, the IVS array and VERA of radio sources with near encounters with Jupiter and Saturn will provide more accurate measurements of the gravito-magnetic effect by measuring the aberrational deflection of the sources. Results are not yet available, although the VLBA precision may be about 0.01 mas and, thus, produce a more accurate result than the 2002 VLBA experiment (Fomalont & Kopeikin 2003).

The National Radio Astronomy Observatory is a facility of the National Science Foundation operated under cooperative agreement by Associated Universities, Inc. Part of this research was performed at the Jet Propulsion Laboratory, California Institute of Technology, under contract with the National Aeronautics and Space Administration. The Parkes 64-meter is operated by the Australian Telescope National Facility (ATNF). The scheduling of the both IVS sessions was kindly done by Dirk Behrend and John Gipson (NVI GSFC). Sergei Kopeikin thanks the Research Council (Grant No. C1669103) and Alumni Organization of the University of Missouri-Columbia (2009 Faculty Incentive Grant) for support of this work.

References

Bertotti, B., Iess, L., & Tortora, P. 2003, *Nature*, 425, 374
Bertotti, B., Ashby, N., & Iess, L. 2008, Classical and Quantum Gravity, 25, 045013
Fomalont, E. & Kopeikin, S. 2003, *ApJ* 598,704
Fomalont, E., Kopeikin, S., Lanyi, G., & Benson, J. (2009) *Ap.J*, 697, 000
Honma, M, Fuji, T., Hirota, T. *et al.* (2003), *PASJ*, 55, L57
Kopeikin, S. 2001, *ApJ*, 556, 1
Kopeikin, S. M., Polnarev, A. G., Schäfer, G. & Vlasov, I. Y. 2007, *Phys Let A*, 367, 276
Kopeikin, S. 2009, *Phys. Lett. A*, in press (e-print arXiv:0901.4818)
Lebach, D. E., Corey, B. E., Shapiro. I. I., Ratner, M. I., Webber, J. C., Rogers, A. E. E., Davis, J. L., & Herring, T. A. 1995, *Phys Rev Lett*, 75, 1439
Shapiro, I. I. 1964, *Phys. Rev. Lett.*, 13, 789
Shapiro, S. S., Davis, J. L, Lebach, D. E., & Gregory, J. S. 2004, *Phys Rev Lett*, 92, 121101

Relativity in Fundamental Astronomy
Proceedings IAU Symposium No. 261, 2009
S. A. Klioner, P. K. Seidelman & M. H. Soffel, eds.

doi:10.1017/S1743921309990548

Gaia: Astrometric performance and current status of the project

Lennart Lindegren[1]

[1]Lund Observatory, Lund University,
Box 43, SE-22100 Lund, Sweden
email: lennart@astro.lu.se

Abstract. The scientific objectives of the Gaia mission cover areas of galactic structure and evolution, stellar astrophysics, exoplanets, solar system physics, and fundamental physics. Astrometrically, its main contribution will be the determination of millions of absolute stellar parallaxes and the establishment of a very accurate, dense and faint non-rotating optical reference frame. With a planned launch in spring 2012, the project is in its advanced implementation phase. In parallel, preparations for the scientific data processing are well under way within the Gaia Data Processing and Analysis Consortium. Final mission results are expected around 2021, but early releases of preliminary data are expected. This review summarizes the main science goals and overall organisation of the project, the measurement principle and core astrometric solution, and provide an updated overview of the expected astrometric performance.

Keywords. methods: data analysis, space vehicles: instruments, catalogs, surveys, astrometry, reference systems, stars: distances, stars: kinematics

---

1. Science goals of Gaia

The ESA Gaia mission is an all-sky astrometric and spectrophotometric survey of point-like objects between 6th and 20th magnitude to be carried out over 5–6 years starting in 2012. Some 1 billion stars, a few million galaxies, half a million quasars, and a few hundred thousand asteroids will be observed. Expected accuracies for stellar parallaxes, annual proper motions and positions at mean epoch ($\simeq$ 2015.0) are 7–25 μas (microarcsec) down to 15th magnitude and a few hundred μas at 20th magnitude (see Sect. 5). The astrometric data are complemented by low-resolution spectrophotometric data in the 330–1000 nm wavelength range and, for the brighter stars, radial velocity measurements using a wavelength region around the near-infrared Ca II triplet. Each object will cross one of the instrument's two fields of view some 12–25 times per year at quasi-irregular intervals, providing a good temporal sampling of variability and orbital motion on all time scales from a few hours to years. A very accurate, dense and faint non-rotating optical reference frame will be established through the quasar observations.

The extremely rich and varied science goals of Gaia rest on its unique combination of very accurate astrometric measurements, the capability to survey large flux limited samples of objects, and the collection of synoptic, multi-epoch spectrophotometric and radial-velocity measurements. The proceedings of the symposium *The Three-Dimensional Universe with Gaia* (Turon *et al.* 2005) provides a broad and detailed overview of the science expectations of Gaia.

A primary goal is to study *galactic structure and evolution* by correlating the spatial distributions and kinematics of stars with their astrophysical properties. The determination of number densities and space motions for large, volume-complete samples of stars allows to trace the galactic potential and hence the distribution of matter (including

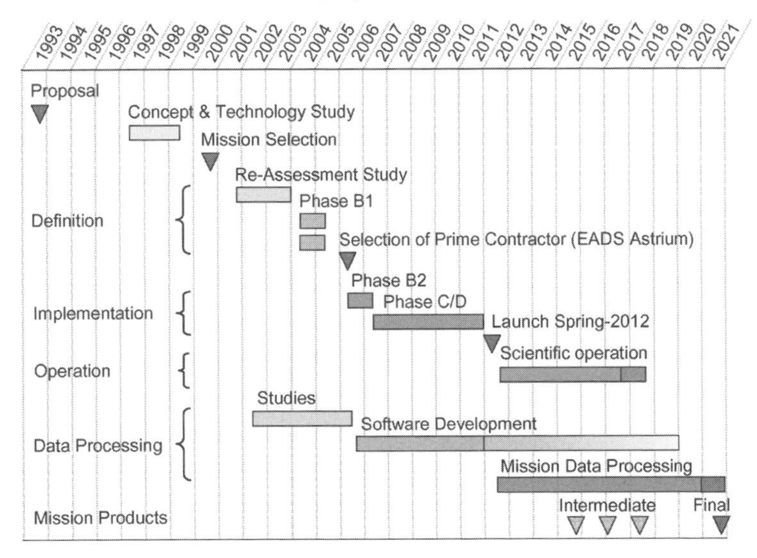

Figure 1. Overview of the main phases of the development of the Gaia project from the original proposal in 1993 to the expected publication of the final catalogue around 2021. The number and dates of the early releases of intermediate mission products are only indicative.

dark matter) in greater detail than has been possible before. The combination of luminosity and colour information for large samples allows to determine the history of star formation, which in combination with kinematic data may provide new insight into how the Galaxy was assembled and evolved. In *stellar astrophysics*, the determination of very accurate ($< 1\%$) parallax distances to literally millions of stars will boost all kinds of investigations based on the intrinsic properties of stars, provide stringent tests of stellar structure models and thereby drive the development of improved theoretical models of stellar atmospheres, interiors, and evolution. Gaia will survey millions of nearby stars for possible giant planetary companions, determine masses and orbital parameters for the detected companions and thereby provide reliable statistics for the occurrence of exoplanetary systems. A huge contribution will be made to *solar system physics* through the systematic observation of all ~300 000 asteroids brighter than Gaia's flux limit. Relatively few new discoveries are expected, but the large number of observations (some 60 epochs per object) and the high accuracy per observation (some 35–1000 μas per epoch) will allow the determination of extremely accurate osculating elements and hence the study of orbit families, their dynamical evolution, and the masses of individual asteroids. Gaia's contributions in the areas of the *reference frame* and *fundamental physics* are discussed in several other papers presented at this symposium; see Mignard & Klioner (2009), Hestroffer *et al.* (2009), Hobbs *et al.* (2009), and others.

2. Project organization and schedule

The Gaia project is the result of a large and complex collaboration between many individuals, teams, and organizations. It involves several formal structures as well as numerous informal groups. The main partners are:

– the European Space Agency (ESA), which has the overall project responsibility for the funding and procurement of the satellite (including the payload and its scientific instruments – there are no PI teams), launch and operations;

– EADS Astrium, who was selected in 2006 as the prime industrial contractor for designing and building the satellite according to the scientific and technical requirements formulated by ESA in consultation with the scientists;

– the Gaia Data Processing and Analysis Consortium (DPAC), charged with designing, implementing and running a complete software system for the scientific processing of the satellite data, resulting in the 'Gaia Catalogue' a few years after the end of the observation phase;

– national funding agencies and other organizations throughout Europe, who support the many individuals, research teams and data processing centres that constitute DPAC;

– the Marie Curie Research Training Network ELSA, a four-year EU-supported project (2006–2010) addressing certain key issues of the mission (Lindegren *et al.* 2007);

– the international scientific community, being the end users of the Gaia Catalogue, and represented at project level by the ESA-appointed Gaia Science Team (GST) chaired by the Gaia Project Scientist, Dr. Timo Prusti.

The current composition of the GST and other organizational details can be found on the Gaia web pages at `http://www.rssd.esa.int/gaia/`.

Figure 1 is an overview of the main phases of the development of the Gaia project, from the original proposal in 1993 to the expected publication of the final catalogue around 2021. Currently (May 2009) the project is in the middle of the implementation phase, on schedule for a targeted launch in spring 2012. Critical payload elements such as the silicon carbide (SiC) telescope mirrors and optical bench, CCDs and associated electronic units, have been designed and are in the process of being manufactured and tested.

The Gaia Data Processing and Analysis Consortium (DPAC) was formed in 2006 in response to an 'Announcement of Opportunity' issued by ESA (Mignard *et al.* 2008). It consists of individuals and institutes organized in nine 'coordination units' (each responsible for the development of one part of the software: overall architecture, simulations, core processing, photometry, etc.). Currently DPAC has nearly 400 individual members in more than 20 countries, including a substantial team of experts at the European Space Astronomy Centre (ESAC) near Madrid in Spain. Six data processing centres participate in the activities of the consortium. Financial support is provided by the various national agencies, ESA, universities and other participating organizations.

The development of the software system for the scientific processing of the satellite data is progressing in parallel with the hardware activities, with the aim to have a complete processing chain in place at the end of 2011. However, considerable further development and fine tuning of the software is expected to happen after the launch, especially based on the experience of the real data. The final catalogue should be ready three years after the end of the observation phase. Intermediate releases of some provisional results are planned after a few years of observations. Science alerts triggered by the Gaia observations (e.g., detection of extragalactic supernovae and near-Earth objects) require the organization of a programme of ground-based follow-up observations.

3. Observation principle

The Hipparcos mission (ESA 1997) established a new paradigm for efficient optical astrometric observations by a free-flying satellite, benefiting from the full-sky visibility and stable environment in space. The basic principles were formulated already in the earliest descriptions of the Gaia concept (e.g., Lindegren *et al.* 1994):

• Two fields of view, set at a large angle to each other, permit to link widely different areas of the sky in a single measurement.

Table 1. Summary of the main mission parameters of Gaia and characteristics of the astrometric instrument. Where applicable, dimensions are given as AL × AC, where AL is the along-scan dimension and AC the across-scan dimension.

targeted launch date	spring 2012, Soyuz/Fregat launcher (Kourou)
operational orbit	Lissajous orbit at Sun–Earth L2 (1.5×10^6 km from Earth)
transfer orbit to L2	1 month
nominal mission length	5 years ($\geqslant 90\%$ available for science)
extended mission length (for consumables)	6 years
solar aspect angle (Fig. 2)	$45 \pm 0.1°$
spin rate	60 arcsec s^{-1} (4 rev day^{-1})
no. of astrometric viewing directions (fields of view)	2
basic angle (between viewing dir.)	106.5°
pupil dimension (per viewing dir.)	1.45×0.5 m^2
focal length	35 m
full size of focal plane	0.93×0.42 m^2 (1.52×0.68 deg^2)
active solid angle per astrometric field (AF)	0.44 deg^2
number of astrometric CCDs	62
number of pixels per CCD	4500×1966
pixel size	10×30 μm^2 (58.9×176.8 mas^2)
integration time	4.417 s per CCD (39.75 s per AF transit)
mean number of AF transits per object in 5 years	78 (see Table 3)
down-linked CCD samples	(6–18 pix AL) × (12 pix binned AC)
wavelength region (G magnitude band)	$\lambda \simeq 330$–1000 nm

• The angular measurements are basically one-dimensional, namely, on the great circle connecting the two fields (Fig. 4).

• Continuous scanning permits to measure all programme stars as they pass through the two fields, while minimizing the need for attitude manoeuvres and keeping external influences as constant as possible.

• Rapid rotation of the whole instrument about an axis perpendicular to the two fields permits to determine critical parameters, in particular the scale value and the 'basic angle' between the two fields, from the 360° closure condition on a full rotation. Instrument stability is uncritical on all time-scales longer than the rotation period.

• Two-dimensional positions, and eventually the parallaxes and proper motions of the stars, are built up by slowly changing the orientation of the rotation axis so that each point on the sky is scanned many times, in different directions, over the mission.

• The data processing consists of a simultaneous least-squares adjustment of all the unknowns, whether they represent the instrument, its scanning motion, or the stars. Thus the mission is self-calibrating in the sense that critical calibration parameters, such as the basic angle, are derived from the scientific observations themselves – no special calibration observations are required.

Although the Gaia mission as it is now being realized is entirely different from the concept described in 1994 (cf. Høg 2007), the above principles remain valid. The main mission parameters are given in Table 1.

Figure 2 (left) shows the precessional motion of the spin axis causing the two fields of view (represented by the two viewing directions at the field centres) to successively cover the sky. The precession period is about two months, and after six months every part of the sky is covered by at least two scans intersecting at a large angle. With the ten-fold sky coverage obtained in a five-year mission it is clear that at least the positions and proper motions of the stars can be determined. That their *absolute* parallaxes can also be determined (i.e., without making use of distant 'background' objects assumed to have zero parallax) is shown in the right part of Fig. 2. The large angle between the Sun and the spin axis (45°) and the large basic angle between the viewing directions (106.5°) are essential for Gaia's capability to do global astrometry and determine absolute parallaxes.

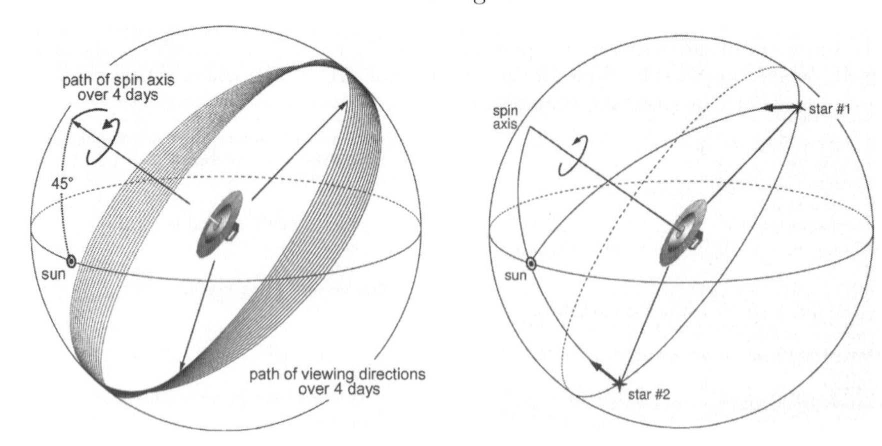

Figure 2. *Left:* the Gaia scanning law over a time interval of four days. The spin axis shifts by some 17° relative to the stars, remaining at a constant angle of 45° from the Sun, while each of the two viewing directions trace sixteen revolutions on the sky (only the path of the preceding viewing direction is shown). *Right:* the principle of absolute parallax measurement. The parallactic displacement of a star is directed along the great-circle arc from the star to the Sun (actually the Solar-System Barycentre). The observed angle between the indicated stars (#1 and #2) therefore depends on the parallax of #1, but is independent of the parallax of #2. At some other time during the mission, the same pair may be observed with the Sun between the spin axis and star #1, in which case the observed angle will depend only on the parallax of #2. In general the observed angle will depend on both parallaxes, but their different projection factors will allow the absolute values to be disentangled.

The Gaia telescope is an off-axis three-mirror anastigmat with 35 m focal length, two entrance pupils of 1.45×0.5 m^2 each, and a combined focal plane of nearly 0.4 m^2 area; the astrometric part of the focal plane consists of a mosaic of 62 CCDs occupying a field of about 40×40 arcmin^2 in each viewing direction. Plane folding mirrors and a beam combiner at the intermediate pupil complete the optical trains of the astrometric instrument. Separate CCDs preceded by filters and dispersive optics form the two photometers and the radial-velocity spectrometer. All the CCDs have a pixel size of 10×30 μm^2 matched to the optical resolution of the telescope with its rectangular pupil.

The satellite is continuously rotating according to the pre-defined scanning law described above at a nominal rate of 1 arcmin s^{-1}; thus the images of the stars in either field of view cross the astrometric CCDs in about 40 s. The autonomous on-board processing system detects any object brighter than $\simeq 20$ mag as it enters the special skymapper CCDs next to the astrometric field, then tracks the object across the various CCDs dedicated to the astrometric, photometric and radial-velocity measurements. The CCDs sample a small window centred on the projected path of the optical image of each detected object (Fig. 3). The pixel values, usually binned in the across-scan direction to reduce the readout noise and data rate, are transmitted to ground for the off-line data analysis. In the astrometric CCDs, the basic measurements are therefore the one-dimensional (along-scan) locations of the image centres in the pixel stream. The across-scan coordinates of the image centres are also determined to a lower accuracy in the skymappers, as part of the real-time object detection, and for a subset of the stars (in particular those brighter than 13th magnitude) in the astrometric CCDs. Although the astrometric measurements are essentially one-dimensional, viz., in the along-scan (AL) direction, the across-scan (AC) coordinates are needed for the attitude determination and certain instrument calibrations. However, it can be seen from elementary geometry that the sensitivity of the AL measurements to errors in the AC direction is reduced roughly

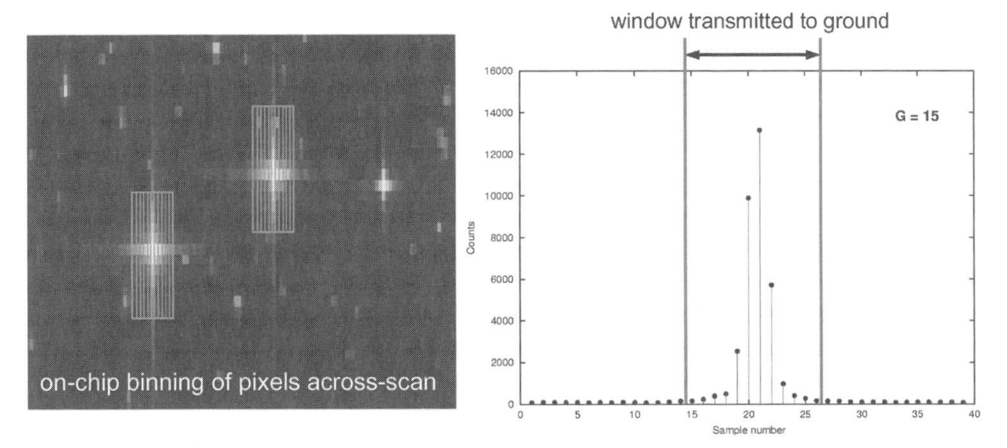

Figure 3. The left image is a simulated CCD image of a stellar field as seen with the astrometric instrument of Gaia (courtesy DPAC/CU2). Horizontal is the along-scan direction. Note the diffraction spikes corresponding to the rectangular telescope pupil. The on-board software causes a rectangular window to be read out around each detected image. The right graph shows the simulated output for a 16th magnitude star. Only values in the indicated window (6 to 18 samples, depending on the magnitude) are transmitted to the ground for further analysis.

by a factor equal to the transverse size of the field of view, or $\simeq 10^{-2}$ (Fig. 4). The AC measurements made by the skymappers for all stars (and by the astrometric CCDs for bright stars) have errors that are about a factor 10 worse than in the AL direction, but this is more than sufficient to provide the necessary information in the AC direction.

4. Astrometric data analysis

It is not possible to give here even a sketchy outline of the complete Gaia data analysis chain. Instead, only some basic principles of the core astrometric data analysis, known as AGIS, will be explained. AGIS is an acronym for Astrometric Global Iterative Solution, which implies the use of an iterative solution method. However, the essential point is rather the *global* nature of the solution: it is a single least-squares adjustment of the global set of parameters needed to describe (a) a subset of the objects (known as 'primary stars'), (b) the celestial pointing (attitude) of the telescope as function of time, and (c) various instrument calibration parameters. However, the resulting system of equations is so large that in practice only iterative solution methods are feasible.

As previously explained, the main observational data resulting from an astrometric CCD crossing is the AL image location in the pixel stream, which may be converted into a time of observation, $t^{\rm obs}$. For some observations also the corresponding AC image coordinate, $\mu^{\rm obs}$ (measured in AC pixels, and fractions thereof, on a particular CCD) is obtained. The time of observation is that of the image crossing a fiducial 'observation line' half-way along the CCD. It is initially expressed on the time scale of the on-board clock, but later transformed to the barycentric coordinate time (TCB) through correlation with an atomic clock at the receiver station. The spatial coordinates in the BCRS of the event are also known from the orbit determination mainly using ranging data. The orbit also gives the barycentric velocity of Gaia at the time of observation.

The field angles η (AL) and ζ (AC) give the proper direction of the object in a local reference system rotating with the satellite. The fiducial observation line for a particular CCD can be thought of as the central line of pixels, half-way through the CCD, traced out in the (η, ζ) angles. Formally, the observation line is defined by the functions $\eta_{fn}(\mu|\mathbf{c})$,

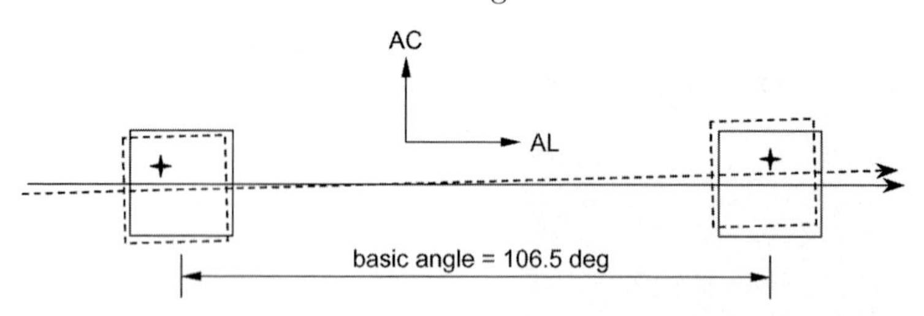

Figure 4. The use of primarily one-dimensional measurements (in the along-scan, AL, direction) is essential for the Gaia observation principle. It means that the measured angle between any two stars is to first order independent of the instrument's orientation in space (the attitude). Attitude errors in the across-scan (AC) direction are multiplied by a factor which is of the order of the size of the field of view, or $\simeq 0.01$ rad. The geometric calibration in the AC direction is similarly less critical than the AL calibration by the same factor.

$\zeta_{fn}(\mu|\mathbf{c})$, describing the field angles as functions of the AC pixel coordinate μ. These functions are different for each CCD (distinguished by the index $n = 1$ through 62 for the different astrometric CCDs) and for each field of view (index $f = 1$ and 2 for the preceding and following fields). They also depend on the calibration parameters in the array $\mathbf{c}$. In the simplest case, the calibration parameters could be the coefficients of polynomials in μ, different for each combination of indices fn. Temporal evolution of the instrument can be taken into account by having a different set of calibration parameters for each time interval (of days or weeks).

On the other hand, the field angles of a particular object at the time of observation can also be calculated from its astrometric parameters and the telescope attitude at that time. This calculation involves the full general-relativistic model for transforming from the astrometric parameters to the proper direction in the co-moving Lorentzian frame, and the subsequent spatial rotation of the reference system to take into account the attitude. This calculation can formally be written as $\eta(t^{\rm obs}|\mathbf{s},\mathbf{a})$, $\zeta(t^{\rm obs}|\mathbf{s},\mathbf{a})$, where the array $\mathbf{s}$ contains the astrometric parameters of all the primary stars and $\mathbf{a}$ contains the complete set of attitude parameters.† The adopted attitude model is a cubic spline (with a knot interval of some 5 to 15 s) for each of the four components of the attitude quaternion as function of time. The array $\mathbf{a}$ then contains the corresponding B-spline coefficients.

The core astrometric data analysis consists in setting up and solving the least-squares problem

$$\min_{\mathbf{s},\mathbf{a},\mathbf{c}} \sum_{\rm AL} \frac{\left[\eta_{fn}(\mu^{\rm obs}|\mathbf{c}) - \eta(t^{\rm obs}|\mathbf{s},\mathbf{a})\right]^2}{\sigma_{\rm AL}^2 + \sigma_0^2} + \sum_{\rm AC} \frac{\left[\zeta_{fn}(\mu^{\rm obs}|\mathbf{c}) - \zeta(t^{\rm obs}|\mathbf{s},\mathbf{a})\right]^2}{\sigma_{\rm AC}^2 + \sigma_1^2}, \tag{4.1}$$

where the sums are taken over all AL and AC observations, respectively, of the primary stars. $\sigma_{\rm AL}$ and $\sigma_{\rm AC}$ are the formal errors (expressed in angular units) of $t^{\rm obs}$ and $\mu^{\rm obs}$ as derived from the photon statistics and CCD readout noise. The additional noises σ_0 and σ_1 represent all other error sources, including modelling errors in the astrometry, attitude, and calibration. σ_0 and σ_1 are not constant numbers but could depend on the object and be functions of time. They have to be modelled and estimated as part of the data analysis process, which includes the selection of primary stars (i.e., relegating objects with consistently large σ_0 to the status of 'secondary stars', to be handled by the off-line object processing) and the systematic detection and down-weighting of outliers.

† These functions may depend on additional 'global parameters' such as the PPN γ parameter (Hobbs *et al.* 2009).

The iterative solution of Eq. (4.1) can be done according to a number of different schemes, the most obvious one being equivalent to a Gauss–Seidel block-iterative solution of the normal equations. In this scheme, which was originally proposed for AGIS, the astrometric, attitude and calibration parameters are cyclically adjusted, so that for example the astrometric parameters are adjusted, one star at a time, while the attitude and calibration parameters are kept fixed at their most recent estimates. As demonstrated by numerical experiments, this simple scheme converges to the correct solution, but rather slowly, requiring over a hundred such cycles (iterations) for full numerical accuracy. An accelerated scheme, requiring about half this number of iterations, is currently used in AGIS. For the final implementation of AGIS it is planned to use a conjugate gradient scheme, which will reduce the number of iterations by another factor of two. An important conclusion from numerical experiments carried out with the small-scale simulation tool known as AGISLab (Holl *et al.* 2009) is that the end results of the different iteration schemes are all equivalent to a rigorous solution of the least-squares problem (4.1). For example, the final astrometric parameters are independent of the starting values used to initiate the iterations.

The core astrometric data analysis contains a final step that produces a kinematically non-rotating catalogue aligned with the ICRS. This uses a subset of the primary stars (or quasars) that either have accurate parameters in the VLBI frame or can be assumed to be extragalactic zero-proper motion objects (i.e., quasars). It is expected that about 500 000 quasars can be used for this purpose; they have to be securely identified as such primarily based on photometric information (colours, variability) and ground-based surveys; see Mignard & Klioner (2009) for further details.

The current implementation of AGIS (at ESAC, Spain) routinely handles solutions based on realistic simulations for several million primary stars and tens of millions of secondary stars. The data processing system will be extended in several steps over the coming years, and tested with increasingly larger data sets, so that at the time of launch it will be ready to handle the full-size problem, including the projected astrometric solution for 100 million primary stars.

5. Expected astrometric performance

The predicted astrometric performance of Gaia has been calculated by means of the Gaia Accuracy Analysis Tool (de Bruijne 2005) using current mission and instrument data. A nominal mission length of 5 years is assumed, with an average dead time of 7%. The accuracy analysis takes into account all identified relevant error sources, assuming that the engineering specifications of the individual components (CCD quantum efficiency, wavefront errors, etc) are met. It also makes certain assumptions about the accuracy of the instrument calibration which in some cases are still uncertain. The rms errors therefore include a science margin factor of 1.2. Table 2 summarizes the resulting predictions for single, unperturbed solar-type stars.

The astrometric accuracy is mainly a function of the magnitude of the object and its ecliptic latitude (β). The values in Table 2 are averages over the celestial sphere. For arbitrary spectra and positions on the sky, the accuracy can be estimated as follows.

The relevant magnitude, denoted G, represents the flux in the broad wavelength band ($\simeq$ 330–1000 nm) defined by the mirror coatings and CCD quantum efficiency. Given the V magnitude and the (Johnson–Cousins) colour index $V-I$, the G magnitude is approximately given by

$$G = V - 0.0107 - 0.0879(V-I) - 0.1630(V-I)^2 + 0.0086(V-I)^3 \,, \qquad (5.1)$$

Table 2. Sky-averaged rms errors of the astrometric parameters for G2V stars (no extinction).

V magnitude	6–13	14	15	16	17	18	19	20	unit
parallax	8	13	21	34	54	89	154	300	μas
proper motion	5	7	11	18	29	47	80	158	μas yr^{-1}
position at mid-epoch ($\simeq$2015)	6	10	16	25	40	66	113	223	μas

valid for $V-I \lesssim 7$. This and other relevant photometric relations (e.g., to the Sloan system) can be found in Jordi (2009). The sky-averaged standard errors of the parallaxes of stars in the magnitude range $6 \leqslant G \leqslant 20$ are then approximated by

$$\sigma_{\varpi} = (4 + 515z + 5z^2 + 2 \times 10^{-8} z^6)^{1/2} \ \mu\text{as}, \tag{5.2}$$

where $z = \max(0.11,\ 10^{0.4(G-15)})$. The quantity z is inversely proportional to the number of detected photons from the star per CCD transit, normalized to 1 for $G = 15$ mag. The floor at $z = 0.11$ represents the onset of the CCD gating strategy to avoid pixel saturation for stars brighter than $G \simeq 12.6$. For all magnitudes brighter than $G = 20$, the expression in Eq. (5.2) is dominated by the second term, viz. $\sigma_{\varpi} \propto z^{1/2}$, showing that photon noise is the main contribution to the error budget at all magnitudes. The other terms represent the contributions from attitude and calibration errors at the bright end, and from CCD readout noise, sky background and missed detections at the faint end.

In order to obtain the standard errors of the different astrometric parameters at a particular point on the sky, the sky-averaged parallax error from Eq. (5.2) should be multiplied by factors that depend in a complex way on the number of observations of the object, the geometry of the scans across the object, and their temporal distribution. However, thanks to the symmetry of the Sun-centred scanning law with respect to the ecliptic, these factors depend mainly on the ecliptic latitude of the object, β, which can be approximately calculated from the equatorial (α, δ) or galactic (ℓ, b) coordinates as

$$\sin\beta \simeq 0.9175\sin\delta - 0.3978\cos\delta\sin\alpha \tag{5.3}$$

$$\simeq 0.4971\sin b + 0.8677\cos b\sin(\ell - 6.38^\circ)\,. \tag{5.4}$$

Table 3 gives the average factors within equal-area zones in ecliptic latitude. The sky-averaged standard errors in the position and annual proper motion components are

$$\sigma_{\alpha*} = 0.802\sigma_{\varpi}\,, \quad \sigma_{\delta} = 0.693\sigma_{\varpi}\,, \quad \sigma_{\mu_{\alpha*}} = 0.565\sigma_{\varpi}\,, \quad \sigma_{\mu_{\delta}} = 0.493\sigma_{\varpi}\,, \tag{5.5}$$

where the asterisk indicates true arcs on the sky ($\sigma_{\alpha*} = \sigma_{\alpha}\cos\delta$, etc). The values in Table 2 are approximately reproduced for a star of colour index $V-I = 0.75$.

6. Conclusions

The Gaia mission is well on track towards its targeted launch in early 2012, with both the hardware and software developments halfway through their implementation phases. The core astrometric data analysis will implement a rigorous simultaneous least-squares solution of the astrometric parameters for 100 million primary stars together with the corresponding attitude and instrument calibration parameters. There is good confidence that the mission will achieve the overall astrometric performance projected in Sect. 5.

Acknowledgement. The author thanks the many individuals who, perhaps without knowing it, have contributed to this review. Financial support by the Swedish National Space Board is gratefully acknowledged.

Table 3. Numerical factor to be applied to the sky-averaged parallax error in Eq. (5.2) for the different astrometric parameters as function of ecliptic latitude, β. N_{tr} is the mean number of field-of-view transits over the mission. All values are for a 5 yr mission with 7% dead time.

$\|\sin\beta\|$	N_{tr}	$\alpha*$	δ	ϖ	$\mu_{\alpha*}$	μ_δ
0.0–0.1	59	1.036	0.746	1.176	0.730	0.534
0.1–0.2	60	1.016	0.750	1.167	0.715	0.534
0.2–0.3	62	0.984	0.744	1.154	0.693	0.537
0.3–0.4	65	0.925	0.722	1.121	0.655	0.524
0.4–0.5	70	0.856	0.701	1.082	0.608	0.504
0.5–0.6	80	0.757	0.675	1.030	0.538	0.482
0.6–0.7	106	0.598	0.630	0.950	0.426	0.446
0.7–0.8	120	0.520	0.619	0.831	0.362	0.430
0.8–0.9	86	0.628	0.654	0.764	0.437	0.455
0.9–1.0	74	0.687	0.686	0.721	0.478	0.476
mean sky	78	0.802	0.693	1.000	0.565	0.493

References

de Bruijne, J. H. J. 2005, in: C. Turon, K. S. O'Flaherty, & M. A. C. Perryman (eds.), *The Three-Dimensional Universe with Gaia*, ESA SP-576, p. 35

ESA 1997, *The Hipparcos and Tycho Catalogues*, ESA SP-1200

Hestroffer, D., Mouret, S., Mignard, F., Tanga, P., & Berthier, J. 2009, *this proceedings*, 325

Hobbs, D., Holl, B., Lindegren, L., Raison, F., Klioner, S., & Butkevich, A. 2009, *this proceedings*, 315

Holl, B., Hobbs, D., & Lindegren, L. 2009, *this proceedings*, 320

Høg, E. 2007, in: W. J. Jin, I. Platais, & M. A. C. Perryman (eds.), *A Giant Step: from Milli- to Micro-arcsecond Astrometry*, Proc. IAU Symposium No. 248 (Cambridge), p. 300

Jordi, C. 2009, GAIA-C5-TN-UB-CJ-041, Gaia technical note (unpublished)

Lindegren, L., Bijaoui, A., Brown, A. *et al.* 2007, in: W. J. Jin, I. Platais, & M. A. C. Perryman (eds.), *A Giant Step: from Milli- to Micro-arcsecond Astrometry*, Proc. IAU Symposium No. 248 (Cambridge), p. 529

Lindegren, L., Perryman, M. A. C., Bastian, U. *et al.* 1994, in: J. B. Breckinridge (ed.), *Amplitude and Intensity Spatial Interferometry II*, Proc. SPIE Vol. 2200, p. 599

Mignard, F., Bailer-Jones, C., Bastian, U. *et al.* 2008, in: W. J. Jin, I. Platais, & M. A. C. Perryman (eds.), *A Giant Step: from Milli- to Micro-arcsecond Astrometry*, Proc. IAU Symposium No. 248 (Cambridge), p. 224

Mignard, F. & Klioner, S. 2009, *this proceedings*, 306

Turon, C., O'Flaherty, K. S., & Perryman, M. A. C. (eds.) 2005, *The Three-Dimensional Universe with Gaia*, ESA SP-576

Relativity in Fundamental Astronomy
Proceedings IAU Symposium No. 261, 2009
S. A. Klioner, P. K. Seidelman & M. H. Soffel, eds.

doi:10.1017/S174392130999055X

Gaia: Relativistic modelling and testing

F. Mignard[1], S.A. Klioner[2]

[1]Observatoire de la Côte d'Azur, Nice, France
email: francois.mignard@oca.eu

[2]Lohrmann Observatory, Dresden Technical University, Dresden, Germany
email: Sergei.Klioner@tu-dresden.de

Abstract. Gaia is an ambitious space astrometry mission of ESA with a main objective to map the sky in astrometry and photometry down to a magnitude 20 by the end of the next decade. Given its extreme astrometric accuracy and the repeated observations over five years, the observation modelling is done in a fully relativistic framework and several tests of General Relativity or of its extensions can be carried out during the data processing. The paper presents an overview of the current activities in this area and of the expected performances.

Keywords. astrometry, relativity, data processing

1. Introduction

Gaia is an all-sky, high precision astrometric and photometric satellite of the European Space Agency. This is an ESA approved and fully funded mission in the C/D phase scheduled for launch in early 2012. Its design reuses the main concepts that were proved so successful with Hipparcos, namely a continuously spinning satellite with two widely separated field of views mapping the sky. Over the five years of its mission Gaia will measure the position, parallaxes and proper motions of every object brighter than visual magnitude V = 20 with an end-of-mission astrometric accuracy of 25 μas at V=15. In addition, Gaia will perform multi-band photometry for all the sources and will carry also a dedicated spectrometer to determine radial velocities for at least 100 million stars.

Beyond the sheer measurement accuracy, a major strength of Gaia follows from (i) its capability to perform an all-sky and sensitivity limited absolute astrometric survey, (ii) the unique combination into a single spacecraft of the three major electronic detectors carrying out nearly contemporaneous observations, (iii) the huge number of objects and observations which allow the accuracy on single objects to be achieved on very large samples, thus yield statistical significance. In relation with the relativistic aspect dealt with in this paper, one must also add the extreme care taken to calibrate the instrumental parameters and the availability on-board of a Rb clock regularly monitored from the ground.

2. Astrometric accuracy and relativity

2.1. *Few relevant numbers*

One can quickly recognize with a few relevant numbers the need to model astrometric observations, that is to say primarily the light path between a star and an observer, by accounting first for the finite light-velocity, and then for higher order kinematical effects or for the interaction between gravitation and the light-propagation. For observations carried out in the vicinity of the Earth, the barycentric velocity of the observer is about 30 km s^{-1}, equivalent to $v/c \sim 20$ arcsec. This is not a small angle for astronomers and the

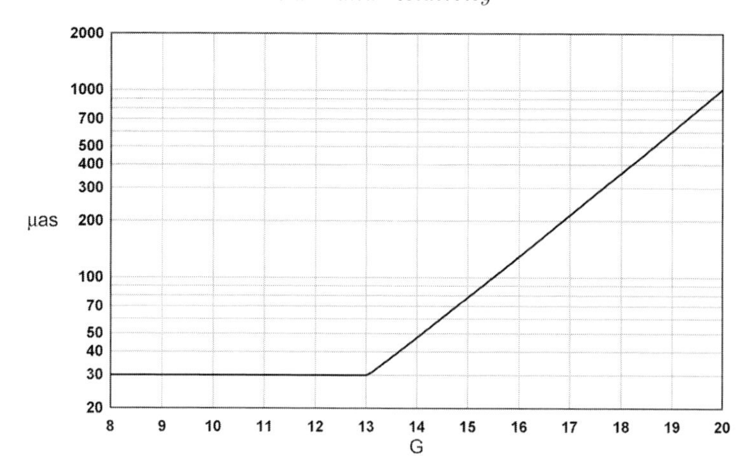

Figure 1. Astrometric accuracy expected with Gaia for a single observation over a 40s transit over the 9 astrometric CCDs. For faint stars, this is primarily determined by the photon noise, while at the bright end, the calibration and attitude determination take over.

astrometric effect was discovered in the early 18th century as the aberration of light. One can then see that the $(v/c)^2$ is of the order of one mas (0.″001) and had to be considered systematically in the Hipparcos data processing. The full special relativity treatment is implemented for Gaia in the nominal astrometric model GREM (Gaia RElativity Model) which aims at an accuracy of 1 μas (Klioner, 2003).

Regarding the gravitation in the solar system, the yardstick is the light deflection by the Sun of a grazing ray, giving

$$\delta\theta = \frac{4GM_\odot}{c^2 b}, \tag{2.1}$$

where b is the impact parameter. For $b \sim R_\odot$ one gets about 1.″75. This was about the best astrometric accuracy with photographic plates in the early years of the 20th century and the detection was actually achieved during the famous eclipse of 1919. While the signal is larger when observing close to the solar surface, this has also several drawbacks due to the photosphere brightness in the visible and the radio wave dispersion by the solar corona. The deflection decreases as $1/b$ and is still 50 mas at 10 degrees from the Sun and 4 mas at 90 degrees. This means that deflection at any angle was within reach of Hipparcos and that for Gaia it must be modelled with great care: at 90° it is already 100 times larger than the single observation accuracy of a bright star. This is also the reason why Gaia is a good tool to test some aspects of General Relativity to an unprecedented accuracy.

There are several ways to define the astrometric performance of Gaia, using the final parallax accuracy on stars or the 1D astrometric accuracy over one transit. The latter is more interesting for the following discussion since it applies to every kind of object, whether stellar, extragalactic or solar system sources and is more relevant to discuss either global or local tests. The expected positional accuracy over one field transit, amounting to a total of 40 s integration over the nine astrometric CCDs is shown in Fig. 1 as a function of the magnitude. There are two obvious regimes, roughly for bright and fainter stars. We have a constant accuracy for sources brighter than $G = 13$ at around 30 μas; then when one goes fainter, the photon noise takes over and one sees the slow degradation in performance, with a single measurement precision of about 1 mas at $G = 20$, yielding about 0.3 mas for the parallax at mission completion. This single transit accuracy is

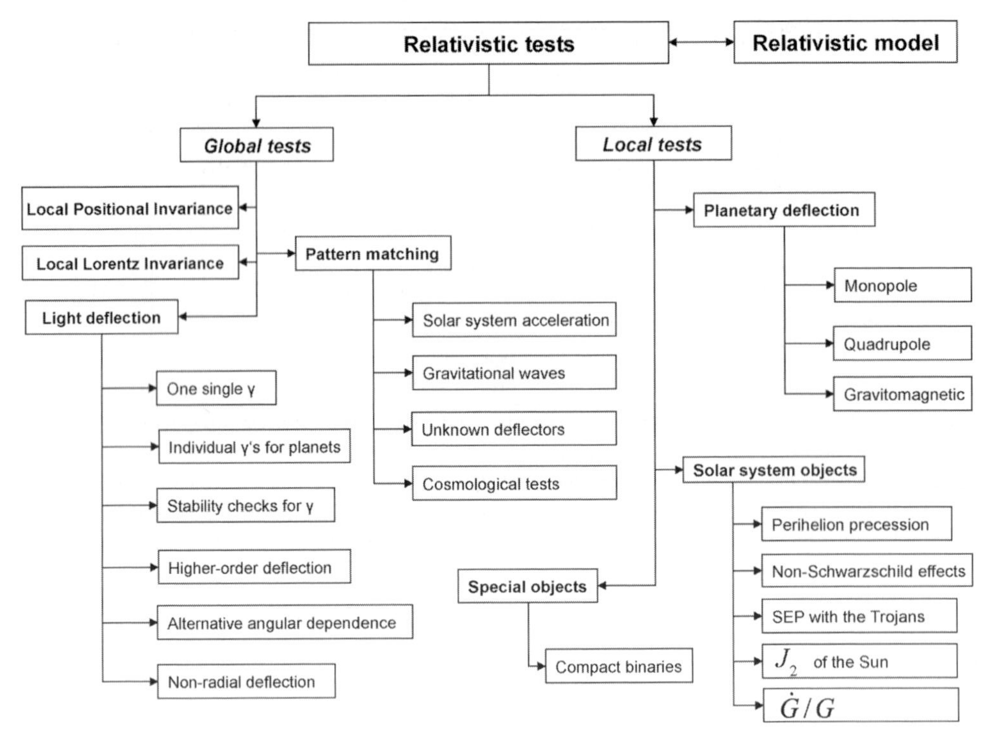

Figure 2. Summary of relativity testing that can be investigated with highly accurate space astrometry, like Gaia.

also the only meaningful one for solar system objects to fit initial conditions, masses or general physical parameters, as shown in (Hestroffer *et al.*, 2010).

2.2. *Relativistic testing with Gaia*

The whole set of relativistic experiments with Gaia can be conveniently divided into two groups:

- global tests which are related to the Gaia global astrometric solution and should use the whole set of Gaia data or at least a sizable fraction of it;
- local tests which are related to some specially designed (e.g., differential) solutions and involve a relatively small amount of selected data, for example collected during observations in the immediate vicinity of a planet.

The variety of testing that can be achieved with highly accurate astrometry is illustrated in the diagram of Fig. 2, where the main division between global and local tests is emphasized by the earlier split into two main branches.

2.2.1. *Global tests*

Global tests are performed by taking advantage of the full sky coverage and use every piece of astrometric information gathered during the mission. These tests take the form of a few additional general parameters included in the global model fitting to the observational data. The most natural and precise test of general relativity expected with Gaia is the measurement of the gravitational light deflection by the Sun with a precision of about 2×10^{-6}. Such a precision for γ is very interesting physically, since it supports or rejects some physical theories (e.g., the hypothetical "built-in" cosmological attractor mechanism of the scalar-tensor theory of gravity towards general relativity (Damour &

Nordvedt, 1993a, Damour & Nordvedt, 1993b). How this test is carried out and how the accuracy is achieved are presented in more detail in Section 3.

Another planned test is to use the relativistic aberration to test the Local Lorentz Invariance. The transformation from the unit vector of the observed direction of the incoming photon to the same vector in the barycentric frame, is basically nothing else than the Lorentz transformation with a specially chosen velocity parameter. That Lorentz transformation will be computed for each of the ≈ 80 observations of each of 10^9 sources and its effect on the observed directions is many orders of magnitude larger than the accuracy of observations. A test theory for special relativity (Robertson, 1949, Mansouri & Sexl, 1977) or any of its modern extensions can be used to add a few additional parameters into the Lorentz transformations. Those parameters can be fitted from the Gaia data with high precision. It is not expected that Gaia could compete with modern laboratory experiments (Laemmerzahl, 2006) verifying Lorentz transformation, but since Gaia is a totally different kind of experiment, this test has its own merit.

Finally, a number of experiments are related to matching certain physically interesting patterns to the measured field of proper motions or individual time-dependent positions. One of these patterns allows one to measure the acceleration of the solar system relative to the quasars. This will be achieved during the process of constructing a Gaia-based ICRF with several 10,000s QSOs as explained in (Mignard & Klioner, 2007). Another pattern allows one to constraint (Pyne *et al.*, 1996, Gwinn *et al.*, 1997) possible gravitational wave flux with a frequency $\omega < 3 \times 10^{-9}$ Hz. The accuracy that can be expected from Gaia is $\Omega_{GW} < (0.001 - 0.005)\, h^{-2}$, h being the normalized Hubble constant $h = H_0/(100\ \mathrm{km/s/Mpc})$. That limit will be several orders of magnitude better than we have from VLBI now. One can also try to fit a pattern produced by an individual gravitational wave to the individual (time-dependent) positions of the quasars. The pattern in positions produced by a plane gravitational wave is uniquely defined by 5 parameters: its amplitude (strain), its frequency, and the direction of its propagation. Although such an approach is difficult to realize (because of the huge amount of data to be processed), with this method Gaia will be sensitive to the gravitational waves with higher frequencies. The same idea of fitting a pattern to the individual positions and proper motions of sources can be used to detect a hypothetical massive body close to the solar system (e.g. a companion of the Sun).

2.2.2. *Local tests*

Three groups of local tests are currently planned for Gaia. First, special solutions for the observations close to planets of the solar system, especially close to Jupiter, as detailed hereafter in Section 3.2.

The second group of local tests is related to the solar system objects – asteroids and Near-Earth Objects (NEO). The most precise test expected here is the relativistic perihelion precession. In the famous Mercury perihelion advance test the parameter fitted from observation depends on the PPN parameters β, γ and on the solar quadrupole J_2, so that an estimate of β is only possible if J_2 is measured independently (usually, γ can be estimated separately from the same observational data from light deflection or Shapiro effect). In case of asteroids one has many bodies with different orbital parameters and this allows us to distinguish the dynamical effects of the solar quadrupole J_2 and the effects of general relativity. Thus, independent estimates of γ, β and J_2 are possible. The expected precisions are $\sim 10^{-3} - 5 \times 10^{-4}$ for β and $\sim 10^{-7} - 10^{-8}$ for solar J_2 (Hestroffer *et al.*, 2007, Hestroffer *et al.*, 2010). The same solution can be used to estimate the time variation of the Newtonian gravitational constant with an accuracy $\dot{G}/G \sim 2 \times 10^{-12}\ \mathrm{yr}^{-1}$.

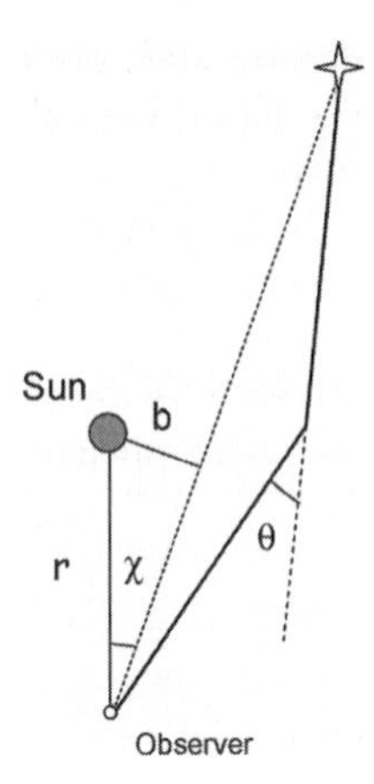

Figure 3. Geometry of the light-deflection by a solar system body at rest with respect to the observer and the source.

Perihelion precession is related to the Schwarzschild field of the Sun. It is interesting that non-Schwarzschild effects related to the post-Newtonian N-body problem will be detected with Gaia. The N-body problem in the post-Newtonian approximation is described by the Einstein-Infeld-Hoffmann (EIH) equations. The dynamical consequences of those equations are well known for major planets with their small eccentricities and inclinations, but not for asteroids. Especially, for resonant asteroids the relativistic N-body effects can be substantially enhanced. Another test is the test of the Strong Equivalence Principle (SEP) with the Trojan asteroids (asteroids orbiting the Lagrange points L_4 and L_5 of the Sun-Jupiter system) and, possibly, other resonant asteroids. Interestingly, the first example of an observable effect due to a possible violation of the SEP (Nordtvedt effect) was a shift of the position of L_4 and L_5 by about 1 arcsecond (for $\eta = 1$) as seen from the Earth. Gaia will be able to provide high-accuracy observations for all Trojan asteroids so that the expected accuracy of η is of the order of 10^{-3}.

The last group of local tests deals with specially selected, relativistic-interesting objects, the processing of which can be improved by special means. As an illustration, there is the case of compact binaries with one component being a black hole candidate. Combining usual Doppler measurements of these objects with Gaia astrometry, one can derive the mass of the invisible companion without any further assumptions (Fuchs and Bastian, 2005). For example, for the well-known system Cyg X-1 the astrometric wobble of the visible companion is expected to be 25 μas and can be measured by Gaia.

3. The light deflection

The grazing-ray expression must be corrected when observing at large angular distance from the deflecting body, which for Gaia would be always the case for the Sun. The star is located at a very large distance compared to the Sun and χ is the angular separation between the Sun and the star (Fig. 3). With the space observations to be carried out by Gaia, χ is not necessarily a small angle. In fact, it can be very small for the planets where grazing observations are feasible, but remains always larger than $\chi_{\min} = 45$ deg for the Sun. The impact parameter of the unperturbed ray is denoted by b and the distance between the observer (on the Earth or space-borne somewhere in the Solar System) is r. The deflector has a mass M and a radius R. To the first order in GM/c^2 and by neglecting any departure from the spherical symmetry (the so-called monopole deflection) the deflection angle is given by ,

$$\delta\theta = \frac{1+\gamma}{2}\frac{2GM}{r\,c^2}\frac{1}{\tan\frac{\chi}{2}}\,, \tag{3.1}$$

while the unit vector in the apparent direction is given by,

$$\mathbf{u} = \mathbf{u_0} + \frac{1+\gamma}{2}\frac{2GM}{c^2}\frac{[1+(\mathbf{u_0}\cdot\mathbf{r})/r]}{b^2}\,\mathbf{b}\,. \tag{3.2}$$

When the angular separation $\chi << 1$, the deflection expression reduces to the classical one with,

$$\delta\theta = \frac{1+\gamma}{2}\frac{4GM}{c^2\,b}\,. \tag{3.3}$$

Although one talks about light deflection or light bending, there is no way to measure this effect directly since the initial direction is not known. By itself the deflection is not an observable quantity and its mathematical expression is coordinate dependant. In fact one has access to the proper direction in the observer frame, and only the variation of this proper direction with time, due to varying geometry with respect to the Sun, is accessible, which eventually permits determining the deflection itself. Relevant astrometric signatures for solar system bodies are given in Table 1, both for the monopole and quadrupole deflection to be considered later. Grazing rays are only meaningful for the planets (all but Mercury which is not observable with Gaia) with a very large monopole deflection by the Gaia standard. The modeling requirement at the level of 1 μas shows that deflection by Jupiter must be included over a wide range of elongations, a non negligible computing effort, even with very optimized ephemeris access, given the number of observations.

Table 1. Relativistic light deflection in the solar system. R_A is the angular radius of body A. The columns with $\delta\theta = 1\mu$as give the planetary or solar elongation where the deflection reaches that value.

Body	Monopole		Quadrupole	
	grazing	χ	grazing	χ
	mas	$\delta\theta = 1\ \mu$as	μas	$\delta\theta = 1\ \mu$as
Sun	17,000	180°		
Mercury	0.083	0.15°		
Venus	0.49	4.5°		
Mars	0.12	0.4°		
Jupiter	16.3	90°	240	8 R_J
Saturn	5.8	17°	95	4 R_S
Uranus	2.1	1.2°	8	2 R_U
Neptune	2.5	0.9°	10	2 R_N

3.1. *The determination of space curvature*

The deflection expression in Eq. (3.1) has been derived in the framework of the PPN extension of the GR and includes γ as free parameter. As said earlier a single observation of a bright star at 90° from the Sun allows potentially to detect deviation from $\gamma = 1$ to 0.01. Given the number of stars (about 10 million stars brighter than $V = 13$) and the repeated observations, one sees immediately that Gaia has a huge potential to evaluate γ to a never achieved accuracy. Consider first the ideal and best case, where the instrument modelling and calibration are fully under control and that the remaining noise is purely random. A straight comparison with Hipparcos provides already a good order of magnitude of the performance achievable with Gaia under these assumptions. This is shown in Table 2 where the different sources of improvement are scaled to Hipparcos. Altogether one can expect with the bright stars something 2000 better than the Hipparcos results, or a determination of γ with the solar deflection as good as 2×10^{-6}.

The estimate seems rather conservative, neglecting the contribution of stars fainter than $V = 13$, but despite that, this is a very challenging goal. It rests upon the statistical improvement with the square root of the number of observations and disregards any adverse unmodelled systematic effect minutely correlated with the deflection astrometric signature as small as $\sim 2 \times 10^{-6} \times 4$mas, or 0.008 μas. The difficulty would not be to solve for γ, but to assess that one has not left any other global parameter, not orthogonal to the subspace determined by the condition equations of γ. Otherwise, the true meaning of the parameter so determined will be questionable. The real challenge to do better than 10^{-5} lies precisely here and extreme care will be exercised to look for correlations with small instrumental parameters, with the zero-point parallax, or the aberration correction, to quote a few of these effects.

Table 2. Determination of the PPN parameter γ with space astrometry scaled on Hipparcos and using only the brightest stars.

Hipparcos	**Gaia**		σ_H/σ_G
10^5 stars $V < 10$	8×10^6 stars $V < 13$		
2.5×10^6 abscissas	6.0×10^8 abscissas	$\Longrightarrow$	15
$\sigma \sim 3$ to 8 mas	$\sigma \sim 40$ μas	$\Longrightarrow$	125
$\chi > 47°$	$\chi > 45°$		
$\Downarrow$	$\Downarrow$		$\Downarrow$
$\sigma_\gamma \sim 3 \times 10^{-3}$	$\sigma_\gamma \sim 2 \times 10^{-6}$	$\Longleftarrow$	2 000

Compared to other experiments or tests related to the measurement of γ, Gaia deals directly with light deflection and not with gravitational signal retardation, as in Viking-like, time dependence of Shapiro delay, as in Cassini-like ones (Bertotti *et al.*, 2003), or differential Shapiro delay (i.e. the difference between the Shapiro delays for signal from the same source received by two spatially separate observing sites), as in VLBI. Before Gaia only direct measurements of stellar positions during total solar eclipses, or the special case of Hipparcos, provided us with measurements of true light deflection. Gaia is going to improve the current best result available from Hipparcos by more than 3 orders of magnitude.

But, as mentioned earlier, Gaia is sensitive to gravitational light deflection in a wide range of angular distances between the observed source and the Sun. This allows one not only to measure γ, but also to map the angular dependence of the light deflection by fitting a number of additional parameters describing alternative light deflection models. The most general way to do this is to fit an expansion of the deflection signal in terms of vector spherical harmonics. The fits of relativistic parameters in Gaia will be done simultaneously with the fits of all other parameters. This is also an advantage compared to many other experiments, when the relativistic fit is a special post-processing on the residuals of the "main solution" that assumes general relativity to be valid. In the latter case the correlations between the relativistic parameters (e.g. γ) and other parameters fitted in the main solution are completely neglected. This leads to a substantial uncertainty in the real accuracy of estimates of the relativistic parameters. One more advantage of Gaia is that full-scale simulations of the observations and data processing are being performed to test the suggested data processing algorithms. This allows to reliably check if the observational data and the selected data processing algorithms are really sensitive to a given signal (see for example Hobbs *et al.*, 2010).

3.2. *Planetary light deflection*

During the course of the Gaia observing program, planets will also enter the fields of view, typically 60 times during the mission (it's below the average at low ecliptic latitude). The planets themselves won't be observable as being both too big in angular size and too bright for the detector. However, faint satellites of Jupiter and Saturn, as well as stars that would happen to be around, will be normally detected and subsequently observed. Observing outside the atmosphere with CCDs will allow to see stars fairly close to the planetary disk, probably as close as 0.1 radius from the surface, at least on the leading side (blooming due to the transit of the bright planet may hinder similar observations on the trailing side). The light bending by the giant planets given in Table 1 is very significant for Gaia. The monopole light deflection by Jupiter will be detected at several degrees from Jupiter or Saturn. Although the deflection is smaller at large distance, there are more stars contributing to the signal. The deflection decreases as the inverse of the impact parameter, while the number of sources grows in the same proportion, meaning that each circular annulus contributes evenly to the final determination of γ. Altogether these observations alone will lead to an independent determination of γ with a precision $\sim 10^{-3}$, that is, with about the same accuracy as Hipparcos for the Solar deflection (Froeschlé *et al.*, 1997). What matters here, is not the accuracy of γ, which is low compared to the main determination with the Sun, but the fact it can be achieved with a different body with no relation to the Sun.

Jupiter is not at rest with respect to the barycenter of the solar system and clearly the deflection pattern of a moving body is not identical to that of a static one and this can be evidenced with Gaia observations. Simulations have shown that the relevant parameter for this translational motion could be determined to an accuracy of 2×10^{-3}, i.e. two orders of magnitude better than current best results (Anglada, Klioner & Torra, 2007).

Given the sensitivity of Gaia astrometry one can hope also to detect the departure of the deflection pattern from a pure isotropic monopole deflection and see the non-radial signature brought about by the quadrupole moment. For Jupiter the magnitude is 240 μas (Table 1) for a grazing ray and falls off rapidly as the cube of the impact parameter. Therefore, it will be really observable on the few instances where a bright star is seen very close to the Jupiter disk, preferentially on the leading side (defined relative to the scan motion). The main properties of the quadrupole deflection and its detection with Gaia have been investigated in (Crosta & Mignard, 2006) and (Anglada, Klioner & Torra, 2007). Simulations have shown that the detection will be very challenging, but could be achieved to better than a 3-sigma level. Since the evidence is based on a few favorable transits of bright stars, this is not a statistical effect and it must be looked at on a case by case basis with realistic simulations of the observations. For a given orbit of Gaia, it is shown that there exist good initial conditions for the scanning law (there are two free parameters) that yield favorable close approaches between Jupiter and bright stars, which at the end increase the detection of the quadrupole deflection up to an 8-sigma level as shown in (de Bruijne, 2010).

However, the selection of these initial conditions may conflict with other mission requirements and the choice will be, at the end, a compromise and no risk will be taken that could impact the mission core science in exchange of an uncertain detection of the quadrupole deflection. Moreover, the deflection is also sensitive to the actual orbit of Gaia, since the apparent direction of Jupiter must be known to better than 0.05 of its radius, or better than 3500 km. In practice, the orbit of Gaia to that accuracy won't be known before the final injection that would take place a few days after the launch.

Acknowledgement

This short review reflects a more collective work undertaken by the DPAC REMAT (standing for **RE**lativistic **M**odels **A**nd **T**ests) task force, in charge of designing, testing and implementing the relativistic astrometric model and the different steps of the data processing necessary to carry out the tests described or briefly mentioned in this paper.

References

Anglada Escudé, G., Klioner, S., & Torra, J., 2007, Relativistic light deflection near giant planets using gaia astrometry, in *MG11 Proceedings.*, H. Kleinert & R. T. Jantzen eds., p. 2588.

Bertotti, B., Iess, L., & Tortora, P., 2003, *Nature* **425**, 374.

de Bruijne, J., 2010, *this proceedings*, 331

Crosta, M. T. & Mignard, F., 2006, *Classical and Quantum Gravity* **23**, 4853.

Damour, T. & Nordtvedt, K., 1993, *Phys. Rev. Lett.* **70**, 2217.

Damour, T. & Nordtvedt, K., 1993, *Phys. Rev. D* **48**, 3436.

Froeschlé, M., Mignard , F., & Arenou, F., 1997, Determination of the PPN parameter γ with the Hipparcos data, in *Proceedings of the ESA Symposium "Hipparcos - Venice 97", ESA SP-402*.

Fuchs, B. & Bastian, U., 2005, Weighing stellarmass black holes with gaia, in *ESA SP-576: The Three-Dimensional Universe with Gaia*, eds. C. Turon, K. S. O'Flaherty and M. A. C. Perryman.

Gwinn, C., Eubanks, T., Pyne, T., Birkinshaw, M., & Matsakis, D., 1997, *Astrophys. J.* **485**, 87.

Hestroffer, D., *et al.*, 2007, Relativistic tests from the motion of the asteroids, in *MG11 Proceedings.*, H. Kleinert & R. T. Jantzen eds., p. 2600.

Hestroffer, D., Mouret, S., Mignard, F., Tanga, P., & Berthier, J. 2010, *this proceedings*, 325

Hobbs, D., Holl, B., Lindegren, L., Raison, F., Klioner, S., & Butkevich, A., 2010, *this proceedings*, 315

Klioner, S. A., 2003, *AJ* **125**, 1580.

Laemmerzahl, C., 2006, Test theories for Lorentz invariance, in *Lect. Notes Physics*, Springer, **702**, 349.

Mansouri, R. & Sexl, R., 1977, *Gen. Rel. Grav.* **8**, p. 497.

Mignard, F. & Klioner, S., 2007, Space Astrometry and Relativity, in *MG11 Proceedings.*, H. Kleinert & R. T. Jantzen eds., p. 245.

Pyne, T., Gwinn, C., Birkinshaw, M., Eubanks, T., & Matsakis, D., 1996, *Astrophys. J.* **465**, 566.

Robertson, H., 1949, *Rev. Mod. Phys.* **21**, 378.

Relativity in Fundamental Astronomy
Proceedings IAU Symposium No. 261, 2009
S. A. Klioner, P. K. Seidelman & M. H. Soffel, eds.

doi:10.1017/S1743921309990561

Determining PPN γ with Gaia's astrometric core solution

David Hobbs[1], Berry Holl[1], Lennart Lindegren[1], Frédéric Raison[2], Sergei Klioner[3] and Alexey Butkevich[3]

[1]Lund Observatory, Lund University, Box 43, SE-22100 Lund, Sweden
email: david@astro.lu.se

[2]European Space Astronomy Center, ESA, Spain
email: frederic.raison@sciops.esa.int

[3]Lohrmann Observatory, Technical University, 01062 Dresden, Germany
email: Sergei.Klioner@tu-dresden.de

Abstract. The ESA space astrometry mission Gaia, due for launch in early 2012, will in addition to its huge output of fundamental astrometric and astrophysical data also provide stringent tests of general relativity. In this paper we present an updated analysis of Gaia's capacity to measure the PPN parameter γ as part of its core astrometric solution. The analysis is based on small-scale astrometric solutions taking into account the simultaneous determination of stellar astrometric parameters and the satellite attitude. In particular, the statistical correlation between PPN γ and the stellar parallaxes is considered. Extrapolating the results to a full-scale solution using some 100 million stars, we find that PPN γ could be obtained to about 10^{-6}, which is significantly better than today's best estimate from the Cassini mission of 2×10^{-5}.

Keywords. Astrometry, relativity, gravitation, data analysis

1. Introduction

The space astrometry mission Gaia, planned for launch by the European Space Agency (ESA) in 2012, will determine accurate astrometric data for about 1 billion objects in the magnitude range from 6 to 20. Accuracies of 8–25 micro-arcseconds are expected for the trigonometric parallaxes, positions at mean epoch and annual proper motions of simple (apparently single) stars down to 15th magnitude, increasing to a few hundred μas for the faintest objects. To meet the challenge of processing the raw data, the Gaia Data Processing and Analysis Consortium (DPAC) was formed in 2006 (Mignard *et al.* 2008). A core part of the processing is to determine the accurate spacecraft attitude, geometric instrument calibration and astrometric model parameters for a well-behaved subset of all the objects (the 'primary stars'). In addition, a small number of global parameters will be estimated, one of these being PPN γ.

Of the 10 parameters used in the PPN formalism (see Will (2006) for a review), Gaia is expected to provide useful constraints on β (Hesteroffer *et al.* 2009), measuring the degree of nonlinearity in the superposition law of gravity, and γ, measuring the curvature of space time (and hence the degree of gravitational light deflection) due to rest mass. Of interest here is that some alternative theories of gravity predict deviations of these parameters from their values in general relativity ($\gamma = \beta = 1$). In this paper the discussion is restricted to γ, where deviations of the order 10^{-5} to 10^{-8} could occur. It is therefore highly interesting to estimate the accuracy by which this parameter could be determined from the Gaia observations. Previous estimates (Mignard 2002; Vecchiato *et al.* 2003) were based on extrapolations from the Hipparcos results (Frœschlé *et al.* 1997) or using

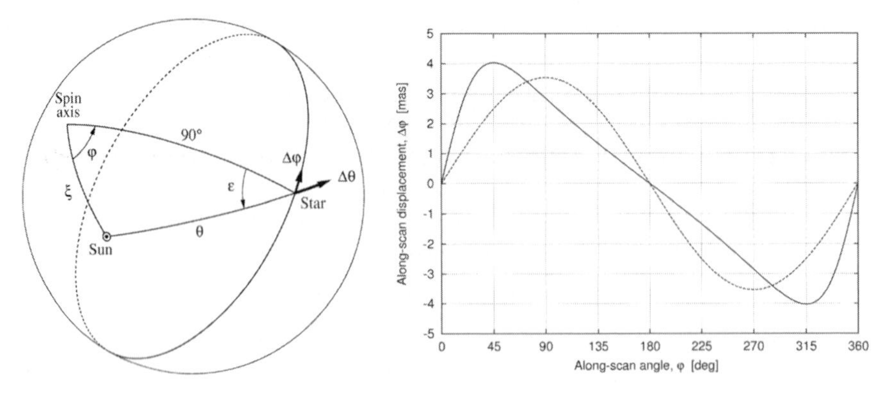

Figure 1. Left: the gravitational deflection by the Sun apparently shifts the star away from the Sun by $\Delta\theta$. Gaia will measure its along-scan component $\Delta\varphi$. Right: the along-scan gravitational deflection (solid curve) together with the effect of a parallax shift by -5 mas (dashed curve).

partly obsolete assumptions about the mission. With the data simulations in DPAC and the core astrometric solution now reaching a high degree of sophistication and realism, more reliable estimates become possible.

2. The core astrometric solution

The core astrometric solution is a simultaneous least-squares adjustment of all the different astrometric, attitude, calibration and global parameters to the measured image positions on the CCDs of the primary stars. In total there are more than 5×10^8 unknowns. Although the resulting system of equations is very sparse, a direct solution is unfeasible by many orders of magnitude, and an iterative method must be used. The basic method, known as the Astrometric Global Iterative Solution (AGIS), consists of four blocks executed cyclically until convergence:

- Source Update determines the five astrometric parameters (α, δ, ϖ, $\mu_{\alpha*}$, μ_δ) for each primary star;
- Attitude Update determines the celestial orientation of the instrument axis as a function of time (using a spline representation of the attitude quaternion);
- Calibration Update determines the geometric instrument calibration parameters (basic angle, CCD geometry, etc.);
- Global Update determines a small number of model parameters that are constant throughout the mission.

In each block, the results of the other three blocks are considered as given. Various schemes can be used to accelerate the convergence of AGIS, but the end result is equivalent to a direct solution of the rigorous normal equations. Since there is no simple way to compute the formal variances of the unknowns, Monte Carlo experiments may be used to 'calibrate' the approximate variances obtained in each block solution.

3. Measuring gravitational deflection and parallax

The direct measurement of large angles (of the order of the basic angle, 106.5°) is fundamental to Gaia's ability to construct a globally consistent reference system as well as for the determination of parallaxes and PPN γ. Gravitational deflection by the Sun causes an apparent shift of a distant object by $\Delta\theta_\gamma = (1+\gamma)(GM_\odot/rc^2)\cot(\theta/2)$ in the direction *away* from the Sun (Fig. 1, left), where $GM_\odot$ is the heliocentric gravitational constant, c

the speed of light, $r \simeq 1.01$ au the heliocentric distance of Gaia, and θ the Sun–star angle. By contrast, the parallax ϖ causes an apparent shift *towards* the Sun (actually towards the Solar System Barycenter but this difference is neglected here) by $\Delta\theta_\varpi = -\varpi r \sin\theta$. Gaia is mainly sensitive to shifts in the along-scan direction, normal to the nominal spin axis. With ξ, φ and ε defined as in Fig. 1 (left), the along-scan gravitational deflection is $\Delta\varphi_\gamma = \Delta\theta_\gamma \sin\varepsilon = (1+\gamma)(GM_\odot/rc^2)\sin\xi\sin\varphi/(1-\sin\xi\cos\varphi)$, while the along-scan parallax shift is $\Delta\varphi_\varpi = \Delta\theta_\varpi \sin\varepsilon = -\varpi r \sin\xi\sin\varphi$. The solar aspect angle $\xi = 45°$ is fixed by the scanning law adopted for Gaia. Resulting variations with φ are shown in Fig. 1 (right). To first order, the effect of the gravitational deflection by the Sun is not unlike a global shift of the parallaxes by $\simeq -5$ mas (dashed curve). This correlation between parallax zero point and γ was discussed by Mignard (2002) who estimated a correlation coefficient $\rho = 0.85$ for the $\xi = 55°$ then foreseen for Gaia. We find that the correlation coefficient can be analytically evaluated to

$$\rho = -\left[\int_0^{2\pi} \Delta\varphi_\gamma^2 \mathrm{d}\varphi \int_0^{2\pi} \Delta\varphi_\varpi^2 \mathrm{d}\varphi\right]^{-1/2} \int_0^{2\pi} \Delta\varphi_\gamma \Delta\varphi_\varpi \mathrm{d}\varphi = \sqrt{\frac{2\cos\xi}{1+\cos\xi}} \tag{3.1}$$

or $\simeq 0.8538$ for $\xi = 55°$ in agreement with Mignard. For the current (definitive) Gaia design with $\xi = 45°$ we find $\rho \simeq 0.9102$. As noted by Vecchiato *et al.* (2003) this correlation increases the formal standard error for PPN γ by a factor $(1-\rho^2)^{-1/2}$ compared with a hypothetical solution in which the parallaxes did not have to be determined; unfortunately neglecting parallax would instead result in a huge bias in γ. With (3.1) we find $(1-\rho^2)^{-1/2} = \cot(\xi/2) \simeq 2.414$ for the current Gaia.

The statistical correlation between PPN γ and stellar parallaxes slows down the convergence of the AGIS iterations considerably. To handle this, a global pseudo-parameter ϖ_0 was introduced in the solution. This ϖ_0 formally represents a global parallax shift but in the solution it is set to zero immediately after fitting, and therefore does not change the final solution vector in any way. However, it allows the correlation to be taken into account in the Global Update block, resulting in much improved convergence, a better estimate of the formal standard error of γ, and a numerical estimate of ρ directly from the solution. In each iteration, ϖ_0 is updated but immediately reset to zero.

4. Monte Carlo experiments

Development and testing of the PPN γ algorithm uses our simulation software AGIS-Lab (Holl *et al.* 2009). This tool allows us to perform many independent astrometric solutions in a reasonable time, based on the same principles as the full-scale AGIS (Sect. 2) but using a much smaller number of primary stars, along-scan measurements only, and several other time-saving simplifications (e.g., Calibration Update is not included for the present experiments). The scaling is designed to allow reliable extrapolation to the full-scale solution by preserving key parameters such as the average number of primary stars in the field of view.

For the present experiments the following simple model (Lindegren *et al.* 1992) was used to calculate the apparent star direction $\hat{\mathbf{u}}$ (ignoring stellar aberration) and its partial derivative with respect to γ (angular brackets denoting vector normalization):

$$\hat{\mathbf{u}} = \left\langle \bar{\mathbf{u}} + \mathbf{r}\frac{(1+\gamma)GM_\odot c^{-2}}{r(r+\bar{\mathbf{u}}\cdot\mathbf{r})} \right\rangle \qquad \frac{\partial\hat{\mathbf{u}}}{\partial\gamma} = (\mathbf{r} - \bar{\mathbf{u}}(\bar{\mathbf{u}}\cdot\mathbf{r}))\frac{GM_\odot c^{-2}}{r(r+\bar{\mathbf{u}}\cdot\mathbf{r})} \tag{4.1}$$

Here $\bar{\mathbf{u}}$ is the star's coordinate direction and $\mathbf{r}$ the heliocentric position of Gaia. Gravitational light deflection due to the planets is not included.

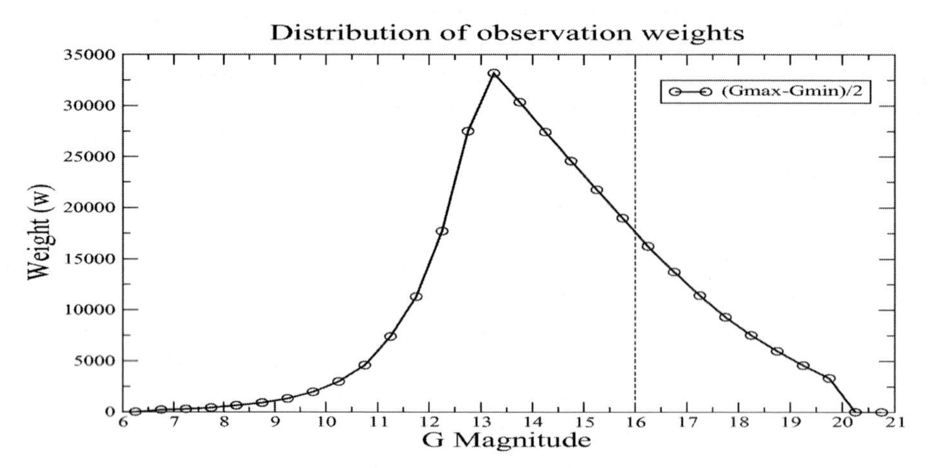

Figure 2. The distribution of observation weights for each G-magnitude bin (0.5) clearly shows that primary stars with $G \leqslant 16$ contribute most to the determination of PPN γ. The figures assume that only 10% of the measured stars are used.

Using AGISLab, we have run large numbers (typically $M = 100$) of solutions with different initial conditions, random star distributions and observation noises, and calculated the resulting PPN γ (assuming $\gamma = 1$ when generating the observations) and its formal standard error $\sigma_\gamma^{\rm Lab}$ in the Global Update (which takes into account the correlation between γ and ϖ_0). Based on simulations with $N = 10\,000$ and $50\,000$ primary stars, it was verified that the formal standard error σ_γ as well as the sample standard deviation of the many different estimates of γ scale exactly as $w^{-1/2}$, where $w = \sum_{\rm obs} \sigma_{\rm obs}^{-2}$ is the total statistical weight of the observations used in the solution. Here, $\sigma_{\rm obs}$ is the assumed along-scan standard error of a single observation, resulting from the crossing of one primary star over one CCD in the astrometric field of Gaia. In the real Gaia, $\sigma_{\rm obs}$ is mainly a function of the G magnitude of the star (with G representing the very broad spectral response of Gaia, $\simeq$ 330–1000 nm), and has been estimated through extensive Monte Carlo simulations of the image location process using the Gaia Accuracy Analysis Tool (de Bruijne 2005). For bright stars ($G = 6$ to 13) this gives $\sigma_{\rm obs} = 75$ μas assuming that the CCDs can be operated near full-well capacity for these stars; the corresponding numbers at $G = 15$ and 20 are 240 μas and 3.1 mas, respectively.

The small-scale simulations using AGISLab are thus used to calculate statistics for the distribution of errors in the estimated PPN γ which can then be extrapolated to the full-scale AGIS solution. The extrapolation uses the following formula:

$$\sigma_\gamma^{\rm Full} = U \times \sigma_\gamma^{\rm Lab} \times \left(\frac{w^{\rm Full}}{w^{\rm Lab}}\right)^{-\frac{1}{2}}, \qquad U = \sqrt{\frac{1}{M}\sum_{i=1}^{M}\left(\frac{\gamma_i - 1}{\sigma_{\gamma i}^{\rm Lab}}\right)^2} \tag{4.2}$$

Here, the labels Full and Lab refer to the full-scale AGIS and the small-scale AGISLab solutions, respectively, and γ_i and $\sigma_{\gamma i}^{\rm Lab}$ are the estimated value in the ith solution and its formal standard error. U is an empirical factor correcting for the neglected correlations between γ and all other unknowns except ϖ_0.

5. Results and discussion

A typical Monte Carlo experiment consisted of 100 simulated solutions of 10 000 primary stars, each using on average 8 768 000 observations with an assumed observation noise of $\sigma_{\rm obs} = 75~\mu$as; thus $w^{\rm Lab} = 1559~\mu{\rm as}^{-2}$. From several such experiments we derived $U = 1.08 \pm 0.08$ with no significant variation between experiments. The mean formal standard error was $\sigma_\gamma^{\rm Lab} = 4.61 \times 10^{-5}$. The mean correlation coefficient between γ and ϖ_0 was $\rho = 0.9086$, in good agreement with (3.1).

The main uncertainty in (4.2) comes from $w^{\rm Full}$, which depends strongly both on the assumed number of primary stars used in AGIS and their distribution in magnitude. The final AGIS solution will use at least 100 million primary stars (10% of the total number of stars with $G \leqslant 20$), but their distribution in magnitude is not fixed. Binaries and other problematic objects must be filtered out, but the fraction of 'well-behaved' stars remaining is likely to be much greater than 10%.

Figure 2 shows the distribution of the weight of all potential observations in half-magnitude bins of G. For $G < 13$, $\sigma_{\rm obs}$ is roughly constant and the weight per bin increases in proportion to the number of stars. For fainter stars, the increased photon noise is only partially offset by the larger number of stars. Since the magnitude range up to $G \simeq 16$ contains about 100 million stars, the maximum possible number of primary stars should be selected in this interval. For example, assuming that 80% of the stars between $G = 6$ and 16 can be used as primary stars, we find $w^{\rm Full} = 2\,420\,000~\mu{\rm as}^{-2}$, leading to $\sigma_\gamma^{\rm Full} = 1.3 \times 10^{-6}$. Including many more fainter stars gives only a marginal improvement. This estimated precision on PPN γ is significantly better than today's best estimate from the Cassini mission (2×10^{-5}).

The small-scale simulations described above will in the future be calibrated against the results of large-scale solutions under development at the Gaia data processing center at ESAC (Madrid, Spain). Current solution runs with simulated observations for 2 million primary stars, based on the Besançon galaxy model (Robin *et al.* 2003), and the number of stars and the realism of the simulated data will successively increase in the next few years. Of greatest importance is that possible sources of systematic errors are carefully evaluated in subsequent studies.

Acknowledgements. The work of DH and LL is supported by the Swedish National Space Board. BH is an ELSA Fellow supported by the Marie Curie FP6 contract MRTN-CT-2006-033481.

References

de Bruijne, J. H. J. 2005, in: C. Turon, K. S. O'Flaherty, & M. A. C. Perryman (eds.), *The Three-Dimensional Universe with Gaia*, ESA SP-576, p. 35

Hesteroffer, D., Mouret, S., Mignard, F., Tanga, P., & Berthier, J. 2009, *this proceedings*, 325

Holl, B., Hobbs, D., & Lindegren, L. 2009, *this proceedings*, 320

Frœschlé, M., Mignard, F., & Arenou, F. 1997, in: B. Battrick, M. A. C. Perryman, & P. L. Bernacca (eds.), *HIPPARCOS Venice '97*, ESA SP-402, p. 49

Lindegren, L., Høg, E., van Leeuwen, F., *et al.* 1992, *A&A*, 258, 18

Mignard, F. 2002, in: O. Bienaymé & C. Turon (eds.), *GAIA: A European Space Project*, EAS Publications Series, Vol. 2 (EDP Sciences), p. 107

Mignard, F., Bailer-Jones, C., Bastian, U., *et al.* 2008, in: W. Jin, I. Platais, & M. A. C. Perryman (eds.), *A Giant Step: From Milli- to Micro-Arcsecond Astrometry*, Proc. IAU Symposium No. 248 (Cambridge), p. 224

Robin, A. C., Reylé, C., Derriére, S., & Picaud, S. 2003, *A&A*, 409, 523

Will, C. M. 2006, *The Confrontation between General Relativity and Experiment*, Living Rev. Relativity 9, 3, URL (27-05-2009): `http://www.livingreviews.org/lrr-2006-3/`

Vecchiato, A., Lattanzi, M. G., Bucciarelli, B., Crosta, M., de Felice, F., & Gai, M. 2003, *A&A* 399, 337

Relativity in Fundamental Astronomy
Proceedings IAU Symposium No. 261, 2009
S. A. Klioner, P. K. Seidelman & M. H. Soffel, eds.

doi:10.1017/S1743921309990573

Spatial correlations in the Gaia astrometric solution

Berry Holl[1], **David Hobbs**[1] **and Lennart Lindegren**[1]

[1]Lund Observatory, Lund University, Box 43, SE-22100 Lund, Sweden
email: berry@astro.lu.se

Abstract. Accurate characterization of the astrometric errors in the forthcoming Gaia catalogue is essential for making optimal use of the data. Using small-scale numerical simulations of the astrometric solution, we investigate the expected spatial correlation between the astrometric errors of stars as function of their angular separation. Extrapolating to the full-scale solution for the final Gaia catalogue, we find that the expected correlations are generally very small, but could reach some fraction of a percent for angular separations smaller than about one degree. The spatial correlation length is related to the size of the field of view of Gaia, while the maximum correlation coefficient is related to the mean number of stars present in the field at any time. Our scalable simulation tool (AGISLab) makes it possible to characterize the astrometric errors and correlations, e.g., as functions of position and magnitude.

Keywords. Astrometry, reference systems, catalogs, methods: data analysis, methods: statistical, space vehicles

1. Introduction

The space astrometry mission Gaia, planned to be launched in 2012 by the European Space Agency (ESA), will provide the most comprehensive and accurate catalogue of astrometric data for galactic and astrophysical research in the coming decades. Accuracies of 8–25 μas are expected for the trigonometric parallaxes, positions at mean epoch and annual proper motions of simple (apparently single) stars down to 15th magnitude with lower accuracy down to 20th magnitude. The astrometric data are complemented by photometric and spectroscopic information collected by the satellite. The resulting catalogue will become available to the scientific community around 2020.

Accurate characterization of the errors in the catalogue is essential for making optimal use of the data. While estimates of the standard errors for individual stars have been reported elsewhere (e.g., Lindegren 2009), little is yet known about what happens when Gaia data are combined for large numbers of objects. This will often be the case in important applications dealing with stellar clusters, nearby dwarf galaxies, galactic stellar populations, and when looking for large-scale patterns e.g. in the apparent proper motions of quasars. In such cases it may be important to know the statistical correlation of the astrometric errors as a function of the angular separation of the objects, which we refer to as *spatial correlations*. In this paper we present the first results of an estimation of spatial correlations in the future Gaia catalogue.

2. The importance of correlations

We examine the estimated value of the generic astrometric parameter x (representing α, δ, π, μ_α, or μ_δ) for star i, denoted x_i. It is assumed that the estimate is unbiased, so $\mathrm{E}[e_i] = 0$, where $e_i = x_i - x_i^{\mathrm{true}}$ is the error. For the two stars $i \neq j$ the estimates are correlated with correlation coefficient ρ_{ij} if

$$\text{Cov}[x_i, x_j] \equiv \text{E}[e_i e_j] = \rho_{ij}\sigma_i\sigma_j \neq 0 \tag{2.1}$$

where $\sigma_i = \sqrt{\text{E}[e_i^2]}$, $\sigma_j = \sqrt{\text{E}[e_j^2]}$ are the standard errors.

Consider now any quantity y calculated from the estimated parameters $x_1 \ldots x_N$ of N different stars. We can generally formulate this as $y = f(\boldsymbol{x})$ where $\boldsymbol{x}$ is the vector of estimates. Assuming that f is linear in the small errors, the variance of y is given by

$$\begin{aligned}\sigma_y^2 &= \left(\frac{\partial y}{\partial \boldsymbol{x}}\right)' \text{Cov}(\boldsymbol{x}) \left(\frac{\partial y}{\partial \boldsymbol{x}}\right) \\ &= \sum_i \left(\frac{\partial y}{\partial x_i}\right)^2 \sigma_i^2 + \sum_i \sum_{j \neq i} \frac{\partial y}{\partial x_i} \frac{\partial y}{\partial x_j} \rho_{ij}\sigma_i\sigma_j \end{aligned} \tag{2.2}$$

The first sum is the computed variance if correlations are neglected; depending on the sign of the correlations this can be an under- or overestimate of the true variance.

As a simple example, consider the calculation of the mean parallax or proper motion of N stars in a cluster, so $y = N^{-1}\sum_i x_i$. If the stars are of approximately the same magnitude, they will have roughly the same standard error, $\sigma_i \simeq \sigma$. If the area on the sky occupied by the cluster is small, it will be found (cf. Sect. 5) that the correlation coefficient is positive and roughly the same for all pairs of stars, $\rho_{ij} \simeq \rho > 0$. Then

$$\sigma_y^2 \simeq \sigma^2 \left(\frac{1}{N} + \frac{N-1}{N}\rho\right) \tag{2.3}$$

In the absence of correlations the improvement in σ_y is by a factor $N^{-1/2}$, as could be expected. However, in the presence of (positive) correlations, $\sigma_y \to \sigma\sqrt{\rho}$ as $N \to \infty$. The limiting accuracy is effectively reached by averaging over some ρ^{-1} stars.

3. Origin of spatial correlations in the Gaia catalog

Although the individual positional measurements in Gaia's focal plane are essentially uncorrelated (the errors are dominated by photon noise), the geometry of the observations and the way they are combined in the astrometric solution will create spatial correlations on different angular scales. Gaia's scanning law and the two fields of view, widely separated by the basic angle of 106.5°, are designed to minimize large-scale correlations, but cannot entirely eliminate them.

The origin of spatial correlations in the Gaia catalog can be understood in terms of errors in the attitude determination (Fig. 1). An error in the attitude at a particular time will 'bias' all observations made at that time, in both fields of view, thus partially correlating stars within each field of view as well as stars separated by the basic angle. This suggests that we can expect spatial correlations to fall off over an angular scale on the order of the field of view size, or $\simeq 0.7°$ for the astrometric instrument of Gaia, possibly reappearing to some degree for separations of about 106.5°.

In a single realization of the Gaia catalogue (which is all that we will have!), the spatially correlated errors may look like localized 'biases' on the sky. However, it is important to realize that they originate from attitude errors that are themselves also a result of the random observation noise and that a different realization of the observation noise would have resulted in a completely different set of 'biases'.

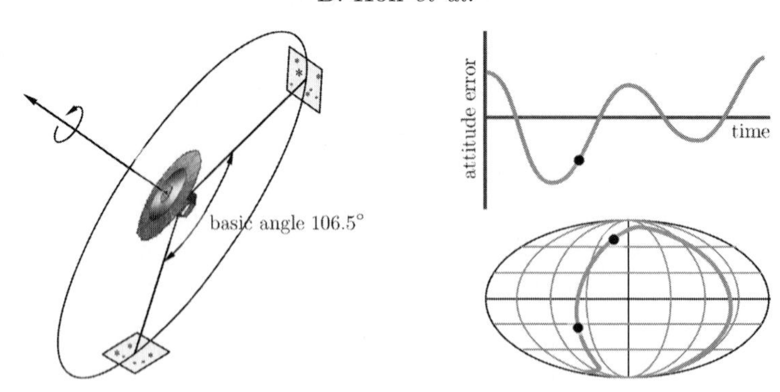

Figure 1. Gaia scans the sky roughly along great circles, observing stars in two fields of view separated by the basic angle. An error in the attitude at a particular time will affect all observations in both fields of view, producing correlations among stars both for small angular separations and for separations of about 106.5°, as illustrated by the all-sky map on the lower right.

4. Monte Carlo experiments with AGISLab

The baseline method for determining the astrometric parameters of Gaia stars is the Astrometric Global Iterative Solution (AGIS). This is an iterative least-squares estimation of the five astrometric parameters for a subset of $\sim 10^8$ well behaved (non-variable, apparently single) 'primary' stars, with additional unknowns for the spacecraft attitude, instrument calibration, and global parameters such as PPN γ (Hobbs *et al.* 2009). The total number of unknowns is $\sim 5 \times 10^8$. This large number prevents a rigorous calculation of the covariance matrix of the solution. To overcome this we estimate the correlations statistically from Monte Carlo experiments with different noise realizations.

While AGIS is currently being tested with 10^6–10^7 primary stars, these simulations take too much time and resources for making a significant number of Monte Carlo experiments. They also depend on externally generated simulated observations which are not always suitable for the tests we want to run. We have therefore developed a scaled version of AGIS called AGISLab, which allows us to run simulations with less than 10^6 stars in a (much) shorter time and using input observations that fit our experiments (e.g., with many different noise realizations but otherwise identical conditions). The scaling uses a single parameter S such that $S = 1$ corresponds to the astrometric solution using approximately the current Gaia design and a minimum of 10^6 primary stars, while $S = 0.1$ would only use 10% as many primary stars. When $S < 1$ the Gaia design used in the simulations is modified to preserve certain key quantities such as the mean number of stars in the focal plane at any time, the mean number of field transits of a given star over the mission, and the mean number of observations per degree of freedom of the attitude model. In practice this is done by formally reducing the focal length of the astrometric telescope and the spin rate of the satellite by the factor $S^{1/2}$, and increasing the interval between attitude spline knots by the factor S^{-1}.

For the present study we made astrometric solutions with AGISLab, using 3 000, 10 000, and 30 000 uniformly distributed primary stars (i.e., for $S = 0.003$, 0.01 and 0.03). In each experiment (A, B, C) many different noise realizations were made and the corresponding solutions computed in order to improve the statistics. As the scaling preserves the mean number of stars per field of view, additional experiments (D, E) were made in which this number could be increased. An overview of the experiments is given in Table 1. All experiments used a noise level of 100 μas per along-scan observation, which is roughly the expected noise for bright stars down to magnitude $V = 13$ (for unreddened G2V stars).

Table 1. Overview of experiments in this study. S = scale parameter; $N_{\rm star}$ = number of stars in the solution; Φ = field of view size (side length); n = mean number of stars per field of view; $N_{\rm run}$ = number of runs in the experiment (with different noise realizations); $\rho_{\rm max}$ = maximum correlation of parallaxes (for separations $\theta \ll \Phi$); $\theta_{1/2}$ = correlation half-length. The reference case for the scaling law (corresponding to $S = 1$) has $N_{\rm star} = 10^6$, $\Phi = 0.7°$ and $n = 12$.

Experiment	S	$N_{\rm star}$	Φ [deg]	n	$N_{\rm run}$	$\rho_{\rm max}$	$\theta_{1/2}$ [deg]	Figure
A	0.030	30 000	4	12	49	0.085	1.7	2, 3
B	0.010	10 000	7	12	112	0.085	3.0	2, 3
C	0.003	3 000	13	12	759	0.085	5.2	2, 3, 4
D	0.003[1]	9 000	13	36	25	0.032	5.2	4
E	0.003[1]	30 000	13	120	17	0.010	5.0	4

[1] In these experiments n is increased with respect to the usual scaling law.

Although the actual noise level is irrelevant for studying correlations, the assumption of a single noise level, as well as the uniform sky distribution of the stars, are of course gross simplifications of the real case. These complications will be addressed in a future paper.

To estimate the correlation as a function of pair separation θ, the range $0 \leqslant \theta \leqslant 180°$ was divided into bins of 0.5° or 1.0° and the sample correlation coefficient calculated in each bin by summing over all relevant pairs in all the runs:

$$\rho(\theta) = \left(\sum_{ij\in\text{bin}} e_i e_j\right)\left[\left(\sum_{ij\in\text{bin}} e_i^2\right)\left(\sum_{ij\in\text{bin}} e_j^2\right)\right]^{-1/2} \tag{4.1}$$

5. Results and discussion

Results are shown graphically in Figs. 2–4. In all cases the strongest correlation is obtained for the smallest separations. For $\theta \simeq 106.5°$ there is also a much weaker positive correlation for α and μ_α, and a similar negative correlation for π, δ and μ_δ (Fig. 2). This is expected if the attitude errors are mainly a rotation offset around the spin axis.

Table 1 gives two key characteristics of the correlation curves for small separations: the maximum correlation coefficient $\rho_{\rm max}$ (obtained in the first 0.5° bin), and the correlation half-length, i.e., the angle $\theta_{1/2}$ such that $\rho(\theta_{1/2}) \simeq \rho_{\rm max}/2$. It is noted that: (i) the correlation length scales with the size of the field of view, $\theta_{1/2} \simeq 0.4\Phi$; and (ii) the maximum correlation depends mainly on the number of stars in the field.

Extrapolating to the real Gaia mission with $\Phi = 0.7°$, the expected correlation half-length is $\simeq 0.3°$. The maximum correlation depends on the assumed number of primary stars and their magnitude distribution. Although the final solution will use about 100 million primary stars, their combined astrometric weight corresponds to a smaller number of perhaps 20 million bright primary stars, suggesting $\rho_{\rm max} \simeq 0.005$ for bright stars ($V < 13$) and smaller for fainter stars. More detailed studies are needed to determine how the magnitude distribution and real-sky non-uniformity affect the correlations. The final goal is to model the covariance of all astrometric parameters for any pair of stars in terms of their magnitudes and positions on the sky, as required for astrophysical applications combining Gaia data for many stars.

Acknowledgements. BH is an ELSA Fellow supported by the Marie Curie FP6 contract MRTN-CT-2006-033481. DH and LL acknowledge support by the Swedish National Space Board.

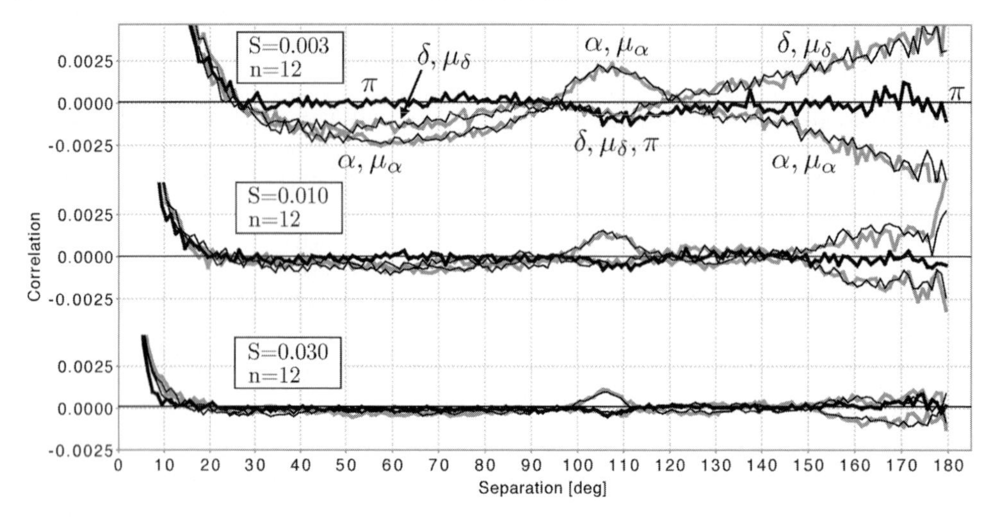

Figure 2. Spatial correlations in experiments A, B and C (see Table 1). Thin black lines are α and δ, the thick black line is π, thick gray lines are μ_α and μ_δ (binsize 1.0°).

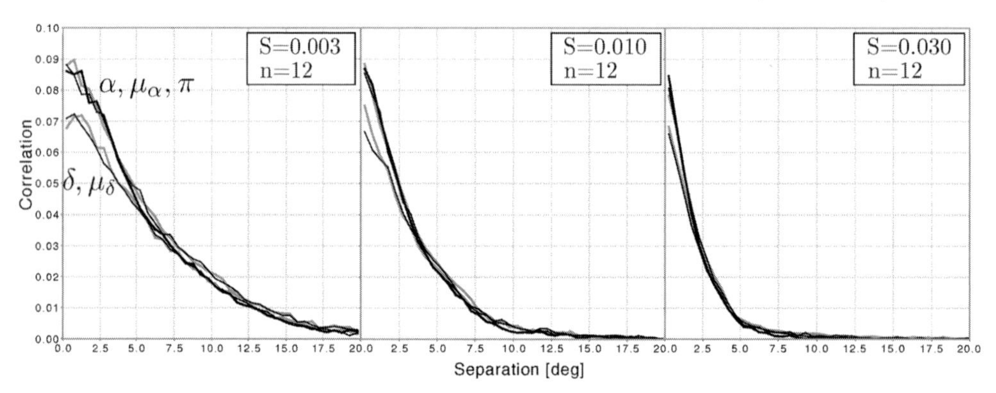

Figure 3. Same as Fig. 2 but plotted for separations $\theta = 0$–$20°$ (binsize 0.5°).

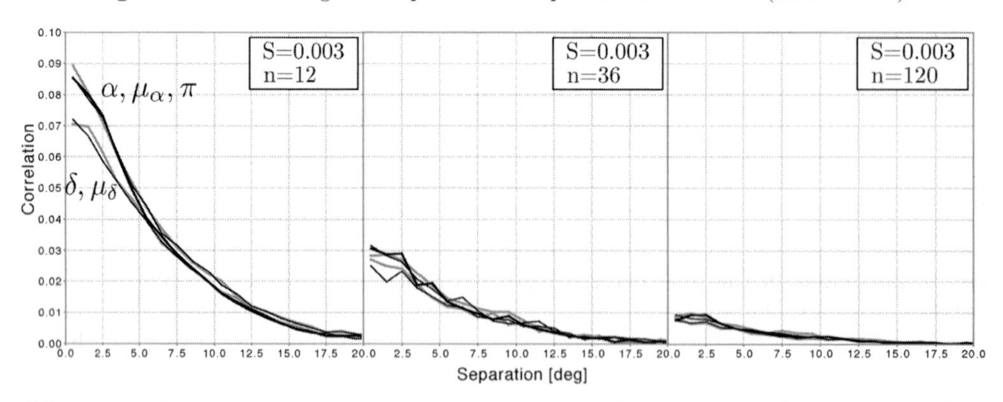

Figure 4. Spatial correlations in experiments C, D and E (see Table 1). Binsize is 1.0°.

References

Hobbs, D., Holl, B., Lindegren, L., Raison, F., Klioner, S., & Butkevich, A. 2009, *this proceedings*, 320

Lindegren, L. 2009, *this proceedings*, 296

Relativity in Fundamental Astronomy
Proceedings IAU Symposium No. 261, 2009
S. A. Klioner, P. K. Seidelman & M. H. Soffel, eds.

doi:10.1017/S1743921309990585

Gaia and the asteroids: Local test of GR

Daniel Hestroffer[1] and S. Mouret[1,2] and F. Mignard[3] and P. Tanga[3] and J. Berthier[1]

[1]IMCCE, Observatoire de Paris, CNRS,
77 Av. Denfert-Rochereau F-75014 Paris, France
email: hestro@imcce.fr, berthier@imcce.fr

[2]Lohrman Observatory, Dresden, Germany; email: serge.mouret@tu-dresden.de

[3]Cassiopée, Observatoire de la Cote d'Azur, CNRS, Mont-Gros F-06300 Nice, France
email: mignard@oca.eu, tanga@oca.eu

Abstract. We present in the following some capabilities of the Gaia mission for performing local test of General Relativity (GR) based on the astrometry of asteroids. This ESA cornerstone mission, to be launched in Spring 2012, will observe—in addition to the stars and QSOs—a large number of small solar system bodies with unprecedented photometric and, mostly, astrometric precisions. Indeed, it is expected that about 250, 000 asteroids will be observed with a nominal precision ranging from a few milli-arcsecond (mas), to sub-mas precision, depending on the target's brightness. While the majority of this sample is constituted of known main-belt asteroids orbiting between Mars and Jupiter, a substantial fraction will be made of near-Earth objects, and possibly some newly discovered inner-Earth or co-orbital objects.

Here we show the results obtained from a simulation of Gaia observations for local tests of GR in the gravitational field of the Sun. The simulation takes into account the time sequences and geometry of the observations that are particular to Gaia observations of solar system objects, as well as the instrument sensitivity and photon noise. We show the results from a variance analysis for the nominal precision of the joint determination of the solar quadrupole J_2 and the PPN parameter β. Additionally we include the link of the dynamical reference frame to the conventional kinematically non-rotating reference frame (as obtained in the visible wavelength by Gaia observations of QSOs). The study is completed by the determination of a possible variation of the gravitational constant $\dot{G}/G$, and deviation from Newtonian $1/r^2$ gravitational law. Comparisons to the results obtained from other techniques are also given.

Keywords. Gaia, asteroids, astrometry, GR, solar quadrupole, reference frame

1. Introduction

One century almost after the appearance of Albert Einstein's "Allgemeine Relativitätstheorie" paper (1915), the theory of general relativity (GR) is still the subject of debates as alternative to the metric theories have been proposed, and tests of the GR in particular in the parameterized post-Newtonian (PPN) frame can still be undertaken (e.g. Will 2006). After the first experiment of Einstein with the perihelion drift of Mercury, Gilvarry (1953) or Dicke (1965) noted that near-Earth asteroids with large eccentricity as (1566) Icarus are also good candidates for such local tests. Past efforts to test the theory of GR with the asteroid Icarus and other solar system objects revealed however unsuccessful (Shapiro *et al.* 1971) or could provide a test at the percent level (Sitarski 1992, Zhan 1994), mainly because of the many systematic or large stochastic errors in the observations, as well as in the dynamical model itself. For instance it was basically uneasy to disentangle relativistic effects from possible unknown non-gravitational effects. With the advent of modern high precision astrometry, from ground-based radar observations (Ostro 2007), or from space with Hipparcos (Hestroffer *et al.* 1998) and Gaia

(Mignard *et al.* 2007), the situation will change drastically and asteroids should be potentially powerful targets as is the Moon from LLR ranging, Mercury with Beppi-Colombo, or other planets with radar ranging. There are some advantages to consider asteroids: they are numerous; probing a wide range in distance in the Solar System[a]; and they are also small in size and mass, making them behave dynamically as free test particles with little shape effects on their astrometry. On another hand, their fast motion and faintness make them difficult to observe with small telescopes[b], and last, non-gravitational effects by perturbing their orbits can mimic some of the foreseen relativistic effects.

We present in the following an analysis of the performance that could be achieved from the astrometry of Gaia asteroids, and comparison to other works. First the mission and instruments characteristics are briefly presented. We then give the formal precision for the determination of the PPN parameter β together with the solar quadrupole J_2, a possible variation of the constant of gravitation $\dot{G}/G$, a possible deviation from the Newtonian force $(1/r^2 + \kappa)$, and last the precision that will be achieved for linking the dynamical reference frame to the kinematical optical-ICRF that will be obtained with Gaia.

2. The Gaia mission

The astrometric Gaia mission will regularly scan the whole celestial sphere down to magnitude $V \leqslant 20$, providing high precision data for a huge number of celestial objects (Lindegren 2009); including not only stars but also solar system objects, mostly asteroids. Compared to its precursor Hipparcos, Gaia will provide a wealth of information on asteroids: much higher precision in astrometry and spectro-photometry, for a number of targets about 4 orders of magnitude larger. In particular Gaia will observe NEOs down to low solar elongation (see Table 1). One can estimate that about 250,000 asteroids will be regularly observed. The average number of observations given here for an MBA can be smaller for a NEO or for an object close to the magnitude limit. The high precision astrometry (at sub-mas level precision) that will be acquired for asteroids should allow to revise the test of GR from analysing their orbit. The present work can be separated in two steps: simulation of the observations for the targets, and variance analysis for the various parameter estimation.

Based on the principle of Hipparcos for global astrometry, the Gaia telescope will not allow a pointing observation of an object. Instead, the target is observed when it is transiting the field-of-view (FOV). We have thus performed a simulation of the observations taking into account the time sequences and geometry of the observations that are particular to Gaia observations of solar system objects, as well as the instrument sensitivity and photon noise. We have also considered two sets of object populations. The first set includes all known asteroids, the second one consists of a population of synthetic NEOs. Indeed, not all NEOs larger than ≈ 500 m have been discovered yet, but many will be known at the time Gaia will operate from current or future ground-based surveys. Among this second set, several synthetic populations have been constructed combining random brightness and orbital elements following the de-biased distributions from Bottke *et al.* (2002). The second step of simulations in the chain gives an estimation of the astrometric precision as a function of the object size, motion and brightness. As given in Table 1,

[a] Going from near-Earth objects - NEO, main-belt asteroids - MBA, Jupiter Trojans, Centaurs and trans-Neptunian objects - TNO, or long-period comets - LPC.

[b] High precision optical astrometry generally necessitates a very good astrometric catalogue free as possible of any zonal error. Radar echo can be obtained only for the closest NEOs.

the formal precision[c] of one astrometric point in the highest resolution direction varies between 0.3 to 5 mas (milli-arcsecond) and mainly depends on the source brightness.

Table 1. General figure of the Gaia mission and observations of Solar System Objects.

Launch date / duration	spring 2012 / 5 years
Celestial sphere coverage	6 month
Limiting magnitude / size	$V \leqslant 20$ / ≈ 250 mas
Number of stars	$\approx 10^9$
Number of asteroids /NEOs	≈ 250.000 / ≈ 2500
Number of observations	≈ 60 transits/target
Solar elongation	$45^\circ \leqslant L \leqslant 135^\circ$
Astrometry (CCD)	AL: 0.3 – 5 mas
	AC: 6 – 12 × AL
Photometry (CCD)	0.001 – 0.05 mag

3. Global parameters determination

Starting from the simulated data of the previous section, we can perform a variance analysis for various parameters to be estimated (or adjusted), in particular global parameters common to all, or a large subset of targets. Since the orbits are already known and all foreseen parameters are small, we can linearise our system of observational equations and solve it by least squares. The vector of unknown parameters $d\mathbf{p} = (d\mathbf{q}_i; d\mathbf{q}_g)$ will contain the correction to the initial conditions $d\mathbf{q}_i$ specific to each asteroid i and the global parameters $d\mathbf{q}_g$ common to all objects. The matrix of partial derivatives:

$$[\partial \mathbf{x}/\partial \mathbf{p}]_i = [\partial x/\partial q_i; \partial x/\partial q_g] \equiv [\mathbf{B}_i; \mathbf{A}_i] \tag{3.1}$$

yields the variation of the target's position at time of observation, and is computed numerically. The variance of the global parameters is then obtained from the inversion of a reduced normal matrix (Söderhjelm & Lindegren 1982) :

$$\mathbf{U} = \sum_i (\mathbf{A}'_i.\mathbf{A}_i) - \mathbf{A}'_i.\mathbf{B}_i(\mathbf{B}'_i.\mathbf{B}_i)^{-1}\mathbf{B}'_i.\mathbf{A}_i \qquad ; \qquad var(d\mathbf{q}_g) = \sigma_o \mathbf{U}^{-1} \tag{3.2}$$

where matrices $(\mathbf{B}'_i.\mathbf{B}_i)$ are of dimension 6×6 as the state vector of the asteroid, and $(\mathbf{A}'_i.\mathbf{A}_i)$ of dimension $n_g \times n_g$, i.e. the number of global parameters, no more than a dozen.

The global parameters that can be estimated are the mass of perturbing asteroids (Mouret *et al.* 2007) or other parameters that would affect all orbits such as the PPN parameter β, the solar J_2. It is well known that both the Sun quadrupole J_2 and GR imply an advance of the perihelion of the orbit that can hardly be separated from the observation of one single target, such as Mercury alone. Indeed, putting $m_\odot = GM_\odot/c^2$ the secular drift[d] of the orbit's argument of periapsis is given by :

$$\Delta\omega \quad = \quad \Delta\omega_{|PPN} + \Delta\omega_{|J_2}$$

[c] A typical star will cross 9 CCDs in a row during one FOV transit, but a fast moving object will not; thus, in order to be more conservative we consider that one transit is reduced to one single CCD exposure. Note that the typical 20 μas precision found in the literature for a star is much smaller because it moreover combines all ≈100 transits gathered over the mission.

[d] Linear part of the orbital elements variation after periodic terms have been averaged, and also after the larger—but well known—secular perturbations of the planets have been removed.

Table 2. Overview of the expected results for the local test of GR with Gaia and the asteroids, and comparison to other experiments in the solar system. (Check mark means parameter determination is possible but not explicitly provided.)

	$\sigma(\beta)$ –	$\sigma(J_2)$ –	Correl –	$\sigma(\dot{G}/G)$ [yr^{-1}]	$\sigma(\kappa)$ [$\mathrm{m.s}^{-2}$]	$\sigma(W_o)$ [μas]	$\sigma(\dot{W})$ [$\mu\mathrm{as.yr}^{-1}$]	Ref.
Gaia	0.6–6×10^{-4}	0.5–10×10^{-8}	0.1–0.9	2×10^{-12}	8×10^{-11}	[5 – 5 – 15]	[1 – 1 – 5]	
LLR[1,2]	1.1×10^{-4}	—	—	3×10^{-13}	—	—	—	[1]
Ephemeris[2,3]	2×10^{-4}	assumed	—	5×10^{-13}	√	√	40	[2]
Bepi Colombo[1]	2×10^{-6}	2×10^{-9}	0.997	√	—	—	—	[3]
NEOs radar[4]	√	√	√	—	—	—	—	[4]
TNOs[5]	—	—	—	—	1.6×10^{-10}	—	—	[5]

Notes:
[1] In the LLR technique as well as for the Bepi-Colombo experiment, the PPN β is derived from Nordvedt $\eta_N = 4\beta - \gamma - 3$ parameter. Also γ is hence assumed to be know with sufficient accuracy.
[2] Precision on $\dot{G}/G$ is improving rapidly with time and increased data span. The same is true for the pulsar timing technique (Deller *et al.* 2008).
[3] Based on model value for the Solar J_2 in Fienga *et al.* (2008), Pitjeva (2005); $\sigma(J_2) = 3\times10^{-8}$ in Pitjeva (2009).
[4] Radar measurements will provide these parameters from a set of ≈ 20 observed targets.
[5] Based on analysis of $\approx$25 TNOs, not directly comparable to the Gaia value (see text).
Ref.: [1] Williams & Folkner (2009) ; [2] Folkner (2009) for $\dot{G}$, κ and W Pitjeva (2009), Fienga *et al.* (2008) for β, J_2 and $\dot{G}$; [3] Milani (2009) ; [4] Margot & Giorgini (2009) ; [5] Wallin *et al.* (2007).

$$= \frac{3m_\odot}{a\,(1-e^2)}\left[\frac{2+2\gamma-\beta}{3} + \frac{R_\odot^2}{4\,a\,m_\odot}\,\frac{(5\cos^2 i - 1)}{(1-e^2)}\,J_2\right] n\,(t-t_0) \tag{3.3}$$

while the other elliptical elements, in particular the drift for the longitude of the node, is driven by the Sun quadrupole only ($\Delta\Omega = \Delta\Omega_{|J_2}$). Here one readily sees that the relativistic and J_2 secular effects – being large for high eccentricities – act differently through the asteroids[e] inclination i and mostly through its semi-major axis a and eccentricity e. Other parameters of the dynamical model that can similarly be estimated are a possible time-variation of the gravitational constant $\dot{G}$, a violation of the Newtonian $1/r^2$ law of gravitation[f], or a rotation[h] $\mathbf{W} = \mathbf{W}_o + \mathbf{W}_1\,(t - t_0)$ between the kinematically or dynamically non-rotating frames associated to either the QSOs or the ephemerides, respectively.

4. Results – Discussion

Solving for the global system (3.2) yields the formal precision given in Table 2. The basic output will be the derivation of the solar J_2 and PPN β with no model assumption for the Sun interior, shape or rotation. The precision is not better than what is achieved today from LLR data, but yet independent of the Nordvedt η_N parameter. Note that the parameter γ in Eq. (3.3) is known with much better accuracy (from Gaia itself, Mignard (2009), Hobbs *et al.* (2009), or other experiments). The formal precision and correlation are nevertheless sensitive by roughly one order of magnitude to the actual number of targets brighter than magnitude $V \leqslant 20$, to the actual number of observations per target, and to the observations of highly eccentric objects orbiting close from the Sun.

[e] In the case of major planets most eccentricities and inclination are small, and similar.

[f] One includes an additional acceleration term in the equation of motion in a way similar to Wallin *et al.* (2007), but here systematically at any distance to the Sun, $\ddot{\mathbf{r}} = -GM_\odot\,(1/r^3 + \kappa)\,\mathbf{r}$.

[h] That is the sum of a constant rigid rotation $\mathbf{W}_o$ at reference epoch t_0 and a rotation rate $\mathbf{W}_1$.

One will be able to measure a possible variation $\dot{G}/G$ at the $10^{-12}\,\mathrm{yr}^{-1}$ level, and validate the Newtonian $1/r^2$ law in the main belt of asteroids at the 10^{-10} level. Note a fundamental difference with the 'Pioneer-anomaly' kind of test, which consider a deviation to the Newtonian law triggered only at some given distance to the Sun, e.g. 5, 10 or 20 AU. If we restrict the analysis to the few Centaurs and TNOs observed by Gaia alone over five years, the precision $\sigma(\kappa)$ drops to $6 \times 10^{-7}\,\mathrm{m/s}^2$. Note also that a possible spatial variation of G is not considered, and that the time variation is coupled with the mass-loss of the Sun, i.e. what is actually measured is $d(GM)/dt$. One will also be able to measure a possible rotation rate $\mathbf{W}_1$ to the order of $6\,\mu\mathrm{as/yr}$ ($\approx 3 \times 10^{-11}\,\mathrm{rad/yr}$) between the dynamical and kinematical reference frames. This is still far much larger than the geodetic precession of the Solar System orbiting around the Galaxy $\dot{\Omega}_{GP} = 3/2(V_\odot/c)(GM_g/c/R_\odot^2) \approx 0.02\mu\mathrm{as/yr}$, or current bounds to a Gödelian rotation of the Universe ($|\omega| \leqslant 10^{-2}\mu\mathrm{as/yr}$). This formal precision is moreover to be balanced by the precision with which the materialization of the kinematically non-rotating frame can be achieved (Lindegren 2009, Zharov *et al.* 2009) and the systematic errors that can enter in the process of asteroids orbit fitting. Another relativistic effect that will perturb the orbits is the gravitomagnetic Lense-Thirring effect from the spinning Sun. Putting $J_\odot = \frac{4}{5}M_\odot R^2\omega_\odot$ where $\omega_\odot$ is the spin rate of the Sun, the precession of the orbit are $\dot{\Omega}_{LT} = 2\frac{G}{c^2}\frac{J_\odot}{a^3(1-e^2)^{3/2}}$ for the node and $\dot{\varpi}_{LT} = -\frac{G}{c^2}\frac{J_\odot(1-3\sin^2 i/2)}{a^3(1-e^2)^{3/2}}$ for the longitude of the periapsis and the mean anomaly. The Lense-Thirring perturbation can be of the order of 10 to $100\,\mu\mathrm{as/yr}$, depending on the target orbit, and must hence be taken into account.

Combining the radar data already acquired by Margot & Giorgini (2009), which are of comparable quality and moreover orthogonal in essence by providing the range of the target, to the Gaia data when available, should also improve these numbers and possibly provide one of the best (direct) measure of the PPN β. Possible test of the Strong Equivalence Principle from analysis of the asteroids orbits (in particular the Trojans around the stable Lagrangian points of Jupiter, Orellana & Vucetich (1988)), influence of non-gravitational effects, and derivation of the orbits precession from the Lense-Thirring effect are under study.

References

Bottke, W. F., Morbidelli, A., Jedicke, R., *et al.* 2002, Icarus, 156, 399
Deller, A. T., Verbiest, J. P. W., Tingay, S. J., & Bailes, M. 2008, ApJL, 685, L67
Dicke, R. H. 1965, AJ, 70, 395
Fienga, A., Manche, H., Laskar, J., & Gastineau, M. 2008, A&A, 477, 315
Folkner, W. M. 2009, *this proceedings*, 155. BAAS, 41, #06.01
Gilvarry, J. J. 1953, Physical Review, 89, 1046
Hestroffer, D., Morando, B., Hog, E., *et al.* 1998, A&A, 334, 325
Hobbs, D., Holl, B., Lindegren, L., *et al.* 2009, *this proceedings*, 315. BAAS, 41, #16.03
Lindegren, L. 2009, *this proceedings*. BAAS, 41, #16.01
Margot, J.-L. & Giorgini, J. D. 2009, *this proceedings*, 183. BAAS, 41, #07.01
Mignard, F. 2009, *this proceedings*, 306. BAAS, 41, #16.02
Mignard, F., Cellino, A., Muinonen, K., *et al.* 2007, Earth Moon and Planets, 101, 97
Milani, A. 2009, *this proceedings*, 356. BAAS, 41, #18.01
Mouret, S., Hestroffer, D., & Mignard, F. 2007, A&A, 472, 1017
Orellana, R. B. & Vucetich, H. 1988, A&A, 200, 248
Ostro, S. J. and Giorgini, J. D. and Benner, L. A. M. 2007, IAU Symposium #236, 143
Pitjeva, E. V. 2005, Astronomy Letters, 31, 340–349

Pitjeva, E. V. 2009, *this proceedings*, 170. BAAS, 41, #06.03
Shapiro, I. I., Smith, W. B., Ash, M. E., & Herrick, S. 1971, AJ, 76, 588
Sitarski, G. 1992, AJ, 104, 1226
Söderhjelm, S. & Lindegren, L. 1982, A&A, 110, 156
Wallin, J. F., Dixon, D. S., & Page, G. L. 2007, ApJ, 666, 1296
Will, C. M. 2006, Living Reviews in Relativity, 9
Williams, J. G. & Folkner, W. M. 2009, *this proceedings*. BAAS, 41, #08.01
Zhang, J.-X. 1994, Chinese Astron. and Astroph., 18, 108–115
Zharov, V., Sazhin, M., Sementsov, V., Kuimov, K., & Sazhina, O. 2009, *this proceedings*, 50. BAAS, 41, #02.06

Relativity in Fundamental Astronomy
Proceedings IAU Symposium No. 261, 2009
S. A. Klioner, P. K. Seidelman & M. H. Soffel, eds.

doi:10.1017/S1743921309990597

Optimising the Gaia scanning law for relativity experiments

Jos de Bruijne[1]†, Hassan Siddiqui[2], Uwe Lammers[2], John Hoar[2], William O'Mullane[2], and Timo Prusti[1]

[1]Research and Scientific Support Department of the European Space Agency, European Space Research and Technology Centre, Keplerlaan 1, 2201 AZ, Noordwijk, The Netherlands

[2]Science Operations Department of the European Space Agency, European Space Astronomy Centre, Villanueva de la Cañada, 28692 Madrid, Spain

Abstract. Gaia is ESA's upcoming astrometry mission, building on the heritage of its predecessor, Hipparcos. The Gaia nominal scanning law (NSL) prescribes the ideal attitude of the spacecraft over the operational phase of the mission. As such, it precisely determines when certain areas of the sky are observed. From theoretical considerations on sky-sampling uniformity, it is easy to show that the optimum scanning law for a space astrometry experiment like Gaia is a revolving scan with uniform rotation around the instrument symmetry axis. Since thermal stability requirements for Gaia's payload require the solar aspect angle to be fixed, the optimum parallax resolving power is obtained by letting the spin axis precess around the solar direction. The precession speed has been selected as compromise, limiting the across-scan smearing of images when they transit the focal plane, providing sufficient overlap between successive "great-circle" scans of the fields of view, and guaranteeing overlap of successive precession loops. With this scanning law, with fixed solar-aspect angle, spin rate, and precession speed, only two free parameters remain: the initial spin phase and the initial precession angle, at the start of science operations. Both angles, and in particular the initial precession angle, can be initialized following various (programmatic) criteria. Examples are optimization/fine-tuning of the Earth-pointing angle, of the number and total duration of Galactic-plane scans, or of the ground-station scheduling. This paper explores various criteria, with particular emphasis on the opportunity to optimise the scanning-law initial conditions to "observe" the most favorable passages of bright stars very close to Jupiter's limb. This would allow a unique determination of the light deflection due to the quadrupole component of the gravitational field of this planet.

Keywords. relativity, space vehicles: instruments, astrometry, planets and satellites: Jupiter

---

1. Jupiter's quadrupole moment

For Jupiter, the magnitude of the gravitational monopole deflection for a grazing ray is 16 mas. Superimposed on this monopole, there is a quadrupole field with an amplitude of 240 μas. This quadrupole deflection, which is within Gaia's reach, has a specific pattern which is a function of (i) the position of the star with respect to the oblate deflector, and (ii) the orientation of Jupiter's spin axis. It is, therefore, possible with Gaia to carry out some kind of eclipse experiment – similar to that of Dyson, Eddington, and Davidson in 1920 – by comparing stellar positions in the immediate vicinity of Jupiter (Crosta & Mignard (2006)). We plan to do so and propose to start Gaia's nominal scanning law after launch with optimum initial conditions, such that the maximum number of passages of bright stars close to Jupiter's limb will be observed. This, in turn, will optimise the end-of-mission detection significance of the quadrupole light deflection.

† This poster describes work which is carried out in the frame of the RElativistic Models And Tests (REMAT) group of the Gaia Data Processing and Analysis Consortium (DPAC).

2. The Gaia Nominal Scanning Law (NSL) and operational strategy

During its 5-year operational life time, Gaia will continuously scan the sky using a uniform, revolving scanning law – the Nominal Scanning Law (NSL) – satisfying the loop-overlap condition. The NSL is determined by fixing five constants and two functions: the inertial spin rate ω (= 60 arcsec s^{-1}; rotation period = 6 h); the revolving angle, also known as solar-aspect angle ξ (= 45°); the precession-speed constant S (= 4.223; 5.8 spin-axis revolutions per year); the initial spin phase Ω_0 (free parameter); the initial revolving phase ν_0 (free parameter); and the ecliptic coordinates of the nominal Sun $\lambda_{\rm s}(t)$, $\beta_{\rm s}(t)$ (fixed by Nature). The spin-axis coordinates $\lambda_{\rm z}(t)$, $\beta_{\rm z}(t)$ and spin phase $\Omega(t)$ follow from solving a set of coupled differential equations. Since ω, ξ, and S have been fixed for Gaia, the NSL is fully determined once the two free parameters (Ω_0 and ν_0) are defined, which is planned for at the start of the mission. We investigate the optimum values for these free parameters from the mission point of view.

During each 6-hour revolution around its spin axis, each Gaia field of view (FoV) sweeps out a thickened great circle which crosses the Galactic plane twice. The typical 6-hour data-collection rate of the two FoVs combined contains 4 peaks. During specific periods of the mission, the spin axis points close to one of the Galactic poles. Due to the relatively slow precession of the spin axis, such situations can be sustained for several consecutive days. During these periods, the FoVs effectively continuously sweep along the Galactic plane, with high sustained data rates as a result. These periods are known as Galactic-plane scans. Typically, a handful of such periods occur during each year of the mission. When they occur depends on the NSL initial conditions.

3. Ground-station support

Gaia observes 24 hours per day. Typically, however, ground-station contact is limited to 8 hours per day as a result of programmatic constraints. During the daily non-visibility periods, collected data are stored on board in a mass memory. Mass-memory saturation – resulting in data loss and science-performance degradation – occurs about two dozen times over the mission. These events are associated with Galactic-plane scan periods. In order to minimize data losses, the systematic support of a second ground station is scheduled during Galactic-plane scan periods. When Galactic-plane scan periods precisely occur, depends on the NSL initial conditions, in particular the initial revolving phase ν_0. The peak-to-valley variation of the total number of Galactic-plane-scan days is 40% over all possible initial angles. The NSL initial conditions, thus, determine the total ground-station support required and, hence, are a cost factor.

4. Ground-station conflicts

ESA's ground-station network includes two Deep Space Antennae: a 35-m dish in Cebreros (Spain) and a 35-m antenna in New Norcia (Australia). These antennae are shared resources. During Gaia's operational life (2012–2017), at least one conflicting user has been identified, namely Rosetta. The Rosetta mission is an interplanetary mission whose objective is to rendezvous with and make in-situ measurements of comet 67 P/Churyumov–Gerasimenko. These events will take place in summer/autumn 2014. Assuming Rosetta's ground-station needs during this period will be met with high priority, limited station hours will be left for Gaia during summer 2014. From a ground-station conflict analysis, it follows that Gaia Galactic-plane scans should be avoided from June through August 2014. This will require dedicated NSL initial conditions.

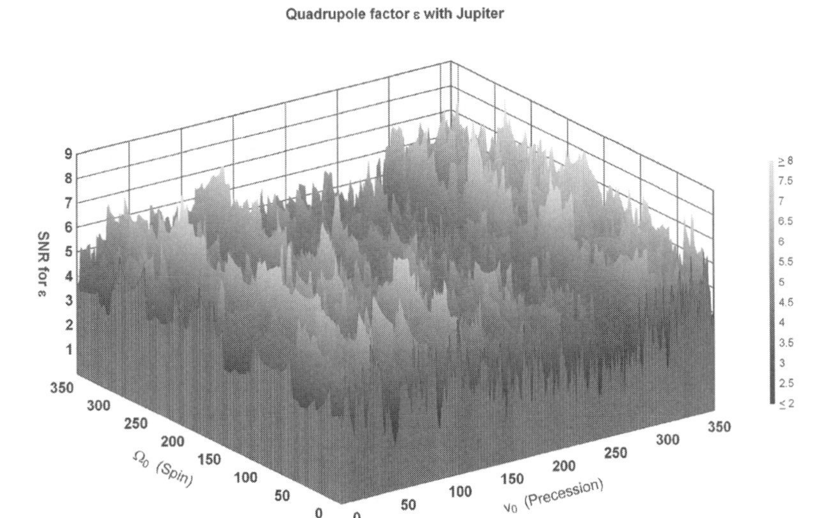

Figure 1. Signal-to-Noise Ratio (SNR) of the detection of the Jupiter quadrupole deflection measurement as function of NSL initial angles. Figure and analysis courtesy François Mignard.

5. Spin-axis – Earth angle

Gaia uses its Phased-Array Antenna (PAA) to transmit its data to the ground. Towards the edges of the usable range (15° away from the central 45° direction), the PAA looses power. It is, hence, favorable for Gaia's overall data return – through an improved link budget and potentially more efficient convolutional encoding of the telemetry stream – to minimise the angle between the spin axis and the Gaia–Earth vector. The evolution of this angle over the mission depends on the NSL initial conditions. Optimised NSL initial conditions can reduce the peak-to-valley variations of the angle by 1° on both extremes, with favorable consequences for the overall science data return.

6. Conclusion

Figure 1 shows the expected Signal-to-Noise Ratio (SNR) of the detection of the Jupiter quadrupole deflection measurement as function of NSL initial angles ν_0 and Ω_0. An actual star catalogue was used to predict close encounters between bright stars and Jupiter and to accumulate the mission-total observable signal. It follows that ν_0 is the most important initial angle to increase the detection significance: up to 8σ detections can be reached for good choices of ν_0. Poor choices for ν_0 exist too (for instance around 225°) which only return 1σ detections. We, therefore, propose to start Gaia's NSL with optimum initial conditions (e.g., $\nu_0 \approx 341^\circ$) such that the quadrupole SNR is optimised.

Performing relativistic tests using a non-dedicated space mission has significant limitations, of both technical and programmatic nature, for instance related to ground-station scheduling and ground-station conflicts with other missions. We are currently performing a system-level trade-off analysis to reveal the (in)compatibilities of the various constraints and will report on the outcome of this study elsewhere.

References

Crosta, M. T. & Mignard, F. 2006, *Classical and Quantum Gravity*, 23, 4853–4871

Relativity in Fundamental Astronomy
Proceedings IAU Symposium No. 261, 2009
S. A. Klioner, P. K. Seidelman & M. H. Soffel, eds.

doi:10.1017/S1743921309990603

Practical relativistic clock synchronization for high-accuracy space astrometry

Christophe Le Poncin-Lafitte[1]

[1]Observatoire de Paris, SYRTE CNRS/UMR8630, Université Pierre et Marie Curie,
77 Avenue Denfert Rochereau, F75014 Paris, France
email: `christophe.leponcin-lafitte@obspm.fr`

Abstract. Future high-accuracy space astrometry missions, such as Gaia and SIM, will need a time-tagging of observations consistent with General Relativity nowadays used as standard background for global data processing scheme. In this work, we are focusing on the realization of the onboard time scale. The onboard clock, being not ideal and consequently tainted with systematic biases, has to be carefully calibrated to the ideal relativistic proper time of the satellite. We present here a modeling of this essential step to provide a reliable relation between the onboard time and TCB, a time scale suitable for global data processing.

1. Introduction

Future space astrometry missions are expected to reach an accuracy of several microarcseconds (μas) for the determination of positions, parallaxes and proper motions of celestial objects. This high accuracy requires subtle relativistic modeling to be used for the data processing. First of all, it is crucial to be able to give a relativistic formulation of astrometric observations, which is usually performed by a resolution of the null geodesics equations for the light propagation from the celestial object to the observer. Then it is also indispensable to control the attitude of the satellite in the four-dimensional spacetime, which requires to construct a particular tetrad or to use a description of the satellite's attitude in a comoving center-of-mass reference system, ideally defined as kinematically non-rotating. Another issue is crucial: the realization and the use of relativistic time scales (Le Poncin-Lafitte 2008).

For instance, the whole data processing of Gaia observations will be done in TCB, the coordinate time scale of the Barycentric Celestial Reference System (Soffel *et al.* 2003). From practical point of view, an onboard clock will produce a time scale, called OnBoard Time OBT, used to tag all kind of observations. In particular, specific tasks, such that observations of variable phenomena, will need precise absolute timing and require that the OBT time scale have to be stabilized with an accuracy of roughly one microsecond. However, if OBT can be viewed as a practical realization of the relativistic ideal Gaia proper time TG along the worldline of the satellite, the onboard clock is not perfect and OBT will be contaminated by some technical clock errors which means that formally OBT and TG time scales will be different. Moreover because of the motion of Gaia and non-zero gravitational potential at the spacecraft location around the Earth/Sun L_2 point, it is not straightforward to relate TG and TCB and a complete relativistic time transformation to go from TG to TCB, and vice versa, is needed (Klioner 1992). The problem is worse when one thinks that in fact we need a relation between OBT and TCB to be able to do data processing in a correct way.

2. Operational strategy

This issue can be achieved by synchronizing the onboard clock with Earth ground clocks. However, it must realized that a satellite is not necessarily continuously observable from the ground during the full time of the mission. Then, the synchronization will be only possible during a period of visibility where many time telemetry procedures are performed at regular intervals. This procedure consists in the interrogation of the onboard clock to create a tag OBT_k. After delay due to the packaging by the onboard computer of that tag into a time telemetry package, the latter is sent to the Earth. After the flight time of the signal between the satellite and the Earth, the package is received by the antenna of a ground station. A new time delay is then necessary to transfer the package from the antenna to a computer where a tag UTC_k *of reception*, in Universal Coordinate Time, is created and stored. We finally obtain a pair (OBT_k, UTC_k) for each procedure. All these pairs constitute the initial data set for the synchronization. The question is now how to deduce from that pairs a new set of (OBT_k, TG_k). A method, illustrated on the next figure, is proposed in this paper to perform this task.

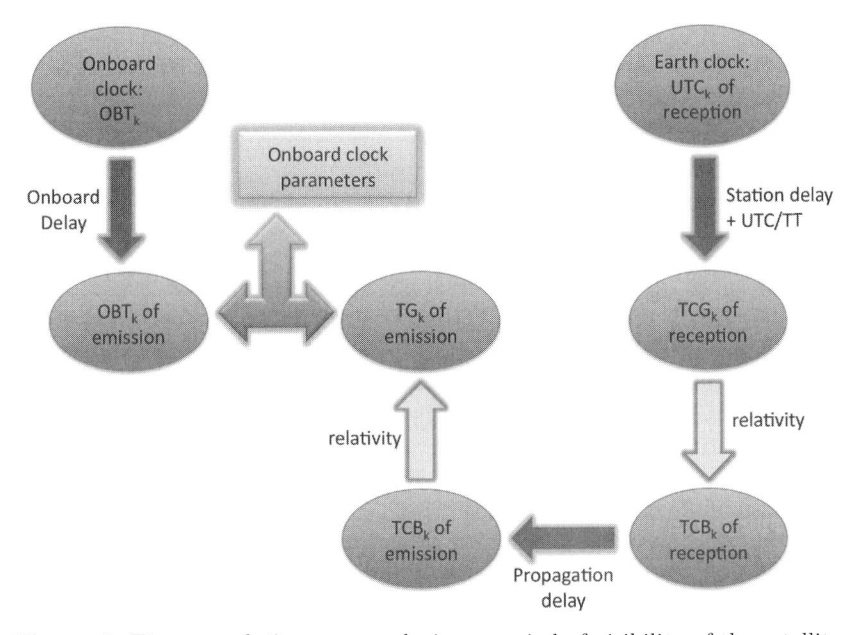

Figure 1. Time correlation process during a period of visibility of the satellite.

Let us detail the steps involved in the procedure:

• UTC_k *of reception* is first transformed into TCG_k *of reception*. Taking into account the instrumental ground delay and using usual relations between terrestrial time scales, it is possible to perform the following transformation $UTC_k \rightarrow TAI_k \rightarrow TT_k \rightarrow TCG_k$ *of reception*.

• TCG_k *of reception* to TCB_k *of reception*. This is a pure relativistic step which is achieved by solving an ordinary differential equation along the wordline of the geocenter.

• TCB_k *of reception* to TCB_k *of emission*. The purpose here is to convert a time tag relative to the reception of the packet on Earth ground station to a time tag relative to the emission time tag from the satellite. This step requires knowing all kinds of time delays during the propagation of the signal between the satellite and the Earth. It can be written as the sum of several contributions involving the instantaneous BCRS distance of the satellite to the Earth based station (it requires precise ephemeris of the probe as well as the knowledge of the position of the Earth based station in the BCRS), the relativistic Shapiro delay, the propagation delay due to the Solar plasma, the troposphere and the ionosphere of the Earth.

• TCB_k *of emission* to TG_k *of emission*. This is the second pure relativistic step of the scheme. It is achieved by solving an ordinary differential equation along the worldline of the satellite.

• OBT_k to OBT_k *of emission*. This step just consists of taking into account some instrumental onboard delays between the interrogation of the clock, producing one tag OBT_k, and the operational instant of sending a time telemetry packet, OBT_k *of emission*, to the Earth ground station.

3. Modeling the errors of the clock

Because the onboard clock can obviously not be ideal for technical reasons, the practical time-scale OBT realized onboard will not be stable. Consequently the time-tagging of all observations will be contaminated with technical errors. It means that the formal difference $OBT - TG$ along the satellite worldline can not be exactly zero ; so for each pair (OBT_k, TG_k) we will get $OBT_k - TG_k = f(OBT_k)$. The whole question is to find a simple expression of the function f. To determine this, a modeling of the clock errors is then indispensable. Usually the frequency of a clock is represented by

$$\frac{d}{d\tau}(\tau - \tau_{Rb}) = A + B\tau + C\sin(2\pi f + \phi) + \frac{D}{\tau_0}F(\tau_0)\,, \tag{3.1}$$

where $d\tau$ is the ideal local proper time interval between two events on the worldline of the onboard clock, $d\tau_{Rb}$ is the time interval between the same two events as measured by the clock, A, B, C, D, f, ϕ and τ_0 are some constants to be determined and $F(\tau_0)$ is a random distribution (white noise) of points at interval τ_0 with unitary standard deviation. Here, it is assumed that the stochastic behaviour of the clock (last term in 3.1) is stationary, *i.e.* independent of which period of data is chosen. That is likely the case, except for malfunctions. The value of A is an arbitrary frequency offset, its value has a priori no effect on the stability and it will have to be estimated in flight. Additionally, we are practically interested in the integral of (3.1), *i. e.* in phase signal, which implies an integration constant (time offset) again to be estimated in flight. The frequency drift B and periodic effect C can be highly correlated and both will also have to be estimated in flight. The synchronization is then achieved when all available pairs (OBT_k, TG_k) have been used to numerically constraint the parameters A, B, C...

References

Klioner, S. A. 1992, *Celestial Mechanics and Dynamical Astronomy*, 53, 81

Le Poncin-Lafitte, C. 2008, *SF2A-2008, 129*

Soffel, M., Klioner, S. A., Petit, G., Wolf, P., Kopeikin, S. M., Bretagnon, P., Brumberg, V. A., Capitaine, N., Damour, T., Fukushima, T., Guinot, B., Huang, T.-Y., Lindegren, L., Ma, C., Nordtvedt, K., Ries, J. C., Seidelmann, P. K., Vokrouhlicky, D., Will, C. M., & Xu, C. 2003, *Astronomical Journal*, 126, 2687

Relativity in Fundamental Astronomy
Proceedings IAU Symposium No. 261, 2009
S. A. Klioner, P. K. Seidelman & M. H. Soffel, eds.

doi:10.1017/S1743921309990615

Global astrometric sphere reconstruction in Gaia: challenges and first results of the Verification Unit

Alberto Vecchiato[1], Ummi Abbas[1], Beatrice Bucciarelli[1], Mario G. Lattanzi[1], Roberto Morbidelli[1]

[1]Osservatorio Astronomico di Torino,
via Osservatorio 20, 10025 Pino Torinese (TO), Italy
email: vecchiato@oato.inaf.it

Abstract. Gaia will estimate the astrometric and physical data of approximately one billion objects. The core of this process, the global sphere reconstruction, is represented by the reduction of a subset of these objects, which will constitute the largest and most precise catalog of absolute astrometry in the history of Astronomy, and will put General Relativity to test by estimating the PPN parameter γ with unprecedented accuracy. As the Hipparcos mission showed, and as it is natural for all kind of absolute measurements, possible errors in the data reduction can hardly be identified at the end of the processing, and can lead to systematic errors in all the works which will use these results. In order to avoid such kind of problems, a Verification Unit was established by the Gaia Data Processing and Analysis Consortium (DPAC). One of its jobs is to implement and perform an independent global sphere reconstruction, parallel to the baseline one, to compare the two results, and to report any significant difference.

Keywords. Astrometry, catalogs, relativity

1. Introduction

The Gaia astrometric catalog, with its $\sim 10^{-9}$ objects at $\sim$10 to $\sim$100 microarcsecond accuracy, will be the richest and most precise ever produced. It will be a milestone of paramount importance for several science topics, going from almost every subject of astrophysics and astronomy to fundamental physics (Turon *et al.*, 2005).

The most evident example comes from the possibility of having parallaxes at the 10% accuracy level at galactic distances, that would result in a complete revision of the cosmic distance ladder since its first step (Webb, 1999).

At the same time, the kind of measurements performed by Gaia belongs to the area of *absolute astrometry*, and the main result of the mission is a *catalog* realizing an astrometric reference frame. A reference frame, as commonly said, is the materialization of a reference system; therefore, it is very difficult to identify possible errors in the measurements or in the data reduction process that brings to the definition of the final catalog, given its nature.

Gaia inherits many ideas from its "parent" mission HIPPARCOS, such as the concept of getting absolute parallaxes from simultaneous observations of two different fields of view separated by a large angle, or that of a scanning law which makes it possible for the satellite instruments to observe the whole celestial sphere. The main goal of HIPPARCOS was also to produce a catalog of absolute positions. Therefore, HIPPARCOS has faced the same kind of problems as above, but at a much smaller scale because of its lower precision and the much smaller size compared to that of the future Gaia catalog. It is then worth learning from that mission in order to tackle these difficulties.

The data reduction process in HIPPARCOS was carried out by two consortia, FAST and NDAC, which operated independently on the same data. Their two results were then compared and appropriately merged in order to obtain the final catalog.

This ideal solution cannot be applied to the case of Gaia. Due to the size of the problem, the data reduction task is much demanding both in terms of the needed resources and manpower. To retain as much as possible the HIPPARCOS approach, without requiring excessive resources, the Gaia Data Processing and Analysis Consortium (DPAC) adopted a strategy that foresees an Astrometric Verification Unit (AVU) within a single data processing pipeline. Before detailing the function of this unit, a brief summary of the structure of the Gaia data reduction pipeline is needed.

2. Overview of the Gaia data processing design

The DPAC is organized in *Coordination Units* (CUs), each of which is in charge of some parts of the whole task. CU3 takes care of the so-called *core processing*, i.e. it will consider a subset of "well-behaved" stars (e.g. single stars, photometrically and astrometrically stable, not too faint, etc.) and will reconstruct very precisely their five astrometric parameters. The number of stars processed by CU3 will be approximately some tens of million, up to 10^8, i.e. $\lesssim 10\%$ of the size of the final Gaia catalog. These will constitute a network to which the position of the other objects will refer. The CU3 reconstruction of the astrometric celestial sphere differs from that involving the other sources because it is *global*, i.e. without any reference to other objects, at least until this global reference frame will be linked to the ICRF.†

The Core Processing is a complex procedure that includes several steps. Since it is here that the "bootstrap" of the Gaia Reference Frame takes place, the DPAC has chosen some of its critical parts to be processed by two independent sub-systems in order to reproduce, for these specific steps, the structure of the two HIPPARCOS consortia. CU3 is designed in such a way that, for each of those steps, there will exist a verification counterpart operating independently from the main data reduction chain. All of these verification sub-systems have been gathered in the so-called AVU, each of them having the task to compare its results with those of its counterpart.

3. The sphere reconstruction at a glance

The main pipeline process which will reconstruct the global sphere is called Astrometric Global Iterative Solution (AGIS). In its bare bones, the sphere reconstruction consists in the solution, in the least-squares sense, of a large and sparse system of linearized equations.

Each equation corresponds to a Gaia observation whose known term contains the measurement, while the other ones are functions of the unknowns to be estimated. Since Gaia is a self-calibrating instrument, the function describing the satellite measurements does not depend just on the (relativistic) *astrometric* model, but also on the *attitude* and the instrument *calibration* parameters for the time of observation. Finally, since one of the expected by-products of the core processing is the estimation of the γ parameter of the Parametrized Post-Newtonian formulation, there is at least one unknown in a last set. Its elements appear in every equation of the system, and therefore these unknowns are called *global parameters*.

† We emphasize that the Gaia reference system will be materialized by all of the $\sim 10^9$ objects of the Gaia catalog, since the stars not processed by the CU3 will be linked to the same reference frame.

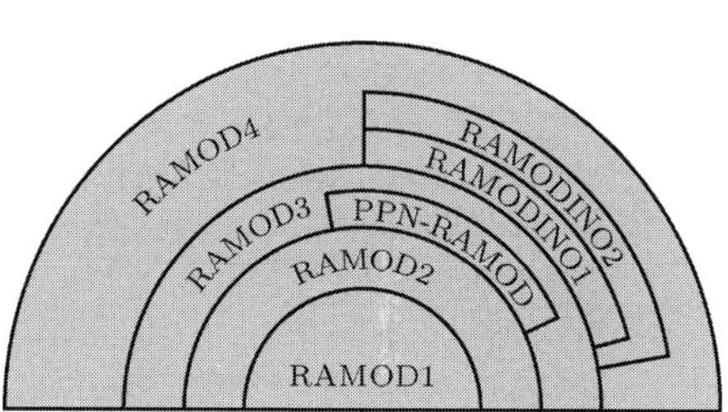

Figure 1. RAMOD identifies a family of astrometric models with increasing accuracies. The attitude models belonging to the project are called RAMODINO1 and RAMODINO2. The present relativistic model implemented in GSR is an adaptation of PPN-RAMOD to the Gaia-type of measurement.

4. The Gaia Sphere Reconstruction in AVU

AGIS has a "duplication" in one of the sub-systems of the AVU called Global Sphere Reconstruction (GSR). The input of both AGIS and GSR is a set of pre-processed data from the Gaia telemetry.

As said in section 2, AGIS will process the data for up to 10^8 well-behaved stars. It is presently foreseen that GSR will rather use a subset of up to 10 million of stars chosen from the AGIS dataset.

To keep the two reductions as independent as possible, GSR will differ from AGIS both from the point of view of the astrometric model and for the algorithm adopted for the sphere reconstruction, i.e. for solving the system of the linearized observation equations.

4.1. *GSR astrometric model*

The astrometric model of AGIS is GREM (Gaia RElativistic Model) (Klioner, 2003), which is an extension of a seminal study (Klioner & Kopeikin, 1992) conducted in the framework of the post-Newtonian (pN) approximation of General Relativity. In GREM this model has been formulated according to a Parametrized Post Newtonian (PPN) scheme accurate to 1 micro-arcsecond.

The astrometric model of GSR is taken from the RAMOD project, which identifies a family of astrometric models with increasing accuracies (see, e.g. Vecchiato *et al.*, 2003, de Felice *et al.*, 2006 and references therein) conceived to solve the inverse ray-tracing problem in a general relativistic framework and to use the tetrad formalism for the description of the observer's reference system.

4.2. *GSR algorithm for the sphere reconstruction*

AGIS takes its name after the method used for solving the system of equations, i.e. the *Global Iterative Solution*. The adopted strategy mainly consists in considering separately each type of parameter: astrometric, attitude, calibration, and global. When, e.g., the astrometric parameters are solved, all of the others are not computed and their present approximate values are used to calculate the known terms. Then the attitude parameters are solved, and the latest estimation for the astrometric ones is used for the known terms, and similarly for the calibration and the global parameters. A complete cycle over all of the parameter types is called *external iteration*, and the process is iterated until convergence is reached (Fig. 2).

This approach allows to easily parallelize the mathematical problem and is probably mandatory when the size of the system of equations to be solved is that of the AGIS.

GSR will use the well-known LSQR algorithm, instead (Paige & Saunders, 1982). LSQR is an iterative algorithm for solving sparse systems of linear equations based on

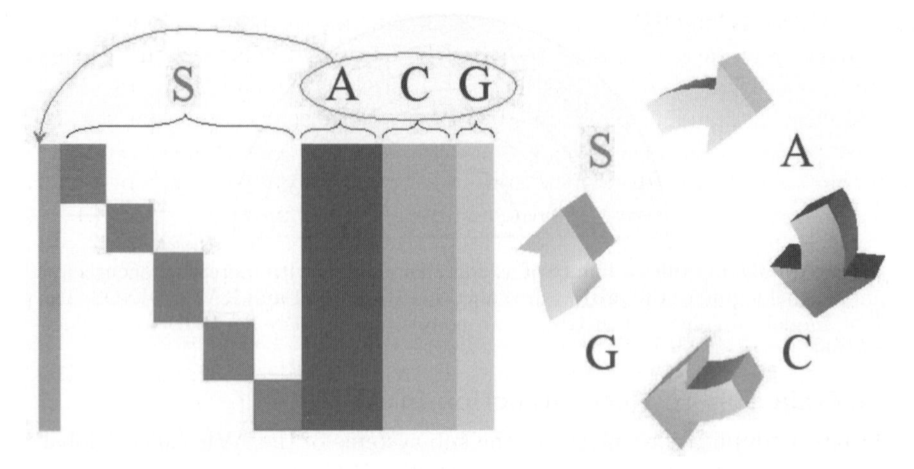

Figure 2. The right panel shows a complete external iteration for AGIS, where the astrometric (S), attitude (A), calibration (C) and global (G) parameters are solved one by one separately. The left panel represents the step of the solution when the astrometric parameters are solved; it shows that, putting the A, C and G parameters to the left-hand-side, which contains the known terms, the design matrix of the system can be arranged in a block-diagonal scheme. This clearly makes the algorithm parallelization easier. A similar arrangement can be used for the other parameters if the rows of the system are ordered by time.

a conjugate-gradients method. Since the procedure can be optimized for the memory requirements, all of the parameters can be estimated in a single iteration, and therefore, given also the smaller number of stars considered, there is no need to resort to the technique of the external iterations. A parallelized version of LSQR can be implemented to improve on the computing time.

4.3. *Algorithms for the comparison*

The AVU-GSR sub-system is also in charge of comparing its sphere solution with that coming from the AGIS.

Up to now the comparison process foresees three different algorithms:

(*a*) the χ^2-test;

(*b*) the Kolmogorov-Smirnov test;

(*c*) the Infinite Overlapping Circle (IOC) test (Bucciarelli *et al.*, 1993).

Using different algorithms to compare the same sets of data is a further precaution to ensure a controlled and error-free solution to the fullest possible extent. If everything behaves according to the expectations, in fact, all of the algorithms will provide quite similar results. On the other hand, each of them is more sensitive to a particular kind of problem. The IOC method, e.g., is very effective at detecting residual regional errors, if any.

Finally, all of these algorithms will run on user-defined subsets of the parameter space(s). This will allow the sub-system to check and isolate differences possibly caused by a single or a limited set of bad observations.

5. Present status of AVU-GSR and future developments

The development path of AVU-GSR includes some successive stages which add refinements to each component of the sub-system. The present version, namely GSR1, is characterized by the first implementation (for RAMOD) of an abscissa-based astrometric model which includes the attitude. The astrometric model uses the PPN-Schwarzschild metric of PPN-RAMOD (Vecchiato *et al.*, 2003), not including the deflection effects caused by the gravitational pull of the other bodies of the Solar System. Therefore, the main limitation of GSR1 is the need to reject approximately half of the observations, i.e. those too close to the giant planets and to the Earth. The accuracy, however, is enough for the purposes of this version, i.e. a first comparison of the two sphere solutions. Such version could also be used during the first period of the Gaia operational phase, when the overall precision of the sphere reconstruction is of ~100 microarcsecond. The attitude model is that of Bini *et al.* (2003) adapted to the PPN-Schwarzschild metric.

Though GSR is able to treat the same input data of AGIS, a complete end-to-end data reduction is still not possible due to the lack of some pieces of software which are still being implemented. Thus, the system is presently running on 1-million-stars self-simulated datasets which allow us to exercise separately the LSQR-algorithm implementation.

The next stage will implement the instrument model (GSR2) while the third and last one will include a fully accurate RAMOD astrometric model (GSR3).

6. Conclusions

The absolute nature of the Gaia measurements calls for an HIPPARCOS-like cross-check of the results of the data reduction; however, the size of the problem produced by Gaia is too large and makes it impossible to provide two independent consortia, as in the Gaia predecessor.

The adopted solution is the duplication of some of the most critical parts of the data reduction, which are gathered in an independent Astrometric Verification Unit (AVU). One of the components of the AVU is GSR that replicates to a smaller scale the global sphere reconstruction (AGIS) produced by the main data processing chain. GSR shall provide the results of a comparison with the AGIS solution, obtained using the same datasets.

GSR1, i.e. the first version of GSR which is presently under completion, will exercise the comparison task. Its future versions (GSR2 and GSR3) will complete the sub-system with the implementation of the instrument model and of the final relativistic astrometric model, respectively.

References

Bini, D., Crosta, M., & de Felice, F. 2003 *Class. Quantum Grav.*, 20, 4695

Bucciarelli, B., Taff, L. G., & Lattanzi, M. G. 1993 *J. Statist. Comput. Simul.*, 48, 29

Klioner, S. A. 2003 *Astron. J.*, 125, 1580

Klioner, S. A. & Kopeikin, S. M. 1992 *Astron. J.*, 104, 897

Paige, C. & Saunders, M. A. 1982 *ACM Trans. Math. Software*, 8, 43

Eds. Turon, C., O'Flaherty, K. S., & Perryman, M. A. C. 2005, *The Three-Dimensional Universe with Gaia, Publ. Astron. Soc. Pac.*, 120, 38

Vecchiato, A., Lattanzi, M. G., Bucciarelli, B., Crosta, M., de Felice, F., & Gai, M. 2003 *Astron. Astrophys.*, 399, 337

de Felice, F., Vecchiato, A., Crosta, M., Bucciarelli, B., & Lattanzi, M. G. 2006 *ApJ*, 653, 1552

Webb, S., 1999, *Measuring the Universe*, Springer-Verlag

Relativity in Fundamental Astronomy
Proceedings IAU Symposium No. 261, 2009
S. A. Klioner, P. K. Seidelman & M. H. Soffel, eds.

doi:10.1017/S1743921309990627

Perspective acceleration and gravitational redshift. Measuring masses of individual white dwarfs using Gaia + SIM astrometry

Guillem Anglada-Escudé[1] and John Debes[2]

[1] Dept. of Terrestrial Magnetism, Carnegie Institution for Science,
5241 Broad Branch Rd. Washington DC 20008, USA,
email:anglada@dtm.ciw.edu

[2] Goddard Space Flight Center, NASA Postdoctoral Program
Greenbelt, Maryland 20771, USA
email:john.h.debes@nasa.gov

Abstract. According to current plans, the SIM/NASA mission will be launched just after the end of operations for the Gaia/ESA mission. This is a new situation which enables long term astrometric projects that could not be achieved by either mission alone. Using the well-known perspective acceleration effect on astrometric measurements, the true heliocentric radial velocity of a nearby star can be measured with great precision if the time baseline of the astrometric measurements is long enough. Since white dwarfs are compact objects, the gravitational redshift can be quite large (40–80 km/s), and is the predominant source of any shift in wavelength. The mismatch of the true radial velocity with the spectroscopic shift thus leads to a direct measure of the Mass–Radius relation for such objects. Using available catalog information about the known nearby white dwarfs, we estimate how many masses/gravitational redshift measurements can be obtained with an accuracy better than 2%. Nearby white dwarfs are relatively faint objects ($10 < V < 15$), which can be easily observed by both missions. We also briefly discuss how the presence of a long period planet can mask the astrometric signal of perspective acceleration.

Keywords. astrometry, (stars:) white dwarfs, relativity, stars: fundamental parameters(masses)

1. Introduction

Perspective acceleration has been previously measured using astrometry for a few nearby stars using ground-based (eg. Gatewood & Russell 1974) and space-based astrometric observations (Dravins *et al.* 1999). The motion of a star on the unit sphere can be obtained by the straight forward normalization of the trajectory vector as a function of time.

$$\mathbf{l}(t) = \frac{\mathbf{r}_0 + \mathbf{v}\,(t - t_0) + \mathbf{d}\,[t, t_0] - \mathbf{x}_{\rm obs}\,[t]}{|\mathbf{r}_0 + \mathbf{v}\,(t - t_0) + \mathbf{d}\,[t, t_0] - \mathbf{x}_{\rm obs}\,[t]|} \tag{1.1}$$

where the squared brackets $[\ldots]$ indicate explicit dependence on t. The vector $\mathbf{d}$ contains any nonlinear contribution to the stellar motion (eg. a Keplerian orbit). To the second astrometric order, the motion of a star on a locally tangent planet (see Anglada-Escudé & Torra 2006) can be written as

$$X_{\rm RA} = (X_0 + \mu^*_{\rm RA}\,(t - t_0) + \delta_{\rm RA}\,[t, t_0] - \pi\, p_{\rm RA})\,(1 - R) \tag{1.2}$$

$$Y_{\rm Dec} = (Y_0 + \mu_{\rm Dec}\,(t - t_0) + \delta_{\rm Dec}\,[t, t_0] - \pi\, p_{\rm Dec})\,(1 - R) \tag{1.3}$$

$$R = R_0 + \frac{1}{r_0} v_{\rm r}\,(t - t_0) + \delta_{\rm r}\,[t, t_0] - \pi\, p_{\rm r} \tag{1.4}$$

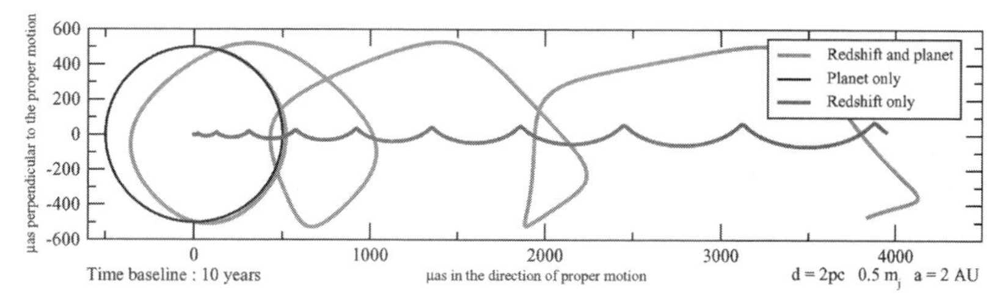

Figure 1. Apparent motion of a star during 10 years due to a planetary companion(black), perspective acceleration(red), and the combined effect(brown). Both axis are in μas.

where $\mu^*_{\rm RA}$ and $\mu_{\rm Dec}$ are the proper motions, the δ's contain the nonlinear motion terms, π is the parallax and $p_{\rm RA}$ & $p_{\rm RA}$ are the corresponding parallax factors. The proper motion in *R.A.* is $\mu^*_{\rm RA} = \mu_{\rm RA} \cos {\rm Dec}$. This differs from the old convention by the $\cos$ Dec term which properly accounts for the deformation of the azimuthal coordinate (R.A) towards the poles. The radial motion is entirely contained in R and contributes as a second order effect. The so-called *perspective acceleration* is the product of the proper motion and the true radial velocity (i.e. v_r/d in R) and grows quadratically with time in the direction of the proper motion.

The true radial velocity of the star is obtained if the perspective acceleration can be observed. Compared to the observed spectroscopic shift, the difference can be attributed to several physical processes, such as the gravitational redshift and the convective outflow. However, the gravitational redshift in white dwarfs (~70 km/s) is the dominant source of non-kinematic shifts. The astrometric signal of the offset induced by the *gravitational redshift* for a white dwarf is

$$\alpha_{\rm GR} = \mu \frac{1}{r_0} v_{\rm GR} \left(t - t_0\right)^2 ; \ \ v_{\rm GR} = \frac{GM}{Rc} \sim 68.9\,{\rm km/s}\,\frac{M_*}{M_{\rm sun}} \frac{R_{\rm earth}}{R_*} \tag{1.5}$$

This method was first demonstrated by Gatewood & Russell (1974) on the Van Maneen 2 star. The gravitational redshift has also already been measured in binary systems such as Sirius B. Gaia+SIM offers the opportunity of applying the method to a statistically significant number of nearby white dwarfs. White dwarfs can be used as absolute stellar candles to obtain distance measures and absolute photometry to old stellar complexes such as globular clusters and nearby galaxies (Davis, Richer, King, *et al.* 2008).

2. Perspective acceleration in the presence of planetary companions

Long term astrometric observations will be extremely sensitive to planets around white dwarfs. Most planetary companions to white dwarfs are expected to lie in regions exterior to a primordial separation of 3-5 AU (Villaver & Livio 2007), which due to central star mass loss on the red giant branch corresponds to final separations of between 6–12 AU or larger. The presence of a long period planet (P > 15 years) should be obvious when combining Gaia+SIM data sets, but its signature will be strongly correlated with the perspective acceleration effect. The combined signal of a planet with a relatively short period compared to the time-span of the observations is shown in Fig. 1. In the short period case the decoupling of the perspective acceleration with the planet signal is feasible. In the case of a long period planet, only the orbital motion perpendicular to the proper motion can be measurable independently. While these systems are interesting from the point of view of planetary systems evolution, they are not useful for gravitational redshift studies.

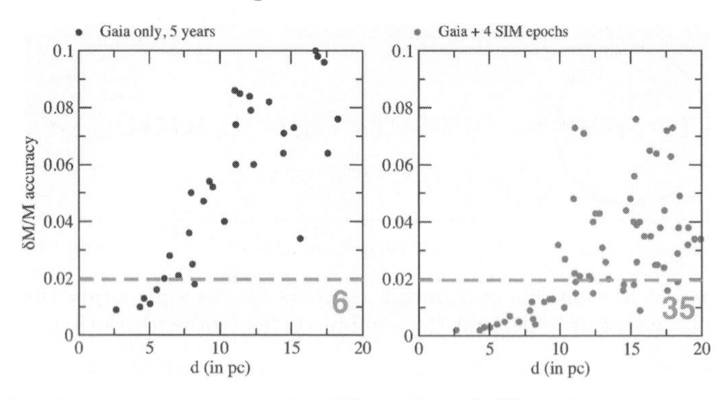

Figure 2. Each point represents a nearby White Dwarf. The relative accuracy at which the mass can be determined is plotted against the distance. The left panel shows Gaia observations alone(5 years). The right panel shows Gaia+SIM(10 years).

3. The nearby white dwarf sample

We have collected a sample of white dwarfs in the Solar neighborhood (d<20 pc). We compiled the sample from Holberg *et al.* (2008b) with proper motions obtained from SIMBAD. In our knowledge, it represents the most complete sample of WD within 20 pc. The distances determined are trigonometric parallaxes if available. Except for a few binaries (eg. Sirius B), the masses are based on model estimates Holberg *et al.* (2008a). The gravitational redshift and its astrometric signal have been simulated in Gaia-alone (5 years), and Gaia+SIM (5 years + 4 SIM epochs at 10μas), obtaining a realistic estimate of what can be obtained for each star.

4. Conclusions

We show how the gravitational redshift term is very significant in the apparent motion of the star and can be measured by comparing the perspective acceleration effect(astrometry) and the Doppler shift (spectroscopy). We find that combining Gaia+SIM observations, the gravitational redshifts of 35 already known objects can be measured with an accuracy of 2%. This number may grow as new nearby white dwarf candidates are identified. The redshift of ALL the known white dwarfs within 20 pc can be obtained with an accuracy better than 10% with 4 additional SIM epochs added to the Gaia ones.

Acknowledgments. This work is part of the SIM Science study : Gaia-SIM Legacy project funded by JPL/Caltech under NASA contract NMO 710776, which explores science cases benefiting from long term astrometric observations combining Gaia and SIM.

References

Anglada-Escudé, G. & Torra, J. 2006, *A&A*, 449, 1281
Davis, D. S., Richer, H. B., King, I. R. *et al.* 2008, *MNRAS*, 383, L20
Debes, J. H. & Sigurdsson, S. 2002, *ApJ*, 572, 556
Dravins, D., Lindegren, L., & Madsen, S. 1999, *A&A*, 348, 1040
Gatewood, G. & Russell, J. 1974, *AJ*, 79, 815
Holberg, J. B., Bergeron, P., & Gianninas, A. 2008a, *AJ*, 135, 1239
Holberg, J. B., Sion, E. M., Oswalt, T., *et al.* 2008b, *AJ*, 135, 1225
Villaver, E. & Livio, M. 2007, *ApJ*, 661, 1192

Relativity in Fundamental Astronomy
Proceedings IAU Symposium No. 261, 2009
S. A. Klioner, P. K. Seidelman & M. H. Soffel, eds.

doi:10.1017/S1743921309990639

Toward inertial reference frames with the SIM observatory

Valeri V. Makarov

NASA Exoplanet Science Institut, Caltech,
770 S. Wilson Ave., MS 100-22, Pasadena, CA, USA
email: vvm@caltech.edu

Abstract. The SIM Lite Observatory is expected to provide a global astrometric reference frame surpassing the 1-μas accuracy threshold in some spherical harmonics. A range of time-varying physical distortions of the reference frame will become observable as large-scale perturbations of the proper motion field. I consider the main sources of the apparent and physical motion of reference objects, such as the aberration of light caused by the acceleration of SIM, long gravitational waves and hypothetical rotation of the Universe, and present some estimates of the astrometric sensitivity to these effects. I argue that a global solution and covariance analysis is of crucial importance for the SIM mission to differentiate the inevitable accidental and systematic zonal errors from real physical phenomena.

Keywords. astrometry, reference systems, gravitational waves, large-scale structure of universe

---

1. Introduction

The standard of inertiality of reference frames has been rising along with the remarkable progress in the accuracy and the density of astrometric systems. The SIM Lite Observatory will approach, for the first time, the level of 10^{-12} rad in accuracy, which poses new problems in the characterization of observable effects related to the accelerating motion of the instrument and the systemic motion of the frame objects. A number of key projects with SIM rely on the rigidity and non-rotation of the SIM reference frame. For example, the dark matter potential of the Milky Way halo can be mapped from ultra-precise measurements of the proper motions of hyper-velocity stars or satellite dwarf galaxies (Unwin *et al.* 2008), but this method requires the residual rotation and other large-scale distortions of the reference frame to be much less than the magnitude of the effects to be measured. Besides the measurement errors, which are accurately estimated in a global astrometric solution (Makarov & Milman 2005), the proper motions of distant quasars will include a number of relativistic effects, making them significantly nonzero.

2. Categories of global reference frame perturbations

An astronomical reference frame is realized through a catalog of angular coordinates of a set of reference objects. Even the most distant astronomical objects are not fixed in position on the celestial sphere; as their angular coordinates change with time, the reference frame undergoes perturbations at various spatial scales. Not all of the predictable perturbations are observable with astrometric techniques. For example, the *physical* motion of extragalactic reference frame objects includes the following categories:

- Rotation of the Universe
- Peculiar motion of quasars and galaxies.

A number of relativistic effects and cosmological phenomena result in the *apparent* motion of reference objects, i.e.

- Spacetime ripples (gravitational waves)
- Gravitational deflection of light
- Acceleration of the Milky Way
- Acceleration of the Sun (secular aberration)
- Acceleration of the observer.

Rotation of the Universe. A generalization of the Friedman model of a homogeneous universe includes a systemic rotation of matter besides the isotropic expansion (Gamow 1946 and Gödel 1949). A global vorticity of the observable matter may result in a number of interesting phenomena, but for an astrometric mission, it mostly results in a rigid rotation of the reference frame. Most of the estimates of the present-day rate of rotation are bounded to $\sim 10^{-13}$ (Li 1998) – 10^{-14} (Hawking 1969) rad s^{-1}, but for certain Bianchi models, rates of up to 10^{-12} can be realized (Ciufolini & Wheeler 1995). The latter estimate is close to the sensitivity of SIM (a few parts in 10^{-12}). However, an astrometric instrument can only measure the angular distances between celestial objects, and any unitary transformation of the reference frame is not directly observable. Speculatively, a global systemic rotation of distant quasars can be detected with respect to an ideal local gyroscope. A mechanical gyroscope matching the required astrometric accuracy is not feasible, but the orbital motion of SIM around the Sun, and the motion of the Sun around the Galactic barycenter, are the substitutes that can be considered. The orbital plane of any free-falling body will slowly tilt with respect to the extragalactic reference frame. This tilt is observable with SIM as a drift in the dipole direction of the global aberration pattern.

Relic gravitational waves. The existence of stochastic, long gravitational waves has not been proven by experiment, but it is consistent with the generic inflationary models (Grishchuk 2006). A monochromatic, single-polarization gravitational wave propagating through the local part of the Universe will cause the apparent direction to a quasar in the transverse direction to oscillate with the frequency of the wave, whereas a quasar observed in the wave direction will not change its position (Fig. 1). On the limited time scale of the SIM mission (5 years), the former quasar will have a net proper motion with respect to the latter quasar. A global grid of quasars will be affected by a smooth pattern of apparent proper motions, represented by the second- and higher-order vector spherical harmonics. The magnitude of these proper motions depends on the amplitude of the relic gravitational waves and on their power spectrum. Different frequencies and modes of polarization may be present today, but most theoretical estimations seem to yield a low energy density $\Omega_{\rm GW} h^2 < 10^{-9}$ (Smith *et al.* 2006). At a signal-to-noise ratio of 10, SIM is expected to be sensitive to energy densities of order 10^{-6}. Given these numbers, the pulsar timing method holds better prospect of the actual experimental discovery of relic gravitational waves. It is noted, however, that compared to other planned and operating experiments in gravitational wave astronomy, such as LIGO, LISA and BBO, SIM will be sensitive to much longer waves, just shortward of the expected Big Bang nucleosynthesis mode ($\nu_{\rm BBN} \approx 10^{-11}$ s).

Aberration. The free fall of the Solar System onto the center of the Galaxy makes the stellar aberration pattern vary with time. The resulting proper motion field is represented by the three first-order electric harmonics (Kopeikin & Makarov 2006). The magnitude of this secular aberration effect is ~ 4 μas/yr, and it will be confidently measured by SIM. Subsequently, this perturbation can be calibrated out of the observed proper motion field of grid stars, and thus, the reference frame can be brought to the Galactic barycenter standard of rest. But the Galaxy is probably involved in a nonlinear motion due to, for example, the gravitational attraction to the other members of the Local Group. If this acceleration is not much smaller than 6 mm s^{-1} yr^{-1}, which is the expected acceleration of the Sun toward the Galactic center, it should result mainly in an offset of the axis of the proper motion dipole from the Galactic center. Other short-term variations in the reference frame positions due to the deflection of light by Solar System bodies, and higher-order relativistic effects, such as the coupling terms between stellar aberration and parallax (Klioner 2003), should be carefully subtracted from the data prior to this analysis.

3. Description of reference frame distortions

The time-varying component of large-scale reference frame distortions is represented by a series of vector spherical harmonics on a unit sphere

$$\mu(\lambda, \beta) = \sum_i e_i \, \mathbf{E}_i + m_i \, \mathbf{M}_i, \tag{3.1}$$

where μ is the proper motion field of reference frame objects, which is a function of the angular coordinates (λ, β), and $\mathbf{E}_i$, $\mathbf{M}_i$ are the electric and magnetic vector spherical harmonics, respectively. The coefficients of the expansion e_i and m_i include the genuine distortions of physical origin, as well as positionally correlated astrometric errors of proper motions, which are for historical reasons called "zonal errors". The latter can be systematic or accidental by origin. The physical effects of scientific interest mostly reside in the low-order vector harmonics:

$$\mu = \sum_{k=-1}^{1} e_1^k \, \mathbf{E}_1^k + \sum_{k=-1}^{1} m_1^k \, \mathbf{M}_1^k + \sum_{k=-2}^{2} e_2^k \, \mathbf{E}_2^k + \sum_{k=-2}^{2} m_2^k \, \mathbf{M}_2^k + \ldots. \tag{3.2}$$

The first term in the right-hand part, which is the set of three first-order electric harmonics or dipoles, includes the effects of secular aberration. The second term, the first-order magnetic harmonics, represents the net rotation of the reference system as a whole. The third and the fourth terms, the second-order harmonics, include most of the perturbation caused by long gravitational waves (Gwinn *et al.* 1997).

It is well known that a global astrometric solution in its canonical form, which includes positions, parallaxes and proper motions of grid stars, is deficient in rank by 6. The 6 indefinite parameters are a rigid rotation and a rigid spin of the reference frame, i.e.,

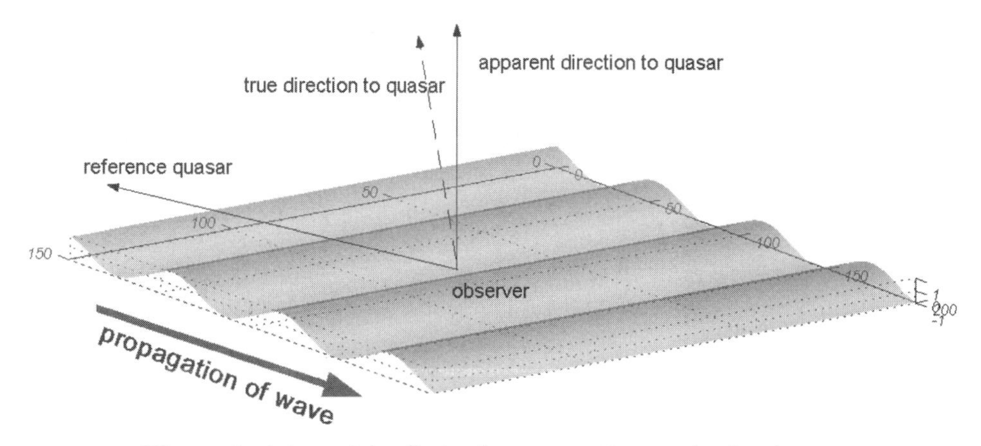

Figure 1. Astrometric effects of a propagating gravitational wave.

the first-order magnetic harmonics of positions and proper motions. The deficiency of rank can be circumvented by constraining the solution for suitably chosen "one and a half" stars in the grid catalog, or by constraining the systemic spin (in the case of proper motions) to zero for the system of grid quasars. The latter method will carry over any accidental or systematic errors of the proper motion harmonics into the reference frame, as well as any genuine rotation of the reference frame, unless we can find other locally inertial systems or bodies available to observation.

4. Galactic velocity field

Hipparcos was not sensitive enough to observe quasars or AGN, and its reference frame is composed of Galactic stars. The stars in the local part of the Galaxy are involved in the differential rotation, azimuthal shear, asymmetric drift, warp, and local dilation or contraction. Each of these patterns is larger in magnitude than the cosmological and relativistic effects considered in this paper. The presence of dynamical streams, associations and the natural dispersion of velocities further complicate the effort to detect more subtle effects at the 1-mas level. Today, some of the phenomena in the local velocity field remain poorly understood or outright mysterious. A fit to the linear Ogorodnikov-Milne model of the local velocity field, reveals at least one unexpected, statistically significant magnetic term (Makarov & Murphy 2007). This term may be related to the Galactic warp, but it has the opposite sign with respect to a static warp model. Furthermore, when a higher-order vector harmonic model is fitted to the data, and all the "known" Ogorodnikov-Milne terms are subtracted, the residual proper velocity field displays a fascinating pattern (Fig. 2), characterized by a few overlapping multipoles and two conspicuous foci, which are close to, but not quite coincident with, the direction of the solar motion with respect to the Local Standard of Rest. Is this field real, or just a

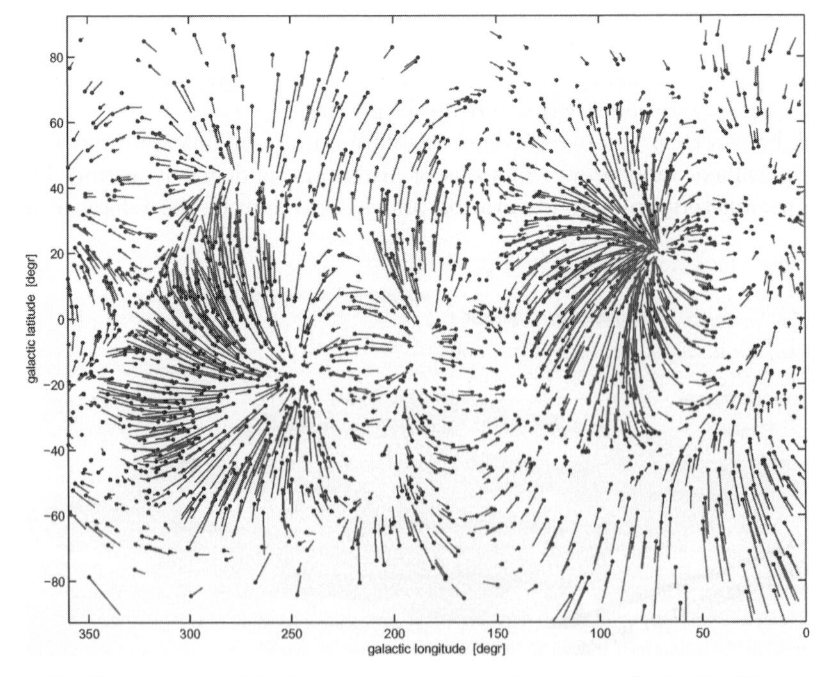

Figure 2. Residual pattern of higher-order proper motion harmonics in the Hipparcos data.

realization of zonal error within the expected variance? It is impossible to answer this question without a rigorous covariance analysis of the zonal error propagation.

5. Benefits of a global solution

A global astrometric problem in the perturbation form can be presented as a system of linearized condition equations

$$\mathbf{A}\,\mathbf{x} = \mathbf{y}, \tag{5.1}$$

where $\mathbf{A}$ is the global design matrix, $\mathbf{x}$ is the vector of unknowns, including the complete set of astrometric parameters for all grid objects, instrument calibration parameters and the attitude corrections, and $\mathbf{y}$ is the vector of observations in the differential form "observed minus calculated". A global solution is a one-step, direct Least-Squares adjustment, yielding $(\mathbf{A}^{\mathrm{T}}\mathbf{A})^{-1}$, the global covariance matrix of all mission unknowns. At first glance, such a solution is intractable due to the large size of the design matrix. Although it is block-structured and sparse, the number of unknowns (of order 10^6 for SIM and 10^7 for JMAPS) makes it impossible to store the entire normal matrix in fast memory, let alone to invert it by standart routines. However, analysis of zonal errors is simplified by the fact that we are interested in a finite set of low-order harmonics. The covariance matrix of a set of harmonics can be computed exactly, and by virtue of their near orthogonality, the covariances with the higher-order terms can be safely neglected. Once the covariance of the harmonics for all 5 astrometric parameters is known, the confidence levels of accidental perturbations in the parameters of interest, (e.g, the residual spin) are readily derived. A full global solution simulation and covariance analysis has been performed for SIM, resulting in an accurate estimation of the zonal error propagation.

References

Ciufolini, I. & Wheeler, J. A. 1995, *Gravitation and Inertia*, Princeton University Press, Princeton

Gamow, G. 1946, *Nature*, 158, 549

Gödel, K. 1949, *Rev. Mod. Phys.*, 21, 447

Grishchuk, L. P. 2005, *Us. Fi. Nauk*, 48, 1235

Gwinn, C. R., *et al.* 1997, *ApJ*, 485, 87

Hawking, S. 1969, *MNRAS*, 142, 129

Klioner, S. A. 2003, *AJ*, 125, 1580

Kopeikin, S. M. & Makarov, V. V. 2006, *AJ*, 131, 1471

Li, L.-X. 1998, *Gen. Rel. Grav.*, 30, 497

Makarov, V. V. & Milman, M. 2005, *PASP*, 117, 757

Makarov, V. V. & Murphy, D. W. 2007, *AJ*, 134, 367

Smith, T. L., *et al.* 2006, *PhRevD*, 73b, 3504

Unwin, S. C., *et al.* 2008, *PASP*, 120, 38

Relativity in Fundamental Astronomy
Proceedings IAU Symposium No. 261, 2009
S. A. Klioner, P. K. Seidelman & M. H. Soffel, eds.

doi:10.1017/S1743921309990640

Space astrometry with the Joint Milliarcsecond Astrometry Pathfinder

Gregory S. Hennessy[1] **and Ralph Gaume**[2]

[1]US Naval Observatory, 3450 Massachusetts Ave NW
Washington, DC 20392, USA
email: gsh@usno.navy.mil

[2]US Naval Observatory, 3450 Massachusetts Ave NW
Washington, DC 20392, USA
email: rgaume@usno.navy.mil

Abstract. The Joint Milliarcsecond Astrometry Pathfinder Survey (JMAPS) is a small, space-based, all sky, visible wavelength, astrometric and photometric survey mission for 0th through 14th I-band magnitude stars with a planned 2013 launch. The primary objective of the JMAPS mission is the generation of an astrometric star catalog with 1 milliarcsecond (mas) positional accuracy or better and photometry to the 1% accuracy level or better at 1st to 12th mag. A 1-mas all–sky survey will have a significant impact on our current understanding of galactic and stellar astrophysics. JMAPS will improve our understanding of the origins of nearby young stars, provide insight into the dynamics of star formation regions and associations, investigate the dynamics and membership of nearby open clusters.

Keywords. astrometry, space vehicles, catalogs, surveys

1. Introduction

JMAPS is a small, single aperture spacecraft funded by the Department of the Navy for launch in 2013. The principal objective of the JMAPS mission is to produce an all–sky, visible wavelength, astrometric and photometric catalog (Zacharias & Dorland 2006). The final JMAPS catalog will be delivered in 2016. Astrometric positions will be reported in ICRS coordinates and tied to the ICRF though direct observations of the visible wavelength counterparts of radio wavelength ICRF sources. An artist's conception of JMAPS, both telescope and bus, is depicted in Figure 1.

2. SPACECRAFT & INSTRUMENT OVERVIEW

2.1. *Spacecraft*

As shown in Figure 1, JMAPS is a single spacecraft consisting of a customized spacecraft bus (bottom) with solar panels and payload deck (top). The solar panels fold against the bus in a stowed and locked position during launch and are deployed on-orbit. In addition to housing the power subsystem, the spacecraft bus contains communications, thermal control, avionics, reaction wheel, and inertial measurement unit subsystems. The Attitude Determination and Control System (ADCS) is split between the bus and instrument deck. While the spacecraft is slewing, the star tracker located on the instrument deck determines spacecraft attitude to approximately 1 arcsecond. During standard observations the ADCS system holds spacecraft pointing stability to a 50 mas specification. This is accomplished by using the primary instrument to generate boresight pointing quaternions at a 5Hz rate as derived from observations of reference stars on the focal plane. The

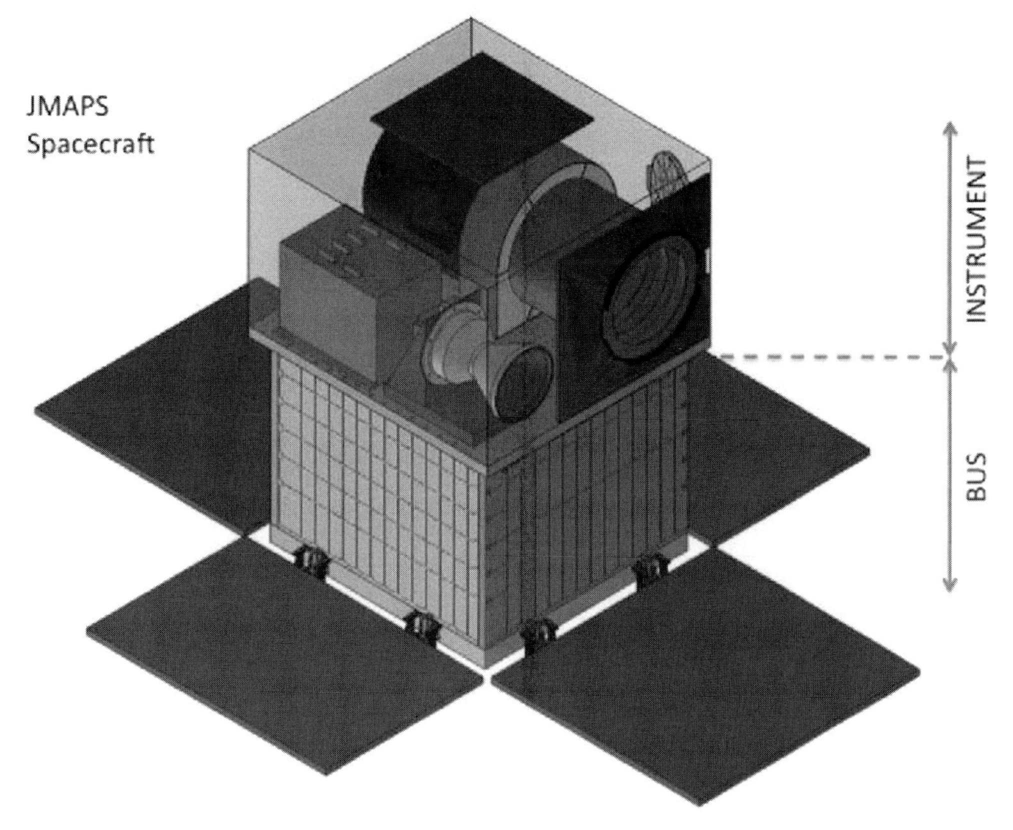

Figure 1. The JMAPS instrument and spacecraft bus.

JMAPS	**Hipparcos**
Single aperture	*two* apertures
Step–Stare	Scanning
Complete to 14th mag	Complete to 7th mag
20 years of proper motion for Hipparcos/Tycho stars	3 years of proper motions
Three years proper motions for others	
Low earth, approximately circular	Highly eccentric orbit

Table 1. Comparison of JMAPS and Hipparcos

total mass of the spacecraft, including contingency, is about 180 kg and the spacecraft occupies a volume of 96.5 cm (h) $\times$ 71 cm $\times$ 61 cm.

Many of the stars JMAPS will observe are stars observed by Hipparcos. One of the major scientific goals of JMAPS is the production of a catalog that has new milliarcsecond precision. The expected mean epoch of 2014.5 for JMAPS combined with the Hipparcos milliarcsecond precisions at 1991.25 will provide proper motions for 118,000 stars at accuracies of about 0.05 mas/year. We provide a comparison between the JMAPS and Hipparcos spacecraft in Table 1. A primary goal for JMAPS is to produce a star catalog in support of next generation star trackers. Multiple non destructive reads of the focal plane, and rapid windowing around the brightest stars will be used to obtain the 15 magnitudes of dynamic range. JMAPS will observe approximately the same dynamic range as Gaia, but will be observing brighter stars.

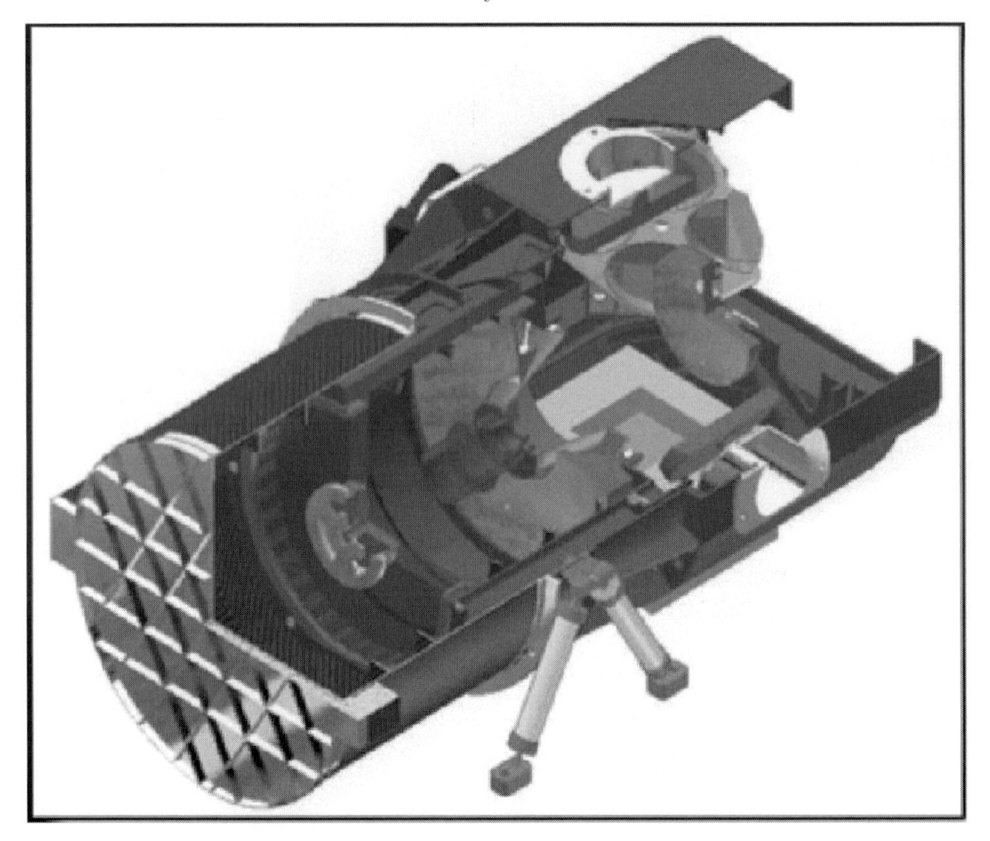

Figure 2. A view of the JMAPS optical system.

2.2. *Instrument*

The Optical Telescope Assembly (OTA) is the largest structure on the payload deck, consisting of a single aperture, 19 cm diameter, f/20 telescope (Dorland and Dudik 2009). A cut–away view of the telescope is visible in Figure 2. A ray trace diagram is shown in Figure 3.

The nominal point spread function (PSF) is 0.87 arcseconds full width half maximum (FWHM). The JMAPS field of view is $1.24° \times 1.24°$. The OTA supports a JMAPS astrometric bandpass of 700–900 nm; spectroscopic observations will be conduced within a wider 450–900nm band. During normal observations, a sun shield behind the OTA (not shown in Figure 1) will protect the OTA from direct exposure to solar heating.

The Focal Plane Assembly (FPA) is located under the OTA (the shaded square in Figure 2). The JMAPS FPA consists of a 2×2 mosaic detectors as shown in Figure 4. The detector is a 10 micron pitch, 4192×4192 pixel CMOS-CCD hybrid. The nominal pixel subtense is 0.55 arcseconds, providing a sampling of approximately 1.6 pixels per PSF FWHM. USNO has sky-tested the first generation detector (Dorland *et al.* 2007). Second generation detectors have been built and are currently undergoing testing at USNO. It is anticipated that JMAPS flight devices will result from the third generation of the H4RG-10 detector.

Major components of the payload deck (also shown in Fig. 3) are the Instrument Electronics Box (IEB)–which houses the primary on-board electronics for both the instrument and bus–the star tracker and FPA radiators. The radiators will be used to maintain the JMAPS FPA at a temperature of 193K.

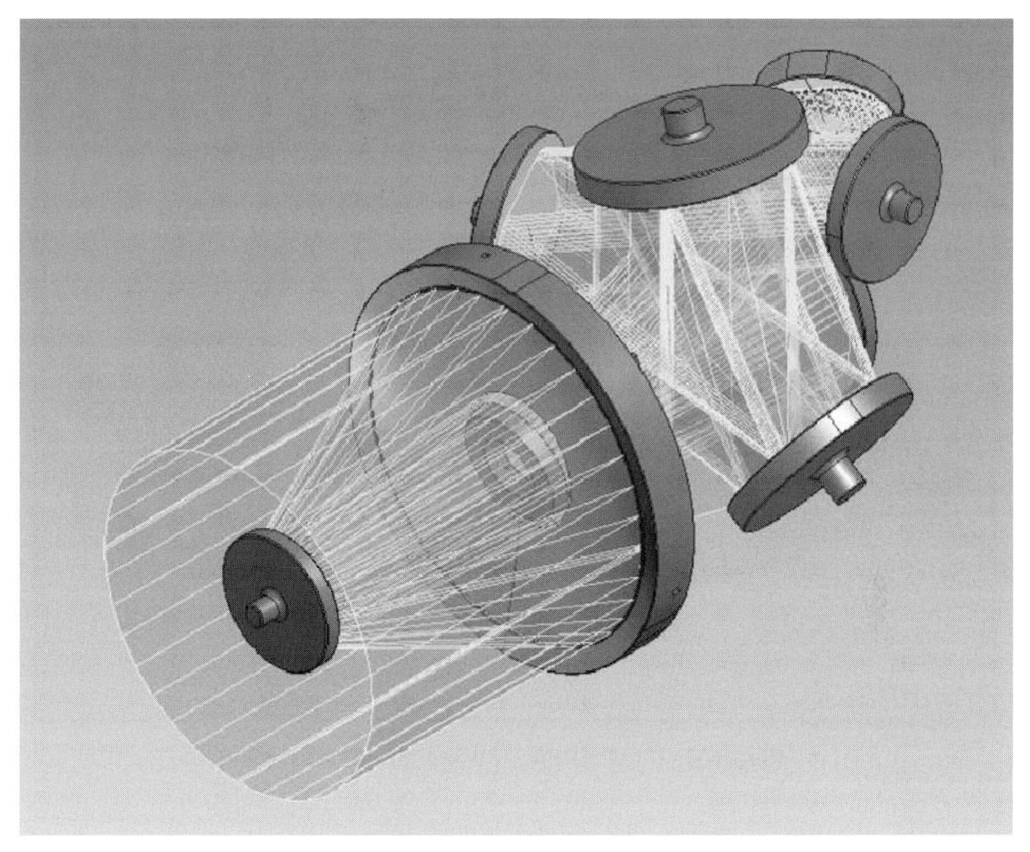

Figure 3. A ray tracing of the JMAPS optical system.

2.3. *JMAPS Concept of Operations*

JMAPS will be launched into a 900 km sun-synchronos terminator orbit and operate in a step-stare mode, typically sweeping out swaths of the sky at approximately a 90° angle to the Earth-Sun line (regions of maximum parallax signal). Windowed FPA integration times of 1, 4.5 and 20 seconds will be used for the majority of stars. Shorter integration times (between 10 msec and 1 sec) will be used for the brightest stars. In order to link the JMAPS reference frame to the ICRF, the optical counterparts of radio wavelength ICRF quasars will be directly observed by JMAPS by employing integration times of up to 500 seconds.

3. Measurements

3.1. *Single field astrometric*

The majority of observations will be taken in quadrature with the sun, in order to maximize the parallax factor. The angle between the boresite and the sun will be allowed to vary by up to 20 degrees from quadrature. After an exposure is taken the boresite will move half a field for the subsequent exposure. Due to the limited on–board data storage capacity, full frames will not be downlinked from the spacecraft. Instead an on–board processor will identify stars, remove cosmic rays and then download a 10×10 pixel window centered on each star. These windows will be processed on the ground to determine the centroid and instrumental magnitude of each star.

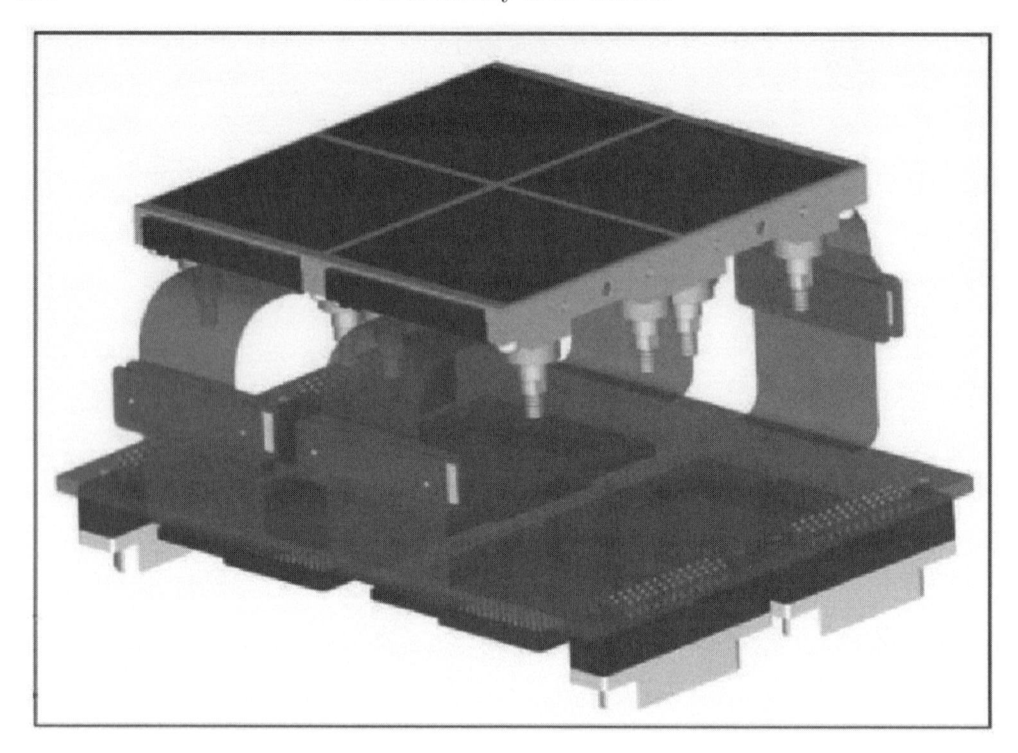

Figure 4. JMAPS FPA 4K 2 × 2 Mosaic Concept

3.2. *Global Reduction*

Global block adjustment algorithms will be used to provide a rigid determination of place centers and rotations for each of the exposures. The use of approximately 100 optically bright quasars that are also part of the International Celestial Reference Frame will be used to place the JMAPS results on an inertial frame. A catalog of 400,000–600,000 "grid stars" will be used to calculate the per frame plate constants. The use of a global reduction is expected to reduce, but not eliminate, zonal errors.

4. JMAPS SCIENCE POSSIBILITIES

The predicted JMAPS single measurement precision (SMP) is shown in Figure 5. The sawtooth nature of the SMP plot is attributed to the various integration times utilized for stars of different magnitudes (see Figure 5). The objective of achieving 1 mas stellar astrometric accuracies is achieved through a global reduction of multiple (approximately 72) observations obtained throughout the 2–3 year mission lifetime. A 1-mas all-sky survey will have a significant impact on our current understanding of galactic and stellar astrophysics. JMAPS will improve our understanding of the origins of nearby young stars, provide insight into the dynamics of star formation regions and associations, investigate the dynamics and membership of nearby open clusters, discover the smallest brown dwarfs at distances up to 5 pc after a 2-year mission and Jupiter-like planets out to 3 pc after 4 years. JMAPS will provide critical milliarcsecond-level parallaxes of tens of millions of stars in the difficult $8\text{–}14^{th}$ magnitude range, which when combined with stellar spectroscopy and relative radii determined from exoplanet transit surveys, allows a determination of stellar radii and exoplanet densities.

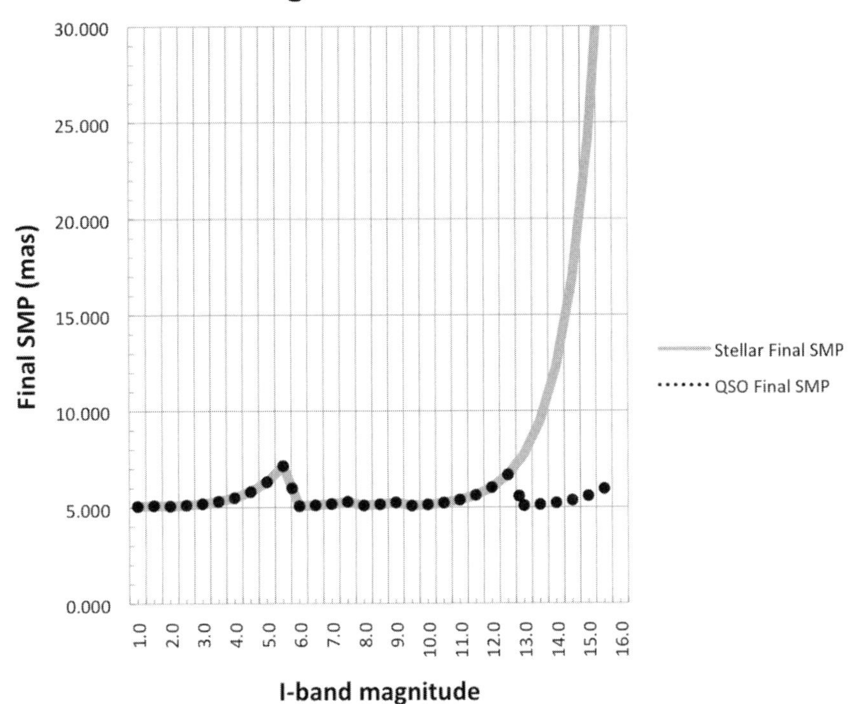

Figure 5. Estimated JMAPS single measurement precision predictions as a function of stellar I-band magnitude

References

Dorland, B. N. & Dudik, R. P. 2009, arXiv:0907.5248

Dorland, B. N., Hennessy, G. S., Zacharias, N., Rollins, C., Huber, D., & Kessel, R. 2007, Proc. SPIE, 6690

Zacharias, N. & Dorland, B. 2006, PASP, 118, 1419

Relativity in Fundamental Astronomy
Proceedings IAU Symposium No. 261, 2009
S. A. Klioner, P. K. Seidelman & M. H. Soffel, eds.

doi:10.1017/S1743921309990652

Relativistic models for the BepiColombo radioscience experiment

Andrea Milani[1], Giacomo Tommei[1], David Vokrouhlický[2], Emanuele Latorre[1] and Stefano Cicalò[1]

[1] Department of Mathematics, University of Pisa, Pisa, Italy
email: milani@dm.unipi.it, tommei@dm.unipi.it, emanuele.latorre@gmail.com, cicalo@mail.dm.unipi.it

[2] Institute of Astronomy, Charles University, Prague, Czech Republic
email: vokrouhl@cesnet.cz

Abstract. To test General Relativity with the tracking data of the BepiColombo Mercury orbiter we need relativistic models for the orbits of Mercury and of the Earth, for the light-time and for all the spatio-temporal reference frames involved, with accuracy corresponding to the measurements: $\simeq$ 10 cm in range, $\simeq$ 2 micron/s in range-rate, over 2 years.

For the dynamics we start from the Lagrangian post-Newtonian (PN) formulation, using a relativistic equation for the solar system barycenter to avoid rank deficiency. In the determination of the PN parameters, the difficulty in disentangling the effects of β from the ones of the Sun's oblateness is confirmed. We have found a consistent formulation for the preferred frame effects, although the center of mass is not an integral. For the identification of strong equivalence principle (SEP) violations we use a formulation containing both direct and indirect effects (through the modified position of the Sun in a barycentric frame).

In the light-time equations, the Shapiro effect is modeled to PN order 1 but with an order 2 correction compatible with (Moyer 2003). The 1.5-PN order corrections containing the Sun's velocity are not relevant at the required level of accuracy.

To model the orbit of the probe, we use a mercury-centric reference frame with its own "Mercury Dynamic Time": this is the largest and the only relativistic correction required, taking into account the major uncertainties introduced by non-gravitational perturbations.

A delicate issue is the compatibility of our solution with the ephemerides for the other planets, and for the Moon, which cannot be improved by the BepiColombo data alone. Conversely, we plan to later export the BepiColombo measurements, as normal points, to contribute with their unprecedented accuracy to the global improvement of the planetary ephemerides.

Keywords. Mercury, radioscience, relativity

1. A radioscience experiment with a Mercury orbiter

BepiColombo is an ESA mission to the planet Mercury, including two spacecrafts (one provided from Japan) to be put in orbit around Mercury; launch is scheduled for 2014, instruments are already being built now. The Mercury Orbiter Radioscience Experiment (MORE) is one of the on board experiments whose goals are:

a) to determine the gravity field of Mercury and its rotation state (to constrain the interior structure of the planet);

b) to determine the orbit of Mercury to constrain the possible theories of gravitation, e.g., by determining the post-Newtonian (PN) parameters;

c) to provide the spacecraft position for geodesy experiments;

d) to contribute to planetary ephemerides improvement.

This is possible thanks to a multi-frequency radio link (in X and Ka bands) allowing to eliminate the uncertainty in the refraction index due to plasma content along the

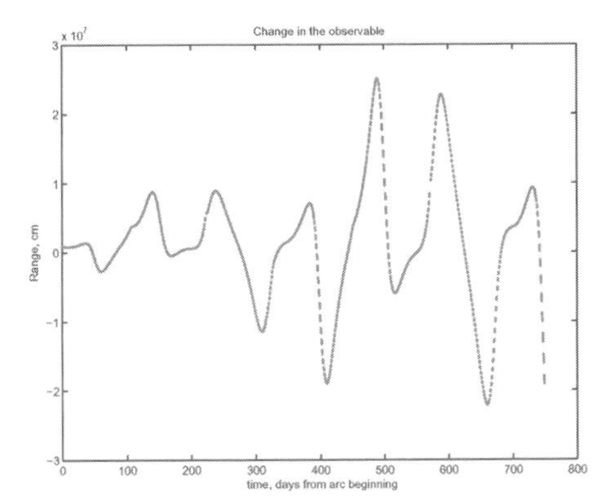

Figure 1. Differences in range using a fully Relativistic and a Newtonian model, over a 2 year Mercury orbiting mission. The total Δr is 4×10^7 cm and $S/N = \Delta r/\sigma(r) \simeq 4 \times 10^6$.

radio waves path. The MORE experiment provides the necessary Ka band transponder and the system to compare the delays in a 5-way link, in combination with instruments installed at the ground stations.

Orders of magnitude for the accuracy which can be achieved in this way are 2 micron/s in range-rate and 10 cm in range: the relative accuracy in range is better than 10^{-12}. This implies the signal to noise ratio (S/N) of all the relativistic effects (both in the dynamics and in the observation equations) is very large, in particular for the range measurements, see Figure 1.

2. The relativistic orbit determination problem

The relativity experiment with MORE needs to solve an orbit determination problem with a full relativistic model (including the terms expressing the violations of general relativity with the PN parameters, such as $\gamma, \beta, \zeta, \eta, \alpha_1, \alpha_2$), not for a generic space-time, but for the one where we are now. Thus we must fit the initial conditions for Mercury and for the Earth-Moon barycenter.

2.1. *Orbit determination with symmetries*

We shall give the equation of motion by using the parametric post-Newtonian approach: the relativistic equation of motion is linearized with respect to the small parameters v_i^2/c^2 and $G\,m_i/(c^2\, r_{ik})$, where v_i is the barycentric velocity for each of the bodies of mass m_i, c the speed of light and r_{ik} a mutual distance, appearing in the metric of the curved space-time, hence in the equations for geodesic motion. This can be formalized by adding to the Lagrangian L_{NEW} of the N-body problem some corrective terms of PN order 1 in the small parameters. By following the notation of (Moyer 2003)

$$\mathbf{r}_{ij} = \mathbf{r}_j - \mathbf{r}_i \ , \ r_{ij} = |\mathbf{r}_{ij}|, \ \ \mathbf{v}_{ij} = \dot{\mathbf{r}_j} - \dot{\mathbf{r}_i} = \mathbf{v}_j - \mathbf{v}_i \ , \ v_{ij} = |\mathbf{v}_{ij}|,$$

for $i, j = 0, \ldots,$ N, where 0 refers to the Sun, the Newtonian Lagrange function is

$$L_{\mathrm{NEW}} = \frac{1}{2} \sum_i \mu_i\, v_i^2 + \frac{1}{2} \sum_i \sum_{j \neq i} \frac{\mu_i \mu_j}{r_{ij}}.$$

The usual Lagrangian is multiplied by G, thus only the gravitational masses $\mu_i = G\,m_i$ appear in the Lagrange function. By Noether's theorem, the 3-parameter group of symmetries $\mathbf{r}_j \longrightarrow \mathbf{r}_j + \mathbf{h}$, $\mathbf{h} \in \mathbf{R}^3$ results in the vector integral of total linear momentum

$$\mathbf{P} = \sum_i \frac{\partial L_{\rm NEW}}{\partial \mathbf{v}_i} = \sum_i \mu_i \, \mathbf{v}_i .$$

The unobservable linear motion of the center of mass of the (N+1)-body system

$$\mathbf{b}(t) = \frac{1}{M} \sum_i \mu_i \, \mathbf{r}_i(t) = \frac{1}{M} \, \mathbf{P}\, t + \mathbf{b}(0)$$

with $M = \sum_i \mu_i$ implies that an orbit determination with mutual observations has a rank deficiency of 6: the normal matrix of the fit has a kernel of dimension 6. There is only one solution to this problem, *descoping* (Milani & Gronchi 2009, Chapter 6), which can be obtained in two ways: either (1) the center of mass is assumed to be fixed, e.g., $\mathbf{b}(t) = \mathbf{0}$, or (2) it is constrained to remain fixed, by adding a priori observations of the form $\mathbf{b}(0) = \mathbf{0} \pm \sigma_1$ and $\dot{\mathbf{b}}(0) = \mathbf{0} \pm \sigma_2$, with a very small a priori uncertainties σ_i. With solution (1) the equation of motion of the Sun is removed, and the position of the Sun is computed from the center of mass, that is $\mathbf{r}_0$ is replaced by $\mathbf{s}$ with

$$\mathbf{s} = -\frac{1}{\mu_0} \sum_{i=1}^{N} \mu_i \mathbf{r}_i .$$

2.2. *Lagrangian formulation for PN Relativity*

The equations of motion of GR, to 1-PN order, can be deduced from the *relativistic Lagrangian*

$$L = L_{\rm NEW} + L_{\rm GR0} + \beta \, L_\beta + \gamma \, L_\gamma ,$$

where γ, β are the "Eddington parameters", both $= 1$ in GR, and $L_{\rm GR0}$ is the portion without free parameters (apart from G):

$$\begin{aligned} L_{\rm GR0} \quad = \quad & \frac{1}{8c^2} \sum_i \mu_i \, v_i^4 + \frac{1}{2\,c^2} \sum_i \sum_{j \neq i} \sum_{k \neq i} \frac{\mu_i \, \mu_j \, \mu_k}{r_{ij} \, r_{ik}} \\ & + \frac{1}{2\,c^2} \sum_i \sum_{j \neq i} \frac{\mu_i \, \mu_j}{r_{ij}} \left[\frac{1}{2} \left(v_i^2 + v_j^2 \right) - \frac{3}{2} \left(\mathbf{v}_i \cdot \mathbf{v}_j \right) - \frac{1}{2} \left(\mathbf{n}_{ij} \cdot \mathbf{v}_i \right) \left(\mathbf{n}_{ij} \cdot \mathbf{v}_j \right) \right], \end{aligned}$$

where $\mathbf{n}_{ij} = \mathbf{r}_{ij}/r_{ij}$,

$$L_\gamma = \frac{1}{2\,c^2} \sum_i \sum_{j \neq i} \frac{\mu_i \mu_j}{r_{ij}} \left(\mathbf{v}_i - \mathbf{v}_j \right)^2, \quad L_\beta = -\frac{1}{c^2} \sum_i \sum_{j \neq i} \sum_{k \neq i} \frac{\mu_i \, \mu_j \, \mu_k}{r_{ij} \, r_{ik}} .$$

The relativistic Lagrangian L is also invariant by translations, thus by Noether's theorem there is a vector integral

$$\mathbf{P} = \sum_i \frac{\partial L}{\partial \mathbf{v}_i} = \sum_i \mu_i \, \mathbf{v}_i + \sum_i \frac{\partial L_{\rm GR0}}{\partial \mathbf{v}_i},$$

where the contributions from the derivatives of L_β vanish (L_β does not depend on $\mathbf{v}_i$) and the ones from L_γ cancel in the sum over i because they are antisymmetric. Thus

$$\mathbf{P} = \sum_i \mu_i \, \mathbf{v}_i \left[1 + \frac{1}{2} \left(\frac{v_i}{c} \right)^2 - \frac{U_i}{2\,c^2} \right] - \frac{1}{2\,c^2} \sum_i \sum_{k \neq i} \frac{\mu_i \, \mu_k}{r_{ik}} \left(\mathbf{n}_{ik} \cdot \mathbf{v}_k \right) \mathbf{n}_{ik} ,$$

where $U_i = \sum_{k \neq i} \mu_k / r_{ik}$ is the Newtonian potential. To PN order 1, there is a vector integral:

$$\mathbf{B} = \sum_i \mu_i \, \mathbf{r}_i \left[1 + \frac{v_i^2}{2c^2} - \frac{U_i}{2c^2}\right], \quad \frac{d\mathbf{B}}{dt} = \mathbf{P}.$$

The relativistic analog of the total mass

$$\mathcal{M} = \sum_i \mu_i \left[1 + (v_i^2 - U_i)/(2c^2)\right]$$

is an integral to order 1-PN (because the PN order 1 term is the Newtonian energy divided by c^2), thus we can define the relativistic center of mass $\mathbf{b} = \mathbf{B}/\mathcal{M}$ with $d\mathbf{b}/dt$ also a vector integral. The rank deficiency problem is the same as in the Newtonian case. To solve it, we can either set $\mathbf{b}(t) = \mathbf{0}$ and solve for the position of the Sun from the ones of the planets

$$\mathbf{s} = -\frac{1}{\mu_0 \left[1 + (v_0^2 - U_0)/2c^2\right]} \sum_{i=1}^{N} \mu_i \left[1 + \frac{v_i^2 - U_i}{2c^2}\right] \mathbf{r}_i.$$

The difference between the position of the Sun from the above formula and the Newtonian one is $\simeq 200$ m, thus it is very significant for our measurement accuracy. On the contrary, the small (< 1 micron/s) differences in the velocity of the Sun are not important, and even formally the changes they introduce in the equations of motion are of PN order 2.

As an alternative, the equation of motion may include the one for the Sun, and a constraint on the center of mass can be added by means of a priori observations: this is somewhat more complicated because the constraints are nonlinear, but it is possible and the results must be the same.

3. Test for parametric post-Newtonian violations

In this Section we will discuss the possible parametric PN violations to understand if we can investigate them with MORE; note that the term containing J_2 is not, strictly speaking, a violation term, but it is closely related because it is strongly correlated with PN parameter β.

3.1. *Three body effects and oblateness of the Sun*

The contribution of the oblateness of the Sun is $J_2 \, L_{J_2}$ with

$$L_{J_2} = -\frac{1}{2} \sum_{i \neq 0} \frac{\mu_0 \, \mu_i}{r_{0i}} \left(\frac{R_0}{r_{0i}}\right)^2 \left[3(\mathbf{n}_{0i} \cdot \mathbf{e}_0)^2 - 1\right],$$

where R_0 is the Sun's radius, $\mathbf{e}_0$ is the unit vector along the Sun's rotation axis.

J_2 affects the precession of the longitude of the node, that generates a displacement in the plane of the solar equator, while the main orbital effect of β is a precession of the argument of perihelion, that is a displacement taking place in the plane of the orbit of Mercury. The angle between these two planes is only $\epsilon = 3.3°$ and $\cos\epsilon = 0.998$, thus it is easy to understand how the correlation between β and J_2 can be 0.997, as found in the numerical simulations of (Milani *et al.* 2002). Short of using another test body, with an orbit plane much more inclined than the one of Mercury, this correlation cannot be avoided. One possible way to mitigate this effect is to use the equation derived by (Nordvedt 1970) that relates the SEP η parameter to β, γ and possibly preferred frame parameters within the metric theories:

$$\eta = 4\,\beta - \gamma - 3 - \alpha_1 - \frac{2}{3}\,\alpha_2.$$

When the values of γ, η and also of the preferred frame parameters (if included in the solution) are well determined, this equation acts essentially as a strong constraint on the value of β, and, as a result, the variance of both β and J_2 is sharply reduced, see Table II in (Milani *et al.* 2002).

3.2. *Gravitational constant and mass of the Sun*

An interesting goal, especially for cosmologists, would be to measure the time variation of the gravitational constant G. The Lagrange function terms are $(\dot{G}/G)\, L_{\dot{G}/G}$, where

$$L_{\dot{G}/G} = \frac{t - t_0}{2} \sum_{i \neq j} \frac{\mu_i\, \mu_j}{r_{ij}},$$

in practice the only terms with measurable effects contain the mass of the Sun. Hence the parameter which can be determined and the corresponding Lagrange term are

$$\zeta = \frac{d\mu_0}{dt}/\mu_0 \quad , \quad L_\zeta = (t - t_0) \sum_{i \neq 0} \frac{\mu_0 \mu_i}{r_{0i}};$$

we cannot discriminate the change with time of G from change with time of m_0.

Thus this is not a null experiment: what should be measured is $\dot{m}_0/m_0 \simeq -7 \times 10^{-14}$ y^{-1}, due to mass shed as radiation. A smaller contribution is the mass of charged particles emitted by the Sun, but the amount of the latter is not that well constrained (Noerdlinger 2008). If the result of our experiment for ζ was close to 10^{-13} y^{-1}, then it would be hard to discriminate the new physics of a change in G from the standard, but inaccurately known, physical effects.

3.3. *Preferred frame effects*

The preferred frame effects are described by the contribution

$$L_\alpha = \frac{\alpha_2 - \alpha_1}{4\,c^2} \sum_j \sum_{i \neq j} \frac{\mu_i\, \mu_j}{r_{ij}}\, (\mathbf{z}_i \cdot \mathbf{z}_j) - \frac{\alpha_2}{4\,c^2} \sum_j \sum_{i \neq j} \frac{\mu_i\, \mu_j}{r_{ij}}\, [(\mathbf{n}_{ij} \cdot \mathbf{z}_i)\, (\mathbf{n}_{ij} \cdot \mathbf{z}_j)],$$

with two additional post Newtonian parameters α_1, α_2 and with the vector $\mathbf{z}_i = \mathbf{w} + \mathbf{v}_i$, where $\mathbf{w}$ is the velocity of the solar system barycenter with respect to the preferred frame, usually assumed to be the one of the cosmic microwave background, thus $|\mathbf{w}| = 370 \pm 10$ km/s in the direction $(\alpha, \delta) = (168°, 7°)$.

The problem arises from the presence of additional terms in the total linear momentum integral $\mathbf{P}$. By applying again Noether's theorem, after the split of the Lagrangian L into parts with and without the preferred frame effects, $L = L_0 + L_\alpha$, we obtain:

$$\mathbf{P} = \mathbf{P}_0 + \mathbf{P}_\alpha, \quad \mathbf{P}_0 = \sum_j \frac{\partial L_0}{\partial \mathbf{v}_j}, \quad \mathbf{P}_\alpha = \sum_i \frac{\partial L_\alpha}{\partial \mathbf{v}_i}.$$

However, L_α is not invariant with respect to a time-dependent translation with constant velocity, and there is no center of mass integral (Will 1993). That is

$$\frac{d\mathbf{P}}{dt} = \frac{d\mathbf{P}_0}{dt} + \frac{d\mathbf{P}_\alpha}{dt} = \mathbf{0} \implies \mathcal{M}\, \frac{d^2\mathbf{b}}{dt^2} = \frac{d\mathbf{P}_0}{dt} = -\frac{d\mathbf{P}_\alpha}{dt}.$$

The accelerated barycentric frame results in apparent forces, giving the same acceleration $d\mathbf{P}_\alpha/dt \cdot (1/\mathcal{M})$ on all the bodies: these apparent forces are of PN order 1 and

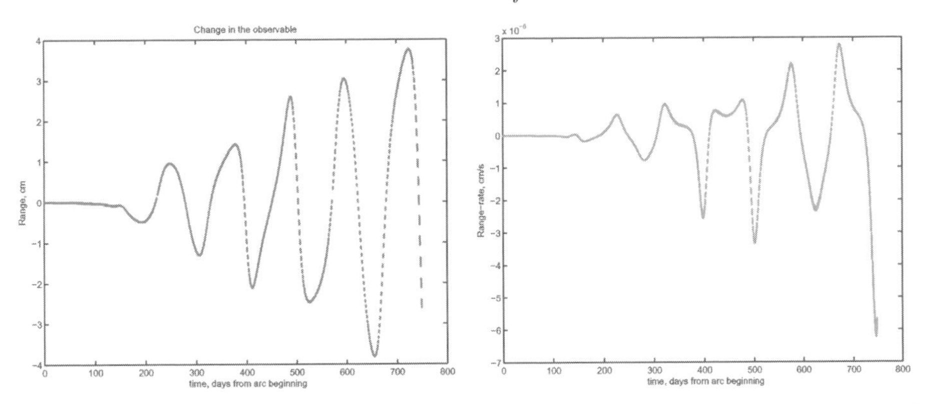

Figure 2. The effect over 2 years of the apparent force associated with preferred frame effects $\alpha_1 = 3 \times 10^{-4}$, $\alpha_2 = 3 \times 10^{-4}$ (range on the left, range rate on the right) has effects small or at the most almost comparable with respect to the measurement accuracy.

are not zero even for $\mathbf{w} = \mathbf{0}$. Even if they results in small effects (Figure 2) they have been included in the model used for simulations. What matters for us is that there is a consistent formulation in "barycentric" coordinates even without a barycenter integral.

3.4. *Violations of the Strong Equivalence Principle*

We can consider that there are for each body i two quantities μ_i and μ_i^I, one in the gravitational potential (including the relativistic part) and the other in the kinetic energy. If there is a violation of the strong equivalence principle involving body i, with a fraction Ω_i of its mass due to gravitational self-energy, then

$$\mu_i = [1 + \eta\Omega_i]\,\mu_i^I \Longleftrightarrow \mu_i^I = [1 - \eta\Omega_i]\,\mu_i + \mathcal{O}(\eta^2),$$

with η a post-Newtonian parameter for this violation. Neglecting $\mathcal{O}(\eta^2)$ terms this is expressed by a Lagrangian term $\eta\, L_\eta$, with an effect on body i:

$$L_\eta = -\frac{1}{2}\sum_i \Omega_i\,\mu_i\,v_i^2 \quad\Longrightarrow\quad \frac{d^2\mathbf{r}_i}{dt^2} = \left.\frac{d^2\mathbf{r}_i}{dt^2}\right|_{\eta=0} [1 + \eta\,\Omega_i].$$

The largest effect of η is a change in the center of mass integral

$$\mathbf{P} = \sum_j \frac{\partial L}{\partial \mathbf{v}_j} = \sum_j [1 - \eta\,\Omega_i]\,\mu_i\,\mathbf{v}_j + \ldots, \qquad \frac{d\mathbf{P}}{dt} = \mathbf{0}$$

and if the center of mass is the origin, the position of the Sun has to be corrected:

$$\mathbf{b} = \frac{1}{\mathcal{M}}\sum_j [1 - \eta\Omega_j]\,\mu_j\,\mathbf{r}_j + \ldots = \mathbf{0} \quad\Longrightarrow\quad \mathbf{s} = \frac{-1}{\mu_0\,[1 - \eta\,\Omega_0]}\sum_{j\neq 0}[1 - \eta\Omega_j]\,\mu_j\,\mathbf{r}_j + \ldots.$$

The partial derivative of the acceleration of the body j with respect to η is

$$\frac{\partial}{\partial\eta}\left[\frac{d^2\mathbf{r}_j}{dt^2}\right] = \Omega_j\left[\frac{\mu_0}{r_{j0}^3}\,\mathbf{r}_{j0} + \sum_{i\neq j,0}\frac{\mu_i}{r_{ji}^3}\,\mathbf{r}_{ji}\right] + \frac{\partial}{\partial\mathbf{s}}\left[\frac{\mu_0}{r_{j0}^3}\right]\frac{\partial\mathbf{s}}{\partial\eta},$$

where the first term is the direct, the second the indirect η-perturbation, and where

$$\frac{\partial\mathbf{s}}{\partial\eta} = \sum_{i\neq 0}(\Omega_j - \Omega_0)\,\frac{\mu_i}{\mu_0}\,\mathbf{r}_i.$$

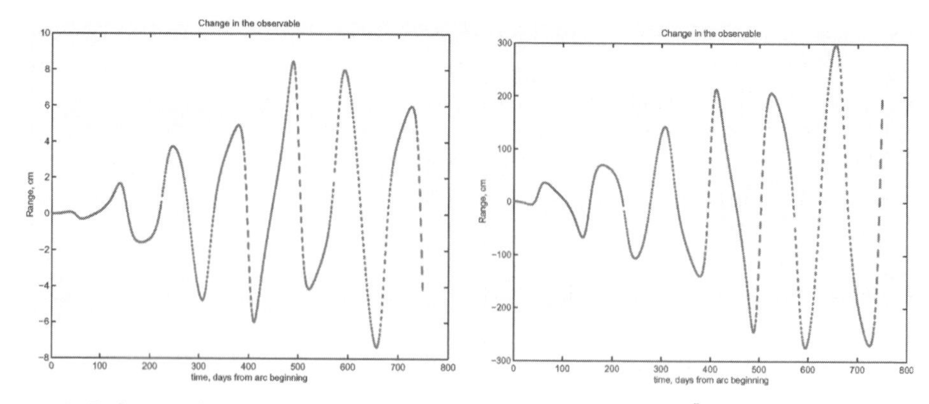

Figure 3. Left: signal in range due to a SEP violation with $\eta = 10^{-5}$, direct part only (acting mostly on Earth); this signal is marginally above measurement accuracy, thus after fitting the initial conditions should not become significant. Right: signal in range due to a SEP violation with $\eta = 10^{-5}$, indirect part only (acting more on Mercury), by assuming the same initial conditions; the fit of initial conditions lowers the signal, but it is anyway significant.

By combining together and omitting smaller terms with $\Omega_i\,\mu_k$ (with $i,k \neq 0$) or η^2

$$\frac{\partial}{\partial\eta}\left[\frac{d^2\mathbf{r}_j}{dt^2}\right] = \Omega_j\,\mu_0\,\frac{\mathbf{r}_{j0}}{r_{j0}^3} - \Omega_0\,\frac{\partial}{\partial\mathbf{r}_0}\left[\frac{1}{r_{j0}^3}\right]\sum_{i\neq 0}\mu_i\,\mathbf{r}_i,$$

with a direct (small parameter $\Omega_j\,\mu_0$) and an indirect (small parameter $\Omega_0\,\mu_i$) part. Figure 3 shows the change in the range due to a SEP violation with $\eta = 10^{-5}$: our experiment should be sensible to the indirect part, not to the direct one.

4. The observables

The observables of our experiment are the distance r between the ground antenna and the spacecraft, and its time derivative $\dot{r}$. The range is computed using 5 state vectors:

$$r = |(\mathbf{x}_{\text{sat}} + \mathbf{x}_M) - (\mathbf{x}_{EM} + \mathbf{x}_E + \mathbf{x}_{\text{ant}})| + S(\gamma),$$

where $\mathbf{x}_{\text{sat}}$ is the mercury-centric position of the orbiter, $\mathbf{x}_M$ the solar system barycentric position of Mercury, $\mathbf{x}_{EM}$ the position of the Earth-Moon center of mass in the same reference system, $\mathbf{x}_E$ the vector from the Earth-Moon barycenter to the center of mass of the Earth, $\mathbf{x}_{\text{ant}}$ the position of the ground antenna center of phase with respect to the center of mass of the Earth. $S(\gamma)$ is the *Shapiro effect*, the difference between distance in a flat space and the geodesic length in the curved space-time, depending upon the post-Newtonian parameter γ, which has a special role in relativistic orbit determination, since it appears in both the dynamics and the equations of observation.

4.1. *Shapiro effect in range*

The Shapiro effect at the 1-PN level is (e.g. Will 1993, Moyer 2003)

$$S(\gamma) = \frac{(1+\gamma)\,\mu_0}{c^2}\,\ln\left(\frac{r_r + \mathbf{k}\cdot\mathbf{r}_\mathrm{r}}{r_t + \mathbf{k}\cdot\mathbf{r}_\mathrm{t}}\right) = \frac{(1+\gamma)\,\mu_0}{c^2}\,\ln\left(\frac{r_t + r_r + r}{r_t + r_r - r}\right),$$

where $\mathbf{r}_\mathrm{r}$ and $\mathbf{r}_\mathrm{t}$ are heliocentric positions of the transmitter and the receiver at the corresponding time instants of photon transmission and reception, $\mathbf{k}$ is the unit vector in the direction of the radio waves. Note that the planetary terms, similar to the solar in $S(\gamma)$ above, can also be included but they prove to be smaller than the accuracy needed for our measurements. However a question arises whether the very high signal to noise in the range requires other terms in the solar gravity influence: (i) motion of the source, or (ii) higher-order PN corrections when the radio waves are passing near the Sun, at just a few solar radii (and thus the denominator in the ln-function of the Shapiro formula is small). The former are of the PN order 1.5, while the latter are of PN order 2.

The PN order 1.5 correction has been widely discussed after the Cassini experiment significantly increased accuracy for the γ determination; see (Bertotti *et al.* 2003) for the Cassini results and (Will 2003, Klioner & Peip 2003, Kopeikin 2008) for further discussion. Figure 4, on the left, shows the incidence of this 1.5-PN term on the range computation: the correction is less than 1 centimeter in range, well below our accuracy.

The important correction is obtained by adding 2-PN terms in the Shapiro formula, due to the bending of the light path. (Moyer 2003) proposed to add a PN 2nd order term $(1+\gamma)\,\mu_0/c^2$ (the radius of a black hole with the mass of the Sun) both in the numerator and denominator of the argument of the natural logarithm:

$$\ln\left(\frac{r_t + r_r + r}{r_r + r_r - r}\right) \rightarrow \ln\left(\frac{r_t + r_r + r + \frac{(1+\gamma)\,\mu_0}{c^2}}{r_t + r_r - r + \frac{(1+\gamma)\,\mu_0}{c^2}}\right).$$

Evaluating the argument of the logarithm near the conjunction configuration we obtain an estimate for the 2-PN correction to the Shapiro formula:

$$S_{2PN} \approx \frac{(1+\gamma)\,\mu_0}{c^2}\,\ln\left(1 - \frac{2\,r_r\,r_t}{b^2}\,\frac{R_0^S}{r}\right) \approx -(1+\gamma)^2\left(\frac{\mu_0}{c^2\,b}\right)^2\frac{2\,r_r\,r_t}{r_r + r_t},$$

where b is the impact parameter of the radio wave path passing near the Sun. Interestingly, Moyer's heuristic correction provides the same result as much more theoretically rooted recent derivations of the 2-PN Shapiro terms by (Le Poncin-Lafitte & Teyssandier 2004, Klioner & Zschocke 2007, Teyssandier & Le Poncin-Lafitte 2008, Ashby & Bertotti 2008). Figure 4, on the right, shows that the 2-PN correction is relevant for our experiment, especially when there is a superior conjunction with small b.

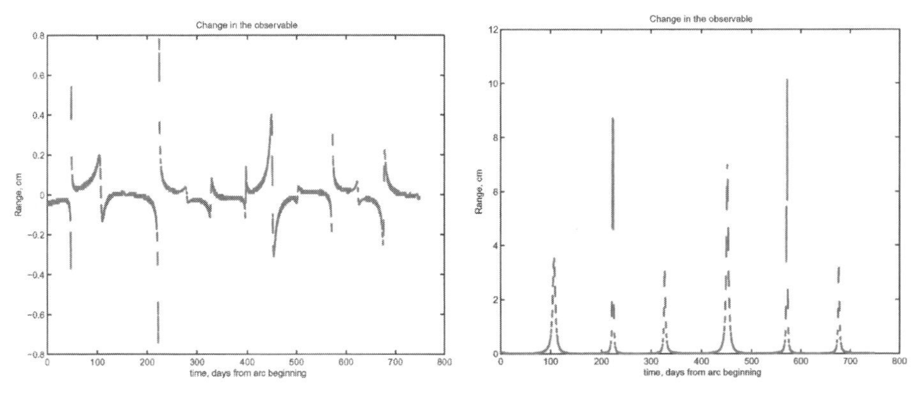

Figure 4. Left: differences in range by using a 1-PN and a 1.5-PN formulation ($\gamma = 1$); the correction has been added with little effort, but does not seem to be important, less than 1 centimeter in range, well below the accuracy level. Right: differences in range by using a 1-PN and a 2-PN formulation ($\gamma = 1$); the correction is relevant for MORE, at least when a superior conjunction results in a small impact parameter b, e.g., in this figure we have plotted data assumed to be available down to $b \simeq 3R_0$. For larger values of b the effect decreases as $1/b^2$.

5. Dynamic Mercury Time

The mercury-centric orbit of the spacecraft is coupled to the orbit of the planet, mostly through the difference between the acceleration from the Sun on the probe and the one on the planet (the Sun tidal term). This coupling is weak because the Sun tide is just 10^{-7} of the monopole acceleration from Mercury. The relativistic perturbations containing the mass of Mercury are small to the point that they are not measurable, being easily absorbed by the much larger non-gravitational perturbations, measured with finite accuracy by the on board accelerometer. Should we conclude that general relativity does not matter in the computation of the mercury-centric orbit? The answer is negative, but the main relativistic effect does not appear in the equation of motion.

There are three different time coordinates to be considered. The dynamics of the planets, as described by the Lagrangian, is the solution of differential equations having a time belonging to a space-time reference frame with origin in the SSB as independent variable. There can be different realizations of such a time coordinate: the currently published planetary ephemerides are provided in a time called TDB (Barycentric Dynamic Time). The observations are based on averages of clocks and frequency scales located on the Earth surface: this corresponds to another time coordinate called TT (Terrestrial Time). Thus for each observation the times of transmission and receiving (t_t, t_r) need to be converted from TT to TDB to find the corresponding positions of the planets, e.g., the Earth and the Moon, by combining information from the precomputed ephemerides and the output of the numerical integration for Mercury and the Earth-Moon barycenter. This time conversion step is necessary for the accurate processing of each set of interplanetary tracking data; the main term in the difference TT-TDB is periodic, with period 1 year and amplitude $\simeq 1.6 \times 10^{-3}$ s, while there is essentially no linear trend, as a result of a suitable definition of the TDB.

The equation of motion of a mercury-centric satellite can be approximated, to the required level of accuracy, by a Newtonian equation provided the independent variable is the proper time of Mercury. Thus, for the BepiColombo radioscience experiment, it is necessary to define a new time coordinate TDM (Mercury Dynamic Time) containing terms of 1-PN order depending mostly upon the distance from the Sun r_{10} and velocity v_1 of Mercury. The relationship with the TDB scale, truncated to 1-PN order (we drop the $O(c^{-4})$ terms on the right hand side, that are in principle known, but certainly not needed for our purposes), is given by a differential equation

$$\frac{dt_{\rm TDM}}{dt_{\rm TDB}} = 1 - \frac{v_1^2}{2\,c^2} - \sum_{k \neq 1} \frac{G\,m_k}{c^2\,r_{1k}},$$

which can be solved by a quadrature formula once the orbits of Mercury, the Sun and the other planets are known. Figure 5 plots the output of such a computation, showing a drift due to the non-zero average of the post-Newtonian term.

The oscillatory term, having the one of Mercury orbit as main period, has an amplitude $\simeq 0.012$ s. In 0.01 s the spacecraft velocity can change by 3 cm/s, $\simeq 10{,}000$ times more than the range-rate measurement accuracy, the position by 30 m, $\simeq 300$ times the range measurement accuracy. Thus this effect has to be accurately taken into account for our experiment†.

† The time scale TDB will be replaced in the planetary ephemerides by the new TCB; when this will happen, we will use a suitably defined Mercury Coordinate Time (TCM), such that the differential equation giving the TCB to TCM conversion will be exactly the same as for TDB to TDM.

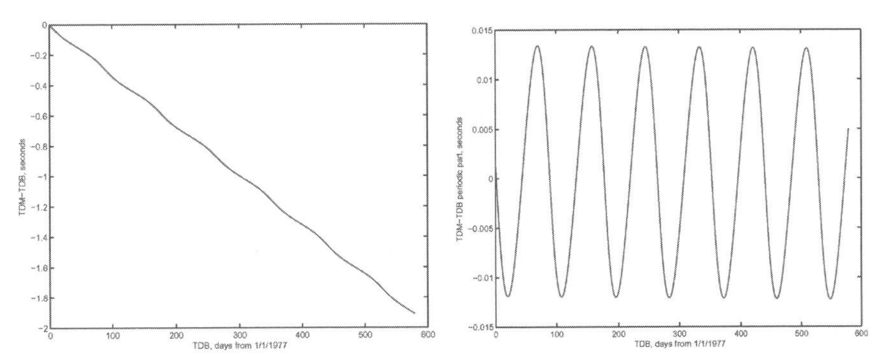

Figure 5. Left: TDM as function of TDB shows a drift due to the non-zero average of the 1-PN term. Right: the oscillatory term is almost an order of magnitude larger than TT-TDB.

6. Restricted ephemerides improvement

The level of accuracy of the measurements of the Mercury orbit radioscience experiment is incompatible with the use of the current planetary ephemerides, which have been solved by using lower accuracy measurements. However the data from BepiColombo by themselves will not allow to improve the ephemerides for all planets. Of the 5 vectors used in the light-time equation, $\mathbf{x}_{ant}$ and $\mathbf{x}_E$ can be assumed known: they cannot be improved by ranging to a Mercury orbiter. For the orbit of the Moon it is more effective to measure the range to the Moon with lunar laser ranging. Navigation satellites and VLBI give more information on the antenna position and on the rotation of the Earth.

The position of the Earth-Moon center of mass and of Mercury can certainly be improved by the range measurements from BepiColombo (the range-rate is less effective: it is more accurate than range only over time scales $\leqslant 50,000$ s). These measurements will also be provided (as normal points) to be used in the planetary ephemerides fit. In this way the BepiColombo data will be available to be used by all the existing and future planetary ephemerides to contribute to more accurate predictions of planetary orbits.

Acknowledgements

This research has been funded by the Italian Space Agency (ASI), contract I/082/06/0. The partecipation to the meeting has been partially funded by an IAU grant.

References

Ashby, N. & Bertotti, B. 2008, *American Astronomical Society, DDA meeting #39*
Bertotti, B., Iess, L., & Tortora, P. 2003, *Nature*, 425, 374
Klioner, S. A. & Peip, M. 2003, *A&A*, 410, 1063
Klioner, S. A. & Zschocke, S. 2007, *GAIA-CA-TN-LO-SK-002-1 report*
Kopeikin, S. 2008, *eprint arXiv:0809.3433*
Le Poncin-Lafitte, C. & Teyssandier, P. 2004, *Class. Quantum Grav.*, 21, 4463
Milani, A. & Gronchi, G. F. 2009, *Theory of Orbit Determination*, Cambridge Univ. Press
Milani, A, Vokrouhlický, D., Villani, D., Bonanno, C., & Rossi, A., 2002, *Phys. Rev. D* **66**
Moyer, T. D. 2003, *Formulation for Observed and Computed Values of Deep Space Network Data Types for Navigation*, Wiley-Interscience
Noerdlinger, P. D. 2008, *eprint arXiv:0801.3807v1*
Nordvedt, K. 1970, *ApJ*, 161, 1059
Teyssandier, P. & Le Poncin-Lafitte, C. 2008, *Class. Quantum Grav.*, 25, 145020
Will, C. M. 1993, *Theory and experiment in gravitational physics*, Cambridge Univ. Press
Will, C. M. 2003, *ApJ*, 590, 683

Relativity in Fundamental Astronomy
Proceedings IAU Symposium No. 261, 2009
S. A. Klioner, P. K. Seidelman & M. H. Soffel, eds.

doi:10.1017/S1743921309990664

Radio astronomy in the future: impact on relativity

Michael Kramer[1,2]

[1]MPI für Radioastronomie, Auf dem Hügel 69, 53121 Bonn, Germany
email: mkramer@mpifr-bonn.mpg.de

[2]University of Manchester, Jodrell Bank Centre for Astrophysics, Alan-Turing Building, Manchster M13 9PL, UK

Abstract. Radio astronomy has played an important part in the study of relativity. Famous examples include the discovery and exploitation of pulsars for precise binary pulsar tests, the proof of the existence of gravitational radiation and the discovery of the Cosmic Microwave background. In the future, radio astronomy will continue to play a decisive role, assisted by new and upcoming instruments like LOFAR and the SKA. In this review, I will present the revolution that is ongoing in radio astronomical techniques and outline the impact expected on the studies of relativity.

Keywords. gravitation, relativity, techniques: radar astronomy, stars: neutron

1. Introduction

Astronomical and astrophysical discoveries are usually driven by technological advances that allow us to probe a previously unexplored region of the discovery phase space (e.g. Harwit 1981). This has been clearly demonstrated by the impressive list of fundamental discoveries achieved in radio astronomy (Wilkinson *et al.* 2004) when a whole new part of the electromagnetic spectrum became available. While we expect similar advances for instance with the anticipated direct detection of gravitational waves using ground-based or space-based detectors, astronomers in other astronomical windows also work towards further technological enhancements. In particular in the electromagnetic spectrum the most notable efforts are clearly those activities towards Extremely Large (optical) Telescopes (ELTs, e.g. Spyromilio 2007) and the Square Kilometre Array (SKA) radio telescope (Schilizzi *et al.* 2008).

The SKA especially tries to benefit from the recent advances in information technology by replacing expensive metal with flexible and upgradable computing and software solutions. These changes are destined to revolutionize the way we do astronomy: astronomers using the SKA will have to deal with data files of (currently) enormous size, in fact, so large that the data hypercubes may still be several TByte large, even after having been pre-processed on the "grid"; astronomers will be able to observe vast areas of the sky in completely different directions at the same time, and they may even be able to "rewind" time by using a data buffer fed with data from all-sky monitors. The real impact, however, will emerge from the science that will become possible. Simply extrapolating from what we know today, the flood of exciting results should be vast, largely enabled by a huge increase in sensitivity and survey speed. This will clearly have a significant impact on the study of relativity which I try to outline and predict here.

2. Radio astronomy: In the past, present and future

Arguably, most of the fundamental astrophysical discoveries of the last century have been made by radio astronomers. The examples are plenty, among which we find the discovery of the Cosmic Microwave Background (CMB), quasars and radio galaxies, the existence of gravitational lenses, jets and super-luminal motion, dark matter, interstellar molecules and powerful masers and mega-masers. Also, very important was the discovery of pulsars, and the subsequent proof of the existence of gravitational waves and the discovery of the first extra-solar planets. All these discoveries were made after the humble beginnings of Karl Jansky and Grote Reber in the early 1930's, which finally opened up a new part of the electromagnetic spectrum and, hence, allowed astronomers to probe a huge new region of phase space.

The rate of discovery accelerated with the construction of the first "big" dishes, clearly marked with the construction of the Lovell radio-telescope (or the "250-feet" or MkI telescope as it was called at the time) at Jodrell Bank, UK. The current state-of-art still includes the Lovell telescope as the 3rd largest fully-steerable telescope in the world, only trumped in size by the 100-m radio-telescopes at Green Bank, USA, and Effelsberg, Germany. In the Southern hemisphere the largest radio-telescope used by astronomers is still the 64-m dish at Parkes. Special types of telescopes like the 300-m Arecibo telescope (a monolithic dish built into a valley on Puerto Rico and, hence, limited in its ability to track sources), or the Very Large Array (VLA), an interferometer of 27 25-m dishes in New Mexico have played their important part in delivering some of the most exciting results in radio astronomy or astrophysics in general.

Currently, new major telescopes are being constructed, all of which will lead to a new era in radio astronomy. The Atacama Large Millimetre Array (ALMA) is being built in Chile and will vastly improve our sensitivity at very high radio frequencies. The most revolutionary new major facility, however, is perhaps the Low Frequency Array (LOFAR, e.g. van Haarlem 2005). LOFAR will observe at the opposite end of the radio spectrum and will enable studies of the largely unexplored low-frequency radio sky. What makes LOFAR special is not only its large sensitivity at low frequencies, but also the way the telescope works. In contrast to most "classical" radio telescopes, LOFAR has no moving parts but uses antenna elements that receive signals from the whole sky overhead. Combining the signals of the antennae digitally and in software, many independent beams

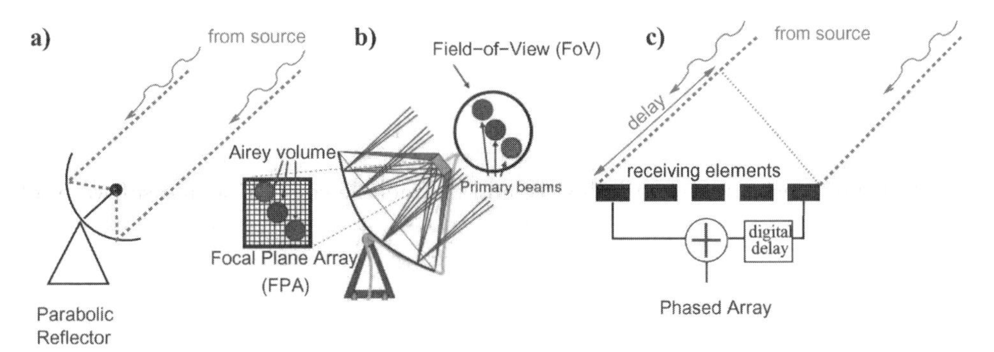

Figure 1. Comparison of pointing a telescope beam towards a source (a) via a conventional parabolic reflector, (b) a parabolic dish equipped with a focal plane array (FPA) or (c) with an aperture array where signals from different receiving elements are combined electronically with appropriate delays to steer a beam to the source direction. Note that in both (b) and (c) the FoV of the telescope is significantly greater than with a conventional telescope. (Figure b courtesy of Andrew Faulkner)

can be formed that allow the astronomers to observe phenomena and sources in different directions of the sky. In essence, radio astronomers start to replace metal and hardware by computer and software solutions.

LOFAR is an important stepping stone into an era that will not only revolutionize radio astronomy itself, but astronomy and astrophysics – and hence physics – as a whole. The goal of this evolution of our radio eyes is the construction of the largest telescope the Earth has ever seen – the Square Kilometre Array (SKA). I will summarise its main features further below, but it is clear that this enormous step from our current facilities to the gigantic global SKA telescope will require some intermediate steps in order to develop technology and, in particular, affordable techniques and production methods. Therefore, a number of SKA "pathfinder" projects are being undertaken that will explore the technology as well as give a taste of the science that will be enabled by the SKA. Apart from LOFAR, the furthest developed project is the Allen-Telescope-Array (ATA, Welch *et al.* 2009) in the USA which consists in its initial phase of 42 6-m dishes, testing the so called large-N–small-D concept. Very significant increases in bandwidths, data transport, data processing and sensitivity are seen in the extensions of MERLIN (the Multi-Element Radio-Linked Interferometer Network) and the VLA to eMERLIN (Garrington 2007) and eVLA (Ulvestad 2007), respectively. In China, a project is underway to build the Five-hundred meter Aperture Spherical Telescope (FAST, Nan 2008) whose 500-m diameter will dwarf even the Arecibo telescope. On the two short-listed SKA sites, two pathfinder telescopes will be completed over the next few years, i.e. MeerKAT in South Africa (Jonas 2007) and ASKAP in Western Australia (Johnston *et al.* 2008), both of which will have collecting areas of the order of a 100-m dish.

All these exciting new facilities will test aspects of SKA technology but are, excitingly, world-class facilities in their own right and are expected to deliver science (and valuable lessons!) over the next 4–5 years. With this knowledge, construction of the SKA could start around 2013. As an interferometer, the SKA will already be able to deliver science while it is being constructed so that early science with about 10% of the final collecting area (Phase I) is expected from 2017, while full-SKA operations are anticipated for the early 2020s.

2.1. *Revolution!*

An important key to achieve the outlined expectations is the increase in telescope sensitivity. For radio astronomy, this is governed by the "radiometer" equation (e.g. Burke & Smith 2002) which specifies the equivalent flux density for a receiving system of temperature $T_{\rm sys}$ on a telescope with the equivalent size of $A_{\rm eff}$ as

$$S_{\rm min} = \frac{2kT_{\rm sys}}{A_{\rm eff}\sqrt{\tau\Delta\nu}} = \frac{T_{\rm sys}}{G}\frac{1}{\sqrt{\tau\Delta\nu}}. \qquad (2.1)$$

Here, k is the Boltzmann constant, τ the integration time and $\Delta\nu$ the observing bandwidth. In the last equation, we have defined the telescope gain as $G = A_{\rm eff}/2k$. As most modern receivers are already at the quantum noise limit, the sensitivity cannot be improved further by reducing the system temperature. Hence, an improvement is only possible if we can increase the gain, G, (i.e. collecting area), the bandwidth or the integration time. All of this is being addressed by the new telescopes by either increasing the collecting area (e.g. LOFAR, SKA), bandwidth (e.g. eMERLIN, eVLA), or the instantaneous Field-of-View (FoV) which allows one to cover a larger area of sky so that one can integrate longer for the same rate of sky coverage (e.g. ASKAP).

These changes, in particular the increase in bandwidth and FoV, are only possible due to the advances in signal and data processing. Larger bandwidths can be sampled, larger data rates can be transferred and larger data volumes can be processed. This is possible using programmable processing units (i.e. Field Programmable Gate Arrays, FPGAs) but also commodity computing power concentrated in "Beowulf Clusters" and even increasingly due to the use of game console hardware, so called GPUs. In particular the latter shows that radio astronomers try to utilise technical developments that are driven outside science but in consumer electronics or telecommunication. These advances indeed allow astronomers to observe large fractions of the sky at once using so-called "phased arrays" where receiving elements are sensitive to vastly different directions or even the whole sky. Their signals are added in hard- or software using particular delays to achieve directionality (see Figure 1). The ultimate number of beams is only limited by the available computing power.

Phased arrays laid flat on the ground are also called "aperture arrays" (AAs), and they are indeed sensitive to the whole sky overhead, although the effective area drops as the cosine of the zenith angle. AAs are an essential part of the European plans to build the SKA. Phased arrays can also be installed in the prime focus of conventional dish telescopes in which case they are referred to as "focal plane arrays" (FPAs). An FPA allows one to detect light rays via paths that are inclined to the geometrical axis of the telescope (see Figure 1). The result is a significant enlargement of the telescope's FoV, sometimes up to an order of magnitude or more. With the combination of the signals from different FPA elements, the enlarged FoV can be fully filled. First examples of both AA and FPA type are very promising, while the remaining technological challenges lie in achieving low system temperatures and in the required data handling and processing.

2.2. *What is it? A telescope, a supercomputer? Both!*

The described efforts to increase the effective sensitivity of new radio telescopes all have one feature in common: they all require an early digitization of the signals and a very significant increase in the processing requirements. This goes in hand with an increase in the number of bits used for digitization in order to be more resilient to man-made radio interference signals. Moreover, a significant increase in collecting area is hard or impossible to achieve with a single monolithic structure, but most new radio telescopes are built as interferometers, i.e. an array of small telescopes that are combined to synthesize a filled aperture. For such instruments, the spatial resolution is given by the largest "baseline", $b_{\rm max}$, i.e. the largest distance between a pair of telescopes. Small "pencil beams" are then synthesized to fill the full FoV of the telescope which in case of single dish elements is as large as the FoV of a single element ("primary beam") or of different size in case of AAs and their combination. Forming an astronomical image with an interferometer is done by combining the signals of the array elements using a correlator. As the signal of each telescope needs to be combined or correlated with every other element, the processing requirements increase with the square of number of antennas, the cube of baseline length and the FoV solid angle.

The actual configuration of the telescope, such as type, size and number of receiving elements, their relative spacing and the observing bandwidth and frequency, will be the result of a complex optimization process regarding costs, technical feasibility and science drivers, but in order to fill a FoV of a few square degree one needs $N_{\rm pix} \sim (b_{\rm max}/D)^2 \sim 10^4 - 10^9$ pixels and about $N_{\rm ops} \sim 10^{15}$ operations per second for a primary beam of a telescope of diameter D. In addition, the obtained data need to be processed on-the-fly, as data rates of several TByte per second as estimated for the SKA cannot be transferred over large distances. All in all, these estimates put a telescope like the SKA safely in the region of the most powerful Exaflop supercomputers anticipated for the next decade.

2.3. *The Square-Kilometre Array (SKA)*

Essentially all upcoming and future telescopes will contribute in one way or the other to advances in the field of relativity. LOFAR will find a large number of neutron stars that are potential sources for tests of relativity or pulsar timing array experiments that attempt to directly detect gravitational waves Stappers *et al.* (2007). Even though these experiments will be performed at high frequencies, LOFAR can find appropriate sources and also monitor the interstellar weather that needs to be corrected for in high precision timing (see contribution by Matthew Bailes). FAST will have a collecting area that allows us to find and time pulsars that will be significantly better than currently achieved with Arecibo (Smits *et al.* 2009). The eVLA promises to give us a first good glance at the pulsar population orbiting the super-massive black hole in the Galactic Centre. However, the real big advance in studies of gravitational theories will be achieved with the SKA, so that we concentrate on this science instrument.

The exact configuration of the SKA will be determined after the ongoing global R&D-phase is completed and a SKA design is presented in 2012. The current efforts for this project that includes 19 countries at the time of writing cover all possible areas, ranging from developing receiving elements to amplifiers, beam-formers, software and governance structure and much more. The results will be compared to the current reference design (Schilizzi 2007) which is outlined in the following.

The defining feature of the SKA will be overwhelming sensitivity due to enormous collecting area. As the name suggests, the total collecting area for the SKA is planned to be one square-kilometer. However, for practical purposes it makes sense to define the sensitivity of the SKA in terms of collecting area and system temperature, as $S_{\rm min}$ directly depends on them (see Eqn. 2.1). The target sensitivity is 2×10^4 m^2 K^{-1} which we will refer to as one SKA unit. The SKA is likely to consist of a sparse AA of tiled dipoles in the frequency range of 70 to 500 MHz, and above 500 MHz the following three implementations are considered:

A) 3 000 15-m dishes with a single pixel feed, a sensitivity of 0.6 SKA units, $T_{\rm sys}$=30 K and 70% efficiency covering the frequency range of 500 MHz to 10 GHz.

B) 2 000 15-m dishes with FPAs from 500 MHz to 1.5 GHz, a sensitivity of 0.35 SKA units, a FoV of 20 deg^2, $T_{\rm sys}$=35 K and 70% efficiency and a single pixel feed from 1.5 to 10 GHz, with $T_{\rm sys}$=30 K.

C) A combination of dense AAs with a FoV of 250 deg^2, a sensitivity of 0.5 SKA units, covering the frequency range of 500 to 800 MHz and 2 400 15-m dishes with a single pixel feed covering the frequency range of 800 MHz to 10 GHz, a sensitivity of 0.5 SKA units, $T_{\rm sys}$=30 K and 70% efficiency.

The signals of these receiving elements can be combined to form independent FoVs.

Most of the collecting area will be concentrated in a central core of the SKA, i.e. 20% of the collecting area will be located within a 1 km radius and 50% within a 5 km radius. The remainder of the receiving elements will be distributed in stations along, probably, log-spiral arms with baselines extending to 3000 km distance in order to enable also high resolution imaging and high precision astrometry. Further resolution will be available via intercontinental baselines between Australia and Southern Africa.

The SKA will be built in three phases. In Phase I about 0.1 SKA unit will be available in the core region covering the low and mid frequencies by 2017. In Phase II, the full array will be completed by 2022. Finally, in Phase III high frequency capability for observations potentially up to 40 GHz will be added from 2022.

The SKA will be a flexible, multi-purpose observatory that will be able to serve the whole astronomy community as *the* premier imaging and surveying instrument. The

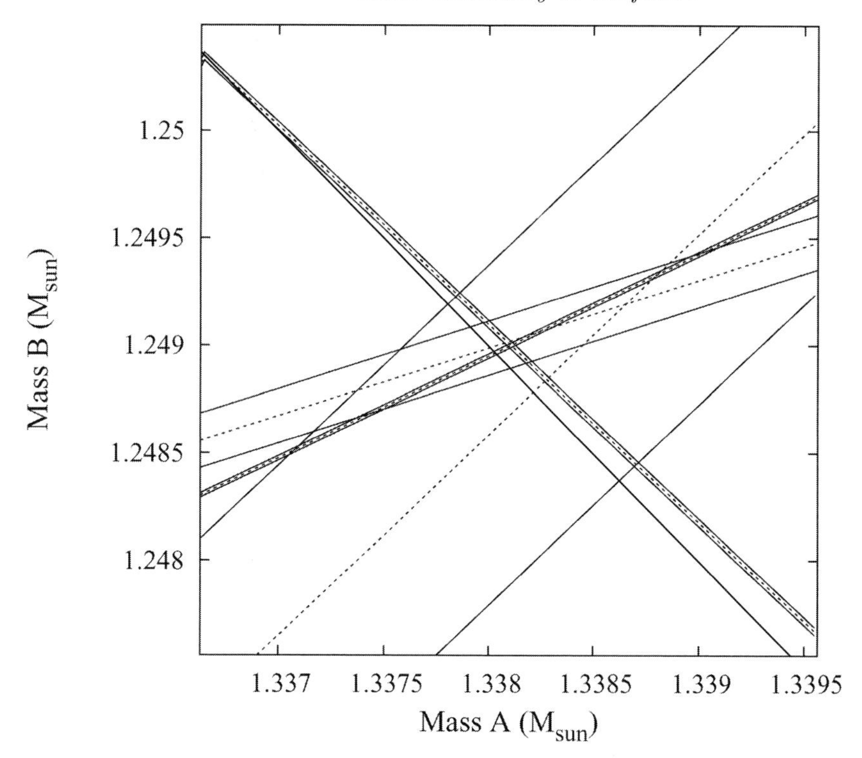

Figure 2. Mass–mass diagram for the Double Pulsar in 20 years from now, based on simulated TOAs (see Kramer & Wex 2009 for details). The solid lines represent the two-sigma errors of the post-Keplerian parameters ($\dot{\omega}$, γ, s, $\dot{P}_b$) and the mass ratio ($R \equiv m_A/m_B$). The $\dot{\omega}$ line was assuming a vanishing moment-of-inertia for pulsar A. Hence, the contribution of the spin-orbit coupling to the total precession of periastron, $\dot{\omega}_{SO}$, can be read off the offset between the $\dot{\omega}$-line and the point of intersection of the $\dot{P}_b$ and the s curves. As the masses of pulsar A and B are roughly equal, the slopes of the $\dot{\omega}$ and the $\dot{P}_b$ curve are roughly the same close to the P_b-s intersection. Hence, the determination of $\dot{\omega}_{SO}$ is much less sensitive to errors in s than to errors in $\dot{P}_b$.

design will ensure that it will be a flexible instrument rather than a fine-tuned experiment, even though five Key Science Projects (KSPs) have been selected by the international community. These KSPs drive essential parts of the design in order to ensure that all current science goals can be achieved. Still, versatility is a key element in the construction of the SKA which also includes natural upgrade paths: Eventually the SKA's power will only be constrained by the available computing power, so that the estimated operation costs budget includes items to constantly replace and upgrade SKA processing capabilities.

The KSPs cover a wide area of topics from fundamental physics to general astrophysics and cosmology. These are summarized in the SKA Science Book edited by Carilli & Rawlings (2004) as *Extreme tests of general relativity* (Kramer *et al.* 2004, see also below), *Galaxies, cosmology, dark matter and dark energy* (Rawlings *et al.* 2004), . *Probing the dark ages the first black holes and stars* (Carilli *et al.* 2004), *The origin and evolution of cosmic magnetism* (Gaensler *et al.* 2004), and the *Cradle of life* (Lazio *et al.* 2004). The versatility of the SKA will also allow a huge range of (astro-)physics to be conducted in addition to the these KSP. This includes obviously the likely possibility to discover unknown phenomena and physics (Wilkinson *et al.* 2004).

3. Applications in Relativity

It is clear that the SKA will have a huge impact on the study of relativity. Observations of weak lensing, the cosmic web, and baryonic oscillations via a billion galaxy survey will firmly establish the nature of dark energy. Also, spectroscopy of highly redshifted quasers will probe the possible variation of fundamental constants, while the monitoring of water mega-masers allow us to probe the expansion of the local Universe. Most importantly, however, will be the search for and timing of radio pulsars. With SKA sensitivity one can essentially find all active Galactic pulsars beaming towards Earth, providing a "Galactic Census" of pulsars (Cordes *et al.* 2004). We expect to find about 20,000 to 30,000 pulsars including 1,000 millisecond pulsars and 100 relativistic binaries (Smits *et al.* 2009).

3.1. *Astrometry & Binary Pulsar Tests*

Astrometry and in particular pulsar timing of millisecond pulsars should allow us to measure proper motions with a precision of about 100 nas/yr and parallaxes of the order of 10–20 μas, i.e. distances of up to 50–100 kpc. Astrometric measurements are not only useful to potentially trace the pulsars' movement in the gravitational field of the Galaxy or globular clusters, but parallax measurements are also needed to correct for relative acceleration between the pulsar and the solar system barycentre which affects the observed period derivatives (for spin and orbit) of relativistic binaries (e.g. Lorimer & Kramer 2005 and contributions by Bailes and Stairs). This is essential if radiative aspects of a theory of gravity are to be tested. The prospects for tests when using the SKA to study the Double Pulsar (Lyne *et al.* 2004) are discussed in detail by Kramer & Wex (2009). Essentially, with the SKA we expect an improvement in timing precision by a factor of $\sim$100, leading to precision tests of general relativity that even exceed the precise tests done in the solar system's weak field regime today. (For a detailed description of tests with pulsars see Ingrid Stairs' contribution.) Moreover, at the same time, the Gaia satellite mission is expected to significantly improve our knowledge of the galactic potential (Brown 2008), and hence lead to a much better understanding of the galactic contributions to the observed period derivative. All this will allow us to measure the moment-of-inertia of pulsar A in the Double Pulsar system (see Figure 2). Furthermore, we expect to measure the previously unmeasured relativistic deformation of the orbit, providing an additional test for general relativity and other theories of gravity (Kramer & Wex 2009). Unfortunately, FAST will not be able to see the Double Pulsar but improvements in timing this system could already be achieved with the Large European Array for Pulsars (LEAP, Kramer priv. comm.). In LEAP the major European telescope are added together to form the equivalent of an illuminated Arecibo dish, increasing the timing precision accordingly.

3.2. *Alternative theories & Black hole properties*

What makes a binary pulsar with a black-hole companion so interesting is that it has the potential of providing a superb new probe of relativistic gravity. As pointed out by Damour & Esposito-Farese (1998), the discriminating power of this probe might supersede all its present and foreseeable competitors. The reason lies in the fact that such a system would be very sensitive to strong gravitational self-field effects, making it for instance an excellent probe for tensor-scalar theories. Moreover, Wex & Kopeikin (1999) showed that the measurement of classical and relativistic spin-orbit coupling in a pulsar-black hole binary, in principle, allows us to determine the spin and the quadrupole moment of the black hole. This would test the "cosmic censorship conjecture" and the "no-hair theorem". While Wex & Kopeikin (1999) pointed out that with current telescopes such an experiment would be almost impossible to perform (with the possible exception of pulsars about the Galactic center black hole), Kramer *et al.* (2004) demonstrated that the SKA sensitivity should be sufficient. Indeed, this experiment benefits from the SKA sensitivity in multiple ways. On one hand, it provides the required timing precision but it also allows to perform the Galactic Census which should eventually deliver the sample of pulsars with a black hole companion.

3.3. *Gravitational wave astronomy*

While pulsars already provide the indirect evidence for the existence of gravitational waves (GW), they can also be used to detect and study them directly, since the timing residuals are affected by a passing GW wave as each pulsar and the Earth can be considered as free masses, whose positions respond to changes in the space-time metric (e.g., Sazhin 1978, Detweiler 1979). An update on the current efforts and details of such experiments is given in the contribution by Hobbs. For our purposes it is important to note that a perturbation of the pulsar and Earth position by a GW would lead to timing residuals of the order of $\sigma \sim h_c(f) \times T^{1.5}$ where $h_c(f)$ describes the characteristic amplitude of the GW per unit logarithmic interval of frequency, and T is the total observing time. Importantly, the sensitivity of GW detection using pulsars scales directly with the achieved timing precision.

Despite the apparent simplicity of the experiment, the timing precision required for the detection of GWs is very much at the limit of what is technically possible today. Only a stochastic background (rather than a single GW source) is likely to be detected today, as it would produce a correlated quadrupole signature among a network or "array" of timed millisecond pulsars on the sky. This facilitates the recognition of a signal originating, for instance, from the orbital motion of binary super-massive black holes in the early Universe (see Hobbs' contribution for more detail). A variety of sources is expected to emit GWs in the nHz-frequency range detectable by a "pulsar timing array" (PTA). Such PTA experiments and, in particular LEAP, harvesting the combined collecting power of Europe's largest telescopes, will have a good chance of detecting GWs, but only with the SKA can a detection be guaranteed and the properties of GWs (e.g. spectrum of a stochastic background, polarisation) be studied. Due to the increased timing precision achievable with the SKA and a large number of suitable sources discovered, a PTA with the SKA would achieve a sensitivity that even exceeds that of LISA (Kramer *et al.* 2004).

Different types of signals can be detected and studied with the SKA, i.e. stochastic, burst, and periodic signals. A stochastic gravitational wave background should arise from a variety of sources. Cosmological sources include inflation, string cosmology, cosmic strings and phase transitions (Kramer *et al.* 2004). The expected correlation signal among the PTA pulsars can be measured with very high precision, so that the polarisation properties of GW can be probed and compared with the predictions of GR (Lee *et al.* 2008). The SKA will also detect single sources of GW emission, such as the periodic GW signal of a binary super-massive black hole in the centre of a nearby Galaxy (e.g. Jenet *et al.* 2004) Here, the SKA and LISA will be complementary: On one hand, both instruments can see similar super-massive binary black hole systems but at different evolutionary stages, i.e. LISA will pick up SKA sources close to their merger. On the other hand, the SKA can observe GW emission of systems that are typically more massive than the LISA sources. We should also be able to discover compact pulsar binaries that emit in radio simultaneously with continuous GW emission detectable with LISA.

Moreover, the black holes studied with the SKA (stellar and the supermassive black hole in the Galactic Centre, see Section 3.2) will be complementary and partly overlapping with the population of black holes to be studied with Advanced LIGO and LISA.

Besides these guaranteed GW signals, the SKA should also be able to measure burst emission from, for instance, explosive events. In the GW window, pulsar glitches may also produce observable events, where the SKA would help to pinpoint the exact time of such an event. Fundamental properties of gravitation can also be tested: if GWs are governed by a massive field propagation law, GWs with frequency less than $m_g c^2/h$ cannot propagate. Hence detecting GWs with a 5 to 15 year period (ie. the frequency range covered by PTAs) reduces the upper bound on gravitation mass m_g by a factor of 15 to 50 (F. Jenet, priv. comm.)

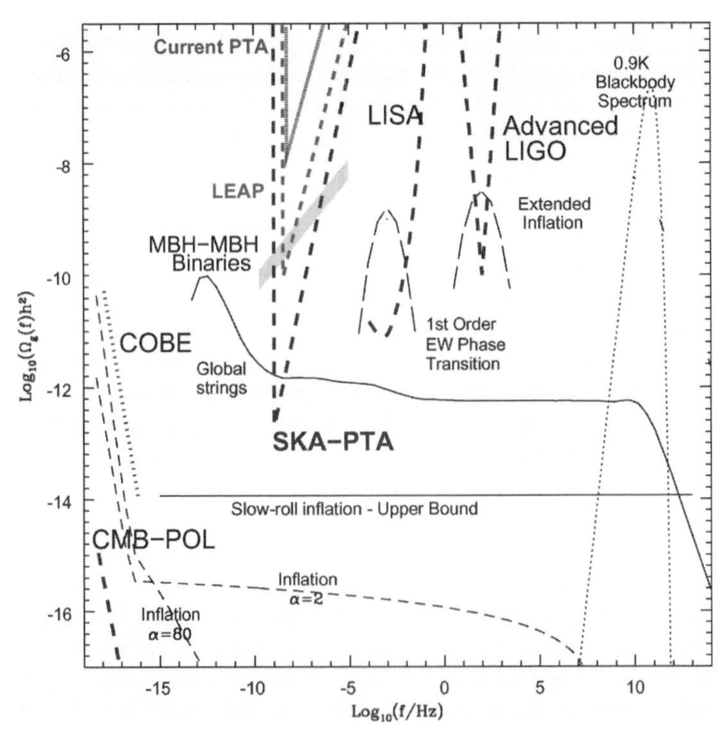

Figure 3. Summary of the potential cosmological sources of a stochastic gravitational background, including inflationary models, first-order phase transitions and cosmic strings and a primordial 0.9K black-body gravitation spectrum as presented by Battye & Shellard (1997). We also overlay bounds from COBE, from current millisecond pulsar timing and the goals from CMB polarization, LISA and Advanced LIGO. The PTA provided by the SKA will improve on the current MSP limit by about four orders of magnitudes. The gray area indicates the spectrum of an additional astrophysical background caused by the merger of massive black holes (MBHs) in early galaxy formation (see text). For this background, $\Omega_{gw} \propto f^{2/3}$, whilst its amplitude depends on the MBH mass function and merger rate. The uncertainty is indicated by the size of the shaded area. The expected sensitivities for LEAP and SKA are marked.

4. Conclusions

The past has shown that large astronomical progress is achieved when new phase space is probed by advances in technology. Radio astronomy, in particular, has benefited greatly from such advances and will continue to do so in the future. In fact, we are experiencing a revolution in the way radio astronomy is conducted as our instruments allow us now to directly "digitize" our photons. This has enormous consequences, since we can greatly benefit from the continuing advances in digital electronics, telecommunication and computing. The results are jumps in achievable bandwidths, FoVs, frequency coverage and also collecting area. The global efforts will culminate in the construction of the SKA as the world's largest and most powerful telescope. Importantly, the SKA, as a versatile observatory, will revolutionize many areas of astrophysics and fundamental physics.

References

Battye, R. A. & Shellard, E. P. S. 1996, *Class. Quantum Grav.*, 13, A239-A246

Brown, A. G. A., in American Institute of Physics Conference Series, Bailer-Jones (ed), *Learning about Galactic structure with Gaia astrometry*, American Institute of Physics Conference Series, 1082, 209

Burke, B. F. & Smith, F. G. 2002, *An introduction to Radio Astronomy*, (Cabridge: Cambridge University Press)

Carilli, C. L. & Rawlings, S. 2004, *New Astronomy Review*, 48, 979

Carilli, C. L., Furlanetto, S., Briggs, F., Jarvis, M., Rawlings, S., & Falcke, H. 2004, *New Astronomy Review*, 48, 1029

Cordes, J. M., Kramer, M., Lazio, T. J. W., Stappers, B. W., Backer, D. C., & Johnston, S. 2004, *New Astronomy Review*, 48, 1413

Damour, T. & Esposito-Farèse, G., *Phys. Rev. D*, 1998, 58(042001), 1

Detweiler, S. 1979, *ApJ*, 234, 1100

Gaensler, B. M., Beck, R., & Feretti, L. 2004, *New Astronomy Review*, 48, 1003

Garrington, S. T., 2007, in From Planets to Dark Energy: the Modern Radio Universe, *e-MERLIN*, The University of Manchester, UK. Published online at SISSA, Proceedings of Science, p.9

Harwit, M., 1981, *Cosmic discovery. The search, scope, and heritage of astronomy*, Brighton: Harvester Press, 1981

Jenet, F. A., Lommen, A., Larson, S. L., & Wen, L. 2004, *ApJ*, 606, 799

Johnston, S. *et al.*, 2008, *Experimental Astronomy*, 22, 151

Jonas, J. 2007, in *From Planets to Dark Energy: the Modern Radio Universe*, The University of Manchester, UK. Published online at SISSA, Proceedings of Science, p.7

Kramer, M. & Wex, N. 2009, *Class. Quant. Grav.*, 26(7), 073001

Kramer, M., Backer, D. C., Cordes, J. M., Lazio, T. J. W., Stappers, B. W., & Johnston, S. 2004, *New Astronomy Review*, 48, 993

Lazio, T. J. W., Tarter, J. C., & Wilner, D. J. 2004, *New Astronomy Review*, 48, 985

Lee, K. J., Jenet, F. A., & Price, R. H. 2008, *ApJ*, 685, 1304

Lorimer, D. R. & Kramer, M. 2005, *Handbook of Pulsar Astronomy*, (Cambridge: Cambridge University Press)

Lyne, A. G. *et al.*, 2004, *Science*, 303, 1153

Nan, R. 2008, in Society of Photo-Optical Instrumentation Engineers (SPIE) Conference Series, *Introduction to FAST: five hundred meter Aperture Spherical radio Telescope*, 7012

Rawlings, S., Abdalla, F. B., Bridle, S. L., Blake, C. A., Baugh, C. M., Greenhill, L. J., & van der Hulst, J. M. 2004, *New Astronomy Review*, 48, 1013

Sazhin, M. V. 1978, *Sov. Ast.*, 22, 36

Schilizzi, R. T., Dewdney, P. E. F., & Lazio, T. J. W. 2008, in Society of Photo-Optical Instrumentation Engineers (SPIE) Conference Series, *The Square Kilometre Array*, 7012

Schilizzi, R. 2007, in From Planets to Dark Energy: the Modern Radio Universe, *The Square Kilometre Array*, The University of Manchester, UK. Published online at SISSA, Proceedings of Science, p.2

Smits, R., Lorimer, D. R., Kramer, M., *et al.*, 2009, *A&A*, in press

Smits, R., Kramer, M., Stappers, B., *et al.*, 2009, *A&A*, 493, 1161

Spyromilio, J. 2007, in From Planets to Dark Energy: the Modern Radio Universe, *Extremely Large Telescopes*, The University of Manchester, UK. Published online at SISSA, Proceedings of Science, p.3

Stappers, B. W., van Leeuwen, A. G. J., Kramer, M., Stinebring, D., & Hessels, J. 2007, arXiv:astro-ph/0701229

Ulvestad, J. 2007, in From Planets to Dark Energy: the Modern Radio Universe. *The EVLA as a scientific and technical pathfinder for the SKA*, The University of Manchester, UK. Published online at SISSA, Proceedings of Science, p.8

van Haarlem, M. P. 2005, in Gurvits L. I., Frey S., Rawlings S. (eds), *LOFAR: The Low Frequency Array*", EAS Publications Series, 15, 431

Welch, J. *et al.*, 2009, arXiv:astro-ph/0904.0762

Wex, N.& Kopeikin, S. 1999, *ApJ*, 513, 388

Wilkinson, P. N., Kellermann, K. I., Ekers, R. D., Cordes, J. M., & Lazio, T. J. W. 2004, *New Astronomy Review*, 48, 1551

Relativity in Fundamental Astronomy
Proceedings IAU Symposium No. 261, 2009
S. A. Klioner, P. K. Seidelman & M. H. Soffel, eds.

doi:10.1017/S1743921309990676

Space clocks to test relativiy: ACES and SAGAS

Peter Wolf[1], Christophe Salomon[2] and Serge Reynaud[3]

[1] LNE-SYRTE, Observatoire de Paris, CNRS, UPMC
61, avenue de l'Observatoire, 75014 Paris, France
[2] Laboratoire Kastler Brossel, ENS, UPMC, CNRS
Département de Physique de l'Ecole Normale Supérieure, 75231 Paris, France
[3] Laboratoire Kastler Brossel, ENS, UPMC, CNRS
Case 74, Campus Jussieu, 75252 Paris, France

Abstract. Atomic clocks are an outstanding tool for the experimental verification of general relativity and more generally for fundamental astronomy (VLBI, pulsar timing, navigation, etc). Recent years have seen a rapid improvement in the performance of such clocks, promising new improved tests of relativity, in particular onboard terrestrial and interplanetary space missions. We present the scientific motivations of such tests taking the ACES Salomon *et al.* (2001) and SAGAS Wolf *et al.* (2009) projects as particular examples.

Keywords. tests of general relativity, atomic clocks, fundamental physics in space

1. Introduction

Gravitation is one of the fundamental interactions of physics. It is well described by General Relativity in the macroscopic world, where the gravitational interaction plays a dominant role. However General Relativity remains a classical theory incompatible with the standard techniques of quantization. Unification models lead to modifications of the theory that have observable consequences. They can become apparent, for example, in violations of Lorentz invariance, of the equivalence principle, of the distance dependence of gravitation, or in the variation of the fundamental constants. Additionally, general relativity is challenged by observations at larger galactic and cosmic scales which are presently taken care of through the introduction of "dark matter" or "dark energy". As long as these components are neither detected through non gravitational means, nor explained as resulting from new physical phenomena, it remains of the uttermost importance to test general relativity at all experimentally accessible scales.

Experimental tests of fundamental physics, with continuously improving precision and in constantly enlarged distance domains, are one of the privileged methods we dispose of to try and uncover the first indications of the "new physics" expected beyond the standard model and general relativity. In this respect, experiments in the solar system provide information of utmost interest, not only because of their excellent precision, but also because of the a priori easier control of the real environment of the experiments. Thus space becomes an ideal laboratory for experiments in fundamental physics.

In this paper, we recall the basic elements of general relativity and briefly review the experimental evidences supporting it. We then discuss perspectives for new tests which are based on recent technological developments as well as new ideas. We focus our attention on tests performed with atomic clocks in particular the ACES Salomon *et al.* (2001) and SAGAS Wolf *et al.* (2009) projects.

2. Tests of general relativity

General relativity (GR) is built up on two basic ideas which have to be distinguished. The first one is the metrical (geometrical) interpretation of gravitation, which identifies the gravitational field with the space-time metric. This is the very core of GR, but it is not sufficient to fix the latter theory. In order to select GR out of the variety of metric theories of gravitation, it is necessary to fix the relation between the geometry of space-time and its matter content. In GR, this relation is given by the Einstein-Hilbert equation which was written only in 1915.

The metrical interpretation of gravity, often coined under the generic name of the "equivalence principle", is one of the most accurately verified properties of nature Will (2001). Freely falling test masses follow the geodesics of the Riemannian space-time, that is also the curves which extremize the integral

$$\Delta s \equiv \int \mathrm{d}s \quad , \quad \mathrm{d}s^2 \equiv g_{\mu\nu} \mathrm{d}x^\mu \mathrm{d}x^\nu , \tag{2.1}$$

$g_{\mu\nu}$ is the metric tensor characterizing the space-time and $\mathrm{d}x^\mu$ the displacements in this space-time. As free motions obey a geometrical definition, they are independent of the compositions of the test masses. The potential violations of this "universality of free fall" property are parametrized by a relative difference in the accelerations undergone by two test bodies in free fall from the same location with the same velocity. Modern experiments constrain this parameter to stay below 10^{-12}, this accuracy being attained in laboratory experiments Adelberger *et al.* (2003), Schlamminger *et al.* (2008) as well as in space tests using lunar laser ranging Williams *et al.* (1996), Williams *et al.* (2004) or planetary probe tracking Anderson *et al.* (1996). These results do not preclude the possibility of small violations of the equivalence principle and such violations are indeed predicted by unification models Damour *et al.* (2002). Large improvements of this accuracy are expected in the future thanks to the existence of dedicated space projects MICROSCOPE Touboul *et al.* (2001) and, on a longer term, STEP Mester *et al.* (2001).

As another consequence of the geometrical interpretation of gravity, ideal atomic clocks operating on different quantum transitions measure the same time, because it is also a geometrical quantity, namely the proper time Δs integrated along the trajectory (see eq. 2.1). This "universality of clock rates" property has also been verified with an extreme accuracy. Its potential variations are measured as a constancy of relative frequency ratios between different clocks at a level of the order of 10^{-16} per year, recently a few 10^{-17} per year Bize *et al.* (2003), Peik *et al.* (2004), Fortier *et al.* (2007), Ashby *et al.*(2007), Rosenband *et al.* (2008). These results can also be interpreted in terms of a potential "variation of fundamental constants" Flambaum (2006), thus opening a window on the "new physics" expected to lie "beyond the standard model". In this domain also, large improvements of the accuracy can be expected in the future with the space project ACES projects using optical clocks Wolf *et al.* (2009), Schiller *et al.* (2009).

As already stated, GR is selected out of the large family of metric theories of gravity by the Einstein-Hilbert equation which fixes the coupling between curvature on one hand, matter on the other one, by setting the Einstein curvature tensor $G_{\mu\nu}$ to be proportional to the stress tensor $T_{\mu\nu}$

$$G_{\mu\nu} \equiv R_{\mu\nu} - \frac{1}{2} g_{\mu\nu} R = \frac{8\pi G}{c^4} T_{\mu\nu} . \tag{2.2}$$

The coefficient is fixed by the Newtonian limit and determined by the Newton constant G and the velocity of light c. At this point, it is worth emphasizing that it is not possible to deduce the relation (2.2) only from the geometrical interpretation of gravity. In other words, GR is one member of the family of metric theories of gravity which has to be selected out of this family by comparing its predictions to the results of observations or experiments. The tests performed in the solar system effectively show a good agreement with GR, as shown in particular by the so-called "parametrized post-Newtonian" (PPN) approach Will (2001).

The main idea of the PPN approach can be described by writing down the solution of GR with a simple model of the solar system where the gravity sources are reduced to the Sun treated as a point-like motion-less mass M (but the formalism can be generalized to the realistic N-body case). Using a specific gauge convention where spatial coordinates are isotropic, the metric element thus takes the following form (coordinates are $x^0 \equiv ct$, the radius r, the colatitude and azimuth angles θ and φ)

$$\mathrm{d}s^2 = g_{00}c^2\mathrm{d}t^2 + g_{rr}\left(\mathrm{d}r^2 + r^2(\mathrm{d}\theta^2 + \sin^2\theta\mathrm{d}\varphi^2)\right). \tag{2.3}$$

With this simple model, an exact solution can be found for the metric. It is convenient to write it as a power series expansion in the reduced Newton potential ϕ (the length scale GM/c^2 has a value of the order of 1.5km, with M the mass of the Sun, so that ϕ is much smaller than unity everywhere in the solar system)

$$g_{00} = 1 + 2\phi + 2\phi^2 + \dots, \quad g_{rr} = -1 + 2\phi + \dots, \quad \phi \equiv -\frac{GM}{rc^2}. \tag{2.4}$$

The family of PPN metrics can then be introduced by inserting a constant β in front of ϕ^2 in g_{00} and a constant γ in front of ϕ in g_{rr} (with $\beta = \gamma = 1$ in GR). The values of these PPN parameters affect the predicted motions, and can therefore be confronted to the observations. Experiments on the propagation of light have led to more and more stringent bounds on $|\gamma - 1|$, with the best current results corresponding to deflection measurements using VLBI astrometry Shapiro *et al.* (2004) and Doppler tracking of the Cassini probe during its 2002 solar occultation Bertotti *et al.* (2003). Meanwhile, analysis of lunar laser ranging data Williams *et al.* (2004) have led to bounds on linear superpositions of β and γ, and then in constraints on $|\beta - 1|$. The current status of these tests clearly favors GR as the best description of gravity in the solar system.

Let us emphasize that the common presentation of this status under the form "general relativity is confirmed by the tests" is a bit too loose. The tests discussed so far do not answer by a final "yes" answer to a mere "yes/no" question. They rather select a vicinity of GR as the best current description of gravity within the family of PPN metrics. This warning is not a mere precaution, it is rather a pointer to possible future progress in the domain. There indeed exist theoretical models (see for example Damour *et al.* (2002)) which deviate from GR while staying within the current observational bounds. Furthermore, as discussed below, extensions of GR do not necessarily belong to the PPN family.

This last point is in particular emphasized by the so-called "fifth force" tests which are focused on a possible scale-dependent deviation from the gravity force law Fischbach (1998). Their main idea is to check the $r-$dependence of the gravity potential, that is also of the component g_{00}. Hypothetical modifications of its standard expression, predicted by unification models, are often parametrized in terms of a Yukawa potential added to the standard g_{00}. This potential depends on two parameters, an amplitude α measured with respect to Newton potential and a range λ related through a Yukawa-like relation to the mass scale of the hypothetical new particle which would mediate the "fifth force". A recent update of the status of such a fifth force is shown on Fig. 1 in Jaekel (2005). It shows that the Yukawa term is excluded with a high accuracy at ranges tested in lunar laser ranging Williams *et al.* (1996) and tracking of martian probes Anderson *et al.*

(1996). At the same time, it also makes it clear that windows remain open for large corrections ($\alpha > 1$) at short ranges as well as long ranges.

The short range window is being actively explored, with laboratory experiments reaching an impressive accuracy at smaller and smaller distances. In the long range window, a test of the gravity force was initiated by NASA as the extension of Pioneer 10/11 missions after their primary planetary objectives had been met. This led to the largest scaled test of gravity ever carried out, with the striking output of a signal that disagreed with the expectations from the physical models (thermal radiation, solar pressure, signal propagation, etc...) and the known laws of gravity Anderson *et al.* (1998), Anderson *et al.* (2002).

This so-called "Pioneer anomaly" was recorded on Doppler tracking data of the Pioneer 10 & 11 probes by the NASA deep space network. It is thus a result of radionavigation techniques, which are based on the performances of the accurate reference clocks located at reception and emission stations Asmar *et al.* (2005). The Doppler observable can equivalently be interpreted as a relative velocity of the probe with respect to the station, which contains not only the effect of motion but also relativistic and gravitational effects. The anomaly has been registered on the two deep space probes showing the best navigation accuracy. This is not an impressive statistics when compared to the large number of tests confirming GR. In particular, when the possibility of an artefact onboard the probe is considered, this artefact could be the same on the two probes. A number of mechanisms have been considered as attempts of explanations of the anomaly as a systematic effect generated by the spacecraft itself or its environment (see the references in Nieto *et al.* (2005)) but they have not led to a satisfactory understanding to date.

The Pioneer anomaly constitutes an intriguing piece of information in a context where the status of gravity theory is challenged by the puzzles of dark matter and dark energy. If confirmed, this signal might reveal an anomalous behavior of gravity at scales of the order of the size of the solar system and thus have a strong impact on fundamental physics, solar system physics, astrophysics and cosmology. It is therefore important to use as many investigation techniques as might be available for gaining new information.

The Pioneer data which have shown an anomaly have been re-analyzed by several independent groups which have confirmed the presence of the anomaly Markwardt (2002), Olsen (2007), Levy *et al.* (2009). As data covering the whole period of Pioneer 10 & 11 missions from launch to the last data point have been saved, it is worth investigating not only the Doppler tracking data, but also the telemetry data Turyshev *et al.* (2006). These efforts should lead to an improved control of systematics and produce new information of importance on several properties of the effect, for example its direction, long-term variation as well as annual or diurnal modulations, spin dependence, a question of particular interest being that of the onset of the anomaly.

In the meantime, new missions have been designed to further test GR eg. Touboul *et al.* (2001), Salomon *et al.* (2001), Christophe *et al.* (2009), Wolf *et al.* (2009). In the following, we will focus our attention on atomic clocks, as used in the projects ACES Salomon *et al.* (2001) and SAGAS Wolf *et al.* (2009). We will also discuss their applications, not only for tests in fundamental physics, but also for other important purposes such as earth sciences, solar system physics or navigation.

3. Atomic Clock Ensemble in Space (ACES)

The measurement of time intervals has experienced spectacular progress over the last centuries. In the middle of the 20th century, the invention of the quartz oscillator and the first atomic clocks opened a new era for time keeping, with inacuracies improving from 10 microseconds per day in 1967 to 100 picoseconds per day for the primary cesium atomic clocks using laser cooled atomic fountains. The best atomic fountains approach 10 picosecond error per day, i.e a frequency stability of 1 part in 10^{16} Bize *et al.* (2005) while the most recent atomic clocks, operating in the optical domain, reach 2 picoseconds per day Rosenband *et al.* (2008), Ludlow *et al.* (2008) and improve at a fast pace.

Because time intervals and frequencies can be measured so precisely, applications of atomic clocks are numerous and diverse. Most precision measurements and units of the SI system can be traced back to frequencies. With the redefinition of the meter 25 years ago and the choice of a conventional value for the speed of light in vacuum, distance measurements have been simply translated into time interval measurements. In other words, the development of quantum technologies has led to large progress in the investigation of our spatio-temporal environment. This is illustrated by the impressive improvement of the measurement of the Einstein redshift effect Vessot *et al.* (1980), Pound (1999), which will be pushed further by the ACES project (see the discussion below). This is also made clear by the measurements of distances in the solar system, with the astronomical unit now connected to atomic units and consequently known with a much better accuracy than previously Shapiro (1999).

The most visible consequence of these ideas is given by the Global Navigation Satellite Systems (GNSS), which are a spectacular and initially unexpected application of the resources of quantum physics (the precision of the system comes from the atomic clocks in the satellites) and general relativity. The US Global Positioning System (GPS) enables today any user with a small receiver or modern cell phone to locate its position on the globe with meter accuracy. In a few years from now, the GALILEO system will be the european contribution to GNSS and its vast uses for various applications, in particular in the scientific domain in geodesy, Earth monitoring, and time metrology.

In the following, we review some of the properties of space clocks in the context of the ACES mission (Atomic Clock Ensemble in Space). ACES is an ESA mission in fundamental physics Salomon *et al.* (2001), Cacciapuoti *et al.* (2007), Salomon *et al.* 2007. ACES aims at flying a new generation of atomic clocks onboard the International Space Station (ISS) and comparing them to a network of ultra-stable clocks on the ground. The ISS is orbiting at a mean elevation of 400 km with 90 min. of rotation period and an inclination angle of 51.6°. ACES will be transported on orbit by the Japanese transfer vehicle HTV, and installed at the external payload facility of the Columbus module using the ISS robotic arm.

The ACES payload accommodates two atomic clocks: PHARAO, a primary frequency standard based on samples of laser cooled cesium atoms, and the active hydrogen maser SHM. The performances of the two clocks are combined to generate an on-board timescale with the short-term stability of SHM and the long-term stability and accuracy of the cesium clock PHARAO. A GNSS receiver installed on the ACES payload and connected to the on-board time scale will provide precise orbit determination of the ACES clocks. The ACES clock signal will be transferred on ground by a time and frequency transfer link in the microwave domain (MWL). MWL compares the ACES frequency reference to a set of ground clocks, enabling fundamental physics tests and applications in different areas of research.

The planned mission duration is 18 months with a possible extension to 3 years. The launch is planned for 2013. In microgravity, the linewidth of the atomic resonance of the PHARAO clock will be reduced by two orders of magnitude, down to sub-Hertz values (from 11 Hz to 110 mHz), 5 times narrower than in Earth based atomic fountains. After clock optimization, performances in the 10^{-16} range are expected both for frequency instability and inaccuracy. Then the on-board clocks will be compared to a network of ground atomic clocks operating both in the microwave and optical domain.

The scientific objectives of the ACES mission cover a wide spectrum. The frequency comparisons between the space clocks and ground clocks will be used to test general relativity to high accuracy. As recalled above identical clocks located at different positions in gravitational fields experience a frequency shift that depends directly on the component g_{00} of the metric, i.e to the Newtonian potential at the clock position. The comparison between the ACES onboard clocks and ground-based atomic clocks will measure the frequency variation due to the gravitational red-shift with a 70-fold improvement on the best previous experiment Vessot *et al.* (1980), testing the Einstein prediction at the 2 ppm uncertainty level. Time variations of fundamental constants will be measured by comparing clocks based on different transitions or different atomic species Fortier *et al.* (2007). Any transition energy in an atom or a molecule can be expressed in terms of the fine structure constant α and the two dimensionless constants m_q/Λ_{QCD} and m_e/Λ_{QCD}, involving the quark mass m_q, the electron mass m_e and the QCD mass scale Λ_{QCD} Flambaum (2006). ACES will perform crossed comparisons of ground clocks both in the microwave and in the optical domain with a resolution of 10^{-17} after a few days of integration time. These comparisons will impose strong and unambiguous constraints on time variations of fundamental constants reaching an uncertainty of 10^{-17}/year in case of a 1-year mission duration and $3 \cdot 10^{-18}$/year after three years. ACES will also perform tests of the Local Lorentz Invariance (LLI), according to which the outcome of any local test experiment is independent of the velocity of the freely falling apparatus. In 1997, LLI tests based on comparing clocks on-board GPS satellites to ground hydrogen masers Wolf and Petit (2007) have been performed. ACES will perform a similar experiment with 10-fold improved uncertainty.

Developed by CNES, the cold atom clock PHARAO will combine laser cooling techniques and micro-gravity conditions to significantly increase the interaction time and consequently reduce the linewidth of the clock transition. Improved stability and better control of systematic effects will be demonstrated in the space environment. PHARAO will reach a fractional frequency instability of $1 \cdot 10^{-13} \cdot \tau^{-1/2}$, where τ is the integration time expressed in seconds, and an inaccuracy of a few parts in 10^{16}. The engineering model of the PHARAO clock has been completed and is presently under test at CNES premises in Toulouse. Design and first results are presented in Laurent *et al.* 2006.

Developed by SPECTRATIME under ESA coordination, SHM provides ACES with a stable fly-wheel oscillator. The main challenge of SHM is represented by the low mass and volume figures required by the space clock with respect to ground H-masers. SHM will provide a clock signal with fractional frequency instability down to $1.5 \cdot 10^{-15}$ after only 10^4 s of integration time. Two servo-loops will lock together the clock signals of PHARAO and SHM generating an on-board time scale combining the short-term stability of the H-maser with the long-term stability and accuracy of the cesium clock. The Frequency Comparison and Distribution Package (FCDP) is the central node of the ACES payload. Developed by ASTRIUM and TIMETECH under ESA coordination, FCDP is the on-board hardware which compares the signals delivered by the two space clocks, measures and optimizes the performances of the ACES frequency reference, and finally distributes it to the ACES microwave link MWL.

With the microwave link MWL, frequency transfer with time deviation better than 0.3 ps at 300 s, 7 ps at 1 day, and 23 ps at 10 days of integration time will be demonstrated. The gravitational shift measurement requires fully relativistic modeling and precise orbit determination and knowledge of the Earth gravitational potential as described in Duchayne *et al.* 2009. MWL is developed by ASTRIUM, KAYSER-THREDE and TIMETECH under ESA coordination. The proposed MWL concept is an upgraded version of the Vessot two-way technique used for the GP-A experiment Vessot *et al.* (1980) and the PRARE geodesy instrument.

The frequency resolution that the ACES microwave link should reach in operational conditions is surpassing by one to two orders of magnitude the existing satellite time transfer comparison methods based on the GPS system and TWSTFT (Two Way Satellite Time and Frequency Transfer) Bauch *et al.* (2006). It may be compared with T2L2 (Time Transfer by Laser Link) Exertier *et al.* (2008) which is now flying onboard the JASON-2 satellite and taking science data since June 2008. The assessment of its time transfer capability is presently ongoing with the interesting prospect to reach a time resolution of $\simeq 10\,\mathrm{ps}$ after one day of averaging for clock comparisons in common view. For the ACES mission, due to the low orbit of the ISS and the limited duration of each pass (300-400 s) over a given ground station, the common view technique will be suitable for comparing ground clocks over continental distances, for instance within Europe or USA or Japan. In a common view comparison, the frequency noise of the onboard clock cancels out to a large degree so that only the link instability remains. ACES mission will also demonstrate the capability to compare ground clocks in non common view with a resolution better than $10^{-13}\Delta t^{1/2}$ for $\Delta t > 1000\,\mathrm{s}$, that is 3 ps and 10 ps for space-ground comparisons separated by 1000 s and 10000 s respectively. This science objective takes full benefit of the excellent onboard time scale realized by the combination of SHM and PHARAO.

These performances will enable common view and non-common view comparisons of ground clocks with 10^{-17} frequency resolution after few days of integration time. ACES will take full advantage of the recent progress of optical clocks Rosenband *et al.* (2008), Ludlow *et al.* (2008), reaching today instability and inaccuracy levels of 2 parts in 10^{17}. Multiple frequency comparisons between a variety of advanced ground clocks will be possible among the 35 institutes which have manifested their interest to participate to the ACES mission. This is important for the tests of the variability of fundamental physical constants, and international comparisons of time scales.

Other applications of the ACES clock signal are currently being developed. ACES will demonstrate a new "relativistic geodesy" based on a differential measurement of the Einstein's gravitational red-shift between distant ground clocks. It will take advantage of the accuracy of ground-based optical clocks to resolve differences in the Earth gravitational potential at the $\simeq 10$ cm level. A 10 cm change in elevation on the ground amounts to a frequency shift of 1 part in 10^{17}, comparable to today's optical clock accuracy of 2 parts in 10^{17}. The GNSS receiver on-board the ACES payload will allow to monitor the GNSS networks and develop interesting applications in Earth remote sensing. This includes studies of the oceans surface via GNSS reflectometry measurements and the analysis of the Earth atmosphere with GNSS radio-occultation experiments.

In addition, ACES will deliver a global atomic time scale with 10^{-16} accuracy, it will allow clocks synchronization at an uncertainty level of 100 ps, and contribute to international atomic time scales. At the 10^{-18} level, ground clocks will be limited by the fluctuations of the Earth potential suggesting that future high precision time references will have to be placed in the space environment where these fluctuations are significantly reduced. From this point of view, ACES is the pioneer of new concepts for global time keeping and positioning based on a reduced set of ultra-stable space clocks in orbit.

4. SAGAS

The SAGAS mission will study all aspects of large scale gravitational phenomena in the Solar System using quantum technology, with science objectives in fundamental physics and Solar System exploration. The large spectrum of science objectives makes SAGAS a unique combination of exploration and science, with a strong basis in both programs. The involved large distances (up to 53 AU) and corresponding large variations of gravitational potential combined with the high sensitivity of SAGAS instruments serve both purposes equally well. For this reason, SAGAS brings together traditionally distant scientific communities ranging from atomic physics through experimental gravitation to planetology and Solar System science.

The payload will include an optical atomic clock optimized for long term performance, an absolute accelerometer based on atom interferometry and a laser link for ranging, frequency comparison and communication. The complementary instruments will allow highly sensitive measurements of all aspects of gravitation via the different effects of gravity on clocks, light, and the free fall of test bodies, thus effectively providing a detailed gravitational map of the outer Solar System whilst testing all aspects of gravitation theory to unprecedented levels.

The SAGAS accelerometer is based on cold Cs atom technology derived to a large extent from the PHARAO space clock built for the ACES mission discussed in the preceding section. The PHARAO engineering model has recently been tested with success, demonstrating the expected performance and robustness of the technology. The accelerometer will only require parts of PHARAO (cooling and trapping region) thereby significantly reducing mass and power requirements. The expected sensitivity of the accelerometer is $1.3 \cdot 10^{-9}\mathrm{m/s^2}/\sqrt{\mathrm{Hz}}$ with an absolute accuracy (bias determination) of $5 \cdot 10^{-12}\mathrm{m/s^2}$.

The SAGAS clock will be an optical clock based on trapped and laser cooled single ion technology as pioneered in numerous laboratories around the world. In the present proposal it will be based on a Sr^+ ion with a clock wavelength of 674 nm. The expected stability of the SAGAS clock is $1 \cdot 10^{-14}/\sqrt{\tau}$ (with τ the integration time), with an accuracy in realising the unperturbed ion frequency of $1 \cdot 10^{-17}$. The best optical single ion ground clocks presently show stabilities below $4 \cdot 10^{-15}/\sqrt{\tau}$, slightly better than the one assumed for the SAGAS clock, and only slightly worse accuracies $2 \cdot 10^{-17}$. So the technology challenges facing SAGAS are not so much the required performance, but the development of reliable and space qualified systems, with reduced mass and power consumption.

The optical link is using a high power (1 W) laser locked to the narrow and stable frequency provided by the optical clock, with coherent heterodyne detection on the ground and on board the spacecraft. It serves the multiple purposes of comparing the SAGAS clock to ground clocks, providing highly sensitive Doppler measurements for navigation and science, and allowing data transmission together with timing and coarse ranging. It is based on a 40 cm space telescope and 1.5 m ground telescopes (similar to lunar laser ranging stations). The short term performance of the link in terms of frequency comparison of distant clocks will be limited by atmospheric turbulence at a few $10^{-13}/\tau$, thus reaching the clock performance after about 1000 s integration time, which is amply sufficient for the SAGAS science objectives. The main challenges of the link will be the required pointing accuracy ($0.3''$) and the availability of space qualified, robust 1 W laser sources at 674 nm. Quite generally, laser availability and reliability will be the key to achieving the required technological performances, for the clock as well as the optical link.

For this reason a number of different options have been considered for the clock/link laser wavelength, with several other ions that could be equally good candidates (e.g. Yb^+ at 435 nm and Ca^+ at 729 nm). Given present laser technology, Sr^+ was preferred, but this choice could be revised depending on laser developments over the next years. We also acknowledge the possibility that femtosecond laser combs might be developed for space applications in the near future, which would open up the option of using either ion with existing space qualified 1064 nm Nd:YAG lasers for the link.

More generally, SAGAS technology takes advantage of the important heritage from cold atom technology used in PHARAO and laser link technology designed for LISA (Laser Interferometric Space Antenna). It will provide an excellent opportunity to develop those technologies for general use, including development of the ground segment (Deep Space Network telescopes and optical clocks), that will allow such technologies to be used in many other mission configurations for precise timing, navigation and broadband data transfer throughout the Solar System.

SAGAS will carry out a large number of tests of fundamental physics, and gravitation in particular, at scales only attainable in a deep space experiment. The unique combination of onboard instruments will allow 2 to 5 orders of magnitude improvement on many tests of special and general relativity, as well as a detailed exploration of a possible anomalous scale dependence of gravitation. It will also provide detailed information on the Kuiper belt mass distribution and determine the mass of Kuiper belt objects and possibly discover new ones. During the transits, the mass and mass distribution of the Jupiter system will be measured with unprecedented accuracy. The science objectives are discussed in the following, based on estimated measurement uncertainties of the different observables.

SAGAS will provide three fundamental measurements: the accelerometer readout and the two frequency differences (measured on ground and on board the satellite) between the incoming laser signal and the local optical clock. Auxiliary measurements are the timing of emitted/received signals on board and on the ground, which are used for ranging and time tagging of data. The high precision science observables will be deduced from the fundamental measurements by combining the measurements to obtain information on either the frequency difference between the clocks or the Doppler shift of the transmitted signals. The latter gives access to the relative satellite-ground velocity, from which the gravitational trajectory of the satellite can be deduced by correcting non-gravitational accelerations using the accelerometer readings.

In the following, we assume that Earth station motion and its local gravitational potential can be known and corrected to uncertainty levels below 10^{-17} in relative frequency (10 cm on geocentric distance), which, although challenging, are within present capabilities. For the Solar System parameters this requires 10^{-9} relative uncertainty for the ground clock parameters (GM and r of Earth), also achieved at present, and less stringent requirements for the satellite.

For long term integration and the determination of an acceleration bias, the limiting factor will then be the accelerometer noise and absolute uncertainty (bias determination). More generally, modelling of non-gravitational accelerations will certainly allow some improvement on the long term limits imposed by the accelerometer noise and absolute uncertainty, but is not taken into account in our preliminary evaluation Wolf *et al.* (2009). Also, depending on the science goal and corresponding signal, the ground and on-board data can be combined in a way to optimise the S/N ratio (see Reynaud *et al.* (2008) for details). For example, in the Doppler observable, the contribution of one of the clocks (ground or space) can be made negligible by combining signals such that one has coincidence of the "up" and "down" signals on board or on the ground.

We will use a mission profile with a nominal mission lifetime of 15 years and the possibility of an extended mission to 20 years if instrument performance and operation allow this. In that time frame, the trajectory allows the satellite to reach a heliocentric distance of 39 AU in nominal mission and 53 AU with extended duration.

In General Relativity (GR), the frequency difference of two ideal clocks is given by (see eq.(2.1) to first order in the weak field approximation, with $g_{00} \simeq 1 - 2w/c^2$ and $g_{rr} \simeq -1$; w is the Newtonian gravitational potential and v the coordinate velocity)

$$\frac{ds_{\rm g}}{cdt} - \frac{ds_{\rm s}}{cdt} \simeq \frac{w_{\rm g} - w_{\rm s}}{c^2} + \frac{v_{\rm g}^2 - v_{\rm s}^2}{2c^2}. \tag{4.1}$$

In theories different from GR this relation is modified, leading to different time and space dependence of the frequency difference. This can be tested by comparing two clocks at distant locations (different values of w and v) via exchange of an electromagnetic signal. The SAGAS trajectory (large potential difference) and low uncertainty on the observable (4.1) allows a relative uncertainty on the redshift determination given by the 10^{-17} clock bias divided by the maximum value of $\Delta w/c^2$. For a distance of 50 AU this corresponds to a test with a relative uncertainty of $1 \cdot 10^{-9}$, an improvement by almost 5 orders of magnitude on the uncertainty obtained by the most sensitive experiment at present Vessot *et al.* (1980).

Additionally, the mission also provides the possibility of testing the velocity term in (4.1), which amounts to a test of Special Relativity (Ives-Stilwell test), and thus of Lorentz invariance. Towards the end of the nominal mission, this term is about $4 \cdot 10^{-9}$, and can therefore be measured by SAGAS with $3 \cdot 10^{-9}$ relative uncertainty. The best present limit on this type of test is $2 \cdot 10^{-7}$ Saathoff *et al.* (2003), so SAGAS will allow an improvement by a factor $\sim$70. Considering a particular preferred frame, usually taken as the frame in which the 3K cosmic background radiation is isotropic, one can set an even more stringent limit. In that case a putative effect will be proportional to $(\mathbf{v}_{\rm s} - \mathbf{v}_{\rm g}).\mathbf{v}_{\rm Sun}/c^2$ (see Saathoff *et al.* (2003)), where $\mathbf{v}_{\rm s}$ and $\mathbf{v}_{\rm g}$ are the velocity vectors of the satellite and ground while $\mathbf{v}_{\rm Sun}$ is the velocity of the Sun through the CMB frame ($\sim$350 km/s). Then SAGAS will allow a measurement with about $5 \cdot 10^{-11}$ relative uncertainty, which corresponds to more than 3 orders of magnitude improvement on the present limit.

Spatial and/or temporal variations of fundamental constants constitute another violation of Local Position Invariance and thus of GR. Over the past few years, there has been great interest in that possibility (see e.g. Uzan (2003) for a review), spurred on the one hand by models for unification theories of the fundamental interactions where such variations appear quite naturally, and on the other hand by recent observational claims of a variation of different constants over cosmological timescales Murphy *et al.* (2003), Reinhold *et al.* (2006). Such variations can be searched for with atomic clocks, as the involved transition frequencies depend on combinations of fundamental constants and in particular, for the optical transition of the SAGAS clock, on the fine structure constant α. Such tests take two forms: searches for a drift in time of fundamental constants, or for a variation of fundamental constants with ambient gravitational field. The latter tests for a non-universal coupling between ambient gravity and non-gravitational interactions (clearly excluded by GR) and is well measured by SAGAS, because of the large change in gravitational potential during the mission.

SAGAS also offers a possibility to improve the test of the PPN family of metric theories of gravitation Will (2001). The two most common parameters of the PPN framework are the Eddington parameters β and γ. Present limits on γ are obtained from measurements on light propagation such as light deflection, Shapiro delay and Doppler velocimetry of the Cassini probe during the 2002 solar occultation Bertotti *et al.* (2003). SAGAS will carry out similar measurements during solar conjunctions, with improved sensitivity and at optical rather than radio frequencies, which significantly minimizes errors due to the solar corona and the Earth's ionosphere. When combining the on board and ground measurements such that the "up" and "down" signals coincide at the satellite (classical Doppler type measurement), the noise from the on-board clock cancels to a large extent and one is left with noise from the accelerometer, the ground clock, and the atmosphere. Details on these noise sources, and on the effects of atmospheric turbulence, variations in temperature, pressure and humidity are presented in Wolf *et al.* (2009). The resulting improvement on the estimation of γ is of the order of 100, with some potential for further improvement if several occultations may be analysed.

As already discussed, experimental tests of gravity have shown a good overall agreement with GR, but most theoretical models aimed at inserting GR within the quantum framework predict observable modifications at large scales. The anomalies observed on the Pioneer probes, as well as the phenomena commonly ascribed to "dark matter" and "dark energy", suggest that it is extremely important to test the laws of gravity at interplanetary distances. This situation has motivated an interest in flying new probes to the distances where the anomaly was first discovered, that is beyond the Saturn orbit, and studying gravity with the largely improved spatial techniques which are available today.

SAGAS has the capability to improve our knowledge of the law of gravity at the scale of the solar system, and to confirm or infirm the presence of a Pioneer anomaly (PA). With one year of integration, all SAGAS observables allow a measurement of any effect of the size of the PA with a relative uncertainty of better than 1%. This will allow a "mapping" of any anomalous scale dependence over the mission duration and corresponding distances. Furthermore, the complementary observables available on SAGAS allow a good discrimination between different hypotheses thereby not only measuring a putative effect but also allowing an identification of its origin. SAGAS thus offers the possibility to constrain a significant number of theoretical approaches to scale dependent modifications of GR. Given the complementary observables available on SAGAS the obtained measurements will provide a rich testing ground for such theories with the potential for major discoveries.

Let us emphasize at this point that this gravity law test is important not only for fundamental physics but also for solar system science. The Newton law of gravity is indeed a key ingredient of the models devoted to a better understanding of the origin of the solar system. The Kuiper belt (KB) is a collection of masses, remnant of the circumsolar disk where giant planets of the solar system formed 4.6 billion years ago. Precise measurements of its mass distribution would significantly improve our understanding of planet formation not only in the solar system but also in recently discovered planetary systems. The exceptional sensitivity and versatility of SAGAS for measuring gravity can be used to study the sources of gravitational fields in the outer solar system, and in particular the class of Trans Neptunian Objects (TNOs), of which those situated in the KB have been the subject of intense interest and study over the last years Morbidelli (2007). Observation of KB objects (KBOs) from the Earth is difficult due to their relatively small size and large distance, and estimates of their masses and distribution are accordingly inaccurate. Estimates of the total KB mass from the discovered objects ($\sim$ 1000 KBOs) range from 0.01 to 0.1 Earth masses, whereas in-situ formation of the observed KBOs would require three orders of magnitude more solid material in a dynamically cold disk.

A dedicated probe like SAGAS will help discriminating different models of the spatial distribution of the KB and for determining its total mass. The relative frequency shift between the ground and space clock due to the KB gravitational potential is indeed a sensitive probe of the spatial distribution of this mass (see Wolf *et al.* (2009) for details). The SAGAS frequency observable is well suited to study the large, diffuse, statistical mass distribution of KBOs essentially due to its sensitivity directly to the gravitational potential (1/r dependence), rather than the acceleration ($1/\mathrm{r}^2$ dependence). The large diffuse signal masks any signal from individual KBOs. When closely approaching one of the objects, the crossover between acceleration sensitivity (given by the $5 \cdot 10^{-12}$ m/s^2 uncertainty on non gravitational acceleration) and the frequency sensitivity (10^{-17} uncertainty on w/c^2) for an individual object is situated at about 1.2 AU. Below that distance the acceleration measurement is more sensitive than the frequency measurement. This suggests a procedure to study individual objects using the SAGAS observables: use the satellite trajectory (corrected for the non gravitational acceleration) to study the gravity from a close object and subtract the diffuse background from all other KBOs using the frequency measurement. Investigating several known KBOs within the reach of SAGAS Bernstein *et al.* (2004) shows that their masses can be determined at the % level when approaching any of them to 0.2 AU or less. Of course, this also opens the way towards the discovery of new such objects, too small to be visible from the Earth. Similarly, during a planetary flyby the trajectory determination (corrected for the non gravitational acceleration) will allow the determination of the gravitational potential of the planetary system. The planned Jupiter flyby with a closest approach of $\sim$ 600000 km will improve present knowledge of Jovian gravity by more than two orders of magnitude.

Doppler ranging to deep space missions provides the best upper limits available at present on gravitational waves (GW) with frequencies of order c/L where L is the spacecraft to ground distance i.e. in the 0.01 to 1 mHz range Armstrong *et al.* (1987), Armstrong *et al.* (2003), and even down to 1 μHz, albeit with lower sensitivity Anderson & Mashoon (1985), Armstrong *et al.* (2003). The corresponding limits on GW are determined by the noise PSD of the Doppler ranging to the spacecraft for stochastic GW backgrounds, filtered by the bandwidth of the observations when looking for GW with known signatures. In the case of SAGAS data, this yields a strain sensitivity of $10^{-14}/\sqrt{\text{Hz}}$ for stochastic sources in the frequency range of 0.06 to 1 mHz with a f^{-1} increase at low frequency due to the accelerometer noise. When searching for GW with particular signatures in this frequency region, optimal filtering using a corresponding GW template will allow reaching strain sensitivities as low as 10^{-18} with one year of data. This will improve on best present upper limits on GW in this frequency range by about four orders of magnitude. Even if GW with sufficiently large amplitudes are not found, still the results might serve as upper bounds for astrophysical models of known GW sources Reynaud *et al.* (2008).

References

Adelberger, E. G., Heckel, B. R., & Nelson, A. E. 2003, *Ann. Rev. Nucl. Part. Sci.* 53, 77.
Anderson, J. D. & Mashoon, B. 1985, *Astrophys. J.* 290, 445
Anderson, J. D. *et al.*, 1996, *Astrophys. J.* 459, 365
Anderson, J. D. *et al.*, 1998, *Phys. Rev. Lett.* 81, 2858
Anderson, J. D. *et al.*, 2002, *Phys. Rev.* D65, 082004
Ashby, N. *et al.*, 2007, *Phys. Rev. Lett.* 98, 070802
Armstrong, J. W. *et al.*, 1987, *Astrophys. J.* 318, 536
Armstrong, J. W. *et al.*, 2003, *Astrophys. J.* 599, 806
Asmar, S. W. *et al.*, 2005, *Radio Science* 40, RS2001
Bauch, A. *et al.*, 2006, *Metrologia* 43, 109120
Bernstein, G. *et al.*, 2004, *Astrophys. J.* 128, 1364
Bertotti, B., Iess, L., & Tortora, P. 2003, *Nature* 425, 374
Bize, S. *et al.*, 2003, *Phys. Rev. Lett.* 90, 150802
Bize, S. *et al.*, 2005, *J. Phys.* B38, S449
Cacciapuoti, L. *et al.*, 2007, *Nuclear Physics* B166, 303
Christophe, B. *et al.* , 2009, *Exp. Astron.* 23, 529

Damour, T., Piazza, F., & Veneziano, G. 2002, *Phys. Rev.* D66, 046007
Duchayne, L., Mercier, F., & Wolf, P., Astron & Astrop, (accepted)
Exertier, P. *et al.*, *IDS workshop*, Nice, November 08, http://ids.cls.fr/documents/report/ids workshop 2008/IDS08 s8 Exertier T2L2.pdf
Fischbach, E. & Talmadge, C., The Search for Non Newtonian Gravity (Springer Verlag, Berlin, 1998).
Flambaum, V. V. & Tedesco, A. F. 2006, *Phys. Rev.* C73, 055501
Fortier, T. M. *et al.*, 2007, *Phys. Rev. Lett.* 98, 070801
Jaekel, M.-T. & Reynaud, S. 2005, *Int. J. Mod. Phys.* A20, 2294
Laurent, P. *et al.*, 2006, *Appl. Phys.* B84, 683
Levy, A. *et al.*, 2009, *Adv. Space Res.* 43, 1538
Ludlow, A. D. *et al.*, 2008, *Science* 319, 1805
Markwardt, C., gr-qc/0208046; http://lheawww.gsfc.nasa.gov/users/craigm/atdf/.
Mester, J. *et al.*, 2001, *Class. Quantum Grav.* 18, 2475
Morbidelli, A., in The Kuiper Belt, A. Barruci *et al.* eds. (Univ. of Arizona Press, 2007).
Murphy, M. T. *et al.*, 2003, Monthly Notic. *Royal Astron. Soc.* 345, 609
Nieto, M. M., Turyshev, S. G., & Anderson, J. D. 2005, *Phys. Lett.* B613, 11
Olsen, O. 2007, *Astron. Astrop.* 463, 393
Peik, E. *et al.*, 2004, *Phys. Rev. Lett.* 93, 170801
Pound, R. V. 1999, *Rev. Mod. Phys.* 71, S54
Reinhold, E. *et al.*, 2006, *Phys. Rev. Lett.* 96, 151101
Reynaud, S. *et al.*, 2008, *Phys. Rev.* D77, 122003
Rosenband, T. *et al.*, 2008, *Science* 319, 1808
Salomon, C. *et al.*, 2001, C. R. Acad. Sci., IV-2, 1313
Saathoff, G. *et al.*, 2003, *Phys. Rev. Lett.* 91, 190403
Salomon, C., Cacciapuoti, L., & Dimarcq, N. 2007, *Int. J. Mod. Phys.* D16, 2511
Schiller, S. *et al.*, 2009, Exp. *Astron.* 23, 573
Schlamminger, S. *et al.*, 2008, *Phys. Rev. Lett.* 100, 041101
Shapiro, I. I. 1999, *Rev. Mod. Phys.* 71, S41
Shapiro, S. S. *et al.*, 2004, *Phys. Rev. Lett.* 92, 121101
Touboul, P. *et al.*, 2001, C. R. Acad. Sci. IV-2, 1271
Turyshev, S. G., Nieto, M. M., Anderson, J. D. 2006, EAS Publ.Ser. 20, 243, (2006); S.G. Turyshev *et al.* , *Int. J. Mod. Phys.* D15, 1
Uzan, J.-P. 2003, *Rev. Mod. Phys.* 75, 403
Vessot, R. F. C. *et al.*, 1980, *Phys. Rev. Lett.* 45, 2081
Will, C. M. 2001, Living Rev. Rel., 4, url: http://www.livingreviews.org/lrr-2001-4 .
Williams, J. G., Newhall, X. X., & Dickey, J. O. 1996, *Phys. Rev.* D53, 6730
Williams, J. G., Turyshev, S. G., & Boggs, D. H. 2004, *Phys. Rev. Lett.* 93, 261101
Wolf, P. & Petit, G. 1997, *Phys. Rev.* A56, 4405
Wolf, P. *et al.*, 2009, Exp. *Astr.*, 23, 651

Relativity in Fundamental Astronomy
Proceedings IAU Symposium No. 261, 2009
S. A. Klioner, P. K. Seidelman & M. H. Soffel, eds.

doi:10.1017/S1743921309990688

Testing the weak equivalence principle

Anna M. Nobili[1,2], Gian Luca Comandi[2], Raffaello Pegna[2], Donato Bramanti[1], Suresh Doravari[3], Francesco Maccarone[1,2] and David M. Lucchesi[2,4]

[1] Department of Physics "E. Fermi", University of Pisa,
Largo Bruno Pontecorvo 3, I-56127 Pisa, Italy
email: nobili@dm.unipi.it

[2]INFN Istituto Nazionale di Fisica Nucleare, Sezione di Pisa,
Largo Bruno Pontecorvo 3, I-56127 Pisa, Italy
[3]Imperial College London, South Kensington Campus, London SW7 2AZ, UK
[4]INAF-IFSI Istituto Nazionale di Fisica dello Spazio Interplanetario,
Via del Fosso del Cavaliere 100, I-00133 Roma, Italy

Abstract. The discovery of Dark Energy and the fact that only about 5% of the mass of the universe can be explained on the basis of the current laws of physics have led to a serious impasse. Based on past history, physics might indeed be on the verge of major discoveries; but the challenge is enormous. The way to tackle it is twofold. On one side, scientists try to perform large scale direct observations and measurements – mostly from space. On the other, they multiply their efforts to put to the most stringent tests ever the physical theories underlying the current view of the physical world, from the very small to the very large. On the extremely small scale very exciting results are expected from one of the most impressive experiments in the history of mankind: the Large Hadron Collider. On the very large scale, the universe is dominated by gravity and the present impasse undoubtedly calls for more powerful tests of General Relativity – the best theory of gravity to date. Experiments testing the *Weak Equivalence Principle*, on which General Relativity ultimately lies, have the strongest probing power of them all; a breakthrough in sensitivity is possible with the "Galileo Galilei" (GG) satellite experiment to fly in low Earth orbit.

Keywords. Gravitation, Large-scale structure of universe

---

1. Dark energy and its challenge

Ever since Newton's *Philosophiae Naturalis Principia Mathematica* was published in London in 1687 gravity is known to govern the physics of the cosmos. In the following two centuries, based on Newton's law of gravity, the best scientists of the time developed sophisticated mathematical tools which allowed them to predict the position of planets in the sky. By comparison with extremely accurate and systematic observations carried out at major astronomical observatories – particularly in Europe – theoretical predictions and observations were found to agree with each other amazingly, superceding Ptolomy's model which had lasted 14 hundred years. Celestial Mechanics became the paradigm of exact science, so much that the existence of Neptune could be inferred, and the planet actually observed in 1846 at the predicted position (though with a bit of luck), on the basis of its gravitational influence on the motion of Uranus which had been found by observations to deviate more and more with time from the theoretical prediction. In point of fact, the contribution from theory was crucial, while the capability to observe Neptune was already there 234 years earlier, when Galileo did indeed see Neptune (Kowal & Drake, 1980; Standish & Nobili, 1997).

Newton's theory of gravity dominated for more than 200 years, even beyond the publication of Einstein's theory of General Relativity (Einstein, 1916). Despite the consistency and beauty of this new theory of gravity, and its profound revolutionary nature with respect to Newton's theory, its observable consequences were minute and hard to measure. Even Einstein's beautiful explanation of the small additional perihelion advance of Mercury predicted by his theory – and until then *missing* in the predictions of Celestial Mechanics as based on Newton's gravity (see, e.g. Nobili & Will, 1986) – was anyway adding only a small contribution to a much larger and astonishingly good prediction of the effects of planetary perturbations to the motion of the perihelion of Mercury.

On a larger scale, since 1929 it became apparent that the universe is expanding. By comparing measurements of velocities (by means of redshifts) and measurements of distances (using Cepheids as standard candles) Edwin Hubble proved that the universe is actually expanding. Gravity would slow down that expansion, and so the question was to establish whether the density of matter is sufficient to "close" the universe or else it will keep expanding forever. About 10 years ago, two teams of astronomers found that there is too little matter in the universe to stop its expansion and, moreover, that the outward motion is indeed speeding up. The conclusion was based on more than 20 years measurements of the distance of extremely far away galaxies using very bright supernovae as standard candles (the Cepheids being too dim at such distances). The discovery was named by Science journal "Breakthrough of the Year for 1998" in Astronomy (see Glanz, 1998 and references therein)

A new, unknown, form of energy – the so called "dark energy" – is required.

Though some skepticism about dark energy remains among scientists, it is quite remarkable that completely different astronomical measurements – namely those of the cosmic microwave background anisotropy performed by BOOMERanG (de Bernardis *et al.*, 2000) and WMAP (Bennet *et al.*, 2003) – lead to the same conclusion. In the future, a dedicated space survey as proposed with the ESA mission EUCLID should provide the scientific community with considerable new insights.

In February 2005 a Dark Energy Task Force (DETF) was established in the US by the Astronomy and Astrophysics Advisory Committee and the High Energy Physics Advisory Panel to advise the Department of Energy, NASA and the National Science Foundation on future dark energy research.

In 2006 DETF published its final report (Report of the Dark Energy Task Force, 2006). The cover page reads:

"Dark energy appears to be the dominant component of the physical Universe, yet there is no persuasive theoretical explanation for its existence or magnitude. The acceleration of the Universe is, along with dark matter, the observed phenomenon that most directly demonstrates that our theories of fundamental particles and gravity are either incorrect or incomplete. Most experts believe that nothing short of a revolution in our understanding of fundamental physics will be required to achieve a full understanding of the cosmic acceleration. For these reasons, the nature of dark energy ranks among the very most compelling of all outstanding problems in physical science. These circumstances demand an ambitious observational program to determine the dark energy properties as well as possible."

And the Executive Summary of the Report begins as follows:

"Over the last several years scientists have accumulated conclusive evidence that the Universe is expanding ever more rapidly. Within the framework of the standard cosmological model, this implies that 70% of the universe is composed of a new, mysterious dark energy, which unlike any known form of matter or energy, counters the attractive force of gravity. Dark energy ranks as one of the most important discoveries in cosmology, with

profound implications for astronomy, high-energy theory, general relativity, and string theory.

One possible explanation for dark energy may be Einstein's famous cosmological constant. Alternatively, dark energy may be an exotic form of matter called quintessence, or the acceleration of the Universe may even signify the breakdown of Einstein's Theory of General Relativity. With any of these options, there are significant implications for fundamental physics."

In addition, the existence of "dark matter" – whose nature is not yet understood – has been postulated long before the discovery of dark energy. Invoked by most astronomers, dark matter probably consists of undiscovered elementary particles whose aggregation produces the gravitational pull capable of holding together galaxies and clusters of galaxies in agreement with observations. The amount required is more than 20% of the total. Hence, only about 5% of the mass of the universe is understood at present.

In this framework it is apparent that the challenge for theoretical physics – especially for General Relativity as the best theory of gravity to date – is enormous.

2. The physical theories: success and problems

The theory of General Relativity (GR) and the Standard Model (SM) of particle physics, taken together, form our current view of the physical world. While the former governs physics in the macroscopic and cosmic scales, the latter governs the physics of the microcosm. According to GR gravity is not a force but a manifestation of space-time curvature. The relation between space-time curvature and space-time content (mass-energy and momentum) being given by Einstein's field equations. The theory has been extensively tested and no astronomical observation or experimental test has been found to deviate from its predictions. Thus it is the best description we have of gravitational phenomena that we observe in nature. The Standard Model of particle physics, since the 1970s gives a unified formalism for the other three fundamental interactions (strong, weak and electromagnetic) between the fundamental particles that make up all matter. It is a quantum field theory which is consistent with both Quantum Mechanics and Special Relativity. It has been spectacularly successful at describing physics down to a distance scale of about 10^{-18} m and no experiment to date contradicts it. Considerable new insights, down to even smaller scales, are expected from the Large Hadron Collider.

However, merging these two very successful theories to form a single unified theory poses significant difficulties. While in SM particle fields are defined on a flat Minkowski space-time, GR postulates a curved space-time which evolves with the motion of mass-energy. The definition of a gravitational field of a particle, whose position and momentum are governed by the Heisenberg Uncertainty Principle, is unclear. In addition quantum mechanics becomes inconsistent with GR near singularities. Current theories break down when gravity plus quantum mechanics both become important.

It is apparent that in spite of their own success, GR and the SM need to be reconciled with each other. Most attempts in this direction indicate that the pure tensor gravity of GR needs modification or augmentation. New physics is needed, involving new interactions which are typically composition dependent. As such, they would violate the Universality of Free Fall (UFF), hence the Weak Equivalence Principle which is the founding pillar of General Relativity (see next Section).

The need to put General Relativity to more and more stringent tests – despite its great success so far – comes therefore not only from facing the challenge of a universe whose mass-energy is mostly unknown, but also from the absence of a quantum theory of gravity.

This need has been clearly identified by the "Committee on the Physics of the Universe" which was appointed by the National Research Council of the US National Academies to investigate the subject and advise the major national research funding agencies. The results of the panel's work have been published in the book "Connecting Quarks with the Cosmos: Eleven Science Questions for the New Century" (2003).

The 3rd of the eleven questions identified in the book is:

"Did Einstein Have the Last Word on Gravity?" and reads:

"Black holes are ubiquitous in the universe, and their intense gravity can be explored. The effects of strong gravity in the early universe have observable consequences. Einstein's theory should work as well in these situations as it does in the solar system. A complete theory of gravity should incorporate quantum effects-Einstein's theory of gravity does not-or explain why they are not relevant."

The last chapter of the book, under the title "Realizing the Opportunities", is devoted to giving recommendations as to how to proceed in order to answer the 11 questions identified. The recommendations focus on very large scientific projects; however, a specific Section is devoted to the importance of setting up an effective program by balancing few big long term projects with more numerous, more affordable, small ones addressing specific crucial issues. The Section is entitled:

"Striking the Right Balance" ("Connecting Quarks with the Cosmos: Eleven Science Questions for the New Century", 2003 p. 162) and reads:

"In discussing the physics of the universe, one is naturally led to the extremes of scale – to the largest scales of the universe as a whole and to the smallest scales of elementary particles. Associated with this is a natural tendency to focus on the most extreme scale of scientific projects: the largest space observatories, the most energetic particle accelerators. However, our study of the physics of the universe repeatedly found instances where the key advances of the past or the most promising opportunities for the future come from work on a very different scale. Examples include laboratory experiments to test gravitational interactions, theoretical work and computer simulations to understand complex astrophysical phenomena, and small-scale detector development for future experiments. These examples are not intended to be exhaustive but to illustrate the need for a balanced program of research on the physics of the universe that provides opportunities for efforts that address the scientific questions but that do not necessarily fit within major program themes and their related large projects.

Two of our scientific questions – "Did Einstein have the last word on gravity?" and "Are there additional space-time dimensions?" – are being addressed by a number of laboratory and solar-system experiments to test the gravitational interaction. Tests of the principle of equivalence using laboratory torsion balances and lunar laser ranging could constrain hypothetical weakly coupled particles with long or intermediate range. These experiments have reached the level of parts in 10^{13} and could be improved by another order of magnitude. Improvement by a factor of around 10^5 could come from an equivalence principle test in space. .. null experimental results provide important constraints on existing theories, and a positive signal would make for a scientific revolution."

In the US the National Research Council of the National Academy of Sciences has recently appointed the "Astronomy and Astrophysics Decadal Survey" (Astro2010) Committee. The Committee on Astro2010 will survey the field of space and ground-based astronomy and astrophysics, recommending priorities for the most important scientific and technical activities of the decade 2010–2020. It would be surprising if further and more stringent tests of General Relativity were not recommended.

3. Tests of the weak equivalence principle as the most powerful probes of General Relativity

According to the so-called Universality of Free Fall (UFF), in a gravitational field all bodies fall with the same acceleration regardless of their mass and composition. First tested by Galileo around 1600 (see e.g. Bramanti *et al.*, 1993) UFF is the most direct experimental consequence of the Weak Equivalence Principle (WEP), in Einstein's own words not a *principle* but rather a *hypothesis* (Einstein, 1907). In this paper Einstein formulates the "*hypothesis of complete physical equivalence*" between a gravitational field and an accelerated reference frame: in a freely falling system all masses fall equally fast, hence gravitational acceleration has no local dynamical effects. Any test mass located inside "Einstein's elevator" – falling with the local acceleration of gravity g near the surface of the Earth – and zero initial velocity with respect to it, remains motionless for the time of fall. An observer inside the elevator will not be able to tell, before hitting the ground, whether he is moving with an acceleration g in empty space, far away from all masses, or else he is falling in the vicinity of a body (the Earth) whose local gravitational acceleration is also g (with the same direction and opposite sign).

This is the WEP, whereby the effect of gravity disappears in a freely falling reference frame: since all bodies fall the same whatever their mass and composition there is no motion relative to each other.

The WEP holds only locally. The elevator is free falling in the vicinity of the Earth, which amounts to saying that the height of fall is much smaller than the radius of the Earth. The cancellation of gravity in a freely falling frame holds locally for each frame, but the direction of free fall is not the same in all of them. Which is a direct consequence of the fact that the gravitational field of a body (like Earth) is non uniform, giving rise to the so called tidal forces between test particles whose centers of mass are not coincident. With the WEP Einstein has moved from Newton's concept of one global reference frame with gravitational forces and the UFF, to many free falling local frames without gravitational forces.

In his further development of the theory of General Relativity (Einstein, 1916), Einstein formulated what is known as the Einstein Equivalence Principle (EEP), which is an even more powerful and far reaching concept. EEP states the following (see e.g. Will, 2006):

- WEP is valid.
- The outcome of any local non-gravitational experiment is independent of the velocity of the freely-falling reference frame in which it is performed (Local Lorentz Invariance).
- The outcome of any local non-gravitational experiment is independent of where and when in the universe it is performed (Local Position Invariance).

EEP is regarded as the "heart and soul" of GR because it is the validity of this "principle" to ensure the fact that in GR the effects of gravity are replaced by a curved 4-dimensional space-time. Since EEP assumes the WEP to be valid, it is apparent that the WEP is the founding pillar of General Relativity. Though all experimental tests of GR are valuable as they contribute to assess its validity and provide further constrains, it is expected that testing the very foundation of GR has a stronger probing power than testing its numerous predictions.

Within string-inspired models, Damour & Polyakov (1994), Damour (1996) and Damour *et al.* (2002) have estimated the level at which a breakdown of the equivalence principle might occur and made quantitative comparisons between the probing power of weak equivalence principle – i.e. composition dependent – tests, and most of the other – composition independent – tests of deviations from GR ("post – Einsteinian" effects in gravitationally interacting systems: solar system, binary pulsars etc. ...). The most remarkable composition independent tests have been performed both in weak-field conditions, by means of radio links with Cassini spacecraft (Bertotti *et al.*, 2003), and in strong-field regime by timing the double pulsar (Kramer *et al.*, 2006). This is a unique system in which both neutron stars are detectable as radio pulsars and is becoming the best available testbed for general relativity and alternative theories of gravity in the strong-field regime (Kramer & Wex, 2009).

The comparison by Damour and colleagues shows the superior probing power (by several orders magnitude) of composition dependent WEP tests.

Most to our amazement, we are led to look back at Galileo's pioneering tests of the UFF in the early 1600, taking advantage of 400 years advance in science and technology, particularly in space science and space technology.

4. WEP tests: state of the art and prospects for improvements

The Weak Equivalence Principle is tested by testing the Universality of Free Fall. In an experiment to test UFF the observable physical quantity is the differential acceleration Δa of two test masses of different composition, relative to each other, while falling in the gravitational field of a source body with an average acceleration a (also referred to as the "driving acceleration"). A deviation from UFF is therefore quantified by the dimensionless parameter:

$$\eta = \frac{\Delta a}{a} \,. \tag{1}$$

The finding of a value $\eta \neq 0$ would disprove the UFF and indicate a violation of WEP on which General Relativity ultimately relies. Instead, $\eta = 0$ (in real experiments η is limited to a minimum value depending, for a given driving signal, on the sensitivity of the experiment to differential accelerations towards the source mass) – as reported by all experiments so far – confirms the basic assumption of General Relativity. By writing the equations of motion of each individual test mass without assuming *a priori* the equivalence of their inertial and gravitational (passive) mass, the parameter η becomes

$$\eta = \frac{2[((m_g/m_i)_A - ((m_g/m_i)_B]}{[((m_g/m_i)_A + ((m_g/m_i)_B]} \tag{2}$$

where subscripts A and B refer to the individual test masses and allow them to be distinguished by their different composition. This parameter is also known as the Eötvös parameter.

It is apparent from (1) that – for any given experimental apparatus – the larger the driving acceleration, the more sensitive the UFF test (hence the WEP test) that it provides. In a Galileo-type mass dropping experiment the driving acceleration is the gravitational acceleration of the Earth along the local vertical ($9.8\,\mathrm{ms}^{-2}$). If the test masses are suspended on a torsion balance the driving acceleration is $0.017\,\mathrm{ms}^{-2}$ (at most) in the field of the Earth – directed along the North-South direction of the local horizontal plane – and $0.006\,\mathrm{ms}^{-2}$ in the field of the Sun (with components along the North-South and East-West directions of the horizontal plane and 24-hr period due to the diurnal rotation of the Earth on which the balance sits). Yet, the first experimental apparatus to provide very accurate UFF tests in the field of the Earth (to 10^{-8}–10^{-9}) was the torsion balance used by Eötvös at the turn of the 20th century, and later on by his students (Eötvös *et al.*, 1922). This is because torsion balances are both extremely sensitive and inherently differential instruments.

The next leap in sensitivity (to 10^{-11}–10^{-12}) came in the 60s and early 70s using again a torsion balance but also recognizing that by taking the Sun as the source mass rather than the Earth, any differential effect on the test masses of the balance would be modulated by the 24-hr rotation of the Earth (Roll *et al.*, 1964; Braginsky & Panov, 1972). Indeed, the modulation frequency should be as high as possible, in order to reduce low-frequency $1/f$ electronic noise.

A turning point came in 1986 with the famous re-analysis of the Eötvös experiment (Fischbach *et al.*, 1986), a paper which made it to the first page of the *New York Times* and attracted the interest of scientists and space agencies alike. Since then, the best tests of UFF (to about 1 part in 10^{13}) have been performed by the "Eöt-Wash" group at the University of Seattle in a systematic series of remarkable experiments using torsion balances placed on a turntable which modulates the signal with a period down to about $20'$ (Su *et al.*, 1994; Baeßler *et al.*, 1994; Schlamminger *et al.*, 2008). Despite the much larger driving acceleration, Galileo-type mass dropping tests of UFF have been unable to compete with rotating torsion balances. The success of torsion balances relies on 3 main properties: i) high sensitivity to differential accelerations; ii) long time duration of the experiment; iii) up-conversion of the signal (DC from the Earth and 24-hr period from the Sun) to higher frequency (the rotation frequency of the balance).

Completely different tests of UFF are performed by laser ranging to the retroreflectors placed on the surface of the Moon by the *Apollo* astronauts and checking if the Moon and the Earth fall any differently in the field of the Sun. With almost 40 years of Lunar Laser Ranging data (currently to cm accuracy), LLR tests have found no deviation from UFF to about 10^{-13} (Williams *et al.*, 2004).

As for the future, rotating torsion balances might still improve by one order of magnitude, though past experience (also at low temperature) indicates how difficult that is. The LLR community has undertaken a considerable effort (Murphy *et al.*, 2007) with the so called "APOLLO" project to employ a bigger telescope and more powerful laser so as to improve laser ranging accuracy from cm to mm level.

For the UFF test to be improved by 1 order of magnitude the physical model used for propagating the orbit of the Moon must be improved to match the improved technology of laser ranging. Which is not an easy task, because of several gravitational (e.g. lunar librations, asteroid perturbations etc...) and non gravitational perturbations (e.g. atmospheric and tidal effects) whose relevant physical parameters are poorly known.

It is worth stressing that ultimately, LLR tests are limited by the non uniformity of the gravity field of the Sun. As shown by Nobili *et al.* (2008) this limitation results in a "classical" dimensionless parameter $\eta_{\rm class}$ indistinguishable from the η parameter of UFF tests as defined above. The authors show that $\eta_{\rm class} = 3\Delta{\rm a}_{\rm sma}/{\rm d}$ ($\Delta{\rm a}_{\rm sma}$ being the measurement error in the semimajor axis of the orbit of the Moon around the Earth and d the orbital distance of the Earth-Moon system from the Sun). For a given value of $\eta_{\rm class}$, typical of the experiment, no deviation from UFF smaller than that can be claimed, no matter how good is the physical model employed. The physical model should be sufficiently good for the UFF test to reach the value of $\eta_{\rm class}$, thus being limited only by the unavoidable non uniformity of the gravitational field. For LLR tests, $1\,cm$ measurement error in the semimajor axis of the moon is consistent with the current level of LLR tests to 10^{-13}; laser ranging to the Moon at $1\,mm$ level with the new APOLLO facility sets the limiting value of $\eta_{\rm class}$ close to 10^{-14}. We note in passing that this kind of limitation is even more severe for laser ranging to LAGEOS (in the field of the Earth) due to its smaller orbital distance from the source body.

High precision gravity measurements can now be performed using atom interferometry (Peters *et al.*, 2001). Experiments of this type are in preparation at Stanford University aiming at measuring the relative gravitational acceleration of falling cold atoms ^{85}Rb and ^{87}Rb – hence testing the UFF for these atoms – to 10^{-15} or even 10^{-17} (Dimopoulos *et al.*, 2007). Such a test of the UFF has already been performed a few years ago – dropping cold atoms ^{85}Rb and ^{87}Rb – at Max-Planck Institute, in Germany, by Fray *et al.* (2004) (with Nobel laureate Ted Hänsch among the authors), yielding a minimum relative difference in the acceleration of the falling atoms $\Delta g/g \simeq 10^{-7}$. The experiment concept is totally new, but the result is not yet competitive with torsion balance tests of the equivalence principle based on macroscopic test bodies, which have achieved 10^{-13} as reported above.

Directly relevant to cold atoms tests of the UFF are experiments devoted to measuring the local gravitational acceleration g by atom interferometry dropping cold atoms. Ten years ago (Peters *et al.*, 1999) (including Nobel laureate Steven Chu) have been able to measure g with an absolute uncertainty of $\Delta g/g \simeq 3 \cdot 10^{-9}$, and this is the best result so far.

Atom interferometry is likely to contribute more and more to precision gravimetry, and a Local Lorentz Invariance test has been performed by Mueller *et al.* (2008). However, from the viewpoint of UFF tests, dropping atoms which differ by 2 neutrons only makes the case for a possible violation depending on composition extremely weak. The choice of different materials is restricted to atoms that can be laser cooled. When comparing with classical tests based on macroscopic test masses, this issue is probably more a matter of concern than the limited statistics and sensitivity.

All other proposals aiming at a considerable improvement over the 10^{-13} level achieved by rotating torsion balances and by LLR refer to experiments to be performed: *i)* inside a capsule dropped during a balloon flight (Iafolla *et al.*, 1998) with target 10^{-14}–10^{-15}; *ii)* in a suborbital flight with a sounding rocket (Reasenberg, 2008) with target 10^{-16}; *iii)* in low Earth orbit inside a spacecraft: μSCOPE (see μSCOPE Website) with target 10^{-15}; MWXG (Ertmer *et al.*, 2009)) with target 10^{-16}; "Galileo Galilei" (GG) (Nobili *et al.*, 2009; GG Website) with target 10^{-17}; STEP (see STEP Website) with target 10^{-18}.

5. A breakthrough possible in space

A spacecraft in low Earth orbit offers major advantages for testing the UFF (hence the WEP) several orders of magnitude more accurately than now, thus probing a completely unknown physics domain and possibly leading to the discovery of a new fundamental force of Nature. The advantages are unquestionable: a driving signal acceleration 3 orders of magnitude stronger; absence of weight (which allows the tests masses to be coupled very weakly, hence with very high sensitivity to differential effects); isolation in space of the "whole lab" (i.e. the spacecraft), thus eliminating a large number of nearby disturbances unavoidable in ground laboratories; possibility to design the spacecraft as co-rotating with the apparatus, to provide the signal modulation (which has made the success of rotating torsion balances) while also passively stabilizing the spacecraft – with no need for a motor.

However, in order to succeed, one should be aware of the difficulties and risks. The lesson of GP-B, which has performed incredibly well, and yet not enough to produce the expected scientific results, should be learned by the entire community.

We can convincingly argue that the "Galileo Galilei" (GG) proposed satellite experiment has been designed to exploit all the advantages of space listed above. GG aims at testing the UFF to 10^{-17} and is now at Phase A-2 Study level by Thales Alenia Space Italy (TAS-I) with ASI (Agenzia Spaziale Italiana) funding. A full scale prototype of the flight instrument ("GG on the Ground" – GGG) is operational in the INFN lab of Pisa – San Piero a Grado and an advanced version of it is under construction (Comandi *et al.*, 2006I; Comandi *et al.*, 2006II).

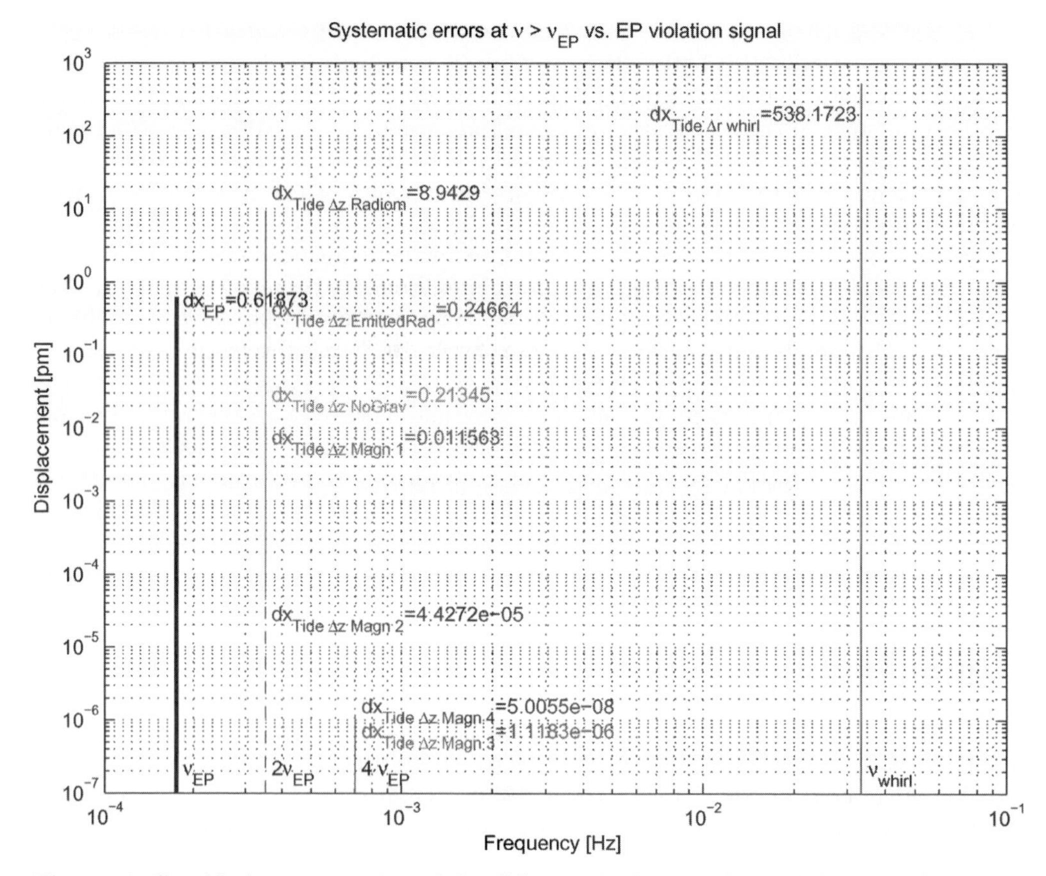

Figure 1. Graphical representation of the GG error budget as obtained from the Space Experiment Simulator. The mission requirements are embedded in the Simulator during a science run; the time series of the relative displacements of the test cylinders (which should be zero if Universality of Free Fall and the Weak Equivalence Principle hold!) allows us to establish systematic and random errors. The plot reports major systematic errors only, caused by various (classical) perturbations, as function of their frequency (in the inertial frame) to be compared with the signal expected for an EP violation to 10^{-17} (indicated by the thick line, first from left). The relative displacement caused on the test masses by the signal and the various perturbations is expressed in picometers. The frequency of the signal is the satellite orbit frequency, indicated as ν_{EP}. It is apparent that even though in some cases the perturbing effects are larger than the signal, they are always sufficiently separated in frequency to be distinguished from it.

GG has been designed to rely on physics laws and physical symmetries more than on brute force, and to be as passive as possible as this is a must for small force gravitational experiments. Very schematically:

- GG has large mass test bodies (10 kg each), to reduce thermal noise and aim at high sensitivity without cryogenics.
- GG has mechanical suspensions (weightlessness makes their stiffness competitive with that of contactless suspensions) in order to guarantee passive electric grounding.
- GG is a cylindrically symmetric co-rotating system, to avoid active attitude control and achieve high frequency signal modulation *for free*.
- The test cylinders are sensitive in the plane perpendicular to the axis – not along the axis (as in STEP and μSCOPE). This is a crucial theoretical condition (which has been known since 1934) in order to guarantee the existence of a physical position of

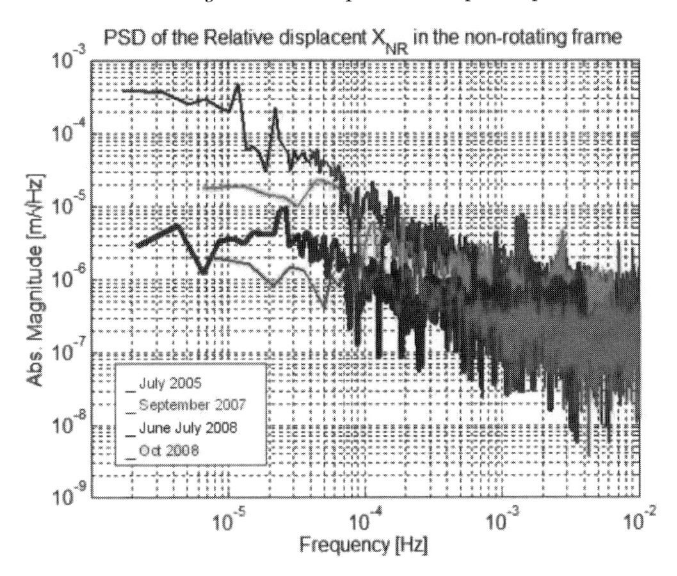

Figure 2. Power Spectral Density of the relative displacements of the GGG test cylinders showing improvements in sensitivity since 2005 (from top to bottom curve). The bottom curve corresponds to the same 2-day run whose Fast Fourier Transform is reported in Figure 3

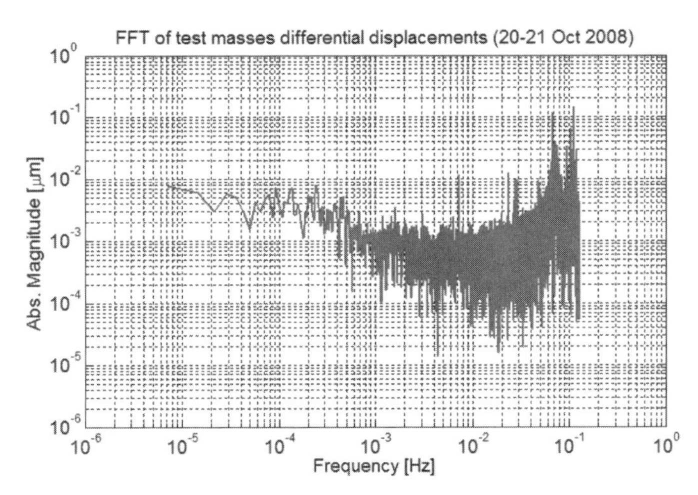

Figure 3. Fast Fourier Transform of the relative displacements of the GGG rotating test cylinders, 10 kg each (in the horizontal plane of the lab). Units are in μm. Even at low frequencies their separation is below 10^{-8} m, reaching a nanometer at higher frequencies. In the GG space experiment, a weak equivalence principle violation to 10^{-17} would produce a displacement of about 10^{-12} m at orbital frequency

relative equilibrium of the test masses at a rapid rotation frequency (1 Hz in the present baseline) well above their natural coupling frequency. In the plane of sensitivity the GG test masses move around that position and a capacitance read-out gives their relative displacements: the masses are not "actively" re-centered, like in the original STEP design and in μSCOPE (the signal being – in those cases – the active force required to re-center them).

• The only active control required on the test masses is that of whirl motions, which require forces much smaller than the coupling ones at frequencies far apart from the signal frequency. Moreover, they will not be applied during science runs (Nobili *et al.*, 1999).

- The spacecraft is capable to provide partial drag compensation with FEEP thrusters.
- Residual electric patch effects are DC or slowly varying (and they are measured in GGG).
- A wide choice of materials for the test masses is possible (rotation makes mass anomalies effects DC) and can be optimized for maximum possible physical effect depending on composition (one test mass might contain H).
- Major radiometer disturbance is negligible by design (otherwise it would be impossible to aim at 10^{-17} at room temperature; see Nobili *et al.*, 2001).

The industrial study team of GG comes from the ESA mission GOCE, for which TAS-I has acted as prime contractor. GOCE has been launched recently and is already in drag-free mode, with a performance a factor 10 better than planned, to the level of 1 to $2 \cdot 10^{-9}\mathrm{ms}^{-2}/\sqrt{\mathrm{Hz}}$ in a range from 2 mHz to 0.1 Hz; see GOCE Website.

The Drag Free Control of GG and a complete Space Experiment Simulator have been built based on expertise from GOCE, on previous GG studies and on inputs from the GGG experimental results. Fig. 1 reports in a graphical manner the error budget (systematic effects) of the GG experiment obtained in April 2009 for a target violation signal of 10^{-17}. Fig. 2 reports the results of the GGG prototype. Platform noise (notably terrain tilt noise and motor noise) needs to be reduced. However, GGG has the same number of degrees of freedom as in GG, the test masses are dominated by their coupling (though stiffer than in space) and not by local gravity and therefore provides a reliable testbed for the instrument to fly in GG.

The technologies required are available, the most recent being the technology of field emission electric propulsion (FEEP) which has been developed for LISA Pathfinder and μSCOPE.

To conclude, a major scientific advance in fundamental physics is possible with current space technology. A null result in testing the Weak Equivalence Principle at the 10^{-17} level that GG is designed to reach would be a milestone for physics theories; a positive result would point at the existence of a new fundamental force of Nature and make for a scientific revolution.

Acknowledgements

We acknowledge support from ASI (grant I/016/07/0) and INFN

References

Baeßler, Heckel, B. R., Adelberger, E. G., Gundlach, J. H., Schimidt, U., & Swanson, H. E. 1999, *Phys. Rev. Lett.*, 83, 3585

Bennet, C. *et al.* 2003, *Ap. J. Suppl.*, 148, 1

Bertotti, B., Iess, L., & Tortora, P. 2003, *Nature*, 425, 374

Braginsky, V. B. & Panov, V. I. 1972, *Sov. Phys. JEPT*, 34, 463

Bramanti, D. *et al.* 1993, *in STEP Symposium, ESA WPP-115*, 319. *http://eotvos.dm.unipi.it/galileo_and_uff*

Comandi, G. L., Toncelli, R., Chiofalo, M. L., Bramanti, D., & Nobili, A. M. 2006, *Rev. Sci. Instrum.*, 77, 034501

Comandi, G. L., Chiofalo, M. L., Toncelli, R., Bramanti, D., Polacco, E., & Nobili, A. M. 2006, *Rev. Sci. Instrum.*, 77, 034502

Connecting Quarks with the Cosmos: Eleven Science Questions for the New Century 2003, *Board on Physics and Astronomy* (The National Academic Press)

Damour, T. 1996, *Class. Quantum Grav.*, 13, A33

Damour, T. & Polyakov, A. M. 1994, *Gen. Relativ. Gravit.*, 26, 1171
Damour, T., Piazza, F., & Veneziano, G. 2002, *Phys. Rev. D*, 66, 046007
de Bernardis, P. *et al.* 2000, *Nature*, 404, 955
Dimopoulos, S., Graham, P. W., Hogan, J. M., & Kasevich, M. A. 2007, *Phys. Rev. Lett.*, 98, 111102
Einstein, A. 1907, *Jahrb. Radioaktiv.*, 4, 411
Einstein, A. 1916, *Annalen der Physik*, 49, 769
Eötvös, R. V., Pekar, D., & Fekete, E. 1916, *Annalen der Physik*, 68, 11
Ertmer, W. *et al.* 2009, *Exp Astron.*, 23, 611
Fischbach, E., Sudarsky, D., Szafer, A., Talmadge, C., & Aronson, S. H. 1986, *Phys. Rev. Lett.*, 56, 2426
Fray, S., Alvarez Diez, C., Hänsch, T. W., & Weitz, M. 2004, *Phys. Rev. Lett.*, 93, 240404
GG Website *http://eotvos.dm.unipi.it*
Glanz, J. 1998, *Science*, 282, 2156
GOCE Website *http://www.esa.int/SPECIALS/GOCE/*
Iafolla, V., Lorenzini, E. C., Milyukov, V., & Nozzoli, S. 1998, *Rev. Sci. Instrum.*, 69, 4146
Kowal, C. T. & Drake, S. 1980, *Nature*, 28, 311
Kramer, M., *et al.* 2006, *Science*, 314, 97
Kramer, M. & Wex, N. 2009, *Class. Quantum Grav.*, 26, 073001
μSCOPE Website *http://smsc.cnes.fr/MICROSCOPE/*
Mueller, H., Chiow, S., Herrmann, S., Chu, S., & Chung, K.-Y. 2008, *Phys. Rev. Lett.*, 100, 021101
Murphy, M. *et al.* 2007, *Int. J. Mod. Phys. D*, 16, 2127
Nobili, A. M. & Will, C. M. 1986, *Nature*, 320, 39
Nobili, A. M. *et al.* 1999, *Class. Quantum Grav.*, 16, 1463
Nobili, A. M., Bramanti, D., Comandi, G. L., Toncelli, R., Polacco, E., & Catastini, G. 2001, *Phys. Rev. D*, 63, 101101(R)
Nobili, A. M., Comandi, G. L., Bramanti, D., Doravari, S., Lucchesi, D. M., & Maccarrone, F. 2008, *Gen. Rel. Grav.*, 40, 1533
Nobili, A. M. *et al.* 2009, *Exp. Astr.*, 23, 689
Peters, A., Chung, K. Y., & Chu, S. 1999, *Nature*, 400, 849
Peters, A., Chung, K. Y., & Chu, S. 2001, *Metrologia*, 38, 25
Report of the Dark Energy Task Force 2006 *http://www.nsf.gov/mps/ast/aaac/dark_energy_task_force/report/detf_final_report.pdf*
Reasenberg, R.D. 2008, *Q2C3 International Workshop, Airlie Center, Virginia USA* *http://funphysics.jpl.nasa.gov/Q2C3/program*
Roll, P. G., Krotkov, R., & Dicke, R. H. 1964, *Annals of Physics*, 26, 442
Schlamminger, S., Choi, K.-Y., Wagner, T. A., Gundlach, J. H., & Adelberger, E. G 2008, *Phys. Rev. Lett.*, 100, 041101
Standish, E. M. & Nobili, A. M. 1997, *Baltic Astronomy*, 6, 97
STEP Website *http://einstein.stanford.edu/STEP/*
Su, Y., Heckel, B. R., Adelberger, E. G., Gundlach, J. H., Harris, M., Smith, G. L., & Swanson, H. E. 1994, *Phys. Rev. D*, 50, 3614
Will, C. M. 2006, *Living Reviews in Relativity*, 9, 3
Williams, J. G., Turyshev, S. G., & Boggs, D. H. 2004, *Phys. Rev. Lett.*, 93, 261101

Relativity in Fundamental Astronomy
Proceedings IAU Symposium No. 261, 2009
S. A. Klioner, P. K. Seidelman & M. H. Soffel, eds.

doi:10.1017/S174392130999069X

Two cylindrical masses in orbit for the test of the equivalence principle

Ratana Chhun, Pierre Touboul and Vincent Lebat

ONERA, BP-72
F-92322 CHATILLON CEDEX, FRANCE
email: ratana.chhun@onera.fr

Abstract. Two pairs of solid test-masses have been considered to perform in space the test of the universality of free fall with an accuracy of at least 10^{-15}. These cylindrical masses are precisely at the heart of the MICROSCOPE mission instrument comprising two differential electrostatic accelerometers. These masses shall exhibit material quality, shapes, positions and alignments in regard to stringent experimental requirements. Indeed the space experiment is based on the control of the two masses submitted to the same gravity acceleration along the same orbit at 810 km altitude with an accuracy of 10^{-11} m. Thus effects of Earth and satellite gravity gradients shall be contained as well as any other disturbances of the mass motions induced by their magnetic susceptibility or electrical dissymmetries, by outgassing of the materials or radiation emissivity. Furthermore, the electrostatic levitation of the two masses depends dramatically on the mass shapes and electrical properties in particular for the definition of the sensitive axes orientation. All these aspects will be presented from the mass characteristics to the space MICROSCOPE experiment performance.

Keywords. MICROSCOPE, Equivalence principle, accelerometer, gravity

---

1. Introduction

For a few decades several unification theories have risen to try and merge general relativity and quantum mechanics. These approaches all call out for the existence of a fifth interaction force (Damour *et al.* 2002, Fayet 2003). Such existence would translate into a violation of the basis of general relativity, the Equivalence Principle which stipulates that all bodies acquire the same acceleration rate in the same uniform gravity field regardless of inertial mass or intrinsic composition.

To this day the Equivalence Principle has been verified to a precision of a few 10^{-13} by ground-based experiments (Schlamminger *et al.* 2008). MICROSCOPE aims to test the Equivalence Principle to better than 10^{-15} (Touboul *et al.* 2001).

In order to attain such precision, the test will be performed in space. The space environment indeed offers several very favorable conditions compared to a ground-based experiment: among those, reduced vibrations from micro to nano level, a very long integration time, highly stable thermal conditions, modulations of the Earth gravity source in the instrument reference frame and the possibility to use a dedicated instrument with sufficiently limited range with a consequent very low noise.

2. Mission description

The MICROSCOPE space mission is now at the production level of the instrument, the satellite and the masses being specified and defined. In the experiment, two concentric test-masses will be placed on the same gravitational trajectory. By keeping the two masses concentric, they are ensured to follow the exact same free fall and to undergo the exact same gravitational field. Instead of trying to detect a difference of free fall trajectory between the two masses, the principle of MICROSCOPE consists in measuring the deviation of electrostatic forces necessary to maintain the two masses on the same orbit along the gravity field direction. A breaking of symmetry between the measured control accelerations, equivalent to the two masses motions, would translate into a violation of the Equivalence Principle. Following an inertial heliosynchronous orbit at an altitude of about 800km, the test is performed at the orbit plus the satellite spin frequency around the normal axis to the orbital plane. Two pairs of test-masses, constituting the two pairs of test-masses of two differential accelerometers, embark the CNES microsatellite for a double differential test, one pair of platinum-rhodium alloy masses and one pair of one platinum rhodium alloy and titanium alloy masses.

The launch of the MICROSCOPE microsatellite is scheduled for the end of 2012 and the mission duration is set to a minimum of twelve months. It will follow a quasi-circular orbit (with an eccentricity inferior to 5×10^{-3}) for a very well known measurement frequency. In order to ensure the mission performance requirements, the satellite will be equipped with twelve micronewton thrusters, a continuous drag-free compensation system and fine orbit and attitude control. It will also feature a passive thermal control.

Currently manufacturing the instrument qualification model, this paper will describe the overall design of the instrument with a focus on the test-masses and the various numerical models and geometry constraints required to reach the experiment performance, i.e. the 10^{-15} accuracy EP test.

3. Instrument description

The payload is composed of two sensor units; each one is composed of one pair of test-masses centered with respect to each other at better than 20 µm after integration. Each test-mass is at the core of an inertial sensor, one half of the differential accelerometer. The ensemble is integrated inside a highly insulated thermal case and a µmetal magnetic shield installed at the center of the satellite so that the orientation of the payload with respect to the star sensor piloting the drag-free and attitude control system is very stable.

The test-masses are cylindrical and their dimensions are designed to obtain spherical inertia properties, trade-off between practicability and relevance for the test, thus minimizing the gravity gradient effects. The two cylinders are concentric and coaxial in addition to having a common center of mass, thus mimicking the behavior of two concentric spheres, with a privileged sensitive measurement axis, the average cylinder axis.

Each test-mass is electrostatically levitated and maintained centered and steady inside its gold-coated silica cage by applying adequate voltages on the electrodes etched on the silica cylinders surrounding the test-mass, which are also used to sense its position and thus its displacement attempts along or around its six degrees of freedom. The only physical contact on the proof mass is a 5 µm gold wire used for charge control and for applying a sinusoid voltage for the capacitive sensing.

Each sensor unit core is maintained under a vacuum less than 10^{-5} Pa thanks to an Invar tight housing and a getter material on top of the sensor unit.

4. Measurement equation

The science measurements of one inertial sensor are the applied voltages needed to maintain the mass centered with respect to its electrode cage, to compensate for both

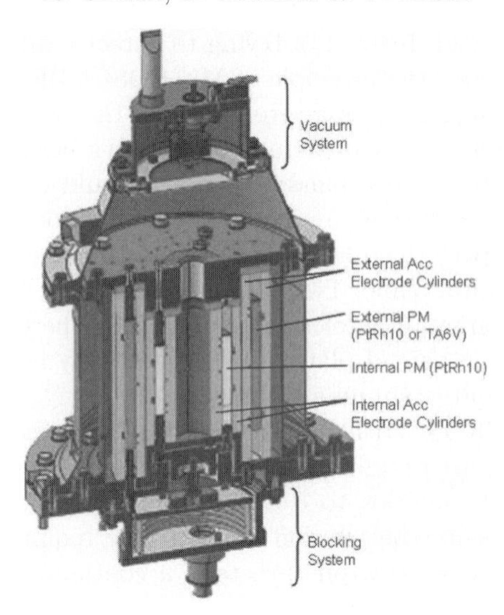

Figure 1. Cross cut view of the differential accelerometer, consisting of the sensor core of test-masses and electrode cylinders, the blocking mechanism below and the vacuum system above.

gravity and surface forces on the satellite. The differential measurement between the two inertial sensors is then expressed as follows:

$$\mathbf{\Gamma}_{\mathbf{app,d}} \approx \left(\delta \cdot \mathbf{g}(\mathbf{O}_{\mathrm{sat}}) + ([T] - [In]) \cdot \mathbf{\Delta} - 2 \cdot [Cor] \cdot \dot{\mathbf{\Delta}} - \ddot{\mathbf{\Delta}}\right)$$

$\delta = m_{G2}/m_{I2} - m_{G1}/m_{I1}$ is the signal to be detected where m_G and m_I are respectively the gravitational mass and the inertial mass, Δ the excentering between the centers of mass of the two test-masses. At about 800 km high, $g(O_{\mathrm{sat}})$ is quite $8\,\mathrm{m/s^2}$, so in order to perform the test at 10^{-15}, the differential acceleration needs to be measured at a resolution better than $8 \times 10^{-15}\,\mathrm{m/s^2}$ at the test frequency. Thus, all other terms of the equation need to be weaker or very well known.

A great part of the error budget is then due to forces resulting directly from the test-masses design, material properties and manufacture.

5. Gravity gradients

The first disturbing effect due to the instrument defects is the effect of the Earth gravity gradient on the test-masses excentering. The Earth gravity gradient is considered uniform over the two masses and processed through the field tensor form. The Earth gravity gradient directly generates major components at DC frequency and at twice the measurement frequency and, due to the eccentricity, at the measurement frequency plus or minus twice the spin frequency of the satellite.

The test-masses relative centering specification is 20 µm but will not be sufficient to limit the Earth gravity gradient effect without an in-orbit calibration and data ground correction to 0.1 µm in the orbital plane and 0.5 µm along its normal through satellite controlled oscillations (Guiu *et al.* 2007).

A second disturbing gravity effect is due to the local gravity gradient of the satellite and the instrument itself, applied on the test-masses. To limit these effects, the masses dimensions are precisely selected, taking into account the fact that the masses actually feature flat areas for axial rotation control and holes at each end for motion limitation using dedicated stops. Equality of the moments of inertia are obtained around all axes at better than 10^{-4} and relative cross products of inertia inferior to 10^{-8}. Furthermore the homogeneity of the mass material is selected in relation. Moreover the satellite and the instrument are in addition designed to exhibit high stiffness and very good thermal stability to get very low disturbances at the measurement frequency due to environment conditions.

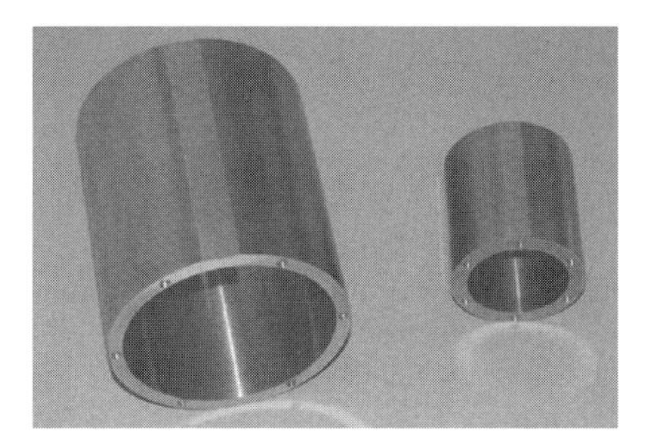

Figure 2. Test-masses featuring flat panels for axial rotation control and holes for displacement limitation.

Finite element models of the instrument on one hand and of the satellite on the other have been computed taking into account various mass motions and positions depending on temperature variations. For the instrument the model results in acceleration sensitivities due to the mass proximity environment thermal conditions of about 2.5×10^{-13} m/s^2/K. Combining the expected thermal environment stability at the measurement frequency of 1 mK, we can expect the instrument self gravity effect to be limited to 2.5×10^{-16} m/s^2. The satellite model shows likewise a satellite self gravity effect of about 10^{-16} m/s^2.

6. Magnetic field

The model for the magnetic field effect which we try so much to protect MICROSCOPE from, using two complementary shieldings, takes into account both the Earth magnetic field and the local magnetic sources, each of those having their own components at various frequencies but all add up to form a global magnetic field. The acceleration effect due to this global field is then $\mathbf{\Gamma}_m = \chi_m/(2\mu_0 \cdot \rho_m)\nabla\left[\mathbf{B}^2\right]$ where χ_m is the magnetic susceptibility of the test-mass material, ρ_m its density and μ_0 the magnetic permittivity in vacuum. χ_m/ρ_m is respectively limited to 1.4×10^{-8} m^3/kg for platinum mass and 1.6×10^{-8} m^3/kg for titanium mass thanks to the selection of the material.

The Earth magnetic field time variations along the MICROSCOPE orbit is derived from Oersted satellite data from which the components of the Earth magnetic field at the various frequencies considered in the global MICROSCOPE error budget, DC, the test measurement frequency, first multiples and random noise, are extracted.

The local sources are considered in a worst case approach as magnetic moments at 30cm from the masses: 1Am^2 at DC, 10^{-3}Am^2 at the test frequency and 0.04Am$^2/\sqrt{Hz}$ random noise.

The magnetic field attenuation due to the shieldings has been measured in CNES facilities and the magnetic field and magnetic field gradients over each test-mass volume has been computed using the FLUX3D software. The maximum acceleration effect at the measurement frequency in case of a purely inertial orbit is 2.08×10^{-16} m/s^2 and only 1.19×10^{-17} m/s^2 in case the satellite is also spinning.

7. Shape dissymmetries and electrical effects

From the electrical point of view, each test-mass is surrounded by a number of gold-coated silica parts, two electrode cylinders, one inside and one outside the test-mass, but also two end disc plates which translate into two planar ring plates regarding the cylindrical mass ends and two sets of three finger stops whose purpose is to block the mass during launch and to limit the mass motion when there is no servo-control after retraction, located also at each end of the mass. These finger stops are at the same potential as the test-mass and sufficiently far away from it to consider that they do not induce any noticeable electrostatic stiffness.

The test-masses and the gold-coated electrode cylinders are all considered as equipotential conductors with limited patch effect thanks to motionless masses with respect to the conductive cages in regard. Since there is no relative displacement of the parts, we can only consider time and thermal variations at the test frequency of the difference of potential and not the DC patch. Not relying only on conjectured models, we also benefit for the evaluation of these disturbance sources from the behavior of previous space accelerometers either in laboratory or in orbit.

With perfect symmetry and coaxiality of all the cylinders and masses, there would then be no electrostatic pressure resultant applied on the mass. This is why the cylinder main symmetry defects must be kept to a minimum, such as cylindricity which translates into a combination of a cone and a barrel shape, or concentricity and coaxiality of the inner and outer diameters.

These defects create electrostatic biasing forces, stiffness and can modify the considered electrode force laws relation to the control electrical potentials. Care has been paid to verify deeply that the manufacturing accuracy of the parts are in agreement with the evaluation of these disturbances.

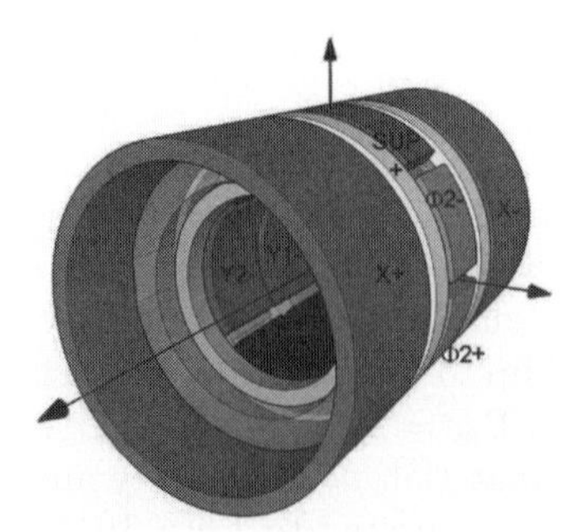

Figure 3. Electrode configuration : the electrode pairs, etched on gold-coated silica cylinders, control the six degrees of freedom of the encaged test-mass.

8. Emissivity and outgassing

We finally consider parasitic effects due to temperature gradients and residual pressure. A prototype model in laboratory under passive vacuum for two years demonstrates by daily measurements a residual pressure kept below 10^{-5} Pa.

With temperature gradients variations inferior to 2 mK/m verified on a thermal breadboard at the test frequency in the case of the satellite rotation mode, the radiometer effect modeled as $\Gamma_{\text{radio}} = 1/2 \cdot PS/mT \cdot \Delta T$ is limited to $\Gamma_{\text{radio}} < 10^{-16}$ m/s^2 at the test frequency.

The outgassing effect depends on the mass material, its cleanliness and creates a difference of forces exerted on each side of the test-mass along any axis by variation of gas pressure induced by surface outgassing fluctuations and is kept below 10^{-17} m/s^2 thanks to the low temperature gradients inside the instrument core.

The radiation pressure model takes into account the emissivity and reflectivity of the mirror-like surfaces in regard in the case of MICROSCOPE and the multiple reflections of photons that would create a differential force on the test-mass resulting in the expression $\Gamma_{\text{radia}} = 8/3 \cdot S\sigma/mc \cdot k\,(\epsilon_c, \epsilon_m)\,T^3 \Delta T$ where $k\,(\epsilon_c, \epsilon_m) = (1 + \epsilon_m)(1 - \epsilon_c)/(1 - \epsilon_m \epsilon_c)$, function of the reflectivity of the test-mass and the cage. Since ϵ_m and ϵ_c are both very close to 1, the value of k can greatly vary from 0 to 2. In that respect, the model we consider is conservative and $k = 2$ for a disturbance of 1.2×10^{-16} m/s^2, still compatible with our needs.

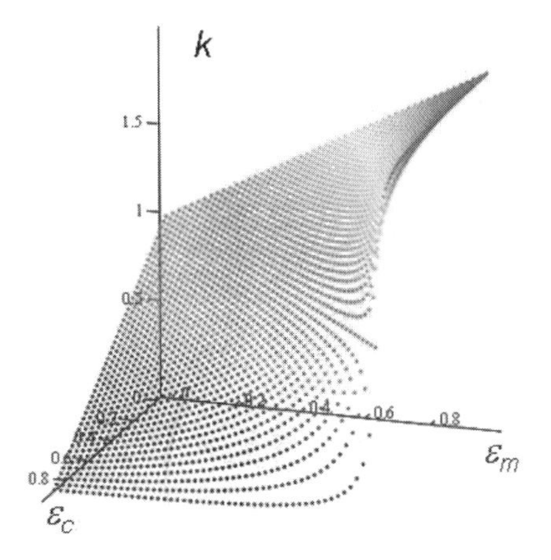

Figure 4. Correction factor of the radiation pressure model, function of the mass and cage reflectivities.

9. Conclusion

Mass geometry, material, magnetic susceptibility, surface outgassing, surface electrical and optical properties all have very mandatory requirements for the objective of the MICROSCOPE space mission, the test of the Equivalence Principle with an accuracy of 10^{-15}. All disturbing effects have been considered to evaluate the instrument performance limitation and to define in consequence the best payload configuration and satellite environment. The present production of the qualification model of the instrument confirms the possibility to meet all requirements. In the near future this instrument will be tested in drop tower and the flight models are foreseen to be delivered at the end of next year for the launch scheduled in 2012.

10. Acknowledgements

The authors would like to thank teams in CNES for their collaboration on the satellite and mission specifications, and funding support, ZARM for their development of a specific free-flyer for the MICROSCOPE instrument test campaign, and the Observatoire de la Cote d'Azur for their contribution to the mission definition and analysis.

References

Damour, T., Piazza, F., & Veneziano, G. (2002), Violations of the equivalence principle in a dilaton-runaway scenario, *Phys. Rev. D*, 66

Fayet, Pierre (2003), Theoretical Motivations for Equivalence Principle Tests, *Adv. Sp. Res.*, 32, 7, pp. 1289–1296

Schlamminger, S., Choi, K., Wagner, T. A., Gundlach, J. H., & Adelberger, E. G. (2008), Test of the equivalence principle using a rotating torsion balance, *Phys. Rev. Letter*, 100, 041101

Touboul, P., Rodrigues, M., Metris, G., & Tatry, B. (2001), MICROSCOPE: Testing the equivalence principle in space, *CRAS*, Paris, 2, IV, 9, pp. 1271–1286

Guiu, E., Rodrigues, M., Touboul, P., & Pradels, G. (2007), Calibration of MICROSCOPE, *Advances in Space Research*, 39, 2, pp. 315–323

Relativity in Fundamental Astronomy
Proceedings IAU Symposium No. 261, 2009
S. A. Klioner, P. K. Seidelman & M. H. Soffel, eds.

doi:10.1017/S1743921309990706

Lorentz violation and gravity

Quentin G. Bailey

Physics Department, Embry-Riddle Aeronautical University,
3700 Willow Creek Road, Prescott, AZ 86301, USA,
email: baileyq@erau.edu

Abstract. In the last decade, a variety of high-precision experiments have searched for miniscule violations of Lorentz symmetry. These searches are largely motivated by the possibility of uncovering experimental signatures from a fundamental unified theory. Experimental results are reported in the framework called the Standard-Model Extension (SME), which describes general Lorentz violation for each particle species in terms of its coefficients for Lorentz violation. Recently, the role of gravitational experiments in probing the SME has been explored in the literature. In this talk, I will summarize theoretical and experimental aspects of these works. I will also discuss recent lunar laser ranging and atom interferometer experiments, which place stringent constraints on gravity coefficients for Lorentz violation.

Keywords. gravitation, relativity

1. Introduction

General Relativity (GR) encompasses all known gravitational phenomena and remains the best known fundamental theory of gravity. No convincing deviations from GR have been detected thus far (Will 2005, Battat *et al.* 2007, Müller *et al.* 2008, and Chung *et al.* 2009). Nonetheless, there remains widespread interest in continuing ever more stringent tests of GR in order to find possible deviations. This is motivated by the idea that small deviations from GR might be the signature of an as yet unknown unified fundamental theory that successfully meshes GR with the Standard Model of particle physics.

The principle of local Lorentz symmetry is a solid foundation of GR. However, the literature abounds with theoretical scenarios in which this symmetry principle might be broken. One promising possibility is that miniscule deviations from perfect local Lorentz symmetry might be observable in high-precision experiments and observations (for reviews see CPT 2008, Bluhm 2006, and Amelino-Camelia *et al.* 2005). The Standard-Model Extension (SME) is a general theoretical framework for tests of Lorentz symmetry, in both gravitational and nongravitational scenarios (Kostelecký & Potting 1995, Colladay & Kostelecký 1997, Colladay & Kostelecký 1998, and Kostelecký 2004). The SME is an effective field theory that includes all possible Lorentz-violating terms in the action while also incorporating the known physics of the Standard Model and GR. The Lorentz-violating terms are constructed from gravitational and matter fields and coefficients for Lorentz violation, which control the degree of Lorentz-symmetry breaking.

A wide variety of experiments with matter have set constraints on many of the coefficients for Lorentz violation in the SME (for a summary and references see the data tables in Kostelecký & Russell 2009). Recently, the gravitational sector and matter-gravity couplings in the SME have been studied (Bailey & Kostelecký 2006, Bailey 2009, and Kostelecký & Tasson 2009). In turns out that some novel effects can occur that are controlled by certain matter sector coefficients in the SME, which are unobservable in the absence of gravity. In addition, experimental work constraining SME coefficients in the pure-gravity sector has begun, including gravimeter experiments (Müller *et al.* 2008) and lunar laser ranging (Battat *et al.* 2007). The theoretical and experimental aspects of the signals for Lorentz violation in gravitational experiments will be discussed in this talk. For convenience, natural units ($\hbar = c = 1$) are used.

2. Theory

In Kostelecký (2004), the SME with both gravitational and nongravitational couplings was presented in the general context of a Riemann-Cartan spacetime. This extended earlier work on the Minkowski spacetime limit. We focus here on the special cases of gravity couplings in the matter sector of the SME and also on the pure-gravity sector.

In the matter sector of the SME, the matter-gravity couplings expected to dominate in many experimental scenarios for ordinary matter (protons, neutrons, and electrons) can be described in terms of Dirac spinor fields for spin-1/2 fermions. The lagrangian density for this limit takes the form

$$\mathcal{L}_m = \tfrac{1}{2} iee^{\mu}{}_{a}\overline{\psi}(\gamma^a - c_{\nu\lambda}e^{\lambda a}e^{\nu}{}_{b}\gamma^b + ...)\overset{\leftrightarrow}{D}_{\mu}\psi - e\overline{\psi}(m + a_{\mu}e^{\mu}{}_{a}\gamma^a + ...)\psi + ..., \qquad (2.1)$$

where the ellipses represent additional terms in the SME, neglected here for brevity. Following standard methods, the spinor fields ψ and the gamma matrices γ^a are incorporated into the tangent space of the spacetime at each point using the vierbein $e_{\mu}{}^{a}$. The symbol e represents the determinant of the vierbein. The coefficients for Lorentz violation appearing in equation (2.1) are $c_{\mu\nu}$ and a_{μ}, which vanish in the limit of perfect local Lorentz symmetry for the matter fields. Note also that the covariant derivative appearing in this lagrangian is both the spacetime covariant derivative and the $U(1)$ covariant derivative, and so contains additional couplings to gravity through the spacetime connection.

In the pure-gravity sector of the SME, and in the Riemann-spacetime limit, the relevant lagrangian is written as

$$\mathcal{L}_g = \frac{1}{2\kappa} e[(1-u)R + s^{\mu\nu}R^T_{\mu\nu} + t^{\kappa\lambda\mu\nu}C_{\kappa\lambda\mu\nu}] + \mathcal{L}'. \qquad (2.2)$$

In this expression, R is the Ricci scalar, $R^T_{\mu\nu}$ is the trace-free Ricci tensor, $C_{\kappa\lambda\mu\nu}$ is the Weyl conformal tensor, and $\kappa = 8\pi G$, where G is Newton's gravitational constant. The leading Lorentz-violating gravitational couplings are controlled by the 20 coefficients for Lorentz violation u, $s^{\mu\nu}$, and $t^{\kappa\lambda\mu\nu}$. The matter sector and possible dynamical terms governing the 20 coefficients are contained in the additional term denoted $\mathcal{L}'$. In the limit of perfect local Lorentz symmetry for gravity these coefficients vanish.

In all sectors, the action in the SME effective field theory maintains general coordinate invariance. However, under local Lorentz transformations and diffeomorphisms of the localized matter and gravitational fields, or what are called *particle transformations,* the action is not invariant. When Lorentz violation is introduced in the context of a general Riemann-Cartan geometry, some interesting geometric constraints arise. For example, introducing the coefficients for Lorentz violation in the matter and gravity sectors as nondynamical or prescribed functions generally conflicts with the Bianchi identities. If instead the coefficients arise through a dynamical process, as occurs in a spontaneous Lorentz-symmetry breaking scenario, conflicts with the geometry are avoided (Kostelecký 2004).

The approach of Kostelecký & Tasson (2009) and Bailey & Kostelecký (2006) is to treat the coefficients for Lorentz violation as arising from spontaneous Lorentz-symmetry breaking. Although specific models with dynamical vector and tensor fields can reproduce the terms in the lagrangians (2.1) and (2.2), it is a challenging task to study the gravitational effects in a generic, model-independent way. However, the weak-field or linearized gravity regime offers some simplifications to the analysis. It is possible, under certain mild assumptions on the dynamics of the coefficients for Lorentz violation, to extract effective linearized equations. In this case the effects of Lorentz violation on gravity and matter involve only the vacuum expectation values of the coefficients for Lorentz violation (denoted $\overline{a}_\mu$, $\overline{s}^{\mu\nu}$, etc.).

By including matter-gravity couplings, the $\overline{a}_\mu$ coefficients, which have remained largely elusive in nongravitational tests, become accessible to experiments. In contrast, stringent constraints on the $c_{\mu\nu}$ coefficients already exist (see the data tables in Kostelecký & Russell 2009). As shown in Kostelecký & Tasson (2009), the dominant Lorentz-violating effects on matter fields arise from an effective vector potential $\widetilde{a}_\mu$ that takes the form

$$\widetilde{a}_\mu = \tfrac{1}{2}\alpha h_{\mu\nu}\overline{a}^\nu - \tfrac{1}{4}\alpha\overline{a}_\mu h^\nu{}_\nu, \tag{2.3}$$

where $h_{\mu\nu}$ are the metric fluctuations around a Minkowski background and α is a constant.

In the pure-gravity sector, the weak-field gravity analysis in Bailey & Kostelecký (2006) reveals that the leading Lorentz-violating effects are controlled by the nine independent coefficients in $\overline{s}^{\mu\nu}$. In the post-newtonian limit, attempting to match the pure-gravity sector of the SME to the standard Parametrized Post-Newtonian (PPN) formalism involves constraining $\overline{s}^{\mu\nu}$ to an isotropic form in a special coordinate system with one independent coefficient $\overline{s}^{00}$. This reveals that there is a partially overlapping relationship between the two approaches (see figure 1 in Bailey & Kostelecký 2006). Thus, the pure-gravity sector of the SME describes effects outside of the PPN while, conversely, the PPN describes effects outside of the pure-gravity sector of the SME. The relationship between the PPN and other sectors of the SME is not known at present.

Many of the dominant effects in the pure-gravity sector of the SME can be studied via an effective post-newtonian lagrangian for a system of point masses given by

$$\begin{aligned} L \;=\; & \tfrac{1}{2}\sum_a m_a \vec{v}_a^{\,2} + \tfrac{1}{2}\sum_{ab}\frac{Gm_a m_b}{r_{ab}}\left(1 + \tfrac{3}{2}\overline{s}^{00} + \tfrac{1}{2}\overline{s}^{jk}\hat{r}^j_{ab}\hat{r}^k_{ab}\right) \\ & -\tfrac{1}{2}\sum_{ab}\frac{Gm_a m_b}{r_{ab}}\left(3\overline{s}^{0j}v^j_a + \overline{s}^{0j}\hat{r}^j_{ab}v^k_a\hat{r}^k_{ab}\right) + \ldots, \end{aligned} \tag{2.4}$$

where v_a is the velocity of mass m_a and r_{ab} is the relative euclidean distance between two masses. In (2.4), the nine coefficients for Lorentz violation $\overline{s}^{\mu\nu}$ are projected into their space and time components $\overline{s}^{00} = \overline{s}^{jj}$, $\overline{s}^{jk}$, and $\overline{s}^{0j}$. For other applications, such as the time-delay effect, it is necessary to directly use the post-newtonian metric which was obtained in Bailey & Kostelecký (2006).

3. Matter-gravity tests

The chief effects from the coefficients $\overline{a}_\mu$ can be described as an effective modification to the gravitational force between two test bodies. Supposing there is a source body S, such as the Earth, and a test body T near the surface, an addition to the usual vertical force arises that takes the form

$$\widetilde{F}_z = -2g(\alpha\overline{a}_t^{\mathrm{T}} + \alpha\overline{a}_t^{\mathrm{S}} m^{\mathrm{T}}/m^{\mathrm{S}}), \tag{3.1}$$

in a laboratory reference frame. In this expression m^{T} and m^{S} are the masses of the test body and source, respectively, and g is the local gravitational acceleration. The force modification in (3.1) depends on the time components of the $\overline{a}_\mu$ coefficients for the test body and source, denoted $\overline{a}_t^{\rm T}$ and $\overline{a}_t^{\rm S}$. For ordinary matter, these coefficients are ultimately related to the 12 $\overline{a}_\mu$ coefficients for the electron, neutron, and proton (denoted $\overline{a}_\mu^e$, $\overline{a}_\mu^n$, and $\overline{a}_\mu^p$).

Two types of effects arise from the signal in (3.1). When the coefficients $\overline{a}_\mu$ take on a flavor dependence, violations of the Weak Equivalence Principle (WEP) occur, as two test bodies of different compositions experience different accelerations. The standard reference frame for reporting measurements of coefficients for Lorentz violation is the Sun-centered celestial equatorial reference frame or SCF for short (Kostelecký & Mewes 2002). When relating the laboratory frame coefficients in (3.1) to the SCF, sidereal day and yearly time dependence from the Earth's rotation and revolution is introduced. Therefore, an additional effect occurs where the signal for Lorentz violation depends on the time of day and season. These effects have direct experimental consequences for both Earth-bound and space-based tests of WEP (Nobili *et al.* 2003), as well as ordinary single-flavor gravimeter-type tests. Estimated sensitivities for specific tests and a single constraint on one of the 12 coefficients $\overline{a}_\mu$ for ordinary matter, implied by existing analysis (Schlamminger *et al.* 2008), are described in more detail in Kostelecký & Tasson (2009).

4. Pure-gravity sector tests

Within the pure-gravity sector of the SME, the primary effects on orbital dynamics, due to the nine coefficients $\overline{s}^{\mu\nu}$, can be obtained from the point-mass lagrangian in equation (2.4). From the post-newtonian metric, modifications to the classic solar-system tests of GR can be determined, such as the gyroscope experiment and the time-delay effect.

One particularly sensitive test of orbital dynamics in the solar system is lunar laser ranging. Highly-sensitive laser pulse timing is achievable using the reflectors on the lunar surface. The dominant Lorentz-violating corrections to the Earth-Moon coordinate acceleration can be calculated from equation (2.4). In the ideal case, a full analysis would incorporate the effects from the pure-gravity sector of the minimal SME, as well as the standard dynamics of the Earth-Moon system, into the orbital determination program. However, it can be shown that the dominant effects can be described as oscillations in the Earth-Moon distance. The frequencies of these oscillations involve the mean orbital and anomalistic frequencies of the lunar orbit, and the mean orbital frequency of the Earth-Moon system (see Table 2 in Bailey & Kostelecký 2006). Using past data spanning over three decades, Battat *et al.* (2007) placed constraints on 6 combinations of the $\overline{s}^{\mu\nu}$ coefficients at levels of 10^{-7} to 10^{-10}. Substantial improvement of these sensitivities may be achievable in the future with APOLLO (Murphy *et al.* 2008). Also of potential interest are Earth-satellite tests, since satellites of differing orientation can pick up sensitivities to distinct coefficients.

For Earth-laboratory experiments, a modified local gravitational acceleration arises due to the $\overline{s}^{\mu\nu}$ coefficients, similar to that which occurred in (3.1). Again, due to the Earth's rotation and revolution relative to the SCF, this acceleration acquires a time variation that can be searched for in appropriate experiments, such as gravimeter tests. Such an experiment was performed by Müller *et al.* (2008) and, more recently, these results were combined with lunar laser ranging analysis to yield new constraints on 8 out of the 9 independent $\overline{s}^{\mu\nu}$ coefficients (Chung *et al.* 2009).

The classic time delay effect of GR becomes modified in the presence of Lorentz violation. The correction to the light travel time for a photon passing near a mass M can be obtained by studying light propagation with the post-newtonian metric of the pure-gravity sector of the minimal SME. If the signal is transmitted from an observer at position r_e, reflected from a planet or spacecraft at position r_p, the time delay of light can be written in a special coordinate system as

$$\Delta T_g \approx 4GM \left[(1 + \overline{s}^{TT}) \ln \left(\frac{r_e + r_p + R}{r_e + r_p - R} \right) + \overline{s}^{JK} \hat{b}^J \hat{b}^K \right]. \tag{4.1}$$

In this expression, R is the distance between the observer and the planet or spacecraft and $\hat{b}$ is the impact parameter unit vector. The coefficients for Lorentz violation are expressed in the SCF, as denoted by capital letters. This result holds for near-conjunction times when the photon passes near the mass M. As discussed in Bailey (2009), time-delay tests can be useful to constrain the isotropic $\overline{s}^{TT}$ coefficient, among others. It would be of definite interest to perform data analysis searching for evidence of nonzero $\overline{s}^{\mu\nu}$ coefficients in sensitive time-delay experiments such as Cassini (Bertotti *et al.* 2003), BepiColombo (Iess & Asmar 2007), and SAGAS (Wolf *et al.* 2009).

In addition to effects discussed above, many standard results of GR receive modifications in the presence of Lorentz violation. This includes classic tests such as the perihelion shifts of the inner planets, the gravitational redshift, and the classic gyroscope experiment. In addition, tests beyond the solar system, such as with binary pulsars (Kramer 2009 and Stairs 2009), can also probe SME coefficients. In the case of binary pulsars, the dominant effects arise from changes in the orbital elements of the pulsar-companion oscillating ellipse. For more details the reader is referred to Bailey (2009) and Bailey & Kostelecký (2006).

5. Acknowledgments

Q.G. Bailey wishes to thank the International Astronomical Union for the travel grant that was provided to attend the symposium.

References

Amelino-Camelia, G. *et al.* 2005, *AIP Conf. Proc.*, 758, 30
Bailey, Q. G. 2009, *Phys. Rev. D*, 80, 044004
Bailey, Q. G. & Kostelecký, V. A. 2006, *Phys. Rev. D*, 74, 045001
Battat, J. B. R., Chandler, J. F., & Stubbs, C. W. 2007, *Phys. Rev. Lett.*, 99, 241103 (2007)
Bertotti, B., Iess, L., & Tortura, P. 2003, *Nature*, 425, 374
Bluhm, R. 2006, *Lect. Notes Phys.*, 702, 191
Chung, K. Y. *et al.* 2009, *Phys. Rev. D*, 80, 016002
Colladay, D. & Kostelecký, V. A. 1997, *Phys. Rev. D*, 55, 6760
Colladay, D. & Kostelecký, V. A. 1998, *Phys. Rev. D*, 58, 11602
Iess, L. & Asmar, S. 2007, *Int. J. Mod. Phys. D*, 16, 2191
Kostelecký, V. A. (ed.) 2008, *CPT and Lorentz Symmetry IV*, (Singapore: World Scientific)
Kostelecký, V. A. 2004, *Phys. Rev. D*, 69, 105009
Kostelecký, V. A. & Mewes, M. 2002, *Phys. Rev. D*, 66, 056005
Kostelecký, V. A. & Potting, R. 1995, *Phys. Rev. D*, 51, 3923
Kostelecký, V. A. & Russell, N. 2009, arXiv:0801.0287
Kostelecký, V. A. & Tasson, J. D. 2009, *Phys. Rev. Lett.*, 102, 010402
Kramer, M. 2009, *this proceedings*, 366
Müller, H. *et al.* 2008, *Phys. Rev. Lett.*, 100, 031101
Murphy, T. W. *et al.* 2008, *PASP*, 120, 20
Nobili, A. M. *et al.* 2003, *Phys. Lett. A*, 318, 172
Schlamminger, S. *et al.* 2008, *Phys. Rev. Lett.*, 100, 041101
Stairs, I. H. 2009, *this proceedings*, 218
Will, C. M. 2005, *Living Rev. Relativity*, 9, 3
Wolf, P. *et al.* 2009, *Exper. Astron.*, 23(2), 651

Relativity in Fundamental Astronomy
Proceedings IAU Symposium No. 261, 2009
S. A. Klioner, P. K. Seidelman & M. H. Soffel, eds.

doi:10.1017/S1743921309990718

Measurement of gravitational time delay using drag-free spacecraft and an optical clock

Neil Ashby[1], Peter L. Bender[2,3], John L. Hall[2,3], Jun Ye[2,3], Scott A. Diddams[3], Steven R. Jefferts[3], Nathan Newbury[3], Chris Oates[3], Rita Dolesi[4], Stefano Vitale[4], & William J. Weber[4]

[1] University of Colorado, Boulder, CO
email: ashby@boulder.nist.gov

[2] JILA, University of Colorado, Boulder, CO

[3] National Institute of Standards & Technology, Boulder, CO

[4] University of Trento, Italy

Abstract. Improved accuracy in measurement of the gravitational time delay of electromagnetic waves passing by the sun may be achieved with two drag-free spacecraft, one with a stable clock and laser transmitter and one with a high-stability transponder. We consider one spacecraft near the Earth-Sun L1 point with an advanced optical clock, and the transponder on a second satellite, which has a 2 year period orbit and eccentricity $e = 0.37$. Superior conjunctions will occur at aphelion 1, 3, and 5 years after launch of the second spacecraft. The measurements can be made using carrier phase comparisons on the laser beam that would be sent to the distant spacecraft and then transponded back. Recent development of clocks based on optical transitions in cooled and trapped ions or atoms indicate that a noise spectral amplitude of about $5 \times 10^{-15}/\sqrt{\text{Hz}}$ at frequencies down to at least 1 microhertz can be achieved in space-borne clocks. An attractive candidate is a clock based on a single laser-cooled Yb^+ trapped ion. Both spacecraft can be drag-free at a level of $1 \times 10^{-13}\text{m/s}^2/\sqrt{\text{Hz}}$ at frequencies down to at least 1 microhertz. The corresponding requirement for the LISA gravitational wave mission is $3 \times 10^{-15}\text{m/s}^2/\sqrt{\text{Hz}}$ at frequencies down to 10^{-4} Hz, and Gravitational Reference Sensors have been developed to meet this goal. They will be tested in the LISA Pathfinder mission, planned by ESA for flight in 2011. The requirements to extend the performance to longer times are mainly thermal. The achievable accuracy for determining the PPN parameter γ is about 1×10^{-8}.

Keywords. gravitation, relativity, time delay

---

1. Introduction

Testing general relativity (GR) more fully is one of the important scientific opportunities for new missions and new observation programs in the next decade. The extra gravitational time delay for two-way measurements of light propagating from Earth to a spacecraft passing behind the Sun can be more than 200 microseconds. In the Parametrized Post-Newtonian (PPN) formulation of gravitational theory, the main contribution to the time delay is proportional to $(1+\gamma)$, where γ is a measure of the curvature of space-time. In GR, $\gamma = 1$.

Recently, a measurement of γ with accuracy $\pm 2.3 \times 10^{-5}$ was made during the Cassini mission (Bertotti *et al.* (2003)). Further improvements in the accuracy for γ to roughly 10^{-6} are expected from two missions of the European Space Agency (ESA): the Gaia astrometric mission, which will measure the gravitational deflection of light rays by the

Sun, and the BepiColombo mission to Mercury, which will make improved measurements of the solar time delay.

Analysis of a time delay mission involving an accurate clock in a spacecraft at L1 began in 2005 (Ashby & Bender (2008)). This proposal was updated recently to include the use of extremely stable new clocks based on optical transitions in cooled atoms or ions (Bender, *et al.*(2008)). We describe here a mission that can reach an accuracy of about 1×10^{-8} for determining γ.

2. Science justification

In view of the well-known lack of a theory that connects general relativity with quantum theory, improvement of high-accuracy tests of the predictions of GR should be the object of research in the coming decade. Many alternatives to general relativity involve additional scalar fields. Studies of the evolution of scalar fields in the matter-dominated era of the universe indicate that the universe's expansion tends to drive the scalar fields toward a state in which the scalar-tensor theory is only slightly different from GR. Some scalar-tensor extensions of GR predict deviations from the GR value of γ in the range from $10^{-5} - 10^{-8}$(Damour & Esposito-Farese (1996), Damour & Nordvedt (1993)). Improved information about γ would provide important insight into the evolution of the universe and directly limit the range of applicability of alternative gravitational theories.

3. Mission orbits and predicted time delay

For the proposed mission, one spacecraft (S1) containing a highly stable optical clock would be placed in an orbit near the L1 point, about 1.5 million km from the Earth in the direction of the Sun. The second spacecraft (S2) would be placed in a 2 year period orbit in the ecliptic plane, with an eccentricity of 0.37. S2 would pass through superior solar conjunction about 1, 3, and 5 years after launch and would be near aphelion at those times. Both spacecraft would have drag-free systems to nearly eliminate the effects of spurious non-gravitational forces. A measurement of γ to a level of 1×10^{-8} would be carried out by observing the time delay of laser signals exchanged between the two spacecraft when the line of sight passed near the Sun's limb. Atmospheric effects would be absent and continuous observation would be possible. With S2 near aphelion, the range rate would be low, and the orbit determination problem would be much reduced.

The crucial measurements of time delay occur within a few days of superior conjunction and are primarily characterized by a logarithmic dependence on the distance of closest approach of the light to the mass source. The predicted gravitational time delay due to a non-rotating mass source (here $\mu = GM_{\odot}/c^2$), expressed in terms of the radii r_A, r_B of the endpoints of the photon path, and the elongation angle Φ between the radius vectors from the source to the endpoints, is (Ashby & Bertotti (2009))

$$c\Delta t_{\mathrm{delay}} = \mu(1+\gamma)\log\left(\frac{r_A + r_B + r_{AB}}{r_A + r_B - r_{AB}}\right) - \frac{\mu^2(1+\gamma)^2 r_{AB}}{r_A r_B(1+\cos\Phi)} + \frac{\mu^2 r_{AB}(8 - 4\beta + 8\gamma + 3\epsilon)}{4 r_A r_B \sin\Phi} \tag{3.1}$$

where r_{AB} is the geometric distance between the endpoints in isotropic coordinates and β and ϵ are PPN parameters measuring the nonlinearity of the time-time and space-space components of the metric tensor. In GR, $(8 - 4\beta + 8\gamma + 3\epsilon)/4 = 15/4$. The time delay in Eq. (3.1) is expressed in terms of observable quantities, and does not involve the unknown impact parameter or distance of closest approach. The non-linear terms are a few nanoseconds so are significant, but do not have to be estimated with great accuracy.

The contributions to time delay due to the solar quadrupole moment are controlled by the parameter $\mu J_2 R^2/(b^2 c) < 10^{-12}$ s, where $J_2 \approx 2 \times 10^{-7}$, R is the sun's radius, and $b \approx R$ is the impact parameter. This effect depends on the relative orientation of the sun's rotation axis and the photon path, but the net effect is small and can be estimated with sufficient accuracy that it will not contribute significantly to the error budget.

The measurements will be made by transmitting a continuous laser beam from S1 to S2, offset phase locking the laser signal from the transponder on the distant spacecraft to the received beam, and then recording the relative phase of the received beam back at S1 with respect to its own laser as a function of time. With 20 cm diameter telescopes, and given the one-way travel time of about 1600 s, the received signal would be roughly 1000 counts/s for 1 W of transmitted power. This is a weak signal, but it is strong enough so that the chances of a cycle slip should be very small, provided that the laser in S2 is well stabilized. If we consider the round-trip delay times $\Delta t_{\rm delay}$ to be the observable, then the change in delay from 0.75 days to 4 days on either side of conjunction is about 64 microseconds.

For initial acquisition of the beat frequency signal on S2, accurate knowledge of the orbits of both spacecraft is needed. Thus K-band tracking of both spacecraft from Earth would be needed over a period of 20 or 30 days around solar conjunction. In addition, a priori knowledge of the frequency of the laser on S2 to a kilohertz or so would be needed to simplify the acquisition procedure. As an example, if NdYAG lasers are used for transmission between the spacecraft, the stabilization of the one on S2 could be achieved by using a fairly simple iodine stabilization scheme. In this case, a frequency comb approach would be needed on S1 to tie the NdYAG laser frequency there to the highly stable optical clock frequency.

4. Signal-to-Noise Analysis

We estimate the uncertainty that can be attained in this experiment on the basis of the optimal Wiener filter, which takes advantage of the known time signature of the signal and includes the expected noise sources (Thorne (1987)). For simplicity, the distance between the spacecraft is assumed to be constant except for changes in the gravitational time delay. The time signature of $\gamma^* = (1+\gamma)/2$ is taken to be represented by the logarithmic function

$$g(t) = -B(\log|Rt| - M) \tag{4.1}$$

where M is the mean value of $\log|Rt|$ over the time periods $-t_2$ to $-t_1$ and t_1 to t_2 (a short time interval around occultation is excluded), and for the proposed experiment $B = 0.97 \times 8\mu/c = 3.82 \times 10^{-5}$ s. The rate at which the line of sight to the distant spacecraft passes across the sun is $R = 1.9$ solar radii per day.

Let $g(f)$ be the Fourier transform of $g(t)$ over the time of the measurements. Then the square of the signal-to-noise ratio is given by

$$\left(\frac{S}{N}\right)^2 = \int_0^\infty \frac{2|g(f)|^2}{n(f)^2} df = \frac{2}{n(f)^2}\int_0^\infty |g(f)|^2 df\,, \tag{4.2}$$

where

$$g(f) = 2\int_{t_1}^{t_2} g(t)\cos(2\pi f t)dt\,, \tag{4.3}$$

and the factor of 2 comes from time symmetry of the time-delay signal. Here we use a

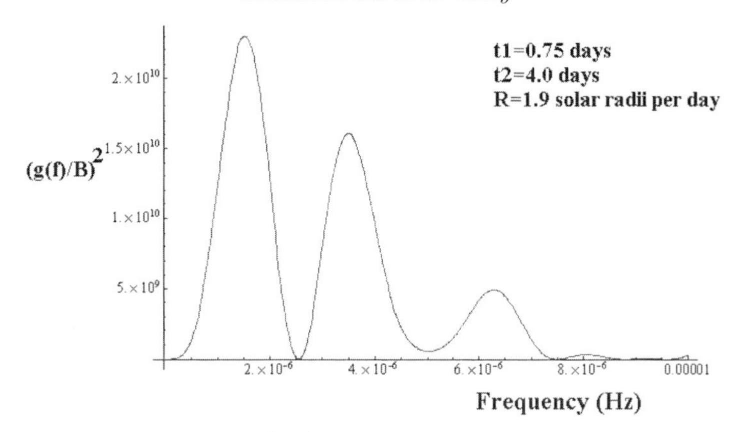

Figure 1. Plot of the function $g(f)^2/B^2$ for frequencies up to 0.01 mHz, for a typical set of measurement parameters.

one-sided Fourier transform, in which negative frequencies are folded into positive frequencies. The mean value M is $M = (t_2 \log(Rt_2) - t_1 \log(Rt_1))/(t_2 - t_1) - 1$. For example, for the case where the time interval of measurements extends from 0.75 days to 4 days, Figure 1 plots the quantity $g(f)^2/B^2$. When combined with estimates of the spectral density of the noise, the function gives us the signal to noise ratio according to Eq. (4.2) above. If the noise has a constant spectral density, only about 5% of the integral in Eq. (4.2) comes from frequencies below 1 microhertz, where the acceleration noise is expected to increase.

So far, we have assumed that time-delay measurements are only made over a total period of 8 days around solar conjunction. This was done in order to make sure that spurious acceleration noise at frequencies below 1×10^{-6} Hz would have little effect. However, simulations for longer observing times are desirable, with full allowance for spurious acceleration noise at the lowest frequencies, as well as for the orbit determination part of the problem.

5. Spacecraft S1 clock

The major requirement for the mission is to fly an optical clock on S1 that has very high stability over a period of at least 8 days around superior conjunction. The nominal design goal for the mission is to achieve a fractional frequency noise power spectral density amplitude of $5 \times 10^{-15}/\sqrt{\mathrm{Hz}}$ from 1 Hz down to at least 10^{-6} Hz. (This is nearly equivalent to an Allan deviation of $3.5 \times 10^{-15}/\sqrt{\tau}$ for times from 1 s to up to 10^6 s.) As an example of the desired performance at frequencies near 3 mHz, a spectral density amplitude of about $2 \times 10^{-15}/\sqrt{\mathrm{Hz}}$ has been achieved for the 698 nm transition in Sr atoms trapped in an optical lattice (Ludlow *et al.* (2008)).

A leading candidate for the optical clock is based on a single trapped cooled Yb^+ ion. The clock transition wavelength is 435 nm. The projected spectral density amplitude is about $3 \times 10^{-15}/\sqrt{\mathrm{Hz}}$ over the necessary frequency range (Peik, E., *et al.* (2006)). An advantage of the Yb^+ clock is that only low power lasers are required. However, substantial development is needed to show that such lasers can be space qualified. Possible alternate choices include trapped ion clocks based on Sr^+ or Al^+.

The relative phase of the transmitted and transponded signals would be strongly affected by the relative motion of the two spacecraft in their orbits, and thus all of the relevant orbit parameters as well as γ would have to be solved for. With the S2 orbit considered here, the S1-S2 distance passes through a maximum at conjunction, when S2 is at aphelion. This reduces the effect of errors due to temperature fluctuations as well as measurement epoch errors. A worst-case drift of 0.1 ps in the clock that is correlated with the predicted time delay would give rise to an error of 3×10^{-9} in determining γ.

In the time delay measurements, care will be needed to reduce the effects of scattered light in the telescopes. Because of the geometry, it would be difficult to avoid having some direct sunlight hit parts of the receiving optics. But, with heterodyne detection, this does not appear to be a serious problem. The plan for such a mission would be to not try to make measurements closer to the limb than about 0.4 solar radii, in order to avoid problems with reacquisition of the signals after solar conjunction.

6. Drag-free system

The required performance builds on that planned for the LISA mission. For frequencies down to 10^{-4} Hz for LISA, the requirement on the acceleration power spectral density amplitude is less than $3 \times 10^{-15} \mathrm{m/s^2}/\sqrt{\mathrm{Hz}}$. However, the performance is expected to degrade at lower frequencies. The main challenge for achieving good performance at low frequencies is minimizing thermal changes, and particularly thermal gradient changes, near the freely floating test mass in the drag-free system. On LISA this is done almost completely by passive thermal isolation. For a time delay mission, a fairly slow active temperature control system would be used at frequencies below 10^{-4} Hz. Changes in solar heat input over the 8 days around conjunction would be quite small for S2, because conjunction occurs near aphelion. The required drag-free performance is roughly $1 \times 10^{-13} \mathrm{m/s^2}/\sqrt{\mathrm{Hz}}$ down to 10^{-6} Hz.

In fact, much of the desired freedom from spurious accelerations needed for LISA has been demonstrated in the laboratory with torsion pendulum measurements (Carbone *et al.* (2007)). But, more important, the overall performance of the drag-free system will be demonstrated in the LISA Pathfinder Mission, which is scheduled for launch by ESA in 2011 (Armano *et al.* (2009); Racca & McNamara (2009)). The requirements are just to demonstrate an acceleration spectral density amplitude of $3 \times 10^{-14} \mathrm{m/s^2}/\sqrt{\mathrm{Hz}}$ performance down to 10^{-3} Hz, because of the less ideal thermal stability expected for LISA Pathfinder, compared with LISA. However, extensive tests of the response to various intentional thermal, electrical, magnetic, etc. disturbances will be carried out, in order to verify the models being used for the disturbances on the test masses. Thus the design of the drag-free systems for a gravitational time delay mission does not appear to be a substantial limitation.

7. Other scientific benefits from the mission

Additional effects such as those arising from non-linear terms in the 00-component of the metric tensor, parameterized by β, as well as other time delay effects originating in the sun's rotation, can also be measured. The clock at the L1 point will experience frequency shifts from the earth's potential, solar tidal effects, and second-order Doppler shifts. Relative to a reference on earth's surface, the fractional frequency shift is about $+6.9 \times 10^{-10}$, and is almost all gravitational. Comparing the clock at L1 with a similar clock on earth's geoid will give accuracies of a few parts per million in a few hours, which is orders of magnitude more accurate than the Vessot-Levine 1976 Gravity Probe A result.

8. Conclusion

Measurement of the time delay of signals passing near the sun is a promising test of the predictions of GR that may provide two orders of magnitude improvement in determination of γ.

References

Armano, M. *et al.*, 2009 in: LISA Pathfinder: the experiment and the route to LISA, *Class. & Quantum Grav.* 26(9), 094001

Ashby, N. & Bender, P. 2008, in: H. Dittus, C. Laemmerzahl, & S. Turyshev, (eds.), *Lasers, Clocks, and Drag-Free Control*, Bremen, Germany, June 2005 Astrophysics and Space Science Library 349, (Springer) 219–230

Ashby, N. & Bertotti, B. 2009, Accurate light-time correction due to a gravitating mass, *to be submitted*

Bertotti, B., Iess, L., & Tortota, P. 2003, *Nature*, 425, 374

Bender, P. *et al.* 2008 in: *Quantum to Cosmos III Workshop*, Warrenton, VA, July 6–10.

Carbone, L. *et al.* 2007 *Phys. Rev.* D75(4), 042001

Damour, T. & Esposito-Farese, G. 1996 *Phys. Rev.* D54, 5541

Damour, T. & Nordtvedt, K. 1993 *Phys. Rev.* D48, 3436

Ludlow, A. D. *et al.* 2008 *Science* 399, 1805

Peik, E. *et al.* 2006 *J. Phys.* B39, 145–158

Racca, G. & McNamara, P. 2009 *Space Science Reviews* in press

Thorne, K. S. 1987 Ch. 9 in: S. W. Hawking and W. Israel, eds. *300 Years of Gravitation* (Cambridge University Press) 330–459

Relativity in Fundamental Astronomy
Proceedings IAU Symposium No. 261, 2009
S. A. Klioner, P. K. Seidelman & M. H. Soffel, eds.

doi:10.1017/S174392130999072X

Modelling and simulation of the space mission MICROSCOPE

Stefanie Bremer[1], Meike List[1], Hanns Selig[1] and Claus Lämmerzahl[1]

[1]ZARM, University of Bremen,
Am Fallturm, 28359 Bremen, Germany
email: bremer@zarm.uni-bremen.de, list@zarm.uni-bremen.de,
selig@zarm.uni-bremen.de, laemmerzahl@zarm.uni-bremen.de

Abstract. The French space mission MICROSCOPE aims at testing the weak Equivalence Principle (EP) with an accuracy of 10^{-15}. The payload, which is developed and built by the French institute ONERA consists of two high-precision capacitive differential accelerometers. The detection of the test mass movement and their control is done via a complex electrode system. The German department ZARM is member of the MICROSCOPE performance team. In addition to drop tower tests, mission simulations and the preparation of the mission data evaluation are realized in close cooperation with the French partners CNES, ONERA and OCA. Therefore a comprehensive simulation of the real system including the science signal and all error sources is built for the development and testing of data reduction and data analysis algorithms to extract the EP violation signal. In this context the focus lays on the correct modeling of the environmental disturbances. Currently new effort to study the influence of the solar radiation and the Earth albedo to the MICROSCOPE mission scenario is underway.

Keywords. Gravitation, relativity

1. Introduction

The French MICROSCOPE mission is designed to test the weak Equivalence Principle (EP) with an accuracy of 10^{-15} (Touboul *et al.* (2001), Chhun *et al.* (2007)). For this purpose two high-precision capacitive differential accelerometers will be applied which are developed and built by the French institute ONERA. A drag-free satellite, that is produced by CNES, will carry out the experiment.

Free-fall tests at the Center of Applied Space Technology and Microgravity (ZARM) are used to study the instrument performance. In addition, the ZARM is also involved in the data evaluation of the MICROSCOPE mission. In this context a comprehensive simulation of the satellite and test mass dynamics is set up, including the instrument modelling and the influence of disturbances on the science signal. The simulation tool HPS will be presented and an overview about different modelling aspects will be given.

2. HPS: Structure and functionality

The High Performance Satellite Dynamics Simulator (HPS) is developed as cooperation project of the Center of Applied Space Technology and Microgravity (ZARM) and the DLR Institute of Space Systems. The software is based on the ZARM Drag-Free simulator (Scheithauer *et al.* (2002), Theil (2002)). It is a tool to support modelling and data analysis of missions that use so-called drag-free techniques. For this purpose the dynamics of a satellite that is equipped with up to four accelerometers each containing two test masses are modelled.

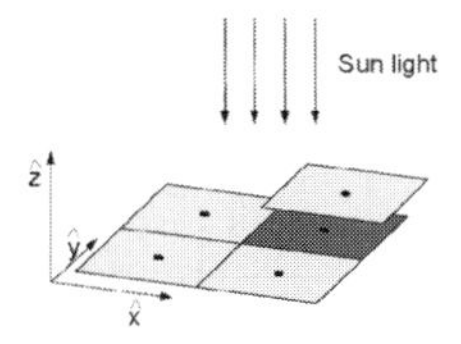

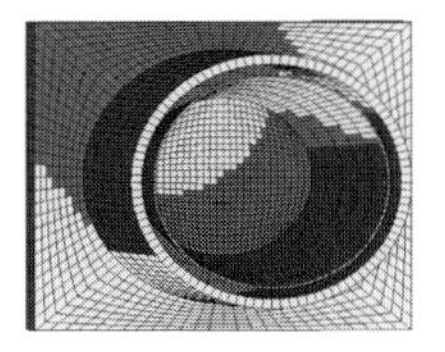

Figure 1. Illustration of shadow algorithm (left); Illumination example for MICROSCOPE radiator (right)

The development environment is Matlab/Simulink which provides the possibility to realize a modular design, i.e. all models can be developed and used separately. The core function calculates the states of the satellite and all test masses by numerical integration of the equations of motion. All effects that are based on the gravitational field are also considered within this main function. Forces and torques that act on the satellite or on its test masses are supplied as time-depending inputs. A multitude of coordinate systems are defined to take into account all technical entities like offsets and alignment errors.

3. Modeling of surface forces by means of finite elements

The correct modelling and understanding of the influence of external forces and torques on the satellite and test mass dynamics is important for the data reduction process. The forces and torques result from the interaction of the satellite with the space environment, basically solar radiation pressure, aerodynamic drag and Earth albedo.

In case of MICROSCOPE a detailed analysis of these effects is intended. The magnitude of all mentioned disturbance sources depends among other things on the satellite geometry and its surface properties. Standard approaches use a reference area for the force computation not taking into account e.g. shadowing. For MICROSCOPE a different procedure is set up, which will be exemplified for the computation of the solar radiation force. The general idea is based on the fact, that the magnitude of the resulting force is well known for a plate. Hence, it is possible to calculate the normalised force for each element of the discretized satellite surface (Wertz (1978), Cicek (2005)):

$$\mathbf{f}_\mathrm{i} = -A_\mathrm{i}[(1 - c_\mathrm{si}) \cdot \mathbf{e}_\mathrm{sun} + 2(c_\mathrm{si}\cos\alpha + \frac{1}{3}c_\mathrm{di}) \cdot \mathbf{e}_\mathrm{N}]\cos\alpha \quad . \tag{3.1}$$

Here A_i is the area of the element i with the reflection coefficients for specular (c_si) and diffuse (c_di) reflexion. The sun vector $\mathbf{e}_\mathrm{sun}$ and the normal vector $\mathbf{e}_\mathrm{N}$ enclose the angle α.

The correct derivation of the illumination conditions is divided into two parts. The identification of all elements that are exposed to the sun is the first step. In a second step, the position of the middle node of the visible elements is compared against each other to ascertain which element is shaded by another one (cf. figure 1 (left)). Figure 1 (right) shows the baseplate of the satellite with the payload's radiator which is prepared for the investigation of the Earth albedo influence. The total force is derived as the sum over all visible elements multiplied by the solar pressure.

4. Modelling of MICROSCOPE payload

Great effort is done to model the MICROSCOPE payload T-SAGE (Twin Space Accelerometer for Gravitation Experimentation). This instrument, developed and built at the French institute ONERA (Guiu (2007)), is composed of two differential accelerometers equipped with two test masses each. The so-called reference sensor contains test masses of the same platinum alloy whereas the test masses of the second accelerometer are made of different material to derive the science signal. The sensor dynamics are

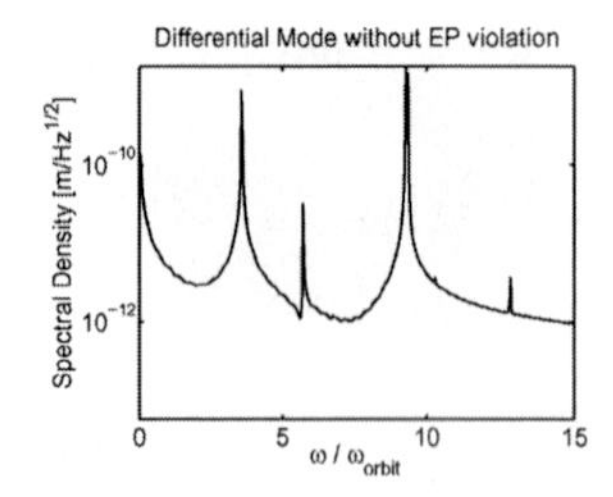

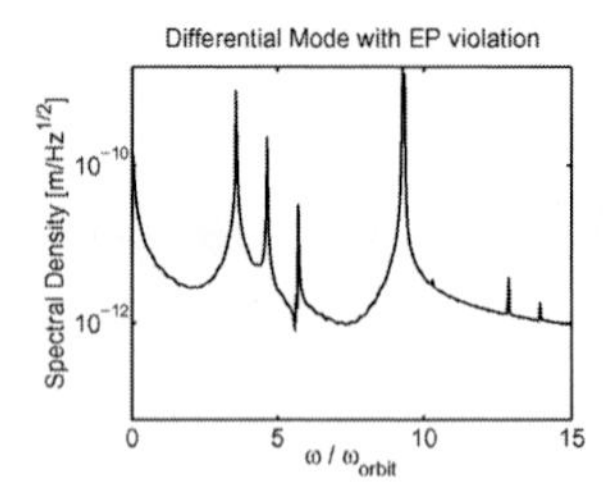

Figure 2. Results for simplified ansatz: Without EP violation (left); With EP violation (right)

linearized in a first approach, i.e. no coupling between the degrees of freedom are considered. The exact test mass position is detected via 18 electrodes. According to the HPS philosophy the sensor dynamics are coded in C and are available for the simulation in the Matlab/Simulink environment via an s-function.

5. Modelling of MICROSCOPE mission: Simplified Ansatz

A simplified ansatz is used to demonstrate an analysis method for the MICROSCOPE mission (Theil (2002)). In this case the satellite follows a circular orbit and rotates with a constant angular velocity with respect to the inertial frame. The test mass motion is constrained to the sensitive x-axis. The satellite and test mass coupling is modelled in terms of a spring-mass system. A spherical potential of the Earth is assumed. Figure 2 shows the results for this approach, on the left hand side without EP violation and on the right with EP violation. Here an additional peak appears at $\omega_{\mathrm{orbit}} + \omega_{\mathrm{spin}} = 4.6\,\omega_{\mathrm{orbit}}$. It can be clearly distinguished from the peaks due to the spin frequency at $3.6\,\omega_{\mathrm{orbit}}$ and the effect due to the gravity gradient at $9.2\,\omega_{\mathrm{orbit}}$. The implementation of other disturbance effects and their detection is underway.

Acknowledgements

The authors like to thank our colleagues from the DLR Institute of Space Systems, in particular Stephan Theil and Hansjörg Dittus. Additionally we like to thank our colleagues from ONERA and from Observatoire de la Cote d'Azur, especially Pierre Touboul, Manuel Rodrigues, Bernard Foulon, Vincent Josselin, Ratana Chhun and Gilles Métris for their support. This work is supported by the German Space Agency of DLR with funds of the BMWi (FKZ 50 OY 0801).

References

Scheithauer, S. & Theil, S. 2002, *Generic Drag-Free Simulator, AIAA Modeling and Simulation Conference, Monterey, California*

Theil, S. 2002, *Satellite and Test Mass Dynamics Modeling and Observation for Drag-free Satellite Control of the STEP Mission, PhD thesis, Department of Production Engineering, University of Bremen*

Touboul, P. & Rodrigues, M. 2001, *Class. Quantum Grav.*, 18, 2487-98

Chhun, R., Hudson, D., Flinoise, P., Rodrigues, M., Touboul, P., & Foulon, B. 2007, *Acta Astronautica*, 60, 873-879

Cicek, M. 2005, *Entwicklung und Validierung einer Methode zur Berechnung der Oberflächenkräfte auf Satelliten, Master Thesis, Hochschule Reutlingen*

Guiu, É. 2007 *Étalonnage de la mission spatiale MICROSCOPE: optimisation des performances, PhD Thesis, École Centrale de Nantes et l'Université de Nantes*

Wertz, J. R. 1978, *Spacecraft Attitude Determination and Control, Kluwer Academic Publishers, Dordrecht, The Netherlands*

Relativity in Fundamental Astronomy
Proceedings IAU Symposium No. 261, 2009
S. A. Klioner, P. K. Seidelman & M. H. Soffel, eds.

doi:10.1017/S174392130999073I

Microscope – A space mission to test the equivalence principle

Meike List[1], Hanns Selig[1], Stefanie Bremer[1] and Claus Lämmerzahl[1]
[1]ZARM – Center of Applied Space Technology and Microgravity,
University of Bremen,
Am Fallturm, D–28359 Bremen, Germany
email: list@zarm.uni-bremen.de, selig@zarm.uni-bremen.de,
bremer@zarm.uni-bremen.de, laemmerzahl@zarm.uni-bremen.de

Abstract. MICROSCOPE is a ESA/CNES space mission for testing the validity of the weak equivalence principle. The mission's goal is to determine the Eötvös parameter η with an accuracy of 10^{-15}. The French space agency CNES is responsible for designing the satellite which is developed and produced within the Myriade series. The satellite's payload T–SAGE (Twin Space Accelerometer for Gravitation Experimentation) consists of two high–precision capacitive differential accelerometers and is developed and built by the French institute ONERA.
As a member of the MICROSCOPE performance team, the German department ZARM performs free fall tests of the MICROSCOPE differential accelerometers at the Bremen drop tower. The project's concepts and current results of the free fall tests are shortly presented.

Keywords. Relativity, Gravitation

1. Overview

MICROSCOPE is a ESA/CNES μ–satellite project which aims at a high precision test of the Weak Equivalence Principle (WEP) which may be violated as predicted by many quantum theories of gravity (Isham 1996, Haugan & Lämmerzahl 2000) as well as within the theoretical framework of string theories (Damour & Polyakov 1994). Thus experiments which test this fundamental principle of general relativity are of big interest, especially concerning the unification of current fundamental theories (quantum chromodynamics, quantum electrodynamics, electroweak theory, general relativity) (e.g. Damour 1996).

The validity of the WEP predicts an equal gravitational acceleration acting on every point mass independent of its composition resulting in identical trajectories in gravitational fields (effects due to the point masses' own gravitational field are neglected). So far this so called *universality of free fall* is confirmed up to an accuracy of 10^{-13} in terms of the Eötvös parameter η defined by

$$\eta = \frac{\mu_1 - \mu_2}{\frac{1}{2}\left(\mu_1 + \mu_2\right)}, \quad \text{with} \quad \mu_a = \left(\frac{m_g}{m_i}\right)_{\text{mass a}} \tag{1.1}$$

where m_g is the gravitational and m_i the inertial mass, respectively. If the WEP holds, the $\eta = 0$. With MICROSCOPE η will be measured with an accuracy of 10^{-15}.
Up to now the best WEP tests have been carried out with the help of torsion balance experiments (Su *et al.* 1994, Smith *et al.* 1999, Boynton 2000, Schlamminger *et al.* 2008) which are strongly influenced by tidal effects as well as effects due to the inhomogenity of the Earth' gravitational field. Thus, there is great effort to put free fall experiments into space (e.g. MICROSCOPE, STEP) (Touboul & Rodrigues 2001, Lockerbie *et al.* 2001, Worden 1993).

Since 2001 ZARM cooperates with CNES and ONERA on the ESA/CNES mission MICROSCOPE. ZARM is in charge of, among other tests, testing and verifying the MICROSCOPE payload in free fall at the Bremen drop tower. These tests are essential for the successful development of the T–SAGE instrument. The behaviour of the instrument can only be tested in free fall with very low disturbance levels. Thus, parabolic flights which reach milli–g conditions only, are far away from the requirements that are given by the limited dynamic range of the instrument. Secondly the development of an end–to–end simulation of the satellite's and payload's dynamics in space is underway. For these purposes several simulation tools are under construction. The simulations will cover the behavior of the differential accelerometer in space as well as the satellite's interaction with the environment (residual atmosphere, solar wind, radiation pressure, etc.). Finally, the mission data analysis will be carried out at ZARM in cooperation with the French project partners.

2. Mission design

MICROSCOPE will be launched in 2012, it is supposed to orbit around the Earth with an altitude of about 800 km. The orbit's eccentricity is planned to be $\epsilon = 5 \cdot 10^{-3}$, the satellite's spinning rate is given by two spinning modes which are stabilised by using a star tracker. MICROSCOPE will be a drag–free mission, the control of μN–thruster application is realised with the help of the common mode signal while scientific data is taken of the differential mode signal.

The science signal will be measured using capacitive sensors. Each sensor unit consists of a slightly modified cylindric test mass (see Figure 1(left)). The motion of the test masses can be detected and stabilised capacitively along six degrees of freedom. Electrostatic forces will compensate disturbances due to environmental and coupling effects, thus ensuring freely falling test masses. Environmental influences which may affect the test masses geodetic motion are e.g. the Earth' atmosphere and albedo, solar pressure, etc.

3. Free fall tests

Since a residual acceleration of $5 \cdot 10^{-7}$ m/s^2 during a drop at the Bremen drop tower is the default value defined by ONERA, disturbances acting on the test masses had to be scaled down using a free-flyer platform. Operation of this platform consists of its disengagement, of damping of vibrations which are a result of the natural oscillation of the system, as well as construction and manufacturing of an appliance for determining the

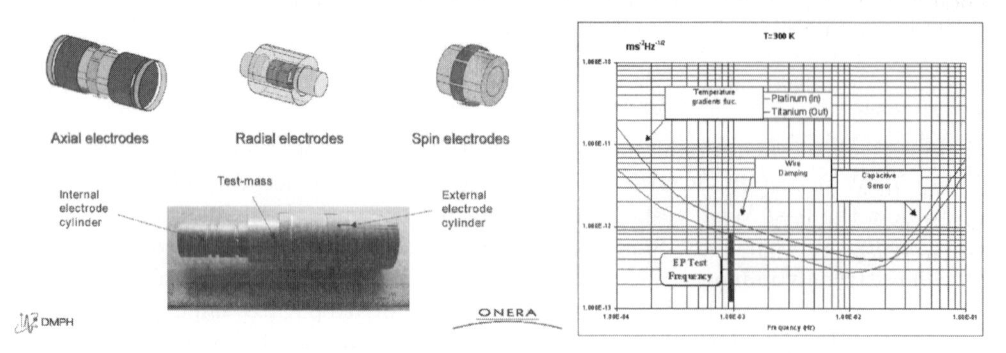

Figure 1. Sensor unit (left) as well as the performance (right) of T–SAGE

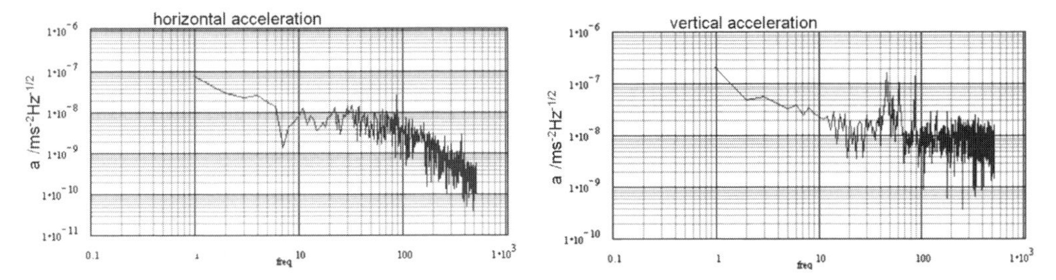

Figure 2. Residual accelerations in horizontal (left) and vertical (right) direction during drop

center of gravity of free–flyer and test masses within an accuracy of 1 mm. For these purposes more than 50 drops were performed and led finally to a successful implementation of a μg–laboratory.

A standard drop using the free–flyer is characterised by:

- μg–duration: 2 – 3 s,
- μg–level: $10^{-7}\text{gHz}^{-1/2}$–$10^{-8}\text{gHz}^{-1/2}$ in horizontal and vertical direction (see Figure 2).

The drop tower test campaigns for the MICROSCOPE instrument will contain 60 drops in 2009 and 2010 for tests and verification of EM and QM hardware. The campaign includes checking of sensor characteristics and performance, verification of the electronic conrol loop behaviour as well as tests of data acquisition and control procedures.

Acknowledgement

The authors like to thank the DLR Space Agency (Deutsches Zentrum für Luft– und Raumfahrt) for funding the project. Additionally we would like to thank P. Touboul, M. Rodrigues, B. Foulon, R. Chhun, V. Josselin, F. Liorzou (ONERA) and G. Metris (OCA) for fruitful discussions. This work is supported by DLR with funds of the BMWi (FKZ 50 OY 0801).

References

Isham, C. J. 1996, *Class. Quantum Grav.*, 13, A5

Haugan, M. P. & Lämmerzahl, C. 2001, *Principles of equivalence: their role in gravitation physics and experiments that test them, Gyros, Clocks, Interferometers, ...: Testing Relativistic Gravity in Space (Lecture notes in Physics)*, Springer Verlag

Damour, T. & Polyakov, A. M. 1994, *Nucl.Phys.*, 423, 532

Damour, T. 1996, *Class. Quantum Grav.*, 13, A33

Touboul, P. & Rodrigues, M. 2001, *Class. Quantum Grav.*, 18, 2487

Lockerbie, N. A., Mester, J. C., Tori, R., Vitale, S., & Worden, P. W. 2001, *STEP: A Status Report, Gyros, Clocks, Interferometers, ...: Testing Relativistic Gravity in Space (Lecture notes in Physics)*, Springer Verlag

Worden, P. W. 1993, *STEP Symposium, Pisa, ESA Publications Division*

Su, Y., Heckel, B. R., Adelberger, E. G., Grundlach, J. H., Harns, M., Smith, G. L., & Swanson, H. E. 1994, *Phys. Rev. D*, 50, 3614

Smith, C. L., Hoyle, C. D., Grundlach, J. H., Adelberger, E. G., Heckel, B. R., & Swanson, H. E. 1999, *Phys. Rev. D*, 61, 2201

Boynton, P. E. 2000, *Class. Quantum Grav.*, 17, 2319

Schlamminger, S., Choi, K.-Y., Wagner, T. A., Gundlach, J. H., & Adelberger, E. G. 2008, *Phys. Rev. Lett.*, 100, 041101

Relativity in Fundamental Astronomy
Proceedings IAU Symposium No. 261, 2009
S. A. Klioner, P. K. Seidelman & M. H. Soffel, eds.

doi:10.1017/S1743921309990743

New precise method for accurate modeling of thermal recoil forces

Benny Rievers[1] & Claus Lämmerzahl[1]

[1]Center of Applied Space Technology and Microgravity (ZARM), University of Bremen
Am Fallturm, D-28359 Bremen, Germany
email: rievers@zarm.uni-bremen.de

Abstract. The exact modeling of external and internal perturbations acting on spacecraft becomes increasingly important as the scientific requirements become more demanding. Disturbance models included in orbit determination and propagation tools need to be improved to account for the needed accuracy. At ZARM (Center of Applied Space Technology and Microgravity) algorithms for the simulation and analysis of thermal perturbations have been developed. The applied methods are based on the inclusion of the actual spacecraft geometry by means of Finite Element (FE) models in the calculation of the disturbance forces. Thus the modeling accuracy is increased considerably and also housekeeping and sensor data can be included in the calculations. Preliminary results for a test case geometry of the Pioneer 10/11 mission are presented and discussed with respect to the Pioneer anomaly. It is shown that thermal effects cannot be neglected for the magnitude scale of the observed anomalous effect.

Keywords. radiation mechanisms: thermal, methods: numerical

---

1. Introduction

For modern spacecraft missions the requirements on perturbation knowledge and modeling accuracy become increasingly demanding. One of the sources for disturbance accelerations is the recoil force resulting from anisotropic heat radiation. For the assessment of this effect at ZARM algorithms have been developed which compute thermal perturbations based on the configuration of the spacecraft with a high level of geometric complexity and numeric accuracy. For this a finite element (FE) model of the craft is developed and a full thermal analysis is processed to calculate the resulting surface temperature distribution including material parameters and thermal boundaries.

2. Basic equations

The resulting recoil from radiation flux $dP_{d\Omega}$ directed to a specific direction characterized by $d\Omega = f(\phi, \beta)$ is

$$F_{d\Omega} = \frac{dP_{d\Omega}}{c}. \tag{2.1}$$

Assuming flat surface and Lambertian radiation the intensity is

$$I(\beta) = I_n \cos(\beta) \tag{2.2}$$

The emission hemisphere can be divided into solid angle elements $d\Omega$ bordered by the angles ϕ and β where $0 \leqslant \phi \leqslant 2\pi$, $0 \leqslant \beta \leqslant \pi/2$ and $d\Omega = \sin\beta\, d\beta\, d\phi$. The radiation flux received by the specific solid angle elements is

$$dP_{d\Omega} = L \cos\beta \sin\beta\, d\beta\, d\phi\, dA, \tag{2.3}$$

where the spectral density L at a given emissivity ϵ is defined as

$$L = \epsilon \frac{\sigma}{\pi} T^4 \quad with \quad \sigma = \frac{2\,\pi^4\,k^4}{15\,h^3 c^3}. \tag{2.4}$$

For the resulting force only flux components perpendicular to the emitting surfaces can contribute. Thus the effective power normal to the emitting surface is

$$P_{\perp} = L \int_0^A \int_0^{\frac{\pi}{2}} \int_0^{2\pi} \cos\beta^2 \sin\beta\, d\beta\, d\phi\, dA = \frac{2}{3} P_{\rm tot} = \frac{2}{3}\epsilon\sigma A T^4, \tag{2.5}$$

which results in a recoil force in normal direction of the emitting plate $\mathbf{e}_{\rm n}$ of

$$\mathbf{F}_{\rm Recoil} = -\frac{2}{3}\frac{P_{\rm tot}}{c}\,\mathbf{e}_{\rm n}. \tag{2.6}$$

3. Computation method

The equations given above are valid for rectangular two-dimensional emitting surfaces. The computation of recoil forces for a detailed spacecraft geometry is much more complicated. At ZARM an elaborated method for the precise computation of thermal recoil forces has been developed. In a first step a complete FE model of the craft is created and thermal FE analysis are conducted to compute equilibrium surface temperatures based on spacecraft configuration, materials, thermal boundaries and loads. In a second step the results and a mathematical model of the craft are exported into a ray tracing algorithm for the computation of the resulting perturbation force including emission, absorption, reflection and shadowing models. The total force $\mathbf{F}_{\rm res}$ is composed of a) an emission component, b) an absorption component and c) a reflection component where

$$\mathbf{F}_{\rm res} = \mathbf{F}_{\rm emis} - \mathbf{F}_{\rm abs} + \mathbf{F}_{\rm ref}. \tag{3.1}$$

The emission component can be acquired with the unit normal vector of each surface and material/temperature data using equation 2.6. For complex model geometries radiation may be absorbed by other model surfaces thus reducing the effective recoil. For this emission is treated by angularly spaced rays emitted into the solid angle elements of the hemisphere, where the number of angular divisions specifies the total number of rays per element. The rays are then traced to detect intersections with other surfaces. For each hit the effective flux is computed by means of view factors. All computations in this respect are performed including shadowing effects. The processing of the reflection component is similar to the absorption computation. The rays hitting a target surface can either be reflected specularly or diffusely thus initialising new sets of reflected rays. The number of reflections or a minimum energy threshold for the processing of rays can be specified.

4. Test case: Pioneer 10/11 mission

For the Pioneer 10/11 spacecrafts a small, constant decrease of velocity has been observed. The residual perturbation acceleration has been computed to $a_{\rm Pio} = 8.74 \cdot 10^{-10} {\rm m/s}^2$ (in Anderson *et al.*(1998)). A conclusive explanation for this effect has not been presented until now. There are many speculations concerning new physics and unmodeled relativistic effects in the codes used for data analysis but also conventional effects may provide an explanation. In Anderson *et al.*(2002) it is stated that an anisotropy of 60 W directed against flight direction may explain a deceleration in the order of magnitude of the PA. This satisfies a closer examination of the effect of anisotropic heat radiation for the Pioneer 10/11 spacecrafts. For this an FE model including the main antenna dish and the two RTG-assemblies (composed of two RTGs each) has been developed. For a first order-of-magnitude assessment only the RTGs are considered as heat sources while the antenna is assumed to be thermally neutral (zero temperature) and only used for absorption/reflection effects. Figure 1 shows the geometry of the test case. The surface temperatures of the RTGs are acquired from a FE thermal analysis based on material, heat load and geometry data. The FE mesh is generated using quadrilateral and hexahedral thermal finite elements to include heat conduction as well as heat radiation.

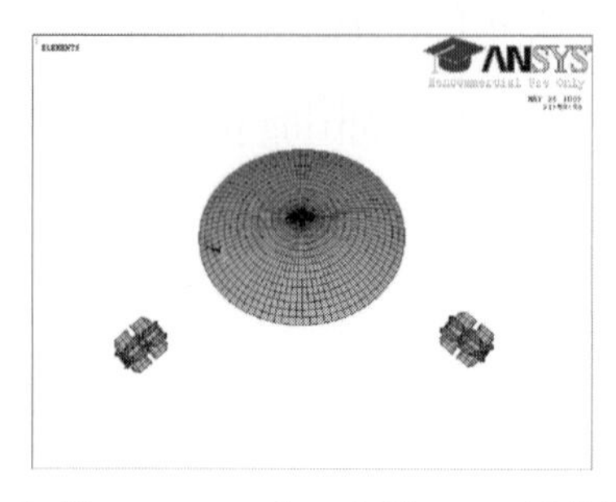

Figure 1. Test case and real Pioneer 10/11 geometry (picture courtesy Craig Markwardt).

The results of the thermal analysis (nodal solutions of the outer quadrilateral surfaces) is then exported into the ray tracing algorithm for computation of the recoil force. The analysis is conducted with varying numbers of emitted rays per surface until the solution converges. The surface model is composed of 1366 individual FE, the computation is performed for BOL power of 2500 W, white surfaces ($\epsilon = 0.9$) and a dry mass of 233 kg.

5. Preliminary results and conclusion

The solution converges for a number of approximately 90000 rays at a model size of approximately 2000 FE to a resulting acceleration component of 35 percent PA oriented against flight direction. Due to the incomplete model geometry (no equipment section, no outer payloads, RTGs only heat source) this result must not be mistaken for an exact value of the total contribution of thermal recoil to the anomalous deceleration. For this a complete model has to be processed also taking into account that the available power decreases over time due to radioactive decay. But the result points out that more detailed analyses are necessary and that thermal effects can not be neglected with respect to PA investigations. Therefore future analysis will include complete geometry, all heat sources as well as dynamic aspects. The presented method for the calculation of thermal recoil forces is of course not limited to the Pioneer case but can be used for any spacecraft mission with high requirements on perturbation knowledge. In particular fundamental physics missions like LISA, LISA pathfinder and Microscope can benefit from the improvement in modeling accuracy.

References

Anderson, J. D., Laing, P. A., Lau, E. L., Liu, A. S., Nieto, M. M., & Turyshev, S. G. 1998, *Phys. Rev. Lett., 1998*, 81, 2858

Anderson, J. D., Laing, P. A., Lau, E. L., Liu, A. S., Nieto, M. M., & Turyshev, S. G. 2002, *Phys. Rev. Lett., 2002*, 65, 082004

Turyshev, S. G., Nieto, M. M., & Anderson, J. D. 2002, *arXiv:physics*, 0502123

Bertolami, O., Franciso, F., Gil, P. J. S., & Paramos, J. 2008, *Phys. Rev. D, 2009*, 78, 103001

Toth, V. T. & Turyshev, S. G. 2009, *Phys. Rev. D, 2009*, 79, 43011

Author Index

Subject Index

Object Index